Michael Vollmer and
Klaus-Peter Möllmann

Infrared Thermal Imaging

Related Titles

Gross, H. (ed.)

Handbook of Optical Systems

6 Volume Set

2011

ISBN: 978-3-527-40382-0

Günzler, H. Gremlich, H.-U.

IR Spectroscopy: An Introduction

2002

ISBN: 978-3-527-28896-0

Gerlach, G., Budzier, H.

Thermische Infrarotsensoren

2010

ISBN: 978-3-527-40960-0

Schuster, N., Kolobrodov, V. G.

Infrarotthermographie

2004

ISBN: 978-3-527-40509-1

Maldague, X. P. V.

Theory and Practice of Infrared Technology for Nondestructive Testing

2001

ISBN: 978-0-471-18190-3

Michael Vollmer and Klaus-Peter Möllmann

Infrared Thermal Imaging

Fundamentals, Research and Applications

WILEY-VCH Verlag GmbH & Co. KGaA

The Authors

Prof. Michael Vollmer
Microsystem and Optical Technologies
University of Applied Sciences
Brandenburg, Germany
vollmer@fh-brandenburg.de

Prof. Dr. Klaus-Peter Möllmann
Microsystem and Optical Technologies
University of Applied Sciences
Brandenburg, Germany
moellmann@fh-brandenburg.de

Cover
Thermal pictures by Vollmer and Möllmann

All books published by **Wiley-VCH** are carefully produced. Nevertheless, authors, editors, and publisher do not warrant the information contained in these books, including this book, to be free of errors. Readers are advised to keep in mind that statements, data, illustrations, procedural details or other items may inadvertently be inaccurate.

Library of Congress Card No.: applied for

British Library Cataloguing-in-Publication Data
A catalogue record for this book is available from the British Library.

Bibliographic information published by the Deutsche Nationalbibliothek
The Deutsche Nationalbibliothek lists this publication in the Deutsche Nationalbibliografie; detailed bibliographic data are available on the Internet at <http://dnb.d-nb.de>.

© 2010 WILEY-VCH Verlag GmbH & Co. KGaA, Boschstr. 12, 69469 Weinheim, Germany

All rights reserved (including those of translation into other languages). No part of this book may be reproduced in any form – by photoprinting, microfilm, or any other means – nor transmitted or translated into a machine language without written permission from the publishers. Registered names, trademarks, etc. used in this book, even when not specifically marked as such, are not to be considered unprotected by law.

Composition Laserwords Private Ltd., Chennai, India
Printing and Bookbinding Strauss GmbH, Mörlenbach
Cover Design Graphik-Design Schulz, Fußgönheim

Printed in the Federal Republic of Germany
Printed on acid-free paper

ISBN: 978-3-527-40717-0

Contents

Preface XV

1   Fundamentals of Infrared Thermal Imaging *1*
1.1   Introduction   *1*
1.2   Infrared Radiation   *6*
1.2.1   Electromagnetic Waves and the Electromagnetic Spectrum   *6*
1.2.2   Basics of Geometrical Optics for Infrared Radiation   *10*
1.2.2.1   Geometric Properties of Reflection and Refraction   *10*
1.2.2.2   Specular and Diffuse Reflection   *12*
1.2.2.3   Portion of Reflected and Transmitted Radiation: Fresnel Equations   *13*
1.3   Radiometry and Thermal Radiation   *14*
1.3.1   Basic Radiometry   *15*
1.3.1.1   Radiant Power, Excitance, Irradiance   *15*
1.3.1.2   Spectral Densities of Radiometric Quantities   *16*
1.3.1.3   Solid Angles   *16*
1.3.1.4   Radiant Intensity, Radiance, and Lambertian Emitters   *17*
1.3.1.5   Radiation Transfer between Surfaces: Fundamental Law of Radiometry and View Factor   *19*
1.3.2   Blackbody Radiation   *21*
1.3.2.1   Definition   *21*
1.3.2.2   Planck Distribution Function for Blackbody Radiation   *21*
1.3.2.3   Different Representations of Planck's Law   *22*
1.3.2.4   Stefan–Boltzmann Law   *25*
1.3.2.5   Band Emission   *26*
1.3.2.6   Order of Magnitude Estimate of Detector Sensitivities of IR Cameras   *27*
1.3.2.7   Fractional Changes of Blackbody Radiation upon Temperature Changes   *29*
1.4   Emissivity   *32*
1.4.1   Definition   *32*
1.4.2   Classification of Objects according to Emissivity   *33*
1.4.3   Emissivity and Kirchhoff's Law   *34*

1.4.4 Parameters Affecting the Value of Emissivity 35
1.4.4.1 Material 35
1.4.4.2 Surface Structure 36
1.4.4.3 Viewing Angle 38
1.4.4.4 Geometry 40
1.4.4.5 Wavelength 42
1.4.4.6 Temperature 44
1.4.4.7 Conclusion 44
1.4.5 Techniques to Measure/Guess Emissivities for Practical Work 45
1.4.6 Blackbody Radiators: Emissivity Standards for Calibration Purposes 47
1.5 Optical Material Properties in the IR 51
1.5.1 Attenuation of IR Radiation while Passing through Matter 52
1.5.2 Transmission of Radiation through the Atmosphere 53
1.5.3 Transmission of Radiation through Slablike Solid Materials 55
1.5.3.1 Nonabsorbing Slabs 55
1.5.3.2 Absorbing Slabs 57
1.5.4 Examples for Transmission Spectra of Optical Materials for IR Thermal Imaging 57
1.5.4.1 Gray Materials in Used IR Spectral Ranges 57
1.5.4.2 Some Selective Absorbers 62
1.6 Thin Film Coatings: Tailoring Desired Optical Properties for IR Optical Components 64
1.6.1 Interference of Waves 64
1.6.2 Interference and Optical Thin Films 65
1.6.3 Examples of AR Coatings 67
1.6.4 Other Optical Components 69
References 71

2 Basic Properties of IR Imaging Systems 73
2.1 Introduction 73
2.2 Detectors and Detector Systems 73
2.2.1 Parameters That Characterize Detector Performance 74
2.2.2 Noise Equivalent Temperature Difference 74
2.2.3 Thermal Detectors 77
2.2.3.1 Temperature Change of Detector 77
2.2.3.2 Temperature-Dependent Resistance of Bolometer 78
2.2.3.3 NEP and D^* for Microbolometer 79
2.2.4 Photon Detectors 83
2.2.4.1 Principle of Operation and Responsivity 83
2.2.4.2 D^* for Signal Noise Limited Detection 85
2.2.4.3 D^* for Background Noise Limited Detection 86
2.2.4.4 Necessity to Cool Photon Detectors 90
2.2.5 Types of Photon Detectors 91
2.2.5.1 Photoconductors 91

2.2.5.2	Photodiodes	92
2.2.5.3	Schottky Barrier Detectors	93
2.2.5.4	Quantum Well Infrared Photodetectors (QWIP)	94
2.3	Basic Measurement Process	97
2.4	Complete Camera Systems	101
2.4.1	Camera Design – Image Formation	102
2.4.1.1	Scanning Systems	103
2.4.1.2	Staring Systems – Focal Plane Arrays (FPA)	103
2.4.1.3	Nonuniformity Correction (NUC)	108
2.4.1.4	Bad Pixel Correction	113
2.4.2	Photon Detector versus Bolometer Cameras	114
2.4.3	Detector Temperature Stabilization and Detector Cooling	116
2.4.4	Optics and Filters	118
2.4.4.1	Spectral Response	118
2.4.4.2	Chromatic Aberrations	118
2.4.4.3	Field of View (FOV)	120
2.4.4.4	Extender Rings	122
2.4.4.5	Narcissus Effect	124
2.4.4.6	Spectral Filters	126
2.4.5	Calibration	129
2.4.6	Camera Operation	133
2.4.6.1	Switch-on Behavior of Cameras	133
2.4.6.2	Thermal Shock Behavior	134
2.4.7	Camera Software – Software Tools	136
2.5	Camera Performance Characterization	137
2.5.1	Temperature Accuracy	137
2.5.2	Temperature Resolution – Noise Equivalent Temperature Difference (NETD)	139
2.5.3	Spatial Resolution – IFOV and Slit Response Function (SRF)	142
2.5.4	Image Quality: MTF, MRTD, and MDTD	145
2.5.5	Time Resolution – Frame Rate and Integration Time	150
	References	154
3	**Advanced Methods in IR Imaging**	**157**
3.1	Introduction	157
3.2	Spectrally Resolved Infrared Thermal Imaging	157
3.2.1	Using Filters	158
3.2.1.1	Glass Filters	159
3.2.1.2	Plastic Filters	160
3.2.1.3	Influence of Filters on Object Signal and NETD	160
3.2.2	Two-Color or Ratio Thermography	163
3.2.2.1	Neglecting Background Reflections	164
3.2.2.2	Approximations of Planck's Radiation Law	166
3.2.2.3	T_{obj} – Error for True Gray Bodies within Wien Approximation	168
3.2.2.4	Additional T_{obj} – Errors due to Nongray Objects	171

3.2.2.5	Ratio versus Single Band Radiation Thermometry	173
3.2.2.6	Application of Two-Color Thermography	174
3.2.2.7	Extension of Ratio Method and Applications	174
3.2.3	Multi- and Hyperspectral Infrared Imaging	176
3.2.3.1	Principal Idea	176
3.2.3.2	Basics of FTIR Spectrometry	178
3.2.3.3	Advantages of FTIR Spectrometers	181
3.2.3.4	Example of a Hyperspectral Imaging Instrument	184
3.3	Superframing	185
3.3.1	Method	186
3.3.2	Example for High-Speed Imaging and Selected Integration Times	189
3.3.3	Cameras with Fixed Integration Time	189
3.4	Processing of IR Images	190
3.4.1	Basic Methods of Image Processing	191
3.4.1.1	Image Fusion	192
3.4.1.2	Image Building	193
3.4.1.3	Image Subtraction	196
3.4.1.4	Consecutive Image Subtraction: Time Derivatives	198
3.4.1.5	Image Derivative in the Spatial Domain	202
3.4.1.6	Digital Detail Enhancement	203
3.4.2	Advanced Methods of Image Processing	205
3.4.2.1	Preprocessing	207
3.4.2.2	Geometrical Transformations	209
3.4.2.3	Segmentation	212
3.4.2.4	Feature Extraction and Reduction	212
3.4.2.5	Pattern Recognition	215
3.5	Active Thermal Imaging	217
3.5.1	Transient Heat Transfer – Thermal Wave Description	219
3.5.2	Pulse Thermography	221
3.5.3	Lock-in Thermography	226
3.5.4	Nondestructive Testing of Metals and Composite Structures	229
3.5.4.1	Solar Cell Inspection with Lock-in Thermography	230
3.5.5	Pulsed Phase Thermography	234
	References	234
4	**Some Basic Concepts of Heat Transfer**	**239**
4.1	Introduction	239
4.2	The Basic Heat Transfer Modes: Conduction, Convection, and Radiation	239
4.2.1	Conduction	240
4.2.2	Convection	243
4.2.3	Radiation	244
4.2.4	Convection Including Latent Heats	245
4.3	Selected Examples for Heat Transfer Problems	247

4.3.1	Overview 247	
4.3.2	Conduction within Solids: the Biot Number 250	
4.3.3	Steady-State Heat Transfer through One-Dimensional Walls and U-Value 252	
4.3.4	Heat Transfer through Windows 257	
4.3.5	Steady-State Heat Transfer in Two- and Three-Dimensional Problems: Thermal Bridges 258	
4.3.6	Dew Point Temperatures 260	
4.4	Transient Effects: Heating and Cooling of Objects 261	
4.4.1	Heat Capacity and Thermal Diffusivity 262	
4.4.2	Short Survey of Quantitative Treatments of Time-Dependent Problems 262	
4.4.3	Typical Time Constants for Transient Thermal Phenomena 267	
4.4.3.1	Cooling Cube Experiment 267	
4.4.3.2	Theoretical Modeling of Cooling of Solid Cubes 267	
4.4.3.3	Time Constants for Different Objects 270	
4.5	Some Thoughts on the Validity of Newton's Law 271	
4.5.1	Theoretical Cooling Curves 272	
4.5.2	Relative Contributions of Radiation and Convection 275	
4.5.3	Experiments: Heating and Cooling of Light Bulbs 276	
	References 279	
5	**Basic Applications for Teaching: Direct Visualization of Physics Phenomena** 281	
5.1	Introduction 281	
5.2	Mechanics: Transformation of Mechanical Energy into Heat 281	
5.2.1	Sliding Friction and Weight 282	
5.2.2	Sliding Friction during Braking of Bicycles and Motorbikes 283	
5.2.3	Sliding Friction: the Finger or Hammer Pencil 286	
5.2.4	Inelastic Collisions: Tennis 286	
5.2.5	Inelastic Collisions: the Human Balance 288	
5.2.6	Temperature Rise of Floor and Feet while Walking 289	
5.2.7	Temperature Rise of Tires during Normal Driving of a Vehicle 290	
5.3	Thermal Physics Phenomena 291	
5.3.1	Conventional Hot-Water-Filled Heaters 292	
5.3.2	Thermal Conductivities 292	
5.3.3	Convections 295	
5.3.4	Evaporative Cooling 298	
5.3.5	Adiabatic Heating and Cooling 300	
5.3.6	Heating of Cheese Cubes 302	
5.3.7	Cooling of Bottles and Cans 305	
5.4	Electromagnetism 308	
5.4.1	Energy and Power in Simple Electric Circuits 308	
5.4.2	Eddy Currents 309	
5.4.3	Thermoelectric Effects 310	

5.4.4	Experiments with Microwave Ovens	*312*
5.4.4.1	Setup	*312*
5.4.4.2	Visualization of Horizontal Modes	*313*
5.4.4.3	Visualization of Vertical Modes	*314*
5.4.4.4	Aluminum Foil in Microwave Oven	*315*
5.5	Optics and Radiation Physics	*316*
5.5.1	Transmission of Window Glass, NaCl, and Silicon Wafer	*316*
5.5.2	From Specular to Diffuse Reflection	*318*
5.5.3	Blackbody Cavities	*320*
5.5.4	Emissivities and Leslie Cube	*322*
5.5.5	From Absorption to Emission of Cavity Radiation	*323*
5.5.6	Selective Absorption and Emission of Gases	*326*
	References	*327*

6	**IR Imaging of Buildings and Infrastructure**	*329*
6.1	Introduction	*329*
6.1.1	Publicity of IR Images of Buildings	*330*
6.1.2	Just Colorful Images?	*331*
6.1.2.1	Level and Span	*331*
6.1.2.2	Sequences of Color and Color Palette	*333*
6.1.3	General Problems Associated with Interpretation of IR Images	*333*
6.1.4	Energy Standard Regulations for Buildings	*336*
6.2	Some Standard Examples for Building Thermography	*338*
6.2.1	Half-Timbered Houses behind Plaster	*338*
6.2.2	Other Examples with Outside Walls	*342*
6.2.3	How to Find Out Whether a Defect is Energetically Relevant?	*342*
6.2.4	The Role of Inside Thermal Insulation	*344*
6.2.5	Floor Heating Systems	*347*
6.3	Geometrical Thermal Bridges versus Structural Problems	*348*
6.3.1	Geometrical Thermal Bridges	*349*
6.3.2	Structural Defects	*352*
6.4	External Influences	*356*
6.4.1	Wind	*356*
6.4.2	The Effect of Moisture in Thermal Images	*358*
6.4.3	Solar Load and Shadows	*361*
6.4.3.1	Modeling Transient Effects Due to Solar Load	*361*
6.4.3.2	Experimental Time Constants	*364*
6.4.3.3	Shadows	*366*
6.4.3.4	Solar Load of Structures within Walls	*367*
6.4.3.5	Direct Solar Reflections	*369*
6.4.4	General View Factor Effects in Building Thermography	*373*
6.4.5	Night Sky Radiant Cooling and View Factor	*375*
6.4.5.1	Cars Parked Outside or Below a Carport	*377*
6.4.5.2	Walls of Houses Facing Clear Sky	*378*
6.4.5.3	View Factor Effects: Partial Shielding of Walls by Carport	*379*

6.4.5.4	View Factor Effects: The Influence of Neighboring Buildings and Roof Overhang *380*	
6.5	Windows *381*	
6.6	Thermography and Blower-Door-Tests *386*	
6.7	Quantitative IR Imaging: Total Heat Transfer Through Building Envelope *391*	
6.8	Conclusions *394*	
	References *394*	
7	**Industrial Application: Detection of Gases** *397*	
7.1	Introduction *397*	
7.2	Spectra of Molecular Gases *397*	
7.3	Influences of Gases on IR Imaging: Absorption, Scattering, and Emission of Radiation *403*	
7.3.1	Introduction *403*	
7.3.2	Interaction of Gases with IR Radiation *403*	
7.3.3	Influence of Gases on IR Signals from Objects *405*	
7.4	Absorption by Cold Gases: Quantitative Aspects *408*	
7.4.1	Attenuation of Radiation by a Cold Gas *408*	
7.4.2	From Transmission Spectra to Absorption Constants *410*	
7.4.3	Transmission Spectra for Arbitrary Gas Conditions and IR Camera Signal Changes *411*	
7.4.4	Calibration Curves for Gas Detection *413*	
7.4.5	Problem: the Enormous Variety of Measurement Conditions *414*	
7.5	Thermal Emission from Hot Gases *416*	
7.6	Practical Applications: Gas Detection with Commercial IR Cameras *417*	
7.6.1	Organic Compounds *417*	
7.6.2	Some Inorganic Compounds *420*	
7.6.3	CO_2 – Gas of the Century *424*	
7.6.3.1	Comparison between Broadband and Narrowband Detection *426*	
7.6.3.2	Detecting Volume Concentration of CO_2 in Exhaled Air *427*	
7.6.3.3	Absorption, Scattering, and Thermal Emission of IR Radiation *427*	
7.6.3.4	Quantitative Result: Detecting Minute Amounts of CO_2 in Air *430*	
7.6.3.5	Quantitative Result: Detection of Well-Defined CO_2 Gas Flows from a Tube *430*	
	Appendix 7.A: Survey of Transmission Spectra of Various Gases *433*	
	References *441*	
8	**Microsystems** *445*	
8.1	Introduction *445*	
8.2	Special Requirements for Thermal Imaging *446*	
8.2.1	Mechanical Stability of Setup *446*	
8.2.2	Microscope Objectives, Close-up Lenses, Extender Rings *446*	
8.2.3	High-Speed Recording *447*	

8.3	Microfluidic Systems	449
8.3.1	Microreactors	449
8.3.1.1	Stainless Steel Falling Film Microreactor	449
8.3.1.2	Glass Microreactor	452
8.3.1.3	Silicon Microreactor	455
8.3.2	Micro Heat Exchangers	456
8.4	Microsensors	459
8.4.1	Thermal IR Sensors	459
8.4.1.1	Infrared Thermopile Sensors	459
8.4.1.2	Infrared Bolometer Sensors	461
8.4.2	Semiconductor Gas Sensors	465
8.5	Microsystems with Electric to Thermal Energy Conversion	467
8.5.1	Miniaturized Infrared Emitters	468
8.5.2	Micro Peltier Elements	470
8.5.3	Cryogenic Actuators	471
	References	474
9	**Selected Topics in Research and Industry**	**477**
9.1	Introduction	477
9.2	Thermal Reflections	477
9.2.1	Transition from Directed to Diffuse Reflections from Surfaces	478
9.2.2	Reflectivities for Selected Materials in the Thermal Infrared	482
9.2.2.1	Metals	483
9.2.2.2	Nonmetals	484
9.2.3	Measuring Reflectivity Spectra: Laboratory Experiments	485
9.2.3.1	Polarizers	485
9.2.3.2	Experimental Setup for Quantitative Experiments	485
9.2.3.3	Example for Measured Reflectivity Curve	486
9.2.4	Identification and Suppression of Thermal Reflections: Practical Examples	487
9.2.4.1	Silicon Wafer	487
9.2.4.2	Glass Plates	488
9.2.4.3	Varnished Wood	489
9.2.4.4	Conclusions	489
9.3	Metal Industry	490
9.3.1	Direct Imaging of Hot Metal Molds	491
9.3.2	Manufacturing Hot Solid Metal Strips: Thermal Reflections	491
9.3.3	Determination of Metal Temperatures if Emissivity is Known	493
9.3.4	Determination of Metal Temperatures for Unknown Emissivity: Gold Cup Method	494
9.3.5	Determination of Metal Temperatures for Unknown Emissivity: Wedge and Black Emitter Method	495
9.3.6	Other Applications of IR Imaging in the Metal Industry	497
9.4	Automobile Industry	498
9.4.1	Quality Control of Heating Systems	498

9.4.2	Active and Passive Infrared Night Vision Systems 500
9.5	Airplane and Spacecraft Industry 503
9.5.1	Imaging of Aircraft 503
9.5.2	Imaging of Spacecraft 505
9.6	Miscellaneous Industrial Applications 508
9.6.1	Predictive Maintenance and Quality Control 508
9.6.2	Pipes and Valves in a Power Plant 509
9.6.3	Levels of Liquids in Tanks in the Petrochemical Industry 510
9.6.4	Polymer Molding 513
9.6.5	Plastic Foils: Selective Emitters 515
9.6.6	Line Scanning Thermometry of Moving Objects 518
9.6.6.1	Line Scans of Fast-Moving Objects with IR Camera 518
9.6.6.2	Line Scanning Thermometry of Slow-Moving Objects 520
9.7	Electrical Applications 522
9.7.1	Microelectronic Boards 522
9.7.2	Old Macroscopic Electric Boards 523
9.7.3	Substation Transformers 524
9.7.4	Overheated High-Voltage Line 527
9.7.5	Electric Fan Defects 528
9.7.6	Oil Levels in High-Voltage Bushings 528
	References 530

10	**Selected Applications in Other Fields** 535
10.1	Medical Applications 535
10.1.1	Introduction 535
10.1.2	Diagnosis and Monitoring of Pain 537
10.1.3	Acupuncture 541
10.1.4	Breast Thermography and Detection of Breast Cancer 544
10.1.5	Other Medical Applications 545
10.1.5.1	Raynaud's Phenomenon 545
10.1.5.2	Pressure Ulcers 546
10.2	Animals and Veterinary Applications 548
10.2.1	Images of Pets 549
10.2.2	Zoo Animals 549
10.2.3	Equine Thermography 550
10.2.4	Others 551
10.3	Sports 553
10.3.1	High-Speed Recording of Tennis Serve 553
10.3.2	Squash and Volleyball 557
10.3.3	Other Applications in Sports 557
10.4	Arts: Music, Contemporary Dancing, Paintings, and Sculpture 559
10.4.1	Musical Instruments 560
10.4.2	Contemporary Dancing 561
10.4.3	Other Applications in the Arts 565
10.5	Surveillance and Security: Range of IR Cameras 566

10.5.1 Applications in Surveillance 566
10.5.2 Range of IR Cameras 568
10.6 Nature 574
10.6.1 Sky and Clouds 574
10.6.2 Wildfires 576
10.6.3 Geothermal Phenomena 579
10.6.3.1 Geysirs 579
10.6.3.2 Infrared Thermal Imaging in Volcanology 580
References 584

Index 589

Preface

The really large steps in the history of thermal imaging took place in intervals of hundred years. First, infrared radiation was discovered in 1800 by Sir William Herschel while studying radiation from the sun. Second, Max Planck was able to quantitatively describe the laws of thermal radiation in 1900. It took more than 50 years thereafter before the first infrared-detecting cameras were developed; initially, these were mostly quite bulky apparatus for military purposes. From about the 1970s, smaller portable systems started to become available; these consisted of liquid nitrogen cooled single photon detector scanning systems. These systems also enabled the use of infrared imaging for commercial and industrial applications. The enormous progress due to microsystem technologies toward the end of the 20th century – the first uncooled micro bolometer cameras appeared in the 1990s – resulted in reliable quantitatively measuring infrared camera systems. This means, that the third large step was taken by about the year 2000. Infrared thermal imaging has now become affordable to a wider public of specialized physicists, technicians and engineers for an ever growing range of applications. Nowadays, mass production of infrared detector arrays leads to comparatively low price cameras which – according to some advertisements – may even become high-end consumer products for everyone.

This rapid technological development leads to the paradoxical situation that there are probably more cameras sold worldwide than there are people who understand the physics behind and who know how to interpret the nice and colorful images of the false color displays: IR cameras easily produce images, but unfortunately, it is sometimes very difficult even for the specialist to quantitatively describe several of the most simple experiments and/or observations.

The present book wants to mitigate this problem by providing an extensive background knowledge on many different aspects of infrared thermal imaging for many different users of IR cameras. We aim at least for three different groups of potential users.

First, this book addresses all technicians and engineers who use IR cameras for their daily work. On the one hand, it will provide extensive and detailed background information not only on detectors and optics but also on practical use of camera systems. On the other hand, a huge variety of different application fields

Infrared Thermal Imaging. Michael Vollmer and Klaus-Peter Möllmann
Copyright © 2010 WILEY-VCH Verlag GmbH & Co. KGaA, Weinheim
ISBN: 978-3-527-40717-0

is presented with many typical examples with hints of how to notice and deal with respective measurement problems.

Second, all physics and science teachers at school or university level can benefit since infrared thermal imaging is an excellent tool for visualization of all phenomena in physics and chemistry related to energy transfer. These readers can particularly benefit from the huge variety of different examples presented from many fields, a lot of them given with qualitative and/or quantitative explanations of the underlying physics.

Third, this text also provides a detailed introduction to the whole field of infrared thermal imaging from basics via applications to up to date research. Thus it can serve as a textbook for newcomers or as a reference handbook for specialists who want to dig deeper. The large number of references to original work can easily help to study certain aspects in more depth and thus get ideas for future research projects.

Obviously, this threefold approach concerning the addressed readers does have some consequences for the structure of the book. We tried to write the ten chapters such that each may be read separately from the others. In order to improve the respective readability, there will be some repetitions and also cross references in each chapter (that more information can be found in other chapters or sections).

For example, teachers or practitioners may initially well skip the introductory more theoretical chapters about detectors or detectors systems and jump right away into the section of their desired applications. Obviously, this sometimes means that not every detail of explanation referring to theory will be understood, but the basic ideas should become clear – and maybe later on, those readers will also get interested in checking topics in the basic introductory sections.

The organization of this book is as follows: the first three chapters will provide extensive background information on radiation physics, single detectors as well as detector arrays, camera systems with optics, and IR image analysis. This is followed by a partly theoretical chapter on the three different heat transfer modes, which will help enable a better understanding of the temperature distribution that can be detected at the surfaces of various objects as e.g., buildings. Chapter 5 then gives a collection of many different experiments concerning phenomena in physics. This chapter was particularly written with teaching applications in mind. The subsequent three chapters discuss three selected application as well as research topics in more detail: building thermography as a very prominent everyday application, the detection of gases as a rather new emerging industrial application with very good future prospects and the analysis of microsystems for research purposes. Finally, the last two chapters give a large number of other examples and discussions of important applications ranging for example from the car industry, sports, electrical, and medical applications via surveillance issues to volcanology.

Our own background is twofold. One of us had originally worked in IR detector design before switching to microsystem technologies whereas the other worked on optics and spectroscopy. Soon after joining our present affiliation, a fruitful collaboration in a common new field, IR imaging, developed, starting with the purchase of our first MW camera in 1996. Meanwhile, our infrared group has access to three different IR camera systems from the extended MW to the LW range including a high speed research camera and a lot of additional equipment such as microscope lenses and so on. Besides applied research, our group focuses also on teaching the basics of IR imaging to students of Microsystem and Optical Technologies at our university.

Obviously, such a book cannot be written without the help of many people, be it by discussions, by providing images, or just by supporting and encouraging us in phases of extreme work load towards the end of this endeavor. We are therefore happy to thank in particular our colleagues Frank Pinno, Detlef Karstädt, and Simone Wolf for help with various tasks that had often to be done at very short notice.

Furthermore, we want to especially thank Bernd Schönbach, Kamayni Agarwal, Gary Orlove, and Robert Madding for fruitful discussions on selected topics and also for permission to use quite a large number of IR images.

We are also grateful to S. Calvari, J. Giesicke, M. Goff, P. Hopkins, A. Mostovoj, M. Ono, M. Ralph, A. Richards, H. Schweiger, D. Sims, S. Simser C. Tanner, and G. Walford for providing IR images and to A. Krabbe & D. Angerhausen, as well as DLR for providing other graphs. Also the following businesses have given permission to reproduce images, which we gratefully acknowledge: Alberta Sustainable Resource development, BMW, Daimler, FLIR systems, IPCC, IRIS, ITC, MoviTHERM, NAISS, NASA, Nature, NRC of Canada, PVflex, Raytek, Telops, Ulis, as well as United Infrared Inc.

Finally we need to especially thank our families for their tolerance and patience, in particular during the final months. Last not least we also need to express special thanks for the effective working together with Mrs. Ulrike Werner from Wiley/VCH.

Brandenburg, June 2010

1
Fundamentals of Infrared Thermal Imaging

1.1
Introduction

Infrared (IR) thermal imaging, also often briefly called *thermography*, is a very rapidly evolving field in science as well as industry owing to the enormous progress made in the last two decades in microsystem technologies of IR detector design, electronics, and computer science. Thermography nowadays is applied in research and development as well as in a variety of different fields in industry such as nondestructive testing, condition monitoring, and predictive maintenance, reducing energy costs of processes and buildings, detection of gaseous species, and many more. In addition, competition in the profitable industry segment of camera manufacturers has recently led to the introduction of low-cost models of the order of just several thousands dollars or euros, which opened up new fields of uses. Besides education (obviously schools are notorious for having problems with financing expensive equipment for science classes), IR cameras will probably soon be advertised in hardware stores as "must have" products for analyzing building insulation, heating pipes, or electrical components of one's own home. This development has its advantages as well as drawbacks.

The advantages may be illustrated by an anecdote based on personal experiences concerning physics teaching in school. Physics was, and still is, considered to be a very difficult subject in school. One of the reasons may be that simple phenomena of physics, for example, friction or the principle of energy conservation in mechanics, are often taught in such an abstract way that rather than being attracted to the subject, students are scared away. One of us clearly remembers a frustrating physics lesson at school dealing first with free-falling objects and next, with the action of walking on a floor. First, the teacher argued that a falling stone would transfer energy to the floor such that the total energy was conserved. He only used mathematical equations but stopped his argument at the conversion of initial potential energy of the stone to kinetic energy just prior to the impact with the floor. The rest was a hand-waving argument that, of course, the energy would be transformed to heat. The last argument was not logically developed it was just one of the typical teacher arguments to be believed (or not). Of course, at those times, it was very difficult at schools to actually measure

Infrared Thermal Imaging. Michael Vollmer and Klaus-Peter Möllmann
Copyright © 2010 WILEY-VCH Verlag GmbH & Co. KGaA, Weinheim
ISBN: 978-3-527-40717-0

the conversion of kinetic energy into heat. Maybe the children would have been more satisfied if he had at least attempted to visualize the process in more detail. The second example – explaining the simple action of walking – was similarly frustrating. The teacher argued that movement was possible owing to the frictional forces between shoe and floor. He then wrote down some equations describing the physics behind and that was all. Again, there were missing arguments: if someone walking has to do work against frictional forces, there must be some conversion of kinetic energy into heat and shoes as well as the floor must heat up. Again, of course, at those times, it was very difficult at schools to actually measure the resulting tiny temperature rises of shoes and floor. Nevertheless not discussing them at all was a good example of bad teaching. And again, maybe some kind of visualization would have helped. But visualizations were not one of the strengths of this old teacher who rather preferred to have Newton's laws recited in Latin.

Visualization is any technique for creating images, diagrams, or animations to communicate an abstract or a concrete argument. It can help bring structure into a complex context, it can make verbal statements clear, and/or give clear and appropriate visual representations of situations or processes. The underlying idea is to provide optical conceptions that help to better understand and better recollect the context. Today, in the computer age, visualization has ever-expanding applications in science, engineering, medicine, and so on. In the natural sciences, visualization techniques are often used to represent data from simulations or experiments in plots or images in order to make analysis of the data as easy as possible. Strong software techniques often enable the user to modify the visualization in real time, thus allowing easy perception of patterns and relations in the abstract data in question.

Thermography is an excellent example of a visualization technique that can be used in many different fields of physics and science. Moreover, it has opened up a totally new realm of physics in terms of visualization. Nowadays, it is possible to visualize easily the (for human eyes) invisible effects of temperature rise of the floor upon impact of a falling object or upon interaction with the shoe of a walking person. This will allow totally new ways of teaching physics and the natural sciences starting in school, and ending in the training of professionals in all kinds of industries. Visualization of "invisible" processes of physics and/or chemistry with thermography can be a major factor creating fascination for and interest in these subjects, not only in students at school and university but also for the layman. Nearly every example described later in this book can be studied in this context.

The drawbacks of promoting IR cameras as mass products for a wide range of consumers are less obvious. Anyone owning an IR camera will be able to produce nice and colorful images, but most will never be able to fully exploit the potentials of such a camera – and most will never be able to correctly use it.

Typically, the first images recorded with any camera will be the faces of people around. Figure 1.1 gives an example of IR images of the two authors. Anyone, confronted with such images for the first time, would normally find them fascinating

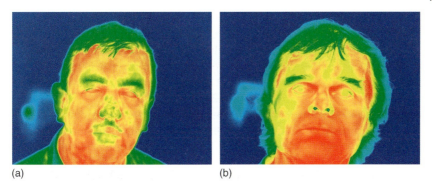

Figure 1.1 IR thermal images of (a) K-P. Möllmann and (b) M. Vollmer.

since they provide a totally new way of looking at people. The faces can still be recognized, but some parts look strange, for example, the eyes. Also, the nostrils (Figure 1.1b) seem to be distinctive and the hair seems to be surrounded by an "aura."

For artists, who want to create new effects, such images are fine, but thermography – if it is to be used for analysis of real problems like building insulation, and so on – is much more than this. Modern IR cameras may give qualitative images, colorful images that look nice but mean nothing, or they can be used as quantitative measuring instruments. The latter use is the original reason for developing these systems. Thermography is a measurement technique, which, in most cases, is able to quantitatively measure surface temperatures of objects. In order to use this technique correctly, professionals must know exactly what the camera does and what they can do to extract useful information from such IR images. This knowledge can only be gathered by professional training. Therefore, the drawback when purchasing an IR camera is that everybody needs professional training before one can correctly use such a camera. A multitude of factors can have an influence on the IR images and, hence, on any interpretation of such images (see Figure 1.2 and Chapters 2, 6).

First, radiation from an object (red) is attenuated via absorption or scattering while traveling through the atmosphere (Section 1.5.2), IR windows, or the camera optics (Section 1.5.4). Second, the atmosphere itself can emit radiation due to its temperature (blue) (this also holds for windows or the camera optics and housing

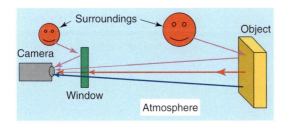

Figure 1.2 Various signal contributions entering an IR camera due to external influences.

itself) and third, warm or hot objects in the surroundings (even the thermographer is such a source) may lead to reflections of additional IR radiation from the object or windows, and so on (pink). The contributions from the object or windows may, furthermore, depend on the material, the surface structure, and so on, which are described by the parameter emissivity. These and other parameters are listed in Table 1.1; they are all discussed in later sections.

Even if all of these parameters are dealt with, there still remain some open questions that need to be answered. Consider for example someone who uses IR imaging in predictive maintenance doing electrical component inspections. Suppose the recording of an IR image shows a component with an elevated temperature. The fundamental problem is the assessment criterion for the analysis of IR images. How hot can a component become and still be okay? What is the criterion for an immediate replacement, or how long can one wait before

Table 1.1 Several parameters and factors that have an influence on images, recorded with modern IR camera systems.

Emissivity of object 　Distance of camera to object (usually in m, in the United States, possibly, ft.) Size of object Relative humidity Ambient temperature (usually in °C or K, in the United States, possibly, °F) Atmospheric temperature External optics temperature External optics transmission	Parameters that influence IR images generated from the raw detector data within a camera and that can usually be adjusted within the camera software. Quantitative results can strongly depend on some of these parameters They can often be changed while analyzing the images (after the recording) if the proper software is used (this may not be possible for the very inexpensive models)
Temperature span ΔT Temperature range + level Color palette	Parameters with influence of how the data are plotted as an image. If chosen unfavorably, important details may be disguised.
Wavelength dependence of emissivity (wavelength range of camera) Angular dependence of emissivity (angle of observation) Temperature dependence of emissivity Optical properties of matter between camera and object Use of filters (high temperature, narrowband, etc.) Thermal reflections Wind speed Solar load Shadow effects of nearby objects Moisture Thermal properties of objects, for example, time constants	Some parameters that can have a strong influence on quantitative analysis and interpretation of IR images.

replacement? These questions involve a lot of money if the component is involved in the power supply of an industrial complex, the failure of which can lead to a shutdown of a facility for a certain period of time.

Obviously, buying a camera and recording IR images sometimes only shifts the problem from not knowing the problem at all to the problem of understanding the used IR technology and all aspects of IR image interpretation. This book deals with the second problem. In this respect, it addresses at least three different groups of people who may benefit. First, it addresses any interested newcomers to the field by giving an introduction to the general topic of IR thermal imaging, by discussing the underlying fundamental physics, and by presenting numerous examples of the technique in research as well as in industry. Second, it addresses educators at all levels who want to include IR imaging into their curriculum in order to help understanding of physics and science topics as well as IR imaging, in general. Third, it addresses all practitioners who own IR cameras and who want to use it as a quantitative or qualitative tool for business. The text shall complement any kind of modern IR camera training/certification course as is offered by nearly every manufacturer of such camera systems.

It is the sincere hope of the authors that this text will help reduce the number of colorful, but often quite wrongly interpreted IR images of buildings and other objects in daily newspapers. In one typical example, a (probably south) wall of a house that had been illuminated by the sun for several hours before the IR image was taken was – of course – showing up to be warmer than the windows and other walls of the house. The interpretation that the wall was obviously very badly insulated was, however, pure nonsense (see Chapter 6). In addition, we hope that in the future, trained specialists will no longer call up their manufacturer and ask, for example, why they are not able to see any fish in their aquarium or in a pond. There will be no more complaints about malfunctioning of the system because the IR camera measures skin temperatures well above 45 °C. Or, to give a last example, people will no longer ask whether they are also able to measure the temperature of hot noble gases or oxygen using IR imaging.

We next come to the reason for bringing out a new book on IR thermal imaging. Of course, there are articles on several related topics in various handbooks [1–6], there is a number of books available on certain aspects like principles of radiation thermometry [7–9], detectors and detector systems and their testing [10–16], IR material properties [17, 18], the fundamentals of heat transfer [19, 20], an overview of the electromagnetic (EM) spectrum in the IR and adjacent regions [21], and there are, finally, also some concise books on practical applications [22–24]. However, not only the detector technologies but also the range of applications has increased enormously during the past decade. Therefore an up to date review of technology as well as applications seems to be overdue.

In this text, the international system of units is used. The only deviation from this rule concerns temperature, probably the most important quantity for thermography. Although temperatures should be given in Kelvin, IR camera manufacturers mostly use the Celsius scale for their images. For the North American customers, there is the option of presenting temperatures in the Fahrenheit scale. Table 1.2 gives a

Table 1.2 Relation between three commonly used temperature scales in thermography.

T (K)	T(°C)	T(°F)
0 (absolute zero)	−273.15	−459.67
273.15	0	32
373.15	100	212
1273.15	1000	1832

$\Delta T(K) = \Delta T(°C); \Delta T(°C) = (5/9) \times \Delta T(°F);$
$T(K) = T(°C) + 273.15; T(°C) = (5/9) \times (T(°F) - 32);$
$T(°F) = (9/5) \times T(°C) + 32.$

short survey of how these temperature readings can be changed from one unit to the other.

1.2
Infrared Radiation

1.2.1
Electromagnetic Waves and the Electromagnetic Spectrum

In physics, visible light, ultraviolet radiation, IR radiation, and so on, can be described as waves – to be more specific, as EM waves (for some properties of IR radiation, e.g., in detectors, a different point of view with the radiation acting like a particle is adopted, but for most applications, the wave description is more useful).

Waves are periodic disturbances (think, e.g., of vertical displacements of a water surface after a stone has been thrown into a puddle or lake) that keep their shape while progressing in space as a function of time. The spatial periodicity is called *wavelength*, λ (given in meters, micrometers, nanometers etc.), the transient periodicity is called *period of oscillation*, T (in seconds), and its reciprocal is the *frequency*, $\nu = 1/T$ (in s^{-1} or hertz). Both are connected via the speed of propagation c of the wave by Eq. (1.1):

$$c = \nu \cdot \lambda \tag{1.1}$$

The speed of propagation of waves depends on the specific type of wave. Sound waves, which exist only if matter is present, have typical speeds of about 340 m s^{-1} in air (think of the familiar thunder and lightning rule: if you hear the thunder 3 s after seeing the lightning, it has struck at a distance of about 1 km). In liquids, this speed is typically at least 3 times higher and in solids, the speed of sound can reach about 5 km s^{-1}. In contrast, EM waves propagate with the much higher speed of

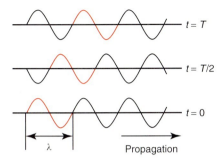

Figure 1.3 Three snapshots of a sinusoidal wave, traveling from left to right. Part of the wave (one wavelength) is marked (thick line) to demonstrate how the wave propagates as a function of time. The snapshots start at $t = 0$ then show the wave after half a period and a full period. In one period T, the wave has traveled one wavelength λ.

light, which is $c = 300\,000$ km s^{-1} in vacuum and $v = c/n$ in matter, with n being the index of refraction, which is a number of the order of unity.

In nature, the geometric form of the disturbances is very often sinusoidal, that is, it can be described by the mathematical sine function. Figure 1.3 schematically depicts snapshots for a sinusoidal wave, which travels from left to right in space. The bottom snapshot refers to a starting time $t = 0$. One wavelength λ is marked with a bold line. After half a period of oscillation (when $t = T/2$), the wave has moved by $\lambda/2$ and when $t = T$, it has moved by one wavelength in the propagation direction.

The disturbances, which resemble waves, can be of great variety. For example, sound waves in gases are due to pressure variations and water waves are vertical displacements of the surface. According to the type of disturbance, two wave types are usually defined: in longitudinal waves (e.g., sound waves in gases), the disturbance is parallel to the propagation direction, in transverse waves (e.g., surface-water waves), it is perpendicular to it. With springs (Figure 1.4), both types of waves are possible (e.g., sound waves in solids). In transverse waves,

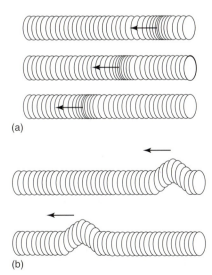

Figure 1.4 Illustration of longitudinal (a) and transverse waves (b). The latter have the additional property of polarization.

1 Fundamentals of Infrared Thermal Imaging

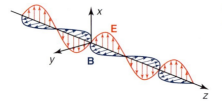

Figure 1.5 IR radiation involves a special kind of electromagnetic waves. In EM waves, the electric field and the magnetic induction field are perpendicular to each other and to the direction of propagation.

the disturbances can oscillate in many different directions. This is described by the property called *polarization*. The *plane of polarization* is the plane defined by the disturbance and the propagation direction.

Light and IR radiation are EM waves. In EM waves, the disturbances are electric and magnetic fields. They are perpendicular to each other and also perpendicular to the propagation direction, that is, EM waves are transverse waves (Figure 1.5). The maximum disturbance (or elongation) is called *amplitude*.

The polarization is defined by the electric field and the propagation direction, that is, the wave is polarized in the x–z-plane in Figure 1.5. Sunlight and light from many other sources like fire, candles, or light bulbs is unpolarized, that is, the plane of polarization of these light waves can have all possible orientations. Such unpolarized radiation may, however, become polarized by reflections from surfaces or on passing a so-called polarizer.

For Vis and IR radiation the simplest polarizer consists of a microscopically small conducting grid (similarly, a metal wire grid can polarize microwave radiation). If unpolarized radiation is incident on such a grid, only those waves whose electric field is oscillating perpendicular to the grid wires are transmitted (Figure 1.6). Such polarizing filters can help suppress reflections when taking photos with a camera (for more details, see [25, 26] and Section 9.2).

Figure 1.7 gives an overview of EM waves, ordered according to their wavelength or frequency. This spectrum consists of a great variety of different waves. All of them can be observed in nature and many have technical applications.

Starting from the top of the figure, for example, γ-rays have the highest frequencies, that is, the shortest wavelengths. X-rays are well known from their medical applications and ultraviolet radiation is important in the context of the ozone hole since less ozone in the upper atmosphere means more UV radiation

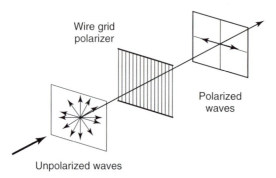

Figure 1.6 Microscopic wire grids can act as polarizers for infrared radiation. The transmitted electric field oscillates perpendicular to the direction of the wires.

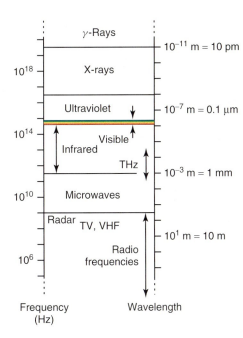

Figure 1.7 Overview of the most common types of electromagnetic waves. The visible spectral range (Vis) covers only a tiny part with wavelengths from 0.38 to 0.78 μm, followed at longer wavelengths by the infrared (IR) from 0.78 μm to 1 mm.

of the sun on earth, which can cause skin cancer. The visible light, defined by the sensitive range of the light receptors in our eyes, only covers a very small range within this spectrum with wavelengths from 380 to 780 nm. The adjacent spectral region with wavelengths from 780 up to 1 mm is usually called *infrared*, which is the topic of this book. This range is followed by microwaves, RADAR, and all EM waves that are used for radio, TV, and so on. Recently, new sensing developments in the frequency range from 0.1 to 10 THz have led to the newly defined range of terahertz radiation, which overlaps part of the IR and microwave ranges.

For IR imaging, only a small range of the IR spectrum is used. It is shown in an expanded view in Figure 1.8. Typically, three spectral ranges are defined for thermography: the long-wave (LW) region from around 7 to 14 μm, the midwave (MW) region from around 3 to 5 μm and the shortwave (SW) region from 0.9 to 1.7 μm. Commercial cameras are available for these three ranges. The restriction to these wavelengths follows from considerations of the amount of thermal radiation to be expected (Section 1.3.2), from the physics of detectors (Chapter 2), and from the transmission properties of the atmosphere (Section 1.5.2).

The origin of naturally occurring EM radiation is manifold. The most important process for thermography is the so-called thermal radiation, which will be discussed in detail in the next section. In brief, the term *thermal radiation* implies that every body at a temperature $T > 0\,K$ ($-273.15\,°C$) emits EM radiation. The amount of radiation and its distribution as a function of wavelength depend on temperature

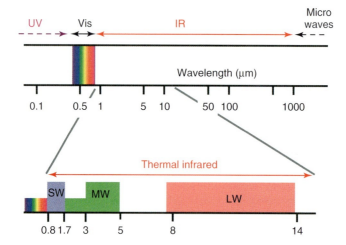

Figure 1.8 Infrared (IR) and adjacent spectral regions and expanded view of the so-called thermal infrared. This is the region where IR imaging systems for shortwave (SW), mid-wave (MW), or long-wave (LW) cameras exist. Special systems have extended MW or SW ranges.

and material properties. For temperatures in the range of natural and technological processes, this radiation is in the IR spectral region.

1.2.2
Basics of Geometrical Optics for Infrared Radiation

1.2.2.1 Geometric Properties of Reflection and Refraction

From the observation of shadows or the use of laser pointers, it is an everyday experience that visible light propagates more or less in straight lines. This behavior is most easily described in terms of geometrical optics. This description is valid if the wavelength of the light is much smaller than the size of the objects/structures onto which the light is incident. IR radiation has a behavior very similar to that of visible light; hence, it can also often be described using geometrical optics:

- In homogeneous materials, IR radiation propagates in straight lines. It can be described as rays, whose propagation follows from geometrical laws. Usually the rays are indicated by arrows.
- At the boundary between two materials, part of the incident radiation is reflected, and part of it is transmitted as refracted IR radiation (Figure 1.9).
- The optical properties of homogeneous materials, for example, those of a lens of an IR camera, are described by the index of refraction n. This index n is a real number larger than unity for nonabsorbing materials and a complex mathematical quantity for absorbing materials. The properties of materials for the thermal IR spectral range are described in detail in Section 1.5.

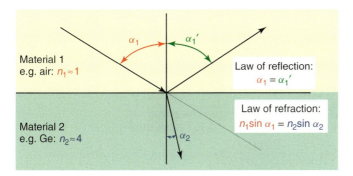

Figure 1.9 The laws of reflection and refraction in geometrical optics. The examples given in the figure refer to IR radiation, which is incident from air onto a germanium surface.

- The orientation between incident radiation and surface normal of the boundary between the two materials is called the *angle of incidence*, α_1. The respective angle between the reflected ray and the surface normal is the angle of reflection α_1'. The law of reflection – which is the basis for mirror optics – states that

$$\alpha_1' = \alpha_1 \tag{1.2}$$

- The orientation between the transmitted refracted radiation and surface normal of the boundary between the two materials is called the *angle of refraction*, α_2. If the index of refraction of the two materials are given by n_1 and n_2, the law of refraction (also called *Snell's law*) – which is the basis for lens optics – states that

$$n_1 \sin \alpha_1 = n_2 \sin \alpha_2 \tag{1.3}$$

The propagation of radiation in matter can be described using these two laws. Nowadays, this is mostly done by using ray-tracing programs, which allow to follow the path of radiation through many different materials and a large number of boundaries, as are usual, for example, in complex lens systems, used in optical instruments such as IR cameras.

Refraction is also the basis for studying spectra and defining the IR spectral region. In the seventeenth century, Isaac Newton demonstrated that the index of refraction of materials depends on the wavelength of visible light. The same holds for IR radiation. Typical optical materials show normal dispersion, that is, the index of refraction decreases with increasing wavelength. If radiation is incident on a prism, the angles of refracted light within the prism are determined from the law of refraction. If the index of refraction decreases with increasing wavelength, the radiation with longer wavelength is refracted less than the short wavelength radiation. The same holds for the second boundary of the prism. As a result, the incident EM radiation is spread out into a spectrum behind the prism. If thermal IR radiation is incident and the prism is made of an IR transparent material the LW

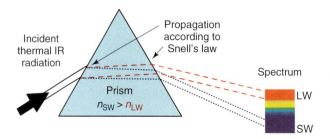

Figure 1.10 Generating a spectrum of EM radiation by using a nonabsorbing prism.

radiation will end up on top and the SW radiation on the bottom of the spectrum in Figure 1.10.

1.2.2.2 Specular and Diffuse Reflection

Finally, IR imaging often studies objects with rough surfaces. In such cases, reflection need not be directed, but can also have a diffuse component. Figure 1.11 illustrates the transition from directed reflection also called specular reflection to diffuse reflection.

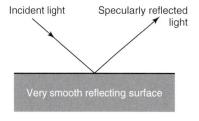

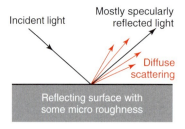

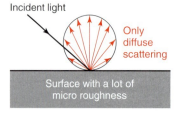

Figure 1.11 During the transition from smooth to rough surfaces, optical reflections change from specular to diffuse reflection. The most common case is a combination of both types of reflection.

1.2.2.3 Portion of Reflected and Transmitted Radiation: Fresnel Equations

Equations (1.2) and (1.3) give the directions of the reflected and transmitted radiation. The portions of reflected and transmitted light, which depend on angle of incidence can also be computed from the so-called Fresnel equations of wave optics [25, 27]. For the purpose of this book, graphical results will be sufficient. Figure 1.12 shows representative results for the reflection of visible light ($\lambda = 0.5\,\mu m$) for a boundary between air and glass and for IR radiation ($\lambda = 10\,\mu m$) for a boundary between air and silicon.

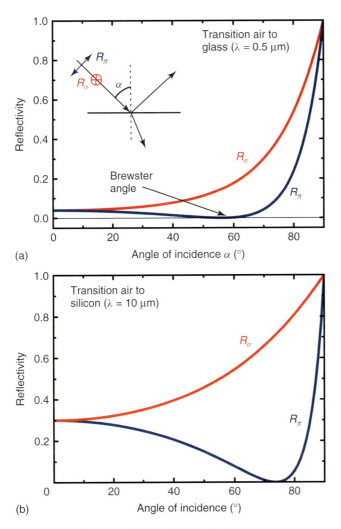

Figure 1.12 The portion of reflected visible light from an air–glass boundary (a) or of IR radiation from an air–silicon boundary (b) depend on the polarization of the radiation.

The reflected light consists of two contributions: the first is light that is polarized parallel (R_π) and the second is light that is polarized perpendicular (R_σ) to the plane of incidence. The latter is defined as the plane made up by the surface normal of the boundary and the propagation direction of the EM radiation. For the perpendicular component, the portion of reflected radiation, called *reflectivity* slowly increases from 0.04 for the Vis case (and about 0.3 for IR) at normal incidence ($\alpha = 0°$) up to the maximum value of 1.0 at grazing incidence ($\alpha = 90°$). In contrast, the parallel component first decreases from the same starting value at $\alpha = 0°$, reaches a specific Brewster angle with zero reflectivity before steeply rising to the maximum value of 1.0 at grazing incidence.

In general, the Brewster angle α_{Br} for a transition from a transparent material A (e.g. air) to a transparent material B (e.g. glass) is defined by the condition

$$\tan(\alpha_{Br}) = \frac{\sin(\alpha_{Br})}{\cos(\alpha_{Br})} = \frac{n_B}{n_A} \tag{1.4}$$

which, for the above examples, gives $\alpha_{Br} = 56.3°$ for air–glass (Vis) and $\alpha_{Br} = 75°$ for air–Si (IR). For absorbing materials, the general form of the curves stays the same, however, the minimum may not reach zero and the Brewster angles may shift.

Thermal reflections are important in many applications of thermography. For quick numerical estimates of transparent materials, we mention Eq. (1.5), which gives the reflectivity R (portion of reflected radiation) in terms of the refractive indices for the case of normal incidence ($\alpha = 0°$). The material A from which radiation is incident is considered to be transparent (e.g., air), whereas the material B from which radiation is reflected can be opaque to the radiation. In this case, the index of refraction, which can be found as a function of wavelength in tables of several handbooks [18] is mathematically a complex number $n_B = n_1 + in_2$ and

$$R(\alpha = 0°, n_A, n_B = n_1 + in_2) = \frac{(n_1 - n_A)^2 + n_2^2}{(n_1 + n_A)^2 + n_2^2} \tag{1.5}$$

In the above examples, $n_A = 1.0$, $n_1 = 1.5$, and $n_2 = 0$, which gives $R = 0.04$ for air–glass at $\lambda = 0.5\,\mu m$ and $n_A = 1.0$, $n_1 = 3.42$ and $n_2 = 6.8 \times 10^{-5}$, which gives $R = 0.30$ for air–Si at $\lambda = 10\,\mu m$. Some applications of these results are discussed in the section on suppression of thermal reflections (Section 9.2).

1.3
Radiometry and Thermal Radiation

In any practical measurement with an IR camera, an object emits radiation into the direction of the camera, where it is focused on the detector and measured quantitatively. Since thermography is mostly done with solid objects, which are, furthermore, opaque to IR radiation, the emission refers to the surfaces of the objects, only (the most important exception – gases – will be treated separately in Chapter 7). Let us consider a small surface area element dA that emits thermal radiation in the specific direction of the detector, which occupies a certain solid angle. In order to characterize the emission, propagation, and irradiance of any kind of radiation – that is, also the thermal radiation discussed here – with respect to the detector, a set of several radiometric quantities has been defined, which will be introduced in Section 1.3.1. As will become evident a certain class of emitters – so-called blackbodies – has unique properties concerning the total amount of emitted radiation as well as their geometrical distribution. This will be discussed in Section 1.3.2.

1.3.1
Basic Radiometry

1.3.1.1 Radiant Power, Excitance, Irradiance

Consider an element dA of the radiating surface of an object. The total energy flux $d\Phi$, from this surface element dA into the hemisphere is called *power*, *radiant power*, or *energy flux* with the SI unit W (watt) (Figure 1.13). This quantity can only be measured directly, if the detector collects radiation from the hemisphere completely. This is usually not the case.

If this radiant power is related to the emitting surface area, we find the excitance M, in W m^{-2}, using

$$M = \frac{d\Phi}{dA} \tag{1.6}$$

Obviously excitance (sometimes also called *emittance* or *emissive power* [16, 20]) characterizes the total radiant power within the hemisphere that is emitted by the surface area. This energy flux contains contributions of all emitted wavelengths (for sake of simplicity, we write total derivatives, we note, however, that these are partial derivatives, since the radiant power depends also on angles and wavelength).

If, in contrast, we consider the total incident power from a hemisphere on a given surface dA, the same definition leads to the irradiance $E = d\Phi/dA$. Obviously, excitance and irradiance refer to the same units of measurement, namely, watts per meter squared (W m^{-2}), but the corresponding energy flux is either emitted or received by a particular surface area dA.

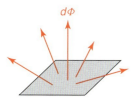

Surface area element dA

Figure 1.13 The total energy flux $d\Phi$, emitted by a surface area element dA is emitted in a hemisphere above dA.

16 | *1 Fundamentals of Infrared Thermal Imaging*

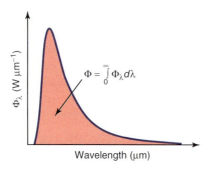

Figure 1.14 Relation between radiant power Φ and its spectral density: The total radiant power Φ (red area) is found by summing up the contribution of Φ_λ (blue curve) over the whole wavelength range.

1.3.1.2 Spectral Densities of Radiometric Quantities

So far, radiant power, excitance, and irradiance for an area dA refer to the total power emitted to or received from a hemisphere. In practice, all radiometric quantities do, however, also depend on wavelength. Therefore one can easily define spectral densities of the various radiometric quantities. As an example, Figure 1.14 illustrates the relation between radiant power Φ and its spectral density Φ_λ.

$$\Phi_\lambda = \frac{d\Phi}{d\lambda} \tag{1.7}$$

Similar relations hold for all other radiometric quantities (Table 1.3). The excitance, that is, emissive power would then be called, for example, *spectral excitance* or *spectral emissive power*.

1.3.1.3 Solid Angles

Most often, surfaces of objects do emit radiation, but not uniformly in the hemisphere. In order to account for this directionality of emitted radiation, we

Table 1.3 Overview of important radiometric quantities.

Name	Symbol	Unit	Definition
Energy flux or radiant power	Φ	W	Emission of energy per time in hemisphere
Excitance	M	W m^{-2}	$M = \dfrac{d\Phi}{dA}$, dA: emitting surface, into hemisphere
Irradiance	E	W m^{-2}	$E = \dfrac{d\Phi}{dA}$, dA: receiving surface, from hemisphere
Radiant intensity	I	W (sr)$^{-1}$	$I = \dfrac{d\Phi}{d\Omega}$
Radiance	L	W (m^{2}sr)$^{-1}$	$L = \dfrac{d^2\Phi}{\cos\delta \, d\Omega \, dA}$
Spectral density X_λ of any chosen radiometric quantity X	X_λ	(Unit of X) (μm)$^{-1}$ or (unit of X) (nm)$^{-1}$ or (unit of X) m^{-1}	$X_\lambda = \dfrac{dX}{d\lambda}$

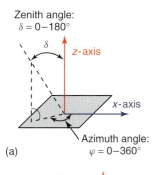

Zenith angle:
δ = 0–180°

(a)

Azimuth angle:
φ = 0–360°

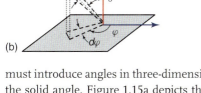

(b)

Figure 1.15 (a) Definition of angles in space and visualization of the solid angle, which, (b) at a given distance from the surface, is related to the area dA_{nor}, which is normal to the chosen direction.

must introduce angles in three-dimensional coordinate systems and the concept of the solid angle. Figure 1.15a depicts the geometry: the emitting area is located in the x–y-plane, that is, the z-axis is perpendicular to the area. Then, any direction in space can be defined by two angles, first, the azimuth angle φ, which is measured from the x-axis to the projection line of the chosen direction onto the x–y-plane, and second the zenith angle δ.

Usually, the direction itself is not important, rather, radiation is emitted toward a detector of given surface area. For simplicity we assume an area dA_{nor}, which is oriented perpendicular to the chosen direction at a distance R from a chosen point on the emitting surface. The area can be characterized by the small increments of angles $d\varphi$ and $d\delta$ as shown in Figure 1.15b. This leads to the definition of the solid angle element $d\Omega$ (for details, see textbooks on mathematics)

$$d\Omega(\delta, \varphi) = \sin\delta \cdot d\delta \cdot d\varphi = \frac{dA_{nor}}{R^2} \tag{1.8}$$

The unit of solid angle is the steradian (sr), similar to the radian (rad) for the planar angle. The full solid angle is 4π. Using δ, φ, and $d\Omega(\delta, \varphi)$, any emission of radiation into any given direction can be characterized, using the quantities radiant intensity and radiance.

1.3.1.4 Radiant Intensity, Radiance, and Lambertian Emitters

The radiant intensity I is the radiant power that is emitted from a point source of a radiating object into a solid angle element $d\Omega$ in a given direction, characterized by (δ, φ). Mathematically it is given by $I = d\Phi/d\Omega$ with unit watts/steradian (W sr^{-1}).

Radiant intensity (which, by the way, is the only quantity in optics where intensity is properly defined [28]) is related to the most often used quantity in radiometry, the radiance L. Radiance is used to characterize extended sources. It is defined as the amount of radiant power per unit of projected source angle and per unit solid angle.

$$L = \frac{d^2\Phi}{\cos\delta \, d\Omega \, dA}, \quad \text{i.e., } d^2\Phi = L\cos\delta \, d\Omega \, dA \tag{1.9a}$$

The significance of this slightly more complicated definition of radiance may become obvious if Eq. (1.9a) is written in a way to calculate the total radiant power from radiance:

$$\Phi = \iint L\cos\delta \, d\Omega \, dA \tag{1.9b}$$

The total radiant power results from summing up radiance contributions over area and solid angle of the hemisphere. If only the integration over solid angle is done, one ends up with the excitance.

$$M = \frac{d\Phi}{dA} = \int_{\text{hemisphere}} L\cos\delta \, d\Omega \tag{1.10a}$$

where as sole integration over surface area results in radiant intensity.

$$I = \frac{d\Phi}{d\Omega} = \int_{\text{source area}} L\cos\delta \, dA \tag{1.10b}$$

The geometrical factor $\cos\delta$ can be easily understood from Figure 1.16. Any emitting surface area dA is observed to be largest for a direction, which is perpendicular to the surface. For any other direction, only the projection of dA perpendicular to it can contribute to the emitted radiation.

Hence, radiance is a measure for the radiant power of an emitter with surface area dA that passes through a surface that is normal to the emission direction. Since this surface defines a solid angle in this direction, radiance is a true measure of the amount of radiation that is emitted into a certain direction and solid angle.

A summary of important radiometric quantities is listed in Table 1.3. SI units are used, however, for the spectral densities the wavelength interval is often given in micrometers or nanometers. This helps to avoid misinterpretations. Consider, for example, the spectral density of excitance, which is the total radiant power per area and wavelength interval. Its unit can be expressed as W (m² μm)$^{-1}$ or also as W m^{-3}. The first choice is much better since it avoids any misunderstanding, for example, when one refers to a power density per volume.

The difference between radiant intensity and radiance will become most obvious for so-called Lambertian radiators (after J.H. Lambert, an eighteenth-century scientist who did a lot of work in the field of photometry). A Lambertian radiator is one that emits or reflects a radiance that is independent of angle, that is,

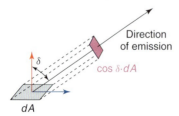

Figure 1.16 For a given direction, only the projection $dA \cdot \cos\delta$ can be seen from the emitting area dA.

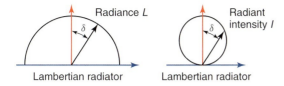

Figure 1.17 Lambertian radiators have constant radiance, but the radiant intensity depends on the direction of emission.

it emits uniformly into the hemisphere. Lambertian radiators are theoretical constructs that can be approximated in the real world by blackbody sources (Section 1.4.6) or perfect diffusely scattering surfaces. In this case, $L = $ constant and $I = L \cdot A \cdot \cos \delta = I_0 \cdot \cos \delta$. This angular dependence of radiant intensity and radiance is schematically depicted in Figure 1.17.

What does constant radiance mean? As stated above, radiance describes the amount of radiation emitted by an emitting area into a given direction and solid angle; therefore, a Lambertian source appears to emit equal amounts of radiation in every direction. A typical example of such behavior is known from visual optics. Diffusely scattering surfaces like, for example, tapestry, reflect the same amount of radiation in every direction. This means that, visually, the illuminated surface has the same brightness irrespective of the direction of observation. Similarly IR cameras detect radiance from objects since the area of the camera lenses define the solid angle, which is used for detection in the direction of the camera.

For Lambertian surfaces, the relation between excitance M and radiance L of a surface is simply given by $M = \pi L$. This also holds for the respective spectral densities, that is, in general, the excitance of Lambertian radiators equals π times their radiance. A summary of properties of Lambertian radiators is given in Table 1.4.

1.3.1.5 Radiation Transfer between Surfaces: Fundamental Law of Radiometry and View Factor

The concept of radiance helps to formulate radiation exchange between two surfaces. This will become important later, when discussing practical examples such as building thermography, where neighboring buildings or objects have a significant influence on the measured surface temperature of a wall or a roof (Section 6.4). The basic relations are introduced here.

Table 1.4 Some relations, holding for Lambertian radiators or reflectors.

Radiance	$L = $ constant	Isotropic
Radiant intensity	$I = I_0 \cos \delta$;	δ angle of direction to surface normal
Relation excitance to radiance	$M = \pi \cdot L$; $M(\lambda) = \pi \cdot L(\lambda)$	

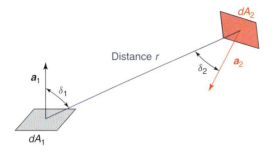

Figure 1.18 Geometry for radiation transfer between two surfaces dA_1 and dA_2.

Consider two surface area elements dA_1 and dA_2 that are arbitrarily positioned and oriented in space (Figure 1.18).

According to Eq. (1.9a), the radiant power $d^2\Phi$ that is emitted by area dA_1 and intercepted by dA_2 can be written as

$$d^2\Phi = L_1 \cos\delta_1 \, dA_1 \, d\Omega_2 \tag{1.11}$$

where L_1 is the radiance from dA_1, and $d\Omega_2$ is the solid angle under which dA_2 is seen from dA_1. The latter is given by $(\cos\delta_2)\, dA_2/r^2$, which leads to the *fundamental law of radiometry*:

$$d^2\Phi = \frac{L_1 \cos\delta_1 \cos\delta_2}{r^2} dA_1 dA_2 \tag{1.12}$$

The radiant power that is emitted by dA_1 and received by dA_2 depends on the distance and relative orientation of the two areas with respect to the connecting line. Eq. (1.12) allows to calculate the portion of the total radiant power of a finite area A_1 of an emitting object which is incident on a finite area A_2 of a receiving object (Figure 1.19).

From Eq. (1.12), the total radiant power Φ_{12} from area A_1 which is incident on area A_2 is given by

$$\Phi_{12} = L_1 \iint_{A_1, A_2} \frac{\cos\delta_1 \cos\delta_2}{r^2} dA_1 dA_2 \tag{1.13}$$

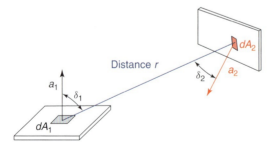

Figure 1.19 Geometry for definition of view factor. For simplicity, areas A_1 and A_2 are plotted as planar, but any kind of arbitrarily curved surface will also work.

Since the total radiant power Φ_1 that is emitted by A_1 into the hemisphere can be computed to be $\Phi_1 = \pi L_1 A_1$, one needs to find the portion of Φ_1 that is intersected by A_2, the so-called view factor.

$$F_{12} = \frac{\Phi_{12}}{\Phi_1} = \frac{1}{\pi A_1} \iint_{A_1,A_2} \frac{\cos \delta_1 \cos \delta_2}{r^2} dA_1 dA_2 \tag{1.14a}$$

Similarly, one may define the view factor F_{21}, which describes the portion of Φ_2 that is intersected by A_1. From these definitions, it follows that

$$A_1 F_{12} = A_2 F_{21} \tag{1.14b}$$

For any practical situation, one usually deals with at least two objects (the object under study and the IR camera itself) at different temperatures, hence, view factors will help analyze the total net energy flux from an object to an IR camera. We return to a the topic of view factors when discussing radiation exchange between objects (Section 6.4.4).

1.3.2
Blackbody Radiation

1.3.2.1 Definition

On the basis of fundamental physics, every object at any given absolute temperature above 0 K emits thermal radiation. The maximum radiant power that can be emitted by any object depends only on the temperature of the object that has led to the term *thermal radiation*. For real bodies, an additional material property, the emissivity, comes into play (Section 1.4).

In this section, we only deal with perfect emitters (see also Section 1.4.5) of thermal radiation. These are called *blackbodies*. Blackbodies resemble ideal surfaces, having the following properties [20]:

1) A blackbody absorbs every incident radiation, regardless of wavelength and direction.
2) For given temperature and wavelength, no surface can emit more energy than a blackbody.
3) Radiation emitted by a blackbody depends on wavelength, however, its radiance does not depend on direction, that is, it behaves like a Lambertian radiator.

As perfect absorbers and emitters, blackbodies serve as standards in radiometry.

Experimentally, blackbodies are most easily realized by cavities whose walls are kept at constant temperature. The notion of black (as defined by property 1) can easily be understood from an optical analog in the visible spectral range. If one observes a distant building with an open window, the inner part of the window looks black, indeed.

1.3.2.2 Planck Distribution Function for Blackbody Radiation

Very precise spectral measurements of thermal radiation of cavities, that is, experimental blackbodies, existed by the end of the nineteenth century. However, it was not before 1900, when Max Planck introduced his famous concept of the Planck constant h, that measured spectra could be satisfactorily explained. Planck's theory was based on thermodynamics but with the quantum nature for emission and absorption of radiation, he introduced a totally new concept not only in the theory of blackbody radiation, but in the whole world of physics. In modern language, the spectral excitance, that is, total radiant power into the hemisphere, of a blackbody of given temperature T in wavelength interval $(\lambda, \lambda + d\lambda)$ can be written as

$$M_\lambda(T)d\lambda = \frac{2\pi hc^2}{\lambda^5} \frac{1}{e^{\frac{hc}{\lambda kT}} - 1} d\lambda \qquad (1.15)$$

The respective radiance is $L_\lambda(T) = M_\lambda(T)/\pi$. Here, $h = 6.626 \times 10^{-34}$ J s is Planck's constant, $c = 2.998 \times 10^8$ m s^{-1} is the speed of light in vacuum, λ is the wavelength of the radiation, and T is the absolute temperature of the blackbody given in Kelvin.

Figure 1.20 depicts a series of blackbody spectra for various temperatures. The spectra refer to either radiance or excitance (the scale has arbitrary units). They have several characteristic features:

1) In contrast to emission from spectral lamps, these spectra are continuous.
2) For any fixed wavelength, radiance increases with temperature (i.e., spectra of different temperature never cross each other).
3) The spectral region of emission depends on temperature. Low temperatures lead to longer wavelengths, high temperatures to shorter wavelength emission.

The wavelength of peak transmission in this representation is found by locating the maximum via the condition $(dM_\lambda(T))/(d\lambda) = 0$. This leads to Wien's displacement law:

$$\lambda_{max} \cdot T = 2897.8 \text{ μm} \cdot \text{K} \qquad (1.16)$$

For blackbodies at 300, 1000, and 6000 K, maximum emission occurs around 10, 3, and 0.5 μm respectively. The first case resembles environmental radiation, the second, for example, a hot plate from an electric stove, and the third, the apparent average temperature of the outer layers of the sun. From daily experience, hot plates start to glow red, because the short wavelength part of the thermal emission enters the red part of the visible spectrum. Radiation from the sun appears white to us since it peaks in the middle of the visible spectrum.

1.3.2.3 Different Representations of Planck's Law

Besides the usual textbook representation of Planck's law in terms of radiance or excitance as a function of wavelengths (Eq. (1.15)), other equations can be found, illustrating the same phenomenon. Many spectrometers that are used to measure spectra of blackbody radiation are, for example, measuring signals as a function of frequency $\nu = c/\lambda$ (in hertz) or wavenumber $\tilde{\nu} = 1/\lambda$ (in cm^{-1}). The corresponding representation of spectra reveals important differences from the

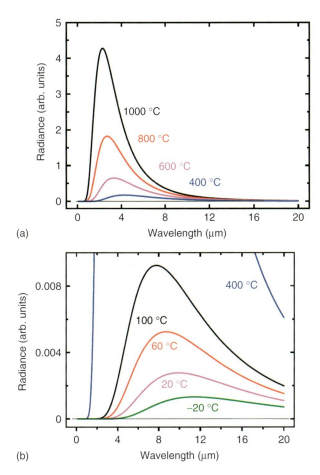

Figure 1.20 Radiance of blackbodies of temperatures between −20 and 1000 °C. Radiance is given in the same arbitrary units for (a) and (b).

wavelength representation. For example, Eq. (1.17) gives the frequency distribution of blackbody radiation:

$$M_v(T)dv = \frac{2\pi h v^3}{c^2} \frac{1}{e^{\frac{hv}{kT}} - 1} dv \quad (1.17)$$

Figure 1.21 compares a set of blackbody spectra in terms of excitance (i.e., spectral emissive power) as a function of wavelength $M_\lambda d\lambda$ and as a function of frequency $M_v dv$ in double log plots. Similar to Wien's displacement law for wavelength (Eq. (1.16)), a displacement law also exists for the frequency representation:

$$\frac{v_{max}}{T} = 5.8785 \times 10^{10} \text{ Hz K}^{-1} \quad (1.18)$$

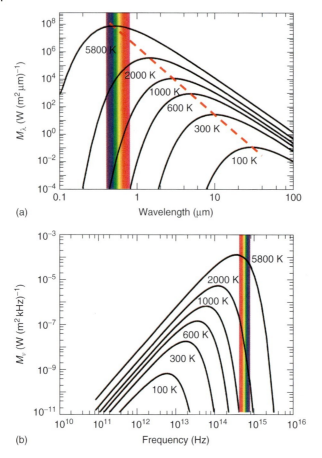

Figure 1.21 (a), (b) Comparison of two representations of Planck's law for blackbody radiation. The maximum of the curves lie at different spectral positions, depending on the choice of the variable. The dotted line in the wavelength representation (a) gives the position of the maxima according to Wien's law.

A note of caution: obviously the two representations have their peaks at different parts of the spectrum for the same temperature, since the distribution functions include $d\lambda$ and $d\nu$ that are related via a nonlinear equation [29]. For example, $M_\lambda d\lambda$ for $T = 5800$ K peaks in the Vis range at 500 nm, whereas $M_\nu d\nu$ peaks at 3.41×10^{14} Hz, which, according to $c = \nu\lambda$, corresponds to a wavelength of 880 nm. This behavior is a consequence of using distribution functions. One needs to be very careful when arguing about maxima of Planck blackbody spectra. The position of a maximum actually depends on the chosen representation. For IR cameras, this has no effect, of course, since we are always interested in the total radiant power within a certain spectral interval; and, of course, in the total radiant power within a certain spectral range, for example, the visible spectral range for

$T = 5800$ K (or the range 7–14 µm for 300 K radiation or any other range) is the same in both representations.

Blackbody radiation is one of the few topics that have inspired several Nobel prizes in physics. First, Wilhelm Wien won the prize in 1911 for his work on thermal radiation although it was Max Planck who finally solved the theoretical puzzle of correctly describing blackbody radiation. Nevertheless, Planck won the prize in 1918 for his general concept of the quantum nature of radiation, which had consequences reaching far beyond thermal radiation in the whole field of physics. The third prize in this field was awarded in 2006 to George Smoot and John Mather. They succeeded in recording the most famous blackbody radiation spectrum in astrophysics. This spectrum of cosmic background radiation is thought to resemble a kind of echo of the big bang of our universe. It was recorded in the early 1990s by the NASA satellite COBE. Figure 1.22 depicts results of this spectrum, which can be fitted with very high accuracy to a Planck function of temperature 2.728 (± 0.002) K.

Usually, only part of the spectrum of blackbody radiation is utilized in IR imaging. In the following two subsections, we deal with the total amount of radiation within certain spectral limits. The Stefan–Boltzmann law deals with the whole spectrum, extending from zero to infinity, whereas band emission is slightly more complicated.

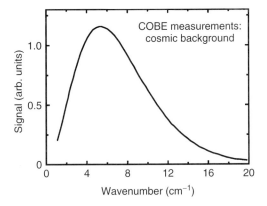

Figure 1.22 Cosmic ray background spectrum measured with the COBE satellite. The theoretical fit refers to $T = 2.728$ K. (Courtesy NASA.)

1.3.2.4 Stefan–Boltzmann Law
The excitance of a blackbody source is calculated from

$$M(T) = \int_0^\infty M_\lambda(T)d\lambda = \int_0^\infty M_\nu(T)d\nu = \sigma T^4 \qquad (1.19)$$

Here, $\sigma = 5.67 \times 10^{-12}$ W m^{-2} K^{-4} denotes the Stefan–Boltzmann constant.

The area under the spectral excitance (spectral emissive power) curve (Figure 1.23) gives the excitance (emissive power), which solely depends on the temperature of

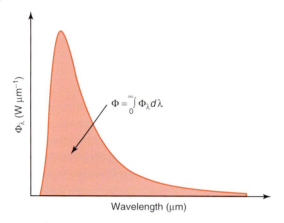

Figure 1.23 Visualization of the Stefan–Boltzmann law.

the blackbody. Hence, the total radiance associated with blackbody radiation is M/π. In astrophysics, the Stefan–Boltzmann law is used to calculate the total energy production of stars, for example, of the sun, from its known surface area and surface temperature.

1.3.2.5 Band Emission

In IR imaging, one never detects radiation of the whole spectrum, but rather the radiation in a predefined spectral range, which is determined by the detector and material properties of the optics and the atmosphere. Unfortunately, the integral of Eq. (1.19) does not have analytical solutions for arbitrary values of lower and upper limit. In order to simplify the results, one defines a blackbody radiation function $F_{(0\rightarrow\lambda)}$ as the fraction of blackbody radiation in the interval from 0 to λ, compared to the total emission from 0 to ∞ (Figure 1.24).

$$F_{(0\rightarrow\lambda)} = \frac{\int_0^\lambda M_\lambda d\lambda}{\int_0^\infty M_\lambda d\lambda} \tag{1.20}$$

We note, that unfortunately, view factors F_{ij} and this blackbody radiation function $F_{(0\rightarrow\lambda)}$ are denoted by the same letter F, the only difference being the subscripts. The context of the two quantities is, however, quite different and can be easily guessed from the subscripts. We therefore adopt this general similar notation for both.

The mathematical analysis shows, that the integrand only depends on the parameter $\lambda \cdot T$; therefore, integrals can be evaluated numerically for this parameter and hence, $F_{(0\rightarrow\lambda)}$ is tabulated as a function of λT (see e.g. [20]). Figure 1.25a depicts the corresponding results.

Obviously, the function $F_{(0\rightarrow\lambda)}$ can be easily used to calculate the fraction of blackbody radiation in an arbitrary wavelength interval (λ_1, λ_2).

$$F_{(\lambda 1\rightarrow\lambda 2)} = F_{(0\rightarrow\lambda 2)} - F_{(0\rightarrow\lambda 1)} \tag{1.21}$$

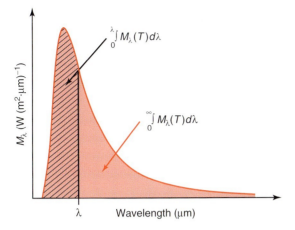

Figure 1.24 Definition of blackbody function: fraction of emitted excitance in spectral band (see text for details).

Figure 1.25b illustrates an example of how quick graphical estimates of $F_{(\lambda 1 \to \lambda 2)}$ are possible for $T = 400$ K and a wavelength range between 8 and 14 µm. In general, a given temperature and wavelength range $\lambda_1 \to \lambda_2$ define $\lambda \cdot T$ values (vertical broken lines). They intercept the $F(0, \lambda)$ curve at two specific values. Their difference (the distance between the horizontal broken lines) gives $F_{(\lambda 1 \to \lambda 2)}$.

As an example, the fraction of blackbody radiation within the hemisphere of an object of 500 K in the LW wavelength interval 7–14 µm is about 42.5%, but only 8.8% in the MW range from 3 to 5 µm. A very hot object, for example, a filament of a light bulb at a temperature of 2800 K leads to a fraction of 10.0% in the Vis wavelength interval from 0.39 to 0.78 µm and, similarly, to about 9.2% in the MW range from 3 to 5 µm.

1.3.2.6 Order of Magnitude Estimate of Detector Sensitivities of IR Cameras

Usually, blackbody radiators are used to calibrate IR cameras (Chapter 2). Using the concepts of radiometry and the laws of blackbody radiation, we can estimate the typical order of magnitude for the sensitivity of thermal radiation detectors, that is, how many watts of input power onto a detector element are needed to detect, for example, a 1 K temperature difference between two objects. Consider, for example, a blackbody radiator of temperature T_{BB} at a distance of $R = 1$ m in front of an LW IR camera at ambient temperature T_{cam}. The blackbody radiator has a circular shape and diameter $2r_{BB} = 5$ cm, the front lens of the IR camera should have a diameter of also $2r_{cam} = 5$ cm. The camera detects radiation in the spectral range from 7 to 14 µm.

The total radiant power that is incident on the camera from the object is given by Eq. (1.11).

$$d^2\Phi = L_1 \cos\delta_1 \, dA_1 \, d\Omega_2$$

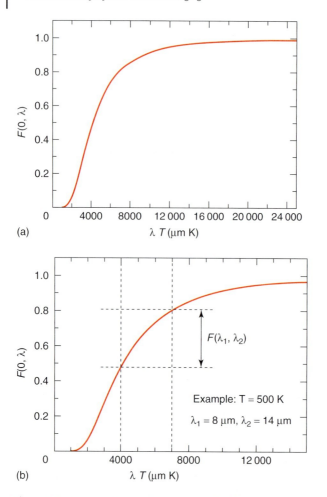

Figure 1.25 Fraction $F(0,\lambda)$ of excitance of blackbody radiation in the wavelength interval $0-\lambda$ as a function of $\lambda \cdot T$ (a), and how it can be used to estimate the fraction $F_{(\lambda 1 \to \lambda 2)}$ in a wavelength interval $\lambda_1 - \lambda_2$ (b).

Here, $\cos\delta_1 \approx 1$, the integral over the blackbody area gives $A_{BB} = \pi r_{BB}^2$, $L_1 = M_1/\pi$, and $d\Omega_2 = \pi r_{cam}^2/R^2$. Therefore the spectrally integrated radiant power incident on the camera is given by

$$\Phi_{BB \to cam} = M_{BB}(T_{BB}) \frac{r_{BB}^2 \cdot r_{cam}^2 \cdot \pi}{R^2} \tag{1.22}$$

Of course, there is also a similar radiant power emitted from the detector toward the blackbody source, however, the respective contributions to Φ due to detector temperature cancel since we deal only with changes of radiant power with object temperature. Therefore, this argument should also hold for all detector types (cooled as well as room-temperature ones).

In the following numerical example, we consider a LW camera system (a similar calculation can also be done for an MW or SW system). The emitted energy flux from a blackbody of temperature T_{BB} in the detector wavelength range is given by

$$\Phi_{BB} = \frac{r_{BB}^2 \cdot r_{cam}^2 \cdot \pi}{R^2} \int_{8\mu m}^{14 nm} M_{\lambda BB}(T_{BB}) d\lambda \tag{1.23}$$

The integral is given by $F_{(\lambda l \to \lambda 2)} \sigma T^4$. Any further calculation requires numerical temperature values. For simplicity, let us assume $T_{BB1} = 303$ K, $T_{BB2} = 302$ K, since most detector sensitivities are rated at $T = 30\,°C$. The integral values can be easily derived from $F_{(\lambda 1 \to \lambda 2)}$. For 303 and 302 K, they are very similar (0.378 and 0.377), therefore a single value of 0.38 is assumed.

In this case, the difference of incident radiative power for variation from 303 to 302 K of the blackbody temperature is given by

$$\frac{\Delta \Phi}{\Delta T} \approx \frac{r_{BB}^2 \cdot r_{cam}^2 \cdot \pi}{R^2} \cdot F_{\lambda 1 \to \lambda 2} \cdot \sigma \left(T_{BB1}^4 - T_{BB2}^4 \right) \tag{1.24}$$

Inserting the above values, we find 2.9×10^{-6} W K^{-1} (the total radiative power due to a blackbody of 303 K is about 2.2×10^{-4} W).

A typical standard lens of an IR camera has an acceptance angle of 24°. At a distance of 1 m, the blackbody source will only occupy an angular diameter of 2.86°. If 24° corresponds to the 320 pixel width of the detector, the blackbody source will be imaged on an angular diameter of about 19 pixels, corresponding to a circular area with about 1140 pixels. This means that each pixel will receive on average a difference in radiant power of 2.54 nW K^{-1}.

These numbers are probably still a factor of at least 3 too large since first, part of the radiation may be attenuated within the atmosphere on its way from the source to the camera (Section 1.5.2); second, the camera optics has a transmission smaller than 100% and third, the active detector area is only about 50% of the complete pixel area. As a final result, one may expect values of the order of 1 nW K^{-1} for the difference in radiative power received by each detector pixel for a 1 K difference. We come back to this estimate when discussing detectors and relate it to the noise equivalent temperatures (Section 2.2.2).

One may gain confidence in this kind of estimate by similarly deriving the solar constant, that is, the total radiant flux onto an area of 1 m^2 outside of the (attenuating) atmosphere of the earth. Starting again with Eq. (1.11) and using $A_{BB} = \pi r_{sun}^2$, $L_1 = M_1/\pi$, and $d\Omega_2 = 1\,m^2/R_{sun-earth}^2$, we find

$$\Phi_{solar\ constant} = M_{BB}(T_{sun}) \frac{r_{sun}^2 \cdot 1}{R_{sun-earth}^2} \tag{1.25}$$

$r_{sun} = 6.96 \times 10^5$ km, $R_{sun-earth} = 149.6 \times 10^6$ km, and $T = 5800$ K gives ≈ 1390 W m^{-2}, which, considering that the sun is not a true blackbody emitter, is a very good approximation for the solar constant.

1.3.2.7 Fractional Changes of Blackbody Radiation upon Temperature Changes

In thermography, IR sensors should be able to detect minute temperature changes. In order to compare the quality of different sensors (Chapter 2), it is useful to study how a fractional change of temperature $\Delta T/T$ of an object leads to a fractional change $\Delta L/L$ of incident radiance. Since radiance is proportional to excitance, we discuss changes of dM/M of a blackbody at temperature T due to temperature changes dT/T.

Differentiating Eq. (1.15) with respect to T and rearranging the quantities, one finds

$$\frac{dM_\lambda(T)}{M_\lambda(T)} \bigg/ \frac{dT}{T} = \frac{c_2}{\lambda T} \frac{e^{\frac{c_2}{\lambda T}}}{e^{\frac{c_2}{\lambda T}} - 1} \qquad (1.26a)$$

where $c_2 = hc/k = 14387.7\,\mu\text{m} \cdot \text{K}$. Figure 1.26 depicts a plot of this quantity as a function of $\lambda \cdot T$.

For most practical applications, Eq. (1.26a) can be simplified. If $\lambda \cdot T < 2898\,\mu\text{m} \cdot \text{K}$ (Eq. 1.16) the exponential is much larger than unity, and the second fraction is close to unity. In this case,

$$\frac{dM_\lambda(T)}{M_\lambda(T)} \bigg/ \frac{dT}{T} \approx \frac{c_2}{\lambda T} \qquad (1.26b)$$

with maximum deviation of 0.7%. For larger values, for example, if $\lambda \cdot T$ reaches 4000, deviations of Eq. (1.26b) from the exact solution of Eq. (1.26a) are less than 2.8%.

Equations (1.26a) or (1.26b) allow to estimate how total radiative power incident onto a detector varies upon a change of temperature as a function of wavelength. As an example, we discuss how a fractional change of 1% in temperature from $T = 400\,\text{K}$ to $T = 404\,\text{K}$ results in different fractional changes of excitance in the MW and the LW spectral region. For simplicity, the MW region should be characterized by a wavelength of $\lambda_1 = 4\,\mu\text{m}$ and the LW region by $\lambda_2 = 10\,\mu\text{m}$.

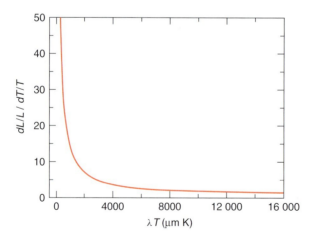

Figure 1.26 Fractional change of excitance of a blackbody due to a fractional change of temperature according to Eq. (1.26a).

Using Eq. (1.26a) with $\lambda_1 \cdot T = 1600\,\mu m\,K$ and $\lambda_2 \cdot T = 4000\,\mu m\,K$, one finds

$$\frac{dM_{4\mu m}(T)}{M_{4\mu m}(T)} \bigg/ \frac{dT}{T} \approx 2.43 \cdot \frac{dM_{10\mu m}(T)}{M_{10\mu m}(T)} \bigg/ \frac{dT}{T} \qquad (1.27)$$

This means that a change of 4 K leads to a 2.43 larger fractional change in MW IR radiation than for LW IR radiation. As a consequence of this fractional changes argument, MW or SW cameras will have advantages compared to LW cameras due to the more pronounced dependance of signal on object temperatures.

Of course, the camera systems only detect certain wavelength ranges of the spectrum, that is, the correct argument must compute the total received power within the respective wavelength ranges 3–5 μm and 8–14 μm. Figure 1.27a depicts the portion of blackbody radiation as a function of temperature. The Planck curves

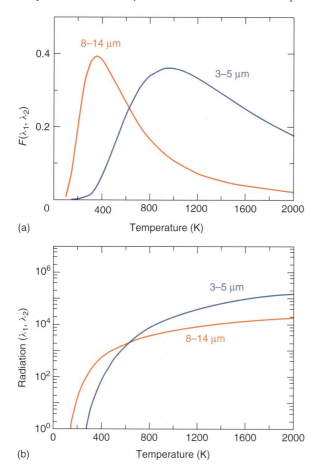

Figure 1.27 (a) Portion of blackbody radiation in the wavelength regions of MW and LW camera systems and (b) respective excitance in these wavelength regions as a function of blackbody temperature.

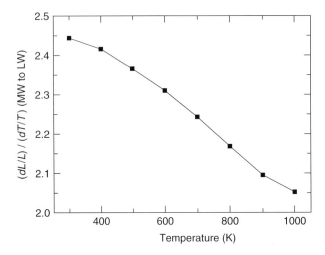

Figure 1.28 Ratio of fractional signal changes with temperature for the MW to the LW range.

(Figures 1.20 and 1.21) shift to shorter wavelengths with increasing temperature. Therefore, the portion in the wavelength window 8–14 μm first increases, reaches a maximum at $T \approx 350$ K, and then decreases. The same happens also for the MW region for higher temperatures (peak at $T \approx 950$ K). The drops are, however, compensated by the much larger increase of radiance with temperature of the Planck curves. This is illustrated in Figure 1.27b, which depicts the excitance within the spectral windows (i.e., $F_{(\lambda 1 \to \lambda 2)} \sigma T^4$).

The fractional signal change with fractional temperature change argument manifests itself in the different slopes of the curves (Figure 1.27b). For any chosen temperature, the slope of the 3–5 μm curve is much steeper than for the 8–14 μm curve. The detailed analysis leads to the result shown in Figure 1.28.

The exact calculation using the wavelength ranges 3–5 μm versus 8–14 μm reveals a slight change from the above simple expectation based upon Eq. (1.26a). For 400 K the result is 2.415 rather than 2.43. In addition, one observes a decrease with temperature. In any case, the overall statement holds that the shorter the wavelength, the more signal change results from any given temperature change.

1.4
Emissivity

1.4.1
Definition

Blackbodies are idealizations and no real object can emit this maximum thermal radiation at a given temperature. The real emission of thermal radiation from any object can, however, be easily computed, by multiplying the blackbody radiation

Table 1.5 Various definitions of emissivity.

Spectral directional emissivity: L, spectral radiance	$\varepsilon(\lambda, \delta, \varphi, T) = \dfrac{L(\lambda, \delta, \varphi, T)}{L_{BB}(\lambda, T)}$	(1.28a)
Spectral hemispherical emissivity, directionally averaged: M, spectral excitance	$\varepsilon(\lambda, T) = \dfrac{M(\lambda, \delta, \varphi, T)}{M_{BB}(\lambda, T)}$ with $M = \int_{\text{hemisphere}} L \cos\delta \, d\Omega$	(1.28b)
Total-directional emissivity: wavelength averaged	$\varepsilon(\delta, \varphi, T) = \dfrac{L(\delta, \varphi, T)}{L_{BB}(T)}$	(1.28c)
Total hemispherical emissivity: wavelength and direction averaged	$\varepsilon(T) = \dfrac{M(T)}{M_{BB}(T)} = \dfrac{M(T)}{\sigma T^4}$	(1.28d)

with a quantity that describes the influence of the object under study, the emissivity ε. In other words: *the emissivity of an object is the ratio of the amount of radiation actually emitted from the surface to that emitted by a blackbody at the same temperature.*

Differing definitions of emissivity are possible, depending on which quantity is used to describe the radiation. In radiometry, four definitions are used (Table 1.5). The relevant radiometric quantities are (i) spectral directional emissivity, (ii) spectral hemispherical emissivity, (iii) directional total emissivity, and (iv) forth total hemispherical emissivity.

Unfortunately, none of these definitions refers to the required conditions for practical IR imaging, where one first usually deals with objects that are observed near normal incidence or at small angles. Hence, a directional quantity, averaged over the desired angular range is needed. Second, IR cameras operate at predefined wavelength ranges. Obviously, the required emissivity needs to be averaged over the desired wavelength range. Symbolically, we need an emissivity $\varepsilon(\Delta\lambda, \Delta\Omega, T)$. We will come back to consequences for practical work at the end of this chapter.

1.4.2
Classification of Objects according to Emissivity

From the definition of emissivity, it is clear that $0 \leq \varepsilon \leq 1$. Figure 1.29 illustrates spectral hemispherical emissivities and corresponding spectra of emission of thermal radiation at given temperature for first a blackbody, second, for a so-called gray body with constant ε-value, that is, ε is independent of wavelength, and, third, a so-called selective emitter where ε varies as a function of wavelength.

For most practical applications in thermography, $\varepsilon(\lambda, T)$ is a constant, which resembles a gray body. Whenever substances are studied, which have absorption and emission bands in the thermal IR spectral range, like, for example, gases or plastic foils, one has to deal with selective emitters, which may complicate the quantitative analysis.

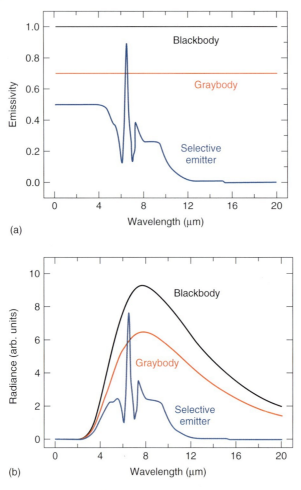

Figure 1.29 Spectral hemispherical emissivities (a) and corresponding thermal radiation (b) spectra for blackbodies, gray bodies, and selective emitters.

1.4.3
Emissivity and Kirchhoff's Law

The emissivity can be guessed from *Kirchhoff's law* (e.g., [20, 30]), which states that the amount of radiation absorbed by any object is equal to the amount of radiation that is emitted by this object. This is usually written in the form

$$\varepsilon = \alpha \tag{1.29}$$

where ε and the so-called absorptivity α denote the fraction of radiation that is either emitted or absorbed. Energy conservation requires that any radiation (Φ_0), incident on any object is either reflected (Φ_R, be it directed according to the law of reflection or diffusely scattered from rough surfaces), transmitted (Φ_T) through the object, or absorbed (Φ_A) within the object.

$$\Phi_0 = \Phi_R + \Phi_T + \Phi_A \tag{1.30a}$$

Considering the fraction of the incident radiation (e.g., in excitance or radiance), this law reads as

$$1 = R + t + \alpha \tag{1.30b}$$

where R and t denote the fraction of radiation that is either reflected or transmitted. Combining Eqs. (1.29) and (1.30b) allows estimates of the emissivity ε. The most simple examples are opaque solids with $t = 0$. In this case,

$$\varepsilon = 1 - R \tag{1.31}$$

that is, the emissivity follows directly from known values for the total reflectivity. It is important to note that R includes not only directed reflectivity (Figure 1.9) as is usual for polished surfaces but also diffuse reflectivity (Figure 1.11), which additionally occurs for rough surfaces. Eq. (1.31) allows to guess ε-values. For example, glass, which is more or less opaque in the IR with reflectivities in the IR range of a few percent, will have emissivities of $\varepsilon > 0.95$. In contrast, metals with high reflectivities above 90% will have emissivities below 0.1. Very well polished metal surfaces can have the lowest possible emissivities of the order 0.01, which practically render IR imaging impossible.

1.4.4
Parameters Affecting the Value of Emissivity

As a material property, the emissivity depends on the following parameters (Table 1.6).

1.4.4.1 Material

The major parameter is the kind of material. Depending on measurement techniques, averages are taken over certain angular and spectral ranges as is useful in thermography. In a simplified classification, one can separately discuss nonmetals and metals, because – fortunately – most nonmetallic materials that are needed for practical thermography applications like skin, paper, paints, stones, glass, and so on are gray emitters and have fairly high emissivity values of above 0.8. In contrast, metals and, in particular, polished metals, pose problems due to their often very low emissivities with values below 0.2.

Table 1.6 Parameters that have an influence on the emissivity ε.

Intrinsic object properties	Variations due to other parameters
Material (metal, insulator, etc.)	Observation direction (viewing angle)
Surface structure (rough/polished)	Wavelength (LW/MW/SW, etc.)
Regular geometry (grooves, cavities, etc.)	Temperature (phase changes, etc.)

1.4.4.2 Surface Structure

For any given material, the emissivity may vary considerably due to surface structure. This leads to the unfavorable situation, that, for the same material, many different values for emissivity are found. This effect is most pronounced for metals. Whereas polished metals can reach values of ε as low as 0.02, the emissivity can be much larger and even reach values above 0.8 if the surfaces are roughened. The highest values for originally metallic parts are found, if their surfaces are modified via oxidation/corrosion over the course of time. Imagine, for example, a metallic bolt of an electrical connection, which is operated for many years, while being exposed to oxygen from the air and water from rain, and so on. A value as high as 0.78 has been reported for strongly oxidized Cu, and 0.90 for some steel compounds due to such chemical modifications of the surface. This factor is most important for any inspection of, for example, bolts, nuts, electrical clamps, and similar parts in electrical components since a thermographer must have criteria for "fail or pass" of the component. Therefore, a quantitative analysis is necessary, that is, one must know the exact temperature difference of these components compared to their surroundings of known emissivity. And the measured temperature does very strongly depend on the actual value of emissivity.

To illustrate the effect of material as well as surface structure on emissivity, we used a so-called Leslie cube. This is a hollow copper metal cube (around 10-cm side length) whose side faces are treated in different ways. One side is covered with a white paint, one with black paint, one is not covered, but just resembles polished copper, and the fourth side consists of roughened copper.

The cube is placed on some Styrofoam as thermal insulation and then filled with hot water. Owing to the good thermal conductivity of the metal, all the side faces of the cube will quickly have the same temperature. The emission of thermal radiation by each side face can easily be analyzed with an IR camera. Figure 1.30 depicts some results (for details of how temperatures are calculated within the used LW camera, see Section 2.3).

For the two painted side faces, the chosen ε-value of 0.96 produced more or less the correct temperature (as can be checked by a contact probe). The rough copper surface had much less emission of thermal radiation and the polished copper surface even less. Using the camera software, one can adjust the emissivity values such that the copper surfaces also give the correct value of the wall temperatures. For our example we find emissivities of the polished copper surface to be around 0.03 and of the roughened copper surface to be about 0.11.

Figure 1.30 illustrates one way of finding correct values for emissivities: measure object temperatures with a contact probe, then adjust the emissivity in the camera

---

Figure 1.30 (a, b) Leslie cube, filled with hot water and viewed at an angle of 45° with respect to the side faces. The emissivity of the whole image was set to $\varepsilon = 0.96$, resulting in the following temperature readings (from left to right). (a) Polished copper ($T = 22.6\,°C$), white paint ($T = 84.3\,°C$). (b) Black paint ($T = 83.4\,°C$) and roughened copper ($T = 29.3\,°C$). The knob on top is a stirrer for reaching thermal equilibrium faster. (c) Visible image of such a cube.

1.4 Emissivity | 37

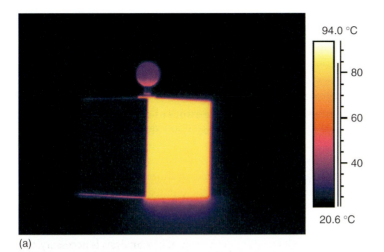
(a)

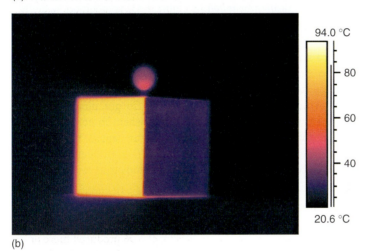
(b)

(c)

until the camera shows the correct temperature reading (assuming, of course, that all other camera parameters are chosen correctly, see Section 2.3).

1.4.4.3 Viewing Angle

We have defined *emissivity* as the ratio of the amount of radiation actually emitted from the surface to that emitted by a blackbody at the same temperature. In terms of radiometric quantities, one defines emissivity as the ratio of the radiance of the radiation, emitted at given wavelength λ and in a direction defined by the two angles δ and φ (Figure 1.15), to the radiance emitted by a blackbody at the same temperature and wavelength.

Blackbodies behave like perfect isotropically diffuse emitters, that is, for any surface emitting radiation, the radiance of the emitted radiation is independent of the direction into which it is emitted (compare the discussion of Lambertian radiators, Figure 1.17). Unfortunately, any real surface shows a different behavior, that is, its radiance shows variations depending on the direction of emission. This is schematically illustrated in Figure 1.31. In addition to the fact that any real surface emits less radiation than a blackbody at the same temperature, the radiance from a real object usually also depends on the angle of emission. Azimuthal symmetry is assumed in Figure 1.31, hence only the angle δ is shown.

This behavior can strongly affect any contactless temperature measurement using IR cameras, since an object that is observed from a direction normal to its surface ($\delta = 0°$) will emit more radiation than when observed at oblique angles. This means that the emissivity depends on the angle of observation with respect to the surface normal. Fortunately, many measurements have been performed and a wide variety of materials has been studied in relation to their directional emissivities. A typical setup for such an experiment is shown in Figure 1.32a.

An angular scale from 0 to 180° is attached to a table with 90° pointing to the IR camera. The (warm or hot) object to be studied is placed on top of the scale

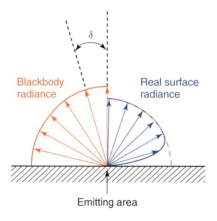

Figure 1.31 Schematic illustration of angular independent radiance for blackbodies (left/red) and directionally dependent radiance of real surfaces (right side/blue).

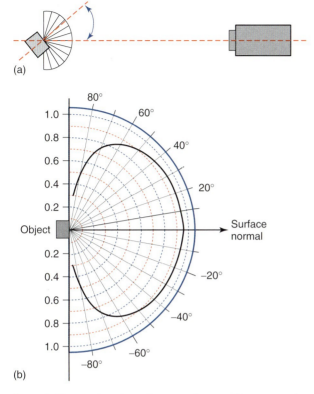

Figure 1.32 A Leslie cube with a painted side face (high emissivity) was used as object to measure the directional emissivity (averaged in the LW range) as a function of viewing angle from 0 to 85° with respect to the surface normal. The cube was rotated with respect to the camera (a) and temperatures recorded. (b) The emissivity as function of viewing angle was then found by varying the emissivity in the camera software until the temperature readings of the rotated side face gave correct values.

with its surface normal facing 90°, that is, the IR camera. Then measurements of the emitted radiation are recorded as a function of angle while rotating the object. In Figure 1.32b, we show an example for the white paint side of a Leslie cube, which was filled with hot water. The actual surface temperature can be measured by contact thermometry, which gives the correct value of the normal direction emissivity. The angular dependent values are found by changing the emissivity value in the camera software until the real temperature is shown (for details of signal processing in the camera, see Chapter 2). Some experiments on $\varepsilon(\phi)$ are discussed in Section 5.5.4.

Figure 1.32b nicely demonstrates an effect that fortunately holds for nearly all practically important surfaces: the emissivity is nearly constant from the normal direction 0° to at least 40 or 45°. The behavior for larger angles differs for metallic and nonmetallic materials (Figure 1.33). For nonconductors, one observes

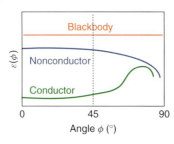

Figure 1.33 Overview of typical directional dependence of emissivities of nonconductors and conductor with respect to blackbodies.

a characteristic drop of ε for larger angles, whereas metallic surfaces usually show first an increase toward larger angles, before decreasing again at grazing incidence [7, 20, 31].

Neglecting diffuse reflections for the moment, this behavior can already be understood from the characteristic properties of directed reflection for different polarizations of the radiation. Figure 1.34 schematically depicts plots for the portion R of directly reflected radiation in the IR spectral region for nonmetals and metals with polished surfaces (similar to Figure 1.12) as a function of angle of incidence.

The directed reflectivity depends on the polarization of the radiation. Unpolarized radiation is characterized by the broken line, which represents the average of both polarizations. Obviously, the reflectivity for nonconductors exhibits a monotonous decrease with angle of incidence, whereas metals show first an increase before decreasing again at larger angles. This characteristic feature (which can be explained theoretically by the Fresnel equations) explains the observed angular plots of emissivity, since $\varepsilon = 1 - R$ for opaque materials according to Eq. (1.31). However, we also note, that most metallic objects have rough surfaces, which induce additional contributions to the emissivity and which can also induce changes in the observed angular distributions.

1.4.4.4 Geometry

The geometry of a surface is related to surface structure; however, here we refer to well-defined structures like grooves, which are used to systematically change emissivity. Cavities are discussed in detail separately in the context of blackbody calibration sources in Section 1.4.6.

Consider a polished metal surface (of e.g., $\varepsilon_{normal} = 0.04$) with a well-defined surface structure in the form of grooves of given slope angle (Figure 1.35, here with apex angle of 60°).

The grooves enhance the emissivity in the direction perpendicular to the macroscopic surface as can be understood from the following argument.

Radiation that is emitted from spot 1 in a direction normal to the macroscopic groove surface is composed of three radiance contributions:

1) Direct radiation emitted from spot 1; this contribution is characterized by $\varepsilon(60°)$ with regard to the real groove surface.

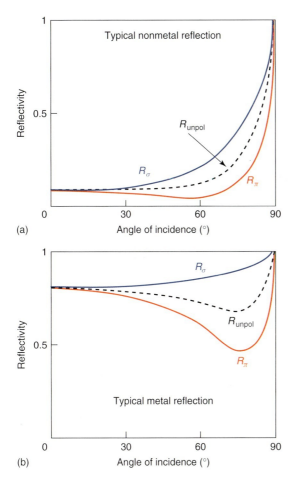

Figure 1.34 Schematic plot of reflectivities of polarized and unpolarized IR radiation of metals (a) and nonmetals (b) as a function of the angle of incidence.

2) Radiation emitted from spot 2, which is then reflected from spot 1 in the normal direction; this contribution is characterized by $\varepsilon(0°) \cdot R(60°) = \varepsilon(0°) \cdot [1 - \varepsilon(60°)]$ with regard to the groove surface.
3) Radiation emitted from spot 1 in direction of spot 2, reflected back from spot 2 to spot 1, and then reflected in the normal direction; this contribution is characterized by $\varepsilon(0°) \cdot R(0°) \cdot R(60°) = \varepsilon(60°) \cdot [1 - \varepsilon(0°)] \cdot [1 - \varepsilon(60°)]$ with regard to the groove surface.

Adding up these radiance contributions and dividing by the blackbody radiance, one can easily see that the normal emissivity of the grooved surface has increased. For a numerical estimate, we assume the polished surface emissivity to be $\varepsilon(0°) = 0.04$ and $\varepsilon(60°) = 0.05$. In this case, the total normal

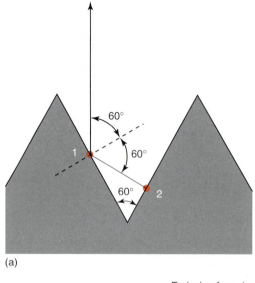
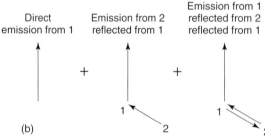

Figure 1.35 V-groove model of a polished metal surface with low emissivity (a). The macroscopic surface is horizontal. Three contributions to radiance and hence emissivity for the groove surface (b).

emissivity of the grooved surface $\varepsilon_{\text{total,normal}} = 0.04 + 0.04 \cdot (1 - 0.05) + 0.04 \cdot (1 - 0.04) \cdot (1 - 0.05) = 0.114$, that is, emissivity has increased by a factor of nearly 3 owing to this surface structure. The basic idea behind this enhancement explains why any rough surface has higher emissivity than polished flat surfaces.

Regular surface structures often lead to nonuniform angular distributions of emissivity. Repeating the above calculation for different emission angles [32] with respect to the macroscopic groove surface (Figure 1.36) reveals strong variations of emissivity as a function of observation angles.

1.4.4.5 Wavelength
As is well known in optics, material properties usually depend on wavelength. Consider, for example, the reflectivity of the noble metals gold (Au), silver (Ag),

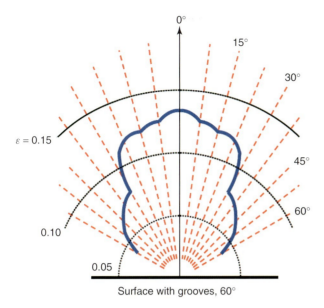

Figure 1.36 Polar diagram for expected dependence of emissivity as a function of observation angle for the 60° V-groove model of a polished metal surface with flat surface emissivity of 0.04 (after [32]).

and copper (Cu). Au and Cu have electronic interband transitions in the Vis range, which give rise to the wavelength-dependent reflectivities, finally resulting in the characteristic golden-yellow as well as red–brown color of these metals. Reflectivity is strongly related to emissivity of materials, hence, any wavelength dependence on reflectivity will also show up in emissivity. Detailed theoretical arguments being beyond the scope of this book, we refer to the respective literature [20, 30] and give schematic diagrams of how emissivities of certain materials changes in general (Figure 1.37).

As can be seen from Figure 1.37, the emissivity of metals usually decreases with wavelength (the opposite effect to their reflectances), whereas oxides and other nonmetals can show increases as well. The examples of Al alloys clearly emphasize the effect that increasing surface roughness from polished surfaces to those roughened by grid paper or finally being sand blasted leads to drastic increases of emissivity. Whenever dealing with substances that have wavelength-dependent emissivity, one must first find out whether emissivities are constant in the IR camera spectral range being used. If not, it is advisable to use narrow band filters or another wavelength band for thermography where emissivities are nearly constant. If this is not possible, one must be aware that any quantitative analysis will be much more complicated since signal evaluation must then use the known variation of emissivity.

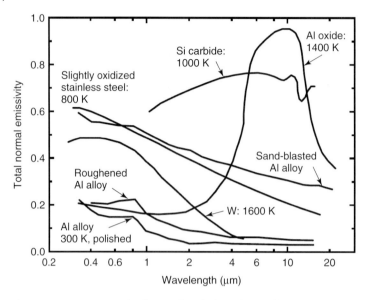

Figure 1.37 Some examples for wavelength dependence of normal emissivity for different materials (after [20, 30]).

We finally mention that in addition to these slowly varying emissivities as a function of wavelength, there are several examples of selective absorbers and emitters such as plastic foils or many gaseous species. These have special applications in IR imaging and are treated in detail in other chapters (Chapter 7, Section 9.6.5).

1.4.4.6 **Temperature**
Material properties usually change with temperature and it is no surprise that this also holds for emissivity. Figure 1.38 gives some examples. Some materials show fairly strong variations, therefore it may be necessary for practical purposes to know whether the temperature of a process under IR observation stays within a certain temperature interval such that the emissivity for this study can be considered to be constant. In addition, if the literature values for emissivity are used, one must know the corresponding temperatures.

1.4.4.7 **Conclusion**
The material property emissivity, which is essential in IR imaging depends on many parameters. Accurate temperature measurement with thermography requires precise knowledge of this quantity. Several sets of tables of emissivities for different materials are available in books [1, 10, 23, 30] and from the manufacturers for such camera systems. Unfortunately, these cannot be used without a word of caution. Measurements always refer to specific experimental

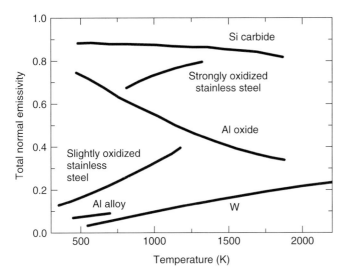

Figure 1.38 Some examples for temperature dependence of emissivity for different materials (after [20, 30]).

conditions, for example, temperature range, wavelength range (LW, MW, SW), or angles (directional or hemispherical measurement). These three factors are usually not critical, since, for most (not all) practical cases in thermography, wavelength and temperature dependencies are not relevant. Furthermore, most practical emitters show directional dependencies only for observation angles larger than 45° with respect to the surface normal. This means that we mostly deal with gray objects, whose emissivity can be guessed within a certain range of accuracy.

Unfortunately, however, metallic objects pose severe problems. Polished metals have very small values of emissivity. Small variations in ε can cause large errors in temperature, therefore the smaller the emissivity, the more precise must its value be known. This poses severe problems, since polished metals have ε values that are quite different compared to roughened or oxidized ones. In electrical inspections, one may often deal with highly oxidized and/or corroded metallic parts. In such cases, guessing emissivity from tables may lead to results that are quite erroneous [33, 34]. In addition, the metal industry (aluminum or steel production, etc.) may also need to consider variations of emissivity with temperature.

1.4.5
Techniques to Measure/Guess Emissivities for Practical Work

Owing to the problems in guessing accurate values for emissivity, it is common practice to directly measure ε. This can be done in various ways, some of those

Table 1.7 Some practical methods of adjusting normal emissivities in thermography.

Method	Tape	Paint, correction fluid, and so on	Contact thermometer	Hole drilling
Equipment needed	Tape	Paint	Thermocouple	Electric drill
Idea behind method	Known emissivity from lab experiments, calibrated with contact probe (usually thermocouple)		Spot measurements with contact probe	Known increased emissivity due to the cavity effect
Advantage/disadvantage	Nondestructive, removable, problem of good thermal contact for very rough surfaces	Nondestructive, problem in removing paint, works also for rough surfaces	Nondestructive, may be time consuming	Destructive, independent of surface structure of object

often used are listed in Table 1.7. In all cases, by "emissivities" we mean directional near normal emissivities, which are integrated over the selected wavelength range of the IR camera.

The easiest method is to attach tape or paint of known emissivity to the object under study. In the analysis, the surface temperatures of the tape or paint follow from their known ε. Assuming good thermal contact and waiting until thermal equilibrium is established, adjacent surface temperatures of the object are assumed to be the same, hence, the object emissivity is found by varying ε in the camera software until the object temperature is equal to known tape surface temperature. The accuracy of this method depends on the accuracy of the known emissivity. Owing to the lab measurements they are related to the temperature accuracy of the contact probe (thermocouple).

One may also directly measure several spot surface temperatures with thermocouples and use them to calibrate the IR images. In this case, one must make sure that good thermal contact is achieved, thermal equilibrium is established and – which is crucial for small objects – that the thermocouple itself does not change the object temperature via conduction of heat. A useful condition is that the heat capacity of the thermocouple must be much smaller than the one of the object. This method need not – but may be more time consuming if objects made of many different materials are investigated. Also, one must make sure that no thermal reflections are present in the IR image (Section 9.2), since these would introduce errors in the analysis.

Sometimes, in building thermography, one may have the chance to drill a hole in the wall. Thereby, one creates a cavity that has very high emissivity values, which can then be used in the image for a correct temperature reading. In a manner similar to the tape method, emissivity can then be estimated for adjacent regions.

1.4.6
Blackbody Radiators: Emissivity Standards for Calibration Purposes

In IR imaging, all commercial camera manufacturers have to calibrate their cameras such that the user will be able to get temperature readings in IR images whenever the proper emissivity is chosen. In addition, some research cameras must be calibrated by the users themselves.

Calibration is usually done by observing the best available experimental approximations to blackbody radiation, so-called blackbody calibration standards. The national institutes that are responsible for standards (in the US, e.g., NIST, in Germany PTB, etc.) have developed standards (for example, $\varepsilon > 0.9996$ in the form of heat pipe cavity type blackbodies [35]). Commercial blackbody radiators, in particular, large area instruments are used as secondary standards by the IR camera manufacturers and other users in laboratory experiments (Figure 1.39). They can be tracked back to the primary cavity type standards but usually have smaller emissivities of about $\varepsilon = 0.98$. They are made of high-emissivity materials with additional surface structures (e.g., pyramids), the surfaces of which are temperature stabilized.

The principle for achieving high emissivities for primary standards is based on Kirchhoff's law (Eq. (1.29)). One needs to construct an object that has high absorptivity. As mentioned above, when introducing blackbody radiation, an open window of a distant building usually looks very black. The reason is illustrated by Figure 1.40.

Radiation is directed into the hole of an otherwise opaque cavity, whose walls are stabilized at a given temperature. There is a certain absorption per interaction with the wall and the radiation will be attenuated successively. Usually, the inner surface is not polished, that is, it is not a reflecting surface according to the law of reflection (Eq. (1.2)) but rather it is diffusely scattering (Figure 1.11). If the hole is small compared to the total surface area of the cavity, the radiation will suffer many interactions with absorption losses before there is a chance of its leaving the cavity again through the entrance hole. For example, if $\alpha = 0.5$, the radiation will be attenuated after 10 reflections already to $(0.5)^{10}$, which is less than 10^{-3}. This means that the overall absorptivity is larger than 99.9% and therefore the respective emissivity has the same value.

The same happens of course to any radiation, which is thermally emitted within the cavity. It interacts with the inner walls many times before leaving the cavity through the only hole. This is the reason, why such cavities are considered to be perfect emitters of thermal radiation.

Many different geometries and sizes of cavities and holes have been studied (Figure 1.41) for a variety of different materials over the years [1, 30]. The simple geometries of sphere and cylinder are usually replaced by either conical shapes or at least cylinders with conical end faces. For high-emissivity sources, the inner surfaces are rough and of materials with high emissivity.

It is now easily possible to generate very high emissivities, but it was also shown, that using low ε wall materials, it is possible to realize cavities with intermediate

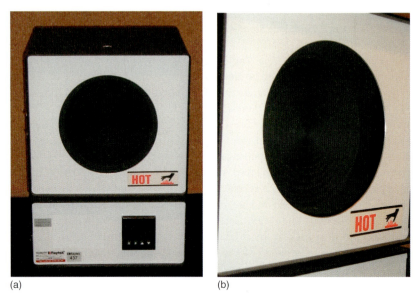

Figure 1.39 Photograph of two commercial secondary standard blackbody source with $\varepsilon = 0.98$. (a, b) With circular symmetry structures, which are visible at grazing incidence. (c) Without structures.

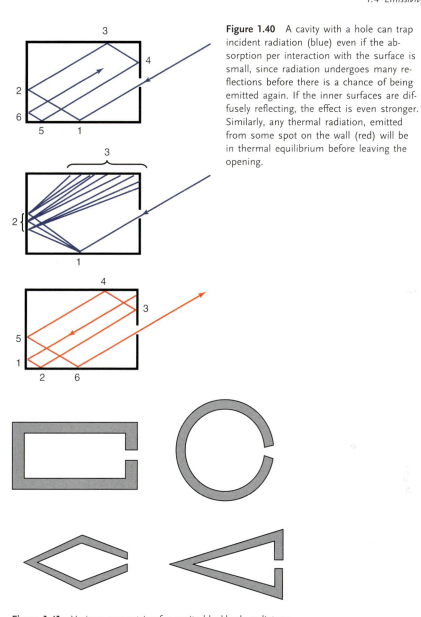

Figure 1.40 A cavity with a hole can trap incident radiation (blue) even if the absorption per interaction with the surface is small, since radiation undergoes many reflections before there is a chance of being emitted again. If the inner surfaces are diffusely reflecting, the effect is even stronger. Similarly, any thermal radiation, emitted from some spot on the wall (red) will be in thermal equilibrium before leaving the opening.

Figure 1.41 Various geometries for cavity blackbody radiators.

emissivity values in the range 0.3–0.7 [36]. As an example, Figure 1.42a shows the geometry of a polished metal cavity. Figure 1.42b depicts the resulting theoretical emissivity as a function of material emissivity and Figure 1.42c demonstrates the effect of cavity geometry on the resulting emissivity. The latter result is easy to understand: the larger the inner surface area, the lower the chance of radiation to hit the small opening hole, hence the larger the emissivity.

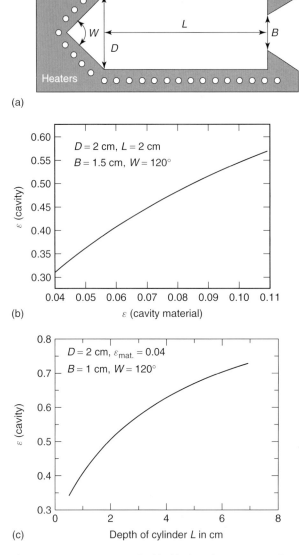

Figure 1.42 (a–c) Design of a blackbody radiator cavity and theoretical expectations for cavity emissivities as function of wall material and geometry.

Polished wall materials within well-defined geometries do, however, have the side effect of angular dependence of emissivity, that is, similar to Figure 1.36, there are specific angles for which more (or less) radiation is emitted with regard to other angles. This is illustrated in Figure 1.43, which depicts IR measurement results from the heated cavity shown in Figure 1.42 for $L = 5$ mm.

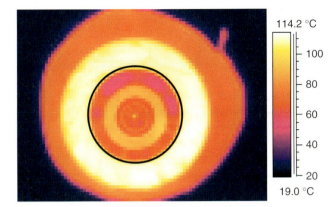

Figure 1.43 Geometrical resonances in the emissivity of metallic cavities. The black circle indicates the size of the aperture.

In conclusion, for cavity blackbody sources, whenever the inner surfaces are made of polished metals, it may be possible to get low and medium emissivities in the range 0.2–0.6; however, the directional character of emissivity usually still shows up, that is, the cavities do not yet resemble Lambertian sources. The best high-emissivity Lambertian blackbody sources are made of cavities with wall materials of high emissivity in the first place, maybe additionally roughened.

1.5
Optical Material Properties in the IR

In any practical contactless temperature measurement with either pyrometry or IR imaging, the radiation from an object under study must reach the detector. However, this requires the IR radiation to pass through the space between object surface and detector within the camera housing (Figure 1.44). The radiation is usually attenuated on this path, since it must pass through various kinds of matter, usually at least the atmosphere and the solid focusing optics materials. In addition, solid materials can lead to additional thermal reflections and – if at

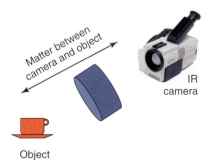

Figure 1.44 Typical setup for thermography. The object is observed through matter, which can be a gas, a liquid, or a solid material. Thereby it can get attenuated. The same holds for the camera optics in front of the detector.

elevated temperatures – to additional emission of thermal radiation, which may contribute to the object signal. Certain special applications may include additional kinds of matter such as other gases, liquids, or additional solid filters.

Here, we summarize mostly the optical material properties due to the atmosphere and solid window or lens materials in the thermal IR spectral range. Additional species, for example, special gases or plastics are dealt with separately in Chapter 7, Section 9.6.5. More information can be found in the literature [1, 4, 12, 13, 17, 30, 37–39].

1.5.1
Attenuation of IR Radiation while Passing through Matter

The most common case is the study of objects that are already contained in the atmosphere, that is, the objects are already immersed in the gas, which then extends to the camera optics. The IR radiation is emitted from the object surface into the atmosphere and one needs to know the scattering/absorption processes within the gas in order to calculate the attenuation of the IR radiation.

For any kind of solid material, incident radiation will be regularly reflected, diffusely scattered, absorbed, or emitted (Figure 1.45). A general treatment of attenuation of IR radiation must therefore be able to compute absorption/scattering within matter as well as reflection losses to the atmosphere at the boundaries.

Attenuation of radiation along its original direction within matter is described by the Bouguer–Lambert–Beer law (Eq. (1.32)), which yields the transmitted portion

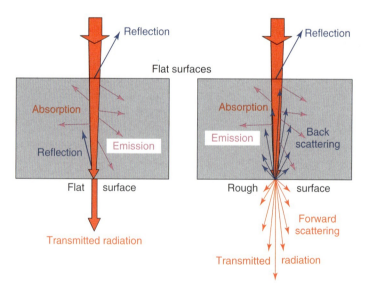

Figure 1.45 IR radiation that is incident on matter with flat surfaces can be reflected, absorbed, or emitted. For rough surfaces, diffuse scattering may take place.

T of the radiance as a function of traveled distance d within matter.

$$T(\lambda, d) = \frac{I(\lambda, d)}{I(\lambda, 0)} = e^{-\gamma(\lambda) \cdot d} \tag{1.32}$$

Here, $\gamma(\lambda) = \alpha(\lambda) + \sigma(\lambda)$ denotes the total attenuation coefficient, which is due to absorption (α) of radiation as well as scattering (σ), that is, a change of direction of radiation. Absorption as well as scattering can be calculated from first principles. In gases, both are due to electronic, vibrational, and/or rotational excitations, and γ can be expressed as

$$\gamma_{\text{gases}}(\lambda) = n \cdot \left(\sigma_{\text{abs}}(\lambda) + \sigma_{\text{sca}}(\lambda) \right) \tag{1.33}$$

where n is the volume concentration of the gas (in number of particles/volume) and σ_{abs}, σ_{sca} are the so-called absorption and scattering cross sections. Results for atmospheric gases are known from many laboratory experiments.

For solid matter, attenuation is due to excitations within electronic bands and lattice vibration processes. The attenuation coefficient in nonscattering solids can be related to a macroscopic well-known quantity, the imaginary part of the index of refraction $n = n_1 + in_2$ (Eq. (1.34)).

$$\gamma_{\text{solids}}(\lambda) = \alpha_{\text{solids}}(\lambda) = \frac{4\pi \cdot n_2}{\lambda} \tag{1.34}$$

The index of refraction has been measured for many solids and is tabulated in a series of handbooks [18].

1.5.2
Transmission of Radiation through the Atmosphere

The dry atmosphere is composed of several natural gases (Table 1.8). In addition, there is a varying amount of water vapor of volume concentration up to several percent. Atomic gases (e.g., Ar) and diatomic gases of the same atomic species (N_2 and O_2) cannot absorb IR radiation in the thermal IR range. However, molecules, made up of two or more different atomic species like, for example, NO, CO, CO_2, CH_4, and of course H_2O, and so on, are, in principle, able to absorb IR

Table 1.8 Composition of dry air. (For CO_2, June 2010: http://www.esrl.noaa.gov/gmd/ccgg/trends/.)

Gas	Symbol	Volume (%)	Concentration (ppm)
Nitrogen	N_2	78.08	—
Oxygen	O_2	20.95	—
Argon	Ar	0.93	—
Carbon dioxide	CO_2	0.0388	388
Neon	Ne	0.0018	18
Helium	He	0.0005	5
Methane	CH_4	0.00018	1.8

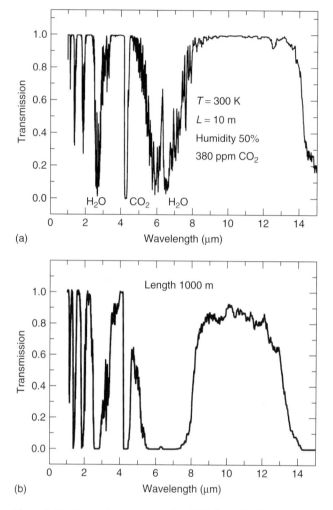

Figure 1.46 Atmospheric transmission $T(\lambda)$ for a (a) 10- and (b) 1000-m horizontal atmospheric path.

radiation. Figure 1.46 depicts two overview transmission spectra of horizontal atmospheric paths of length 10 and 1000 m, demonstrating that mostly CO_2 and H_2O dominate attenuation in otherwise pure air. Aerosols and clouds can induce additional attenuation.

In addition, there are gases with lower concentrations like Krypton (Kr), hydrogen (H_2), nitrous oxides (N_2O, NO), Xenon (Xe), ozone (O_3), and so on. The composition depends on the height above sea level. The current increase of CO_2 is about 1.5 ppm/year, and for methane about 7 ppb/year.

Several characteristic absorption features are present, in particular, some bands around 2.7 µm (H_2O and CO_2), around 4.2 µm (CO_2), between 5.5 and 7 µm (H_2O)

and above 14 μm (H_2O, CO_2). These absorption bands are important for defining the spectral bands of IR cameras (see Figure 1.8).

There are several computer models that compute very precise transmission spectra of IR radiation within the atmosphere, the most well known being the Low Resolution Transmittance (LOWTRAN), Moderate Resolution Transmittance (MODTRAN), or High Resolution Transmittance (HITRAN) codes. These are radiative transfer models for the atmosphere. They incorporate absorption as well as scattering constants for all relevant gaseous species in the atmosphere, and they also include vertical distributions of the constituents, in order to provide an adequate modeling for the gas attenuation. LOWTRAN is a low-resolution propagation model and computer code for predicting atmospheric transmittance and background radiance. MODTRAN is similar to LOWTRAN, but with better spectral resolution. Similarly, HITRAN works with even higher resolution. The 2004 HITRAN model included 1 734 469 spectral lines for as many as 37 different molecules. For IR imaging, however, we mostly deal with two species and a few spectral bands, as shown in Figure 1.46.

Two aspects are particularly important for practical IR thermography. First, attenuation depends on the concentration of the absorbing gas species (Eq. (1.33)). For most gases of the atmosphere, concentrations are constant or slowly varying with time; water vapor, however, can have strong fluctuations, therefore, the relative humidity is an important quantity, which needs to be measured for accurate compensation of water vapor attenuation between object and camera. Second, attenuation depends on the distance from an object to the camera (Eq. 1.32), therefore, this quantity also needs be known. Both, humidity and distance are input parameters within the camera software packages.

We finally mention that for long atmospheric paths (usually not encountered in thermography) additional effects due to aerosol scattering may become important and respective modeling must then include aerosol size and height distributions. If particles are small, as is, for example, typical haze with water droplets of, say 500 nm, visible light is scattered very effectively, whereas longer IR radiation is much less affected (Section 10.5.2).

1.5.3
Transmission of Radiation through Slablike Solid Materials

1.5.3.1 **Nonabsorbing Slabs**
Attenuation of IR radiation in solid materials happens for lenses in the IR cameras, but quite often objects are also observed through some windows. Mostly, flat (polished) surfaces are used such that scattering from surface roughness is neglected. The attenuation is then due to reflection at the boundaries as well as to absorption within the material. For the sake of simplicity we now discuss planar geometry, that is, slabs of material with given thickness and well-defined index of refraction and IR radiation at normal incidence. For camera lenses with curved surfaces, one should redo the argument for finite angles of incidence. In this case, reflection coefficients will change according to the Fresnel equations, giving rise to

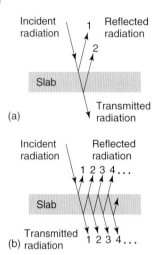

Figure 1.47 Scheme for calculating transmission of an absorbing slab in air. The propagating radiation is drawn as oblique lines for clarity. In the calculation, however, they are assumed to be at normal incidence.

polarization-dependent effects (see Figures 1.12 and 1.34). If the distance from the object to the lens, however, is large compared to the diameter of the lens, then the angles are small, and the general argument will still be valid.

Consider first a slab (coplanar plate) of a nonabsorbing material, surrounded by a nonabsorbing material, usually the atmosphere (Figure 1.47). IR radiation of radiance L_0 is incident at normal incidence (in Figure 1.47, the propagation is drawn at oblique angles in order to separately visualize the various contributions).

In the simplest description (Figure 1.47a), there will be reflection losses L_R (characterized by the reflectivity $R = L_R/L_0$) of the radiation at the first interface. If the losses are small, the wave encounters similar losses at the second interface, that is, the total losses amount to $2R$ and the total transmission $T_{\text{total}} = L_{\text{Trans}}/L_0$ can be written as $T_{\text{total}} \approx 1 - 2R$, with R from Eq. (1.5). A typical example for visible light and a glass plate in air would be $n_{\text{air}} = 1.0$, $n_{\text{glass}} = 1.5 + i0.0$, $R = 0.04$ and $T_{\text{total}} \approx 0.92$. This means that a typical glass plate transmits about 92% of incident visible light, which is why we can see quite clearly through windows. Before turning on to IR windows, let us briefly generalize this result.

If the reflection losses upon hitting the first interface become larger (which will be the case for most IR transparent materials in the IR region), the derivation for the transmission must take into account all contributions of multiple reflections. Figure 1.47b depicts the idea behind the calculation. Part of the incident radiation is reflected at the first encounter with the slab. Using the symbols $R = L_R/L_0$ and $T = L_T/L_0 = (1 - R)$ for reflectivity and transmission at a single interface, this first reflection contribution (1) is given by $L_0 \cdot R$. The transmitted part $L_0 \cdot T$ enters the medium, which is at first assumed to be nonabsorbing. After the second transmission, and exiting the material, the radiance has decreased to $L_0 \cdot T^2$ (transmitted beam number 1). Following the beam, which is internally reflected within the slab, and subsequently studying more reflections and transmissions upon interactions with the slab surfaces, we end up with a number of rays (1, 2, 3, 4, ...) contributing to the total reflected radiance and to the total transmitted radiance. For nonabsorbing materials, $T = 1 - R$, hence, the sum of all reflected contributions can be written as a geometrical sum

$$T_{\text{slab}} = (1 - R)^2 \cdot (1 + R^2 + R^4 + R^6 + \cdots) = \frac{(1 - R)^2}{1 - R^2} \qquad (1.35)$$

Using Eq. (1.5) again for the air–glass example ($n_A = 1.0$, $n_B = 1.5 + i0.0$), the transmission can easily be evaluated to give

$$T_{\text{slab}} = \frac{2n_B}{n_B^2 + 1} \qquad (1.36)$$

For $n_B = 1.5$ this gives $T_{\text{slab}} = 0.923$, which is only slightly higher than the value 0.92 estimated above.

In contrast, a material like germanium at a wavelength of 9 µm has $n_B \approx 4.0 + i0.0$. The simple derivation of a Ge slab in air would give $T_{\text{total}} = 1 - 2 \cdot 0.36 = 0.28$, whereas the correct treatment according to Eq. (1.36) gives $T = 0.47$.

1.5.3.2 Absorbing Slabs

Propagation of radiation within an absorbing slab has the same type of contributions as in Figure 1.47b, the only difference being that during each successive passage through thickness d of the slab, there will be an additional attenuation described by Eq. (1.32).

Repeating the above calculation of the transparent-slab transmission and inserting R from Eq. (1.5). We finally find the transmission of an absorbing slab ($n_B(\lambda) = n_1(\lambda) + in_2(\lambda)$) in air ($n_{1A} = 1$)

$$T_{\text{slab}}(\lambda, d) = \frac{16 n_1^2 \cdot e^{-\frac{4\pi n_2 d}{\lambda}}}{\left[(n_1 + 1)^2 + n_2^2\right]^2 - \left[(n_1 - 1)^2 + n_2^2\right]^2 \cdot e^{-\frac{8\pi n_2 d}{\lambda}}} \qquad (1.37)$$

Equation (1.37) allows to easily compute slab transmission spectra, provided the thickness d of the slab as well as the optical constants $n_1(\lambda)$ and $n_2(\lambda)$ of the slab material are known.

1.5.4
Examples for Transmission Spectra of Optical Materials for IR Thermal Imaging

1.5.4.1 Gray Materials in Used IR Spectral Ranges

There are a number of common materials available in the thermal IR spectral range [4, 9, 17]. These include crystals like BaF_2, NaCl, CdTe, GaAs, Ge, LiF, MgF2, KBr, Si, ZnSe, or ZnS, and inorganic as well as organic glasses like fused silica IR grade or AMTIR-1. The materials can be characterized according to the respective wavelength range where they are used. A number of examples are presented in the following.

Figure 1.48a depicts transmission, reflection, and absorption spectra for a 7.5-mm slab of NaCl. Obviously, NaCl has excellent IR transmission >0.90 up to about 12 µm, and it still has a transmission of about 0.87 at 14 µm. The transmission

58 | *1 Fundamentals of Infrared Thermal Imaging*

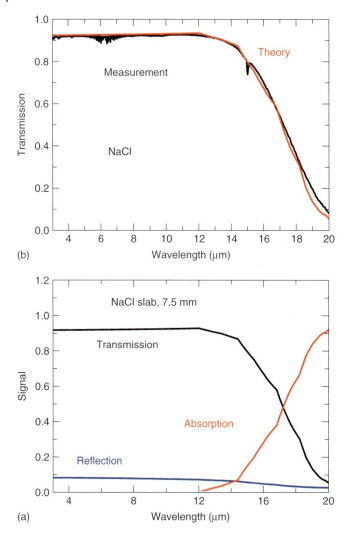

Figure 1.48 Experimental (black) and theoretical (red) transmission spectrum of a 7.5-mm thick slab of NaCl (a) as well as the reflection and absorption contributions (b).

depends on slab thickness wherever absorption plays a role, here for $\lambda > 12\,\mu m$. However, it has the disadvantage of being hygroscopic, that is, it must be protected from water moisture and humidity. Therefore, many alkali halides are not used to manufacture lenses for IR camera systems. They are, however, sometimes used as special windows.

As is the case with all literature data, a word of caution is necessary for the theoretical spectra. Accurate measurements of optical constants of materials are difficult and those collected over the years have later been refined after better sample preparation techniques were available. All of the following theoretical plots are based on collection of data of optical constants from the literature [18]. Often, several slightly different sets of data are available. This is due to the fact that such measurements are usually done under ideal conditions, for example, clean samples, very good crystal or sample film quality, very few surface defects, scratches, and so on with little sideways scattering. Hence, when comparing these data to real-world windows or lenses of the same materials, deviations by several percent are usual; in the case of severe surface damages with a lot of side scattering, even larger deviations are possible. Hence, theoretical spectra should be regarded as order of magnitude expectations and experimental spectra should be recorded for comparison, whenever quantitative analysis is needed. In Figure 1.48b, the comparison of experimental spectra, recorded with Fourier Transform infrared (FT-IR) spectroscopy and theoretical spectra based on tabulated optical constants is very good.

All data presented here refer to normal incidence. Spectra for oblique radiation must take into account the effect of the Fresnel equations (Figure 1.34).

Lenses, windows, and filters for IR cameras are often manufactured from materials like BaF_2, CaF_2, MgF_2, Al_2O_3 (Sapphire), or Si for the wavelength range up to 5 μm and Ge, ZnS, or ZnSe for the LW region. Some of these ideally suited materials for IR imaging are manufactured by special hot-pressing techniques and then have special names like Irtran (acronym for IR transmitting).

Figures 1.49 and 1.50 show some examples for the transmission spectra of materials, which are used in the MW range.

Similar to the NaCl results (Figure 1.48), absorption features become important for longer wavelengths. In this case, the transmission spectra depend on material

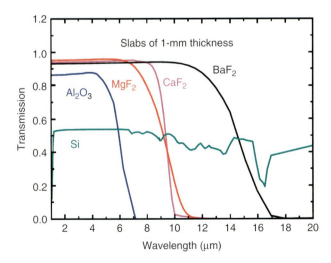

Figure 1.49 Overview of typical theoretical transmission spectra for 1-mm thick slabs of various materials that are used in the MW IR spectral region.

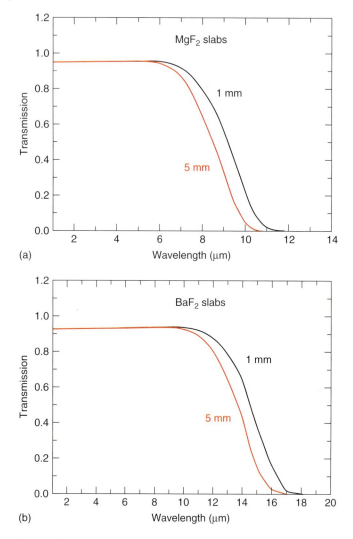

Figure 1.50 Transmission spectra depend on the thickness of the slabs owing to the onset of absorption.

thickness as is shown for BaF_2 and MgF_2 in Figure 1.50. Obviously, it is important to know the exact thickness of windows made of such materials, if the IR camera used is sensitive in the respective spectral range of absorption features. Obviously, BaF_2 may also be used for LW cameras; however, one must take care of keeping the thickness of window materials small.

Figure 1.49 proves that the transmission for Si is rather flat in the MW range; however, absorption features dominate the spectra in the LW range. This is illustrated in Figure 1.51, which depicts a closer look of a Si plate for different thickness.

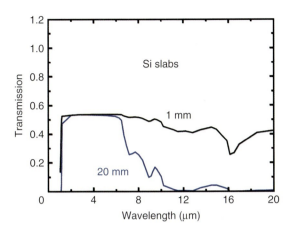

Figure 1.51 Transmission spectra of Si slabs for different thicknesses. Absorption features are prominent for wavelengths above 5 μm.

In the last few decades, the purity of Si crystals has improved considerably. This means that high-quality oxygenfree Si crystals may have appreciably less absorption than the sample shown here, which is based on a compilation of measured optical data from 1985 [18].

Figure 1.52 provides an overview of spectra for materials that can be used in the LW range.

It is should be mentioned that some of these materials have special trade names if manufactured in a certain way. For example, hot-pressed polycrystalline ZnS with about 95% cubic and 5% hexagonal crystals, is also called *Irtran2* by the manufacturer Eastman Kodak. Similar other trade names exist like Irtran 4 for ZnSe, and so on. Since optical properties depend on crystalline structure, mixtures of different crystal forms can lead to differences in optical transmission spectra. ZnSe can be used throughout the LW range; for ZnS or Ge, however, thick samples show appreciable absorption features even below wavelengths of 15 μm

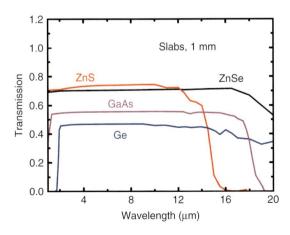

Figure 1.52 Overview of typical theoretical transmission spectra for 1-mm thick slabs of various materials that are used in the LW IR spectral region.

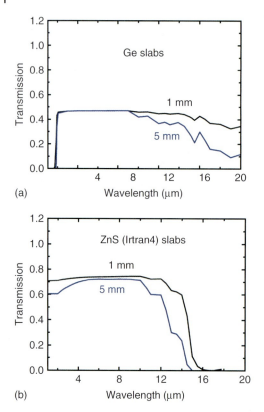

Figure 1.53 Transmission spectra of Ge (a) and ZnS (b) depend on thickness of the slabs owing to the onset of absorption.

(Figure 1.53). This effect is less pronounced for GaAs; however, GaAs is very expensive compared to the other materials, which affects it commercial use for lenses.

For comparison, Figure 1.54 depicts the respective spectra for regular laboratory glass BK7 (Schott) or fused silica, that is, amorphous silicon dioxide. Obviously, common glass cannot be used in either MW or LW IR cameras. Synthetic fused silica can, in principle, be used for MW systems, acting as bandpass filters.

1.5.4.2 Some Selective Absorbers

A common misconception among beginners in IR imaging is the expectation to be able to observe warm or hot objects immersed in water. This is more or less impossible as can be seen from Figure 1.55, which depicts theoretical transmission spectra of thin slabs of water. For direct comparison with Figures 1.48–1.54, the slabs are assumed to be surrounded by air; however, the spectra look very similar if semi-infinite liquids of this thickness are considered. Obviously, only the very thinnest layers of water still allow transmission of radiation in the thermal IR spectra range.

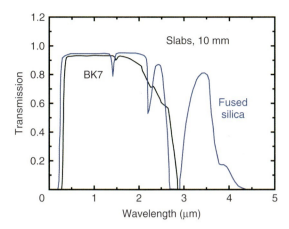

Figure 1.54 Transmission spectra of common laboratory glass BK7 and fused silica. There are usually batch-to-batch variations in the range around the water absorption band at 2.7 μm due to fluctuations in the OH chemical bond contents in these materials.

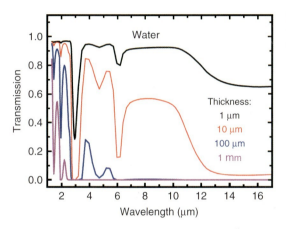

Figure 1.55 Transmission spectra of slabs of water of varying thickness.

Water with 1-mm thickness is sufficient to completely suppress any IR transmission in the MW and LW spectral range. In principle, SW IR imaging ($\lambda < 1.7\,\mu m$) seems possible, however, thermal radiation of objects with less than 600 K is very low in this spectral range (Figures 1.20 and 1.21).

There are, in principle, other liquids like oils or organic compounds that can have finite IR transmission, they are, however, only used for specials applications and are not treated here.

A final example of another often-encountered material in IR imaging is shown in Figure 1.56, which shows IR transmission of a plastic foil. Plastics are complex organic compound materials which, depending on type, show a huge variety in chemical composition. Therefore, they can also have large variations in their IR spectra. Obviously, the chosen example may be used for MW cameras. It has several absorption features in the LW range, but could be, perhaps, used as band filter material for special investigations.

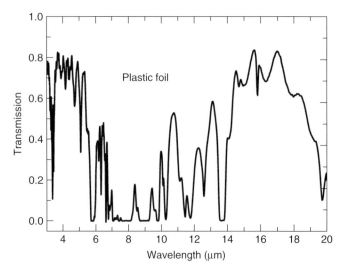

Figure 1.56 Experimental IR transmission spectrum of a plastic foil.

1.6
Thin Film Coatings: Tailoring Desired Optical Properties for IR Optical Components

Many IR transparent optical materials have transmission values much lower than 100%, some of them being in the range of 70% and others like Ge or Si even in the range of only 50%, (Figures 1.49–1.53). These transmission losses are mostly due to the reflection losses from air to the material, with high index of refraction. Obviously, any optical system using such materials would suffer substantial losses of IR radiation reaching the detector. In order to reduce these losses, lenses and optical components of IR cameras are usually treated with antireflection (AR) coatings. In addition, the technique of coatings can also be used to tailor desired optical properties like, for example, band filters for IR radiation. The technique to modify optical properties by deposition with thin film coatings is well known from the visible spectral range [25, 27, 40–42] and has also been successfully applied to the IR spectral range [2, 4, 17].

1.6.1
Interference of Waves

The principle behind AR coatings is based on the wave nature of radiation, in particular, the phenomenon of interference. Figure 1.57 illustrates interference schematically. IR radiation is an EM wave, which may be described by the oscillation of the electric field vector (Section 1.2.1). Whenever two individual waves (light waves, sound waves, water waves, etc.) meet each other at the same time and location, their elongations (here the electric field) superimpose, that is, they add up to give a new total wave elongation and a new maximum elongation, that is, wave amplitude. In Figure 1.57a, the two waves superimpose in such a way, that

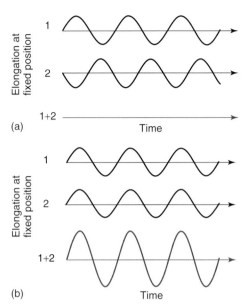

Figure 1.57 Destructive (a) and constructive (b) interference of waves as observed at a fixed location.

the first wave is just shifted by half a wavelength, that is, it is out of phase with the second wave. In this case, the adding up of the two elongations results in complete cancellation of the electric field and the new total elongation is zero. This is called *destructive interference*. In Figure 1.57b, the two waves superimpose in such a way, that the first wave has no shift, that means it is oscillating in phase with the second wave. In this case, the adding up of the two elongations results in a larger elongation of the electric field, that is, the new total amplitude is twice as large as the one from each individual wave. This is called *constructive interference*. The overall energy transported by a wave is then proportional to the square of the amplitude of the wave, that is, in the first case, there is no net energy flux, whereas in the second case, there is twice as much energy flux with regard to the two individual waves. Whenever interference happens, the energy flux of waves is redistributed such that in certain directions, destructive interference reduces the flux, whereas in other directions, there is constructive interference with larger flux. Of course, energy conservation is fulfilled overall.

1.6.2
Interference and Optical Thin Films

Interference can be utilized to tailor optical properties of materials by depositing thin films on top of the desired optical material. The idea behind is schematically illustrated in Figure 1.58. Suppose IR radiation is incident on an optical component made up of a certain material, for example, Ge or Si or ZnSe, and so on. This component material is called *substrate*. For the sake of simplicity, we assume the component to be thick, that is, we only treat the first reflection from the top surface

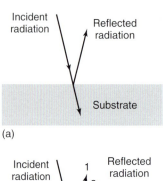
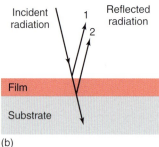

Figure 1.58 (a, b) The idea behind optical coatings (see text for details.).

(Figure 1.58a). We can easily calculate the reflectivity for normal incidence from Eq. (1.5).

If a thin transparent film is deposited onto the substrate (Figure 1.58b), there are at least two dominant reflections, one from the top surface of the film and the second from the interface between film and substrate (for simplicity of the argument, we neglect additional reflections). For normal incidence, the two contributions overlap spatially, and at the top surface of the film, the two reflected waves can interfere with each other. With regard to the top reflection (1), the reflected radiation (2) has traveled an additional distance of twice the film thickness before interfering with the first wave. Thereby a phase shift is induced. For proper film thickness and index of refraction, the phase shift can be chosen in such a way that the two waves interfere destructively. In this case, the thickness of the film needs to be $\lambda/4$, with λ being the wavelength of the EM radiation within the film material. If, in addition, the amplitudes of the two reflected waves can be made equal, destructive interference can lead to a total suppression of reflected radiation. Alternatively, it is possible to enhance the reflected portion of the radiation by using constructive interference. Five parameters determine the optical properties of reflected or transmitted radiation: first, the index of refraction of the substrate material; second, the index of refraction of the film material; third, the thickness of the film material; fourth, the angle of incidence; and finally, the wavelength of the IR radiation.

Optical interference coatings have been studied for many decades in the visible spectral range, the most common and well-known example being AR coatings for glasses. Research has shown that single layers work only for one specific wavelength but are less effective for neighboring wavelengths. If special properties like AR coatings, mirrors, or bandpass filters are needed for broader wavelength ranges,

1.6 Thin Film Coatings: Tailoring Desired Optical Properties for IR Optical Components

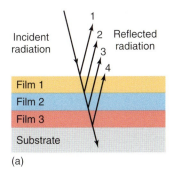
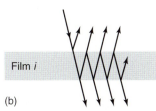

Figure 1.59 Schematic of a multilayer thin film optical coating. In a simplified overview (a), there is a reflection component from each interface; in reality, there are multireflection contributions from each film (b), which need to be summed up.

multilayer coatings are used (Figure 1.59). They can range from a few (like three) to several hundreds of layers. Matrix formulations have been developed to analytically treat such multilayer films [40, 41], it is, however, obvious that such solutions become very complex and, nowadays, computer codes are used to compute the optical properties. We show a few selected results in the following. However, a word of caution is needed when theoretically discussing thin film properties. Besides the above-mentioned theoretical parameters that have an influence on the optical properties, one also needs to consider the availability of transparent film materials, as well as their growth properties on the given substrate materials. If the lattice constants of the selected film material deviate too much from the ones of the substrate, it may be that the coating cannot be manufactured.

1.6.3
Examples of AR Coatings

The effectiveness of thin film optical coatings is first illustrated by an example with visible light. Figure 1.60 depicts the reflectivity for normal glass as well as for the same glass substrate coated with either a single layer of MgF_2 (thickness $\approx$ 90 nm, corresponding to $\lambda/4$ for chosen reference wavelength of 500 nm) or a triple layer using MgF_2, ZrO_2, and Al_2O_3. The typical glass reflectivity of a single interface is reduced from around 4% to a minimum of about 1.5% for a single-layer coating and to less than 0.1% for an appropriate three-layer coating. The multilayer coating has the additional advantage of a very broad band reduction of reflectivity within the visible spectral range. Such AR coatings are of standard use for glasses, nowadays (of course, both glass surfaces must be treated).

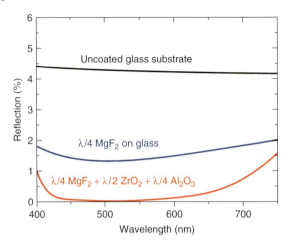

Figure 1.60 Optical antireflection (AR) coatings for glass in the visible spectral range. Multilayer coatings can lead to a broad band reduction of reflectivity (computed with the commercial program Essential McLeod [42]).

Similarly, AR coatings can be applied to materials used for IR optical components like IR cameras. For example, silicon has a reflectivity that is quite high and a corresponding low transmission in the MW range. Again, a simple two-layer coating made of MgF_2 and ZrO_2 could easily reduce the reflectivity to below 10% (Figure 1.61), that is, enhance transmission to over 90% (the sum of both is slightly less than 100% because of absorption). Similar AR coatings can be applied to other

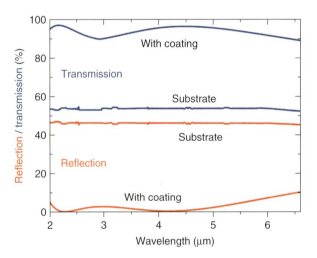

Figure 1.61 Example of antireflection (AR) coatings for Si in the MW IR spectral range (computed with the commercial program Essential McLeod [42]).

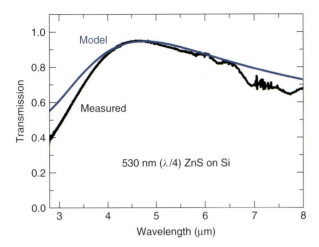

Figure 1.62 Example of a modeled (blue curve) and experimental single-layer antireflection coating of ZnS on top of a silicon wafer.

materials in the LW range. In practice, all component surfaces are treated with AR coatings, not just the front surface.

In conclusion, AR coatings of IR optical components can be easily produced. The exact multilayer composition of manufacturers is usually not known, but the principle behind the method is obvious. Figure 1.62 depicts transmission spectra of a simple single-layer AR coating made of 530 nm ZnS on a silicon wafer. The model result (blue solid line) corresponds quite well with the corresponding real system, which was produced by evaporation techniques in a vacuum chamber and subsequently analyzed using FTIR spectroscopy.

1.6.4
Other Optical Components

Finally, some applications in IR imaging require special optical coatings. There are several common examples. First, one might want to use high-pass or low-pass filters, that is, either only long or only short wavelengths are transmitted with respect to a defined reference wavelength. This may be useful if background radiation of well-defined wavelengths should be suppressed. Second, so-called neutral density filters may be useful if the total radiance, incident on a camera is attenuated in a well-defined way. These filters can have a transmission (which can be chosen more or less arbitrarily over a wide range) that is independent of wavelength. Third, several application require bandpass filters. Such filters are similar to interference filters, known from the visible spectral range. Figure 1.63 depicts a typical commercial filter whose transmission corresponds to the CO_2 absorption lines at $\lambda \approx 4.23\,\mu m$. It is characterized by the maximum transmission, the bandwidth at half maximum, and sometimes the slope of the curve. Commercially, a wide variety of broadband

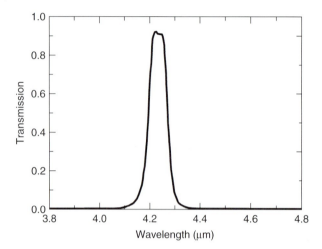

Figure 1.63 Typical example for transmission spectrum of commercially available narrowband IR filter at 4.23 μm.

or narrowband filters are available for nearly every wavelength in the thermal IR range (e.g., [43]).

In conclusion, by using thin optical film coatings, it is possible, within certain limits, to tailor desired optical properties of IR optical components. The only important issue to keep in mind is that the materials for the thin films must not absorb in the respective spectral range. Otherwise, they would behave in a manner similar to absorbing windows, that is, absorb part of the radiation, thereby heat up and emit IR radiation, which introduces an error in the measurement of object temperatures.

Interference filters in the IR are usually a combination of a cut-on and a cutoff filter (Figure 1.64). The use of filters is discussed in Section 3.2 and Chapter 7. Many filter materials transmit short IR waves and start to absorb wavelengths larger than a certain wavelength, that is, filter materials themselves may act as cutoff filters. In this case, it is important to choose the direction of the IR radiation through the filter. It should pass first through the cutoff filter, such that longer wavelengths are

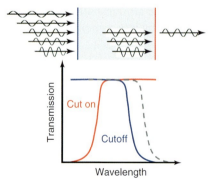

Figure 1.64 An interference filter that has interference coatings on both sides, one acting as cut-on filter (transmitting long wavelengths) and another one acting as a cutoff filter (blocking even longer wavelengths), resulting in a bandpass filter. If the filter material itself is absorbing at long wavelengths, it also resembles a cutoff filter. Only incident IR radiation with suitable wavelength is transmitted, the rest is reflected and none is absorbed.

reflected. If they enter the filter material, they could lead to a heating of the filter, leading to an additional source of thermal radiation, which has an influence on the signal to noise ratio (Chapter 2).

References

1. Wolfe, W.L. and Zissis, G.J. (eds) (1993) *The Infrared Handbook*, revised edition 4th printing, The Infrared Information Analysis Center, Environmental Research Institute of Michigan.
2. Bass, M. (ed.) (1995) *Handbook of Optics*, Sponsored by the Optical Society of America, vol. 1, McGraw Hill, Inc.
3. Dakin, J.R. and Brown, R.G.W. (eds) (2006) *Handbook of Optoelectronics*, vol. 1, Taylor and Francis, New York.
4. (2005) *The Photonics Handbook, Book 3: The Photonics Directory*, 51st edn, Laurin Publishing Company, Pittsfield.
5. Gross, H. (ed.) (2005) *Handbook of Optical Systems*, vol. 1, Wiley-VCH Verlag GmbH, Weinheim.
6. Gross, H. (ed.) (2008) *Handbook of Optical Systems*, vol. 4, Wiley-VCH Verlag GmbH, Weinheim.
7. Bentley, R.E. (ed.) (1998) *Handbook of Temperature Measurement*, Temperature and Humidity Measurement, Vol. 1, Springer, Singapore.
8. Michalski, L., Eckersdorf, K., Kucharski, J., and McGhee, J. (2001) *Temperature Measurement*, 2nd edn, John Wiley & Sons, Ltd, Chichester.
9. Hermann, K. and Walther, L. (eds) (1990) *Wissensspeicher Infrarottechnik (in German)*, Fachbuchverlag, Leipzig.
10. Dereniak, E.L. and Boreman, G.D. (1996) *Infrared Detectors and Systems*, John Wiley & Sons, Inc., New York.
11. Holst, G.C. (1993) *Testing and Evaluation of Infrared Imaging Systems*, JCD Publishing Company, Maitland.
12. Jha, A.R. (2000) *Infrared Technology: Applications to Electro-Optics, Photonic Devices, and Sensors*, John Wiley & Sons, Inc., New York.
13. Schlessinger, M. (1995) *Infrared Technology Fundamentals*, 2nd edn, Marcel Dekker, New York.
14. Schneider, H. and Liu, H.C. (2007) *Quantum Well Infrared Photodetectors*, Springer Series Optical Science, vol. 126, Springer, Heidelberg.
15. Schuster, N. and Kolobrodov, V.G. (2000) *Infrarotthermographie*, Wiley-VCH Verlag GmbH, Berlin.
16. Wolfe, W.L. (1996) Introduction to infrared system design, *Tutorial Texts in Optical Engineering*, vol. TT24, SPIE Press, Bellingham and Washington, DC.
17. Savage, J.A. (1985) *Infrared Optical Materials and Their Antireflection Coatings*, Adam Hilger, Bristol.
18. Palik, E.P. (ed.) (1985) *Handbook of Optical Constants of Solids*, vol. 1, Academic Press, Boston; vol. 2, (1991); vol. 3, (1998).
19. Baehr, H.D. and Karl, S. (2006) *Heat and Mass Transfer*, 2nd revised edn, Springer, Berlin and New York.
20. Incropera, F.P. and DeWitt, D.P. (1996) *Fundamentals of Heat and Mass Transfer*, 4th edn, John Wiley & Sons, Inc., New York.
21. Richards, A. (2001) *Alien Vision, Exploring the Electromagnetic Spectrum with Imaging Technology*, SPIE Press, Bellingham and Washington, DC.
22. Kaplan, H. (1999) *Practical applications of infrared thermal sensing and imaging equipment*, Tutorial Texts in Optical Engineering, vol. TT34, 2nd edn, SPIE Press, Bellingham.
23. Holst, G.C. (2000) *Common Sense Approach to Thermal Imaging*, SPIE Optical Engineering Press, Washington, DC.
24. Moore, P.O. (ed.) (2001) *Nondestructive Testing Handbook*, Infrared and Thermal Testing, vol. 3, 3rd edn, American Society for Nondestructive Testing, Inc., Columbus.
25. Hecht, E. (1998) *Optics*, 3rd edn, Addison-Wesley, Reading.

26. Falk, D.S., Brill, D.R., and Stork, D.G. (1986) *Seeing the Light: Optics in Nature, Photography, Color Vision, and Holography*, Harper & Row, New York.
27. Pedrotti, F. and Pedrotti, L. (1993) *Introduction to Optics*, 2nd edn, Prentice Hall, Upper Saddle River.
28. Palmer, J.M. (1993) Getting intense on intensity. *Metrologia*, **30**, 371–372.
29. Soffer, B.H. and Lynch, D.K. (1999) Some paradoxes, errors, and resolutions concerning the spectral optimization of human vision. *Am. J. Phys.*, **67**, 946–953.
30. De Witt, D.P. and Nutter, G.D. (eds) (1988) *Theory and Practice of Radiation Thermometry*, John Wiley & Sons, Inc., New York.
31. Fronapfel, E.L. and Stolz, B.-J. (2006) Emissivity measurements of common construction materials. Inframation 2006, Proceedings vol. 7, pp. 13–21.
32. Kanayama, K (1972) Apparent directional emittance of V-groove and circular-groove rough surfaces. *Heat Transfer Jpn. Res.*, **1** (1), 11–22.
33. deMonte, J. (2008) Guess the real world emittance. Inframation 2008, Proceedings vol. 9, pp. 111–124.
34. Cronholm, M. (2003) Geometry effects: hedging your bet on emissivity. Inframation 2003, Proceedings vol. 4, pp. 55–68.
35. Hartmann, J. and Fischer, J. (1999) Radiator standards for accurate IR calibrations in remote sensing based on heatpipe blackbodies. Proceedings of the EUROPTO Conference Environmental Sensing and Applications, SPIE vol. 3821, pp. 395–403.
36. Henke, S., Karstädt, D., Möllmann, K.P., Pinno, F., and Vollmer, M. (2004) Challenges in infrared imaging: low emissivities of hot gases, metals, and metallic cavities. Inframation 2004, Proceedings vol. 5, pp. 355–363.
37. Madding, R.P. (2004) IR window transmittance temperature dependence. Inframation 2004, Proceedings vol. 5, pp. 161–169.
38. Richards, A. and Johnson, G. (2005) Radiometric calibration of infrared cameras accounting for atmospheric path effects, in *Thermosense XXVII*, Proceedings of SPIE, Vol. 5782 (eds G.R. Peacock, D.D. Burleigh, and J.J. Miles), SPIE Press, Bellingham, pp. 19–28.
39. Vollmer, M., Möllmann, K.-P., and Pinno, F. (2007) Looking through matter: quantitative IR imaging when observing through IR windows. Inframation 2007, Proceedings vol. 8, pp. 109–127.
40. Kaiser, N. and Pulker, H.K. (eds) (2003) *Optical Interference Coatings*, Springer, Berlin and Heidelberg.
41. Bach, H. and Krause, D. (eds) (1997) *Thin Films on Glass*, Springer, Berlin and Heidelberg.
42. www.thinfilmcenter.com (2010).
43. www.spectrogon.com (2010).

2
Basic Properties of IR Imaging Systems

2.1
Introduction

This chapter presents a brief overview of the radiation detector principles used in thermal imaging systems. It gives the background knowledge of the operation principles, the limiting factors for the detector performance, and the imaging systems. For a practitioner, a detailed knowledge of the detectors is not necessary for using an IR imaging system. However, interpretation of IR images requires some background knowledge of the limiting factors for the camera parameters such as temperature accuracy, temperature resolution (NETD, noise equivalent temperature difference), spatial resolution (MTF, modulation transfer function), and so on.

2.2
Detectors and Detector Systems

The infrared detector or the detector system acts as transducer, which converts radiation into electrical signals. It forms the core of an IR imaging system. The quality of this transduction determines the performance of the imaging system to a great extent.

Infrared detectors can be separated into two groups: photon detectors and thermal detectors.

In photon (or quantum) detectors, a single-step transduction leads to changes of concentration or mobility of the free charge carriers in the detector element upon absorption of photons from the infrared radiation [1]. If the incident radiation generates nonequilibrium charge carriers, the electrical resistance of the detector element is changed (e.g., photoconductors [2, 3]) or an additional photocurrent is generated (e.g., photodiodes [2, 3]).

Thermal detectors can be treated as two-step transducers. First, the incident radiation is absorbed to change the temperature of a material. Second, the electrical output of the thermal sensor is produced by a respective change in some physical property of a material (e.g., temperature-dependent electrical resistance in a bolometer).

Infrared Thermal Imaging. Michael Vollmer and Klaus-Peter Möllmann
Copyright © 2010 WILEY-VCH Verlag GmbH & Co. KGaA, Weinheim
ISBN: 978-3-527-40717-0

2.2.1
Parameters That Characterize Detector Performance

In general, radiation detectors are characterized by a large number of parameters [4]. With respect to the performance of an imaging system, the following detector parameters, listed in Table 2.1, are of particular importance.

The figure of merit for radiation detectors is the value of D_λ^*. With the D_λ^*-value, the performance of all different radiation detectors can be compared (see Figure 2.1 for spectral specific detectivity curves of commercially available infrared detectors.). The D_λ^*-value is determined experimentally from spectral responsivity and frequency-dependent noise measurements.

The knowledge of the D_λ^*-value allows to estimate the noise equivalent power (NEP) for a given detector geometry and bandwidth of signal detection. For example, if we assume an InSb photodiode at 5 µm wavelength with a specific detectivity of about 10^{11} cm $Hz^{1/2}$ W^{-1}, a detector area of 25 µm × 25 µm, and a bandwidth of 10 kHz for signal detection, we find a value of $NEP_\lambda = 2.5$ pW.

As shown in Section 1.3.2, the incident radiant power on a detector element depending on the object temperature can be estimated for any given optics of an IR camera. A change in the blackbody temperature ΔT_{BB} causes a difference in the radiant power $\Delta \Phi_{BB}$ received by the detector according to

$$\Delta \Phi_{BB} = \frac{\partial \Phi_{BB}}{\partial T_{BB}} \Delta T_{BB} \tag{2.1}$$

The object temperature difference ΔT_{BB} can be calculated from the measured radiant power difference $\Delta \Phi_{BB}$

$$\Delta T_{BB} = \left(\frac{\partial \Phi_{BB}}{\partial T_{BB}} \right)^{-1} \Delta \Phi_{BB} \tag{2.2}$$

If the radiant power difference $\Delta \Phi_{BB}$ equals the NEP, the limit for a possible signal detection (with signal-to-noise ratio equal to unity) is achieved.

2.2.2
Noise Equivalent Temperature Difference

The lower limit of the detectable radiant power difference is the NEP. Inserting the NEP in Eq. (2.2) results in the minimum temperature difference of the blackbody ΔT_{BB}^{min} that can be measured.

This minimum temperature difference giving a radiant power difference equal to the NEP is defined as the noise equivalent temperature difference (NETD):

$$\Delta T_{BB}^{min} = NETD = \left(\frac{\partial \Phi_{BB}}{\partial T_{BB}} \right)^{-1} NEP \tag{2.3a}$$

$$NETD = \left(\frac{\partial \Phi_{BB}}{\partial T_{BB}} \right)^{-1} \frac{\sqrt{A_D \Delta f}}{D_\lambda^*} \tag{2.3b}$$

Table 2.1 Several detector parameters that do have a large influence on IR imaging system performance.

Name, symbol, preferred units	Definition, description	Functional relationship
A_D in (cm^2) responsive area of a single pixel	Equals usually the geometric area of a single pixel. Typically 50 µm × 50 µm or 25 µm × 25 µm for thermal detectors and 50 µm × 50 µm down to 15 µm × 15 µm for photon detectors in IR focal plane arrays.	–
τ in (s) Time constant	Characterizes the response time of the detector (see also Figure 2.2).	Equals the decay time or the rise time necessary for the detector signal to reach 1/e-value or (1-1/e)-value, respectively for incident radiation in term of rectangular wave pulse.
R_λ^U in (V W^{-1}) or R_λ^I in (A W^{-1}) Spectral responsivity	Ratio of the detector signal voltage or current to the incident monochromatic radiant flux on the detector area at wavelength λ.	$R_\lambda^U = \frac{U_{Signal}}{\Phi_{radiation}}$ or $R_\lambda^I = \frac{I_{Signal}}{\Phi_{radiation}}$
$\frac{U_N}{\sqrt{\Delta f}}$ in (V Hz$^{-1/2}$) or $\frac{I_N}{\sqrt{\Delta f}}$ in (A Hz$^{-1/2}$) RMS – noise voltage or current density	Detector output noise voltage U_N or current I_N with respect to the square root of the applied bandwidth $\sqrt{\Delta f}$ of the signal measurement.	Depends on noise mechanism. Example: $\frac{U_N}{\sqrt{\Delta f}} = \sqrt{4kT_D R_D}$ for dominating Johnson-Nyquist-noise (R_D-detector resistance)
NEP$_\lambda$ in (W) Spectral noise equivalent power	Necessary incident monochromatic radiant power on the detector area to produce a signal-to-noise ratio of unity. This value mostly depends on the detector area and the bandwidth of signal detection [3, 5].	$NEP_\lambda = \frac{U_N}{R_\lambda^U}$ or $NEP_\lambda = \frac{I_N}{R_\lambda^I}$
Specific spectral detectivity D_λ^* in (cm Hz$^{1/2}$ W^{-1})	Reciprocal of the spectral noise equivalent power normalized to eliminate the dependence of detector area and the bandwidth of signal detection.	$D_\lambda^* = \frac{\sqrt{A_D \Delta f}}{NEP_\lambda}$
T_D (K) Operating temperature	Equals the temperature of the detector.	–

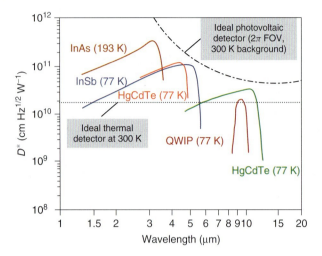

Figure 2.1 Specific spectral detectivity D_λ^* of various photon and thermal detectors for IR radiation in comparison to the theoretical limits of D^* for ideal photon and thermal detectors, respectively after [6] (operating detector temperature as indicated, all detectors with a hemispherical field of view to 300 K background, a chopping frequency of 1 kHz for photon detectors, and 10 Hz for thermal detectors is assumed).

For given D_λ^*, the NETD of the detector depends on the square root of the detector area and the bandwidth of signal detection. For example, a detector is assumed with a very low time constant, allowing the signal detection at 10 kHz and a frequency-independent D_λ^*. If the bandwidth is increased from 100 Hz to 10 kHz, the necessary D_λ^*-value of the detector to achieve the same NETD (i.e., temperature resolution of an IR camera) has to be increased by a factor of 10. If the detector area is decreased by a factor of 4 (e.g., detector size from 50 µm × 50 µm to 25 µm × 25 µm) the D_λ^*-value has to be doubled to achieve the same NETD. Furthermore, the NETD is influenced by the camera optics and all attenuation factors for the incident radiation (limited transmittance of the atmosphere and the optics, filling factor, etc.) will increase the NETD.

If D^* is wavelength dependent (Figure 2.1) and the detector receives the blackbody radiation within the wavelength region $\lambda_1 \leq \lambda \leq \lambda_2$ according to Eqs. (1.23) and (2.3), we find

$$\text{NETD} = \frac{1}{\text{FOV} \int_{\lambda_1}^{\lambda_2} \tau_{\text{optics}}(\lambda)\, D_\lambda^*(\lambda)\, \frac{\partial M_{\lambda\,\text{BB}}(\lambda,T_{\text{BB}})}{\partial T_{\text{BB}}} d\lambda} \sqrt{\frac{\Delta f}{A_D}} \quad (2.4)$$

with the field of view (FOV) related to the hemisphere FOV and the spectral dependent attenuation factor of camera optics $\tau_{\text{optics}}(\lambda)$.

According to the order of magnitude estimate of detector sensitivities of IR cameras (Section 1.3.2.6), all attenuation factors for the incident radiation on the NETD have been neglected.

A simple estimate of the necessary D^* of a detector to achieve a given NETD value uses the results of Section 1.3.2.6. For a long wavelength camera system (7–14 µm) with a standard 24° objective and an object temperature change from 302 to 303 K, the difference of incident radiant power received by each single detector pixel amounts 2.54 nW.

If one assumes this difference in radiant power as the NEP for a detector size of 50 µm × 50 µm and a signal detection with an effective noise bandwidth of 100 Hz, the necessary average D^*-value (average spectral value of $D^*(\lambda)$ within the 7–14 µm wavelength region) can be calculated according to the functional relationship for the D^*-value in Table 2.1. The necessary D^*-value from this estimate amounts to $2 \cdot 10^7$ cm Hz$^{1/2}$ W^{-1}. If the object temperature difference is decreased to NETD $= 0.1$ K, the necessary D^* of the detector increases to $2 \cdot 10^8$ cm Hz$^{1/2}$ W^{-1}. For a more realistic situation, only a decreased value of radiant power will generate the detector signal due to the attenuation factors. In this case, minimum D^* must exceed $5-8 \cdot 10^8$ cm Hz$^{1/2}$ W^{-1}.

2.2.3
Thermal Detectors

2.2.3.1 Temperature Change of Detector

Thermal detectors convert the absorbed electromagnetic radiation into thermal energy causing a rise of the detector temperature. The efficiency of this energy conversion is determined by the absorptance α, that is, the portion of absorbed radiation $\Phi_{absorbed}/\Phi_{incident}$.

Therefore, the spectral dependence of the responsivity is determined by the spectral distribution of the absorptance $\alpha(\lambda)$. The energy conversion can be expressed by the following differential equation resembling energy conservation:

$$\alpha \Phi_0 = C_{th} \frac{d\Delta T}{dt} + G_{th} \Delta T \tag{2.5}$$

with the heat capacitance C_{th} and the heat conductance G_{th} of the detector. The heat conductance G_{th} contains all heat exchange mechanism of the detector as conduction, convection, and radiation. The radiation power $\Phi_o = \Phi(T_{object}) - \Phi(T_{detector})$ in Eq. (2.5) represents the net radiation power transferred to the detector, which is given by the difference between the radiation power received from the object $\Phi(T_{object})$ and the radiation power emitted by the detector $\Phi(T_{detector})$ itself. The absorbed radiant power $\alpha \Phi_o$ leads to an increase in the detector temperature ΔT.

For a square-wave pulse of radiation, the change in detector temperature ΔT exhibits an exponential rise and decay with a time constant τ (Figure 2.2).

Equation (2.5) can be readily solved. One finds for the temperature response $\Delta T(t)$ of the sensor during the rise period:

$$\Delta T = \frac{\alpha \Phi_o}{G_{th}} \left(1 - e^{-\frac{t}{\tau}}\right) \tag{2.6}$$

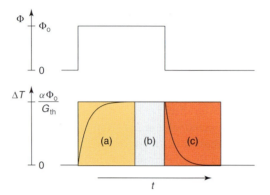

Figure 2.2 Time-dependent temperature change for square-wave pulse of radiation.

with time constant:

$$\tau = \frac{C_{th}}{G_{th}} \qquad (2.7)$$

for the steady-state conditions ($t \to \infty$):

$$\Delta T = \frac{\alpha \Phi_o}{G_{th}} \qquad (2.8)$$

and for the decay period:

$$\Delta T = \frac{\alpha \Phi_o}{G_{th}} e^{-\frac{t}{\tau}} \qquad (2.9)$$

The time constant τ for thermal detectors is a measure of the detector's speed of response. It is determined by the ratio of the heat capacitance and the heat conductance of the sensor. If the temperature difference ΔT due to absorbed radiation is linearly transformed into an electrical signal, then the responsivity of the detector is determined by the ratio of the absorptance α and the heat conductance G_{th} of the detector. Obviously a fast and sensitive thermal detector requires low heat conductance for an optimum temperature increase and, as a consequence, also low heat capacitance (or mass) to exhibit a small time constant. For measurements, see Section 8.4.1.2.

2.2.3.2 Temperature-Dependent Resistance of Bolometer

For the complete detector operation, the transduction of the temperature rise to an electrical output signal is necessary. This can be done by using any temperature-dependent physical property, such as the temperature dependence of the electrical resistance (bolometer), the creation of a voltage by a temperature difference (thermoelectricity or Seebeck effect in thermocouples and thermopiles), the temperature dependence of electrical polarization (pyroelectric detectors), and others. At present, the bolometer principle is the only detector used for IR cameras. Therefore, only the bolometer is discussed here in detail.

The bolometric effect is a change of the electrical resistance of a material due to the temperature increase caused by the absorbed radiation in the detector element.

The temperature dependence of the electrical resistance $R(T)$ is given by the temperature coefficient β, defined by:

$$\beta = \frac{1}{R}\frac{\partial R}{\partial T} \tag{2.10}$$

If the bolometer temperature is increased by ΔT, we find for the resistance change

$$\Delta R = \frac{\partial R}{\partial T}\Delta T = \overline{\beta R}\Delta T \tag{2.11}$$

where $\overline{\beta R}$ is the average value in the temperature interval ΔT.

This resistance change for steady-state conditions can be expressed with the absorbed radiant power $\alpha \Phi_o$ using Eq. (2.8):

$$\Delta R = \frac{\beta R \alpha \Phi_o}{G_{th}} \tag{2.12}$$

2.2.3.3 NEP and D* for Microbolometer

If a steady-state bias current I is applied to the bolometer, the signal voltage $U_{signal} = I\Delta R$ can be measured. According to the definition of R_λ^U, the bolometer responsivity equation is given by

$$R_\lambda^U = \frac{\beta I R \alpha(\lambda)}{G_{th}} \tag{2.13}$$

The responsivity is strongly influenced by the temperature coefficient β of the material used. Therefore, today semiconductors with $\beta = -2$ to -3% K^{-1}, such as vanadium oxide VO$_x$ or amorphous silicon α-Si, are used as bolometer materials instead of metals with β values on the order of 0.1% K^{-1}. The responsivity increases with the current applied to the bolometer. However, with increasing current an additional self-heating of the bolometer occurs associated with a bolometer resistance decrease due to the negative temperature coefficient of the bolometer material. This behavior defines an optimum operating point of the detector.

For a discussion of the specific detectivity D^*, it is necessary to fix the dominant noise mechanisms in bolometer detectors. If no 1/f-noise occurs, we can assume two dominating noise mechanism: the Johnson-noise and the noise resulting from temperature fluctuations [4, 5]. Neglecting an additional amplifier noise, these two dominating noise mechanisms of the NEP lead to

$$\text{NEP} = \sqrt{4kT_D^2 G_{th}\left(1 + \frac{I^2 R}{T_D G_{th}}\right)\Delta f + \frac{4kT_D G_{th}^2 \Delta f}{\beta^2 I^2 R}} \tag{2.14}$$

where the current heating of the bolometer is included. For Eq. (2.14), we have used the fact that the temperature differential due to current heating is small compared to the detector temperature before applying a current flow. The first term in Eq. (2.14) resulting from the temperature fluctuation noise indicates an increasing NEP with current heating. The second term resulting from the Johnson-noise describes a decreasing NEP with increasing current associated with the increasing signal response of the detector (Eq. 2.13). These opposite tendencies result in an optimum current flow with a minimum NEP. The minimum NEP is found by differentiating the NEP with respect to the current. The minimum NEP is found for optimum bolometer current $I(\text{NEP}_{\min})$:

$$I(\text{NEP}_{\min}) = \sqrt{\frac{G_{\text{th}}}{|\beta| R}} \tag{2.15}$$

which defines the minimum NEP to be

$$\text{NEP}_{\min} = \sqrt{4kT_D^2 G_{\text{th}} \left(1 + \frac{2}{|\beta| T_D}\right) \Delta f} \tag{2.16}$$

This expression shows that the temperature coefficient β, the heat conductance G_{th}, and the detector temperature T_D represent optimization parameters for the NEP of a bolometer detector. In IR cameras with bolometer arrays, the focal plane array is operated near room temperature ($T_D \approx 300$ K). The β-value for the materials used is about -2 to -3% K^{-1} as mentioned above; hence, the factor $\frac{2}{|\beta| T_D}$ varies between 0.22 and 0.33. The most interesting parameter in bolometer detector technology is the thermal conductance G_{th}, which directly affects the NEP (Eq. 2.16). For a bolometer detector, the heat transfer can be described by the three fundamental heat transfer processes: conduction, convection, and radiation. Bolometer arrays in IR cameras are operated under vacuum conditions. Therefore, convection and conduction due to a surrounding gas atmosphere can be excluded. The only remaining heat transfer processes are radiant heat exchange and the heat conduction via the solid state material of the bolometer (support legs; Figure 2.3). The total thermal conductance can be calculated from the conductance caused by radiation $G_{\text{th}}^{\text{radiation}}$ and conduction $G_{\text{th}}^{\text{conduction}}$:

$$G_{\text{th}} = G_{\text{th}}^{\text{radiation}} + G_{\text{th}}^{\text{conduction}} \tag{2.17}$$

The thermal conductance caused by the radiant loss of a detector with area A_D at temperature T_D assuming unit emissivity and hemispherical FOV can be calculated by differentiating the Stefan–Boltzmann equation with respect to the

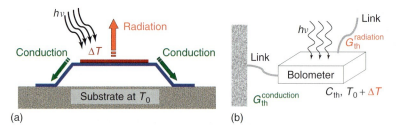

Figure 2.3 (a) Heat transfer from a bolometer membrane by radiative losses from the membrane and conductive losses via the support legs of the membrane and (b) the equivalent circuit for energy transfer.

temperatures:

$$G_{th}^{radiation} = \frac{d(A_D \sigma T_D^4)}{dT} = 4 A \sigma T_D^3 \qquad (2.18)$$

The conduction contribution G_{th}^{cond} can be estimated using a simplified model of one dimensional heat transport (Eq. 2.19):

$$G_{th}^{conduction} = \lambda \frac{A_{cross}}{l} \qquad (2.19)$$

where λ is the thermal conductivity of the material suspending the bolometer membrane, A_{cross} the cross-sectional area, and l the length of the support legs of the bolometer.

To get low NEP values and hence low NETD values, the total heat conductance has to be as small as possible. The minimum value of G_{th} results if conduction can be neglected and the radiation dominates the heat transfer. In this case and if only thermal fluctuation noise occurs in the thermal detector (Johnson-noise is neglected), we will get the minimum NETD possible or in other words the theoretical limit of the specific detectivity D^* (see Eqs. 2.20 and 2.21, respectively):

$$NEP_{ideal} = 4\sqrt{A_D \sigma k T_D^5 \Delta f} \qquad (2.20)$$

$$D^*_{ideal} = \frac{1}{4\sqrt{\sigma k T_D^5}} \qquad (2.21)$$

As a numerical example for detector temperature $T_D = 300$ K, the specific detectivity becomes $1.8 \cdot 10^{10}$ cm Hz$^{1/2}$ W^{-1}. If this value is compared to the 2π FOV and 300 K background background limited infrared photodetection (BLIP) limit for quantum detectors (Figure 2.1), one can conclude that thermal detectors are only competitive in the LW region (7 – 14 µm).

For a 35 µm × 35 µm bolometer, this corresponds to a thermal conductance value of $7.5 \cdot 10^{-9}$ W K^{-1}.

For a real bolometer, the thermal conductance is also influenced by the heat conduction via the suspension legs of the membrane connecting the bolometer with the substrate, acting as a heat sink.

To achieve only small additional contributions to the thermal conductance, a small cross-sectional area and a comparatively large length of the insulation legs are required. The limits for the length of the legs are given by the area of a single pixel (Figure 2.4a). The insulation legs are responsible for a decrease of the filling factor below 100%. The improvements in the micromechanical technologies have increased the filling factor to about 80%.

For a simple estimate (neglecting the temperature coefficient of resistivity (TCR) material) of the heat conductance of the legs, we assume a heat conductivity of the material Si_3N_4 used of about 2 W m^{-1} K^{-1}, a length of the legs of 50 µm, with a cross-sectional area of 2 µm × 0.5 µm [7]. For these parameters, the thermal conductance caused by the thermal conduction via the two bolometer legs is

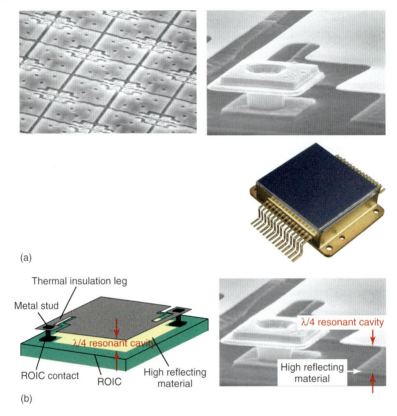

Figure 2.4 (a) Uncooled bolometer FPAs. Left: bolometer array; top right: contact of the bolometer detector; and bottom right: complete bolometer 640 × 480 pixels FPA chip. Image courtesy: ULIS, France. (b) Bolometer pixel with $\lambda/4$ resonant cavity. Image courtesy: ULIS, France.

$8 \cdot 10^{-8}$ W K^{-1}. The vacuum operated bolometer will, therefore, have a total heat conductance of $8.8 \cdot 10^{-8}$ W K^{-1}. According to Eq. (2.9), the time constant equals the ratio of the heat capacitance and the heat conductance. For a bolometer with the heat conductance of $8.8 \cdot 10^{-8}$ W K^{-1} and a time constant of 10 ms, a heat capacitance of $8.8 \cdot 10^{-10}$ J K^{-1} has to be achieved. Assuming a specific heat capacity of 500 J kg^{-1} K^{-1} of the main bolometer material Si$_3$N$_4$, this results in a mass of the bolometer pixel of about 0.6 ng. For a Si$_3$N$_4$ density of 3.2 g cm^{-3} and a bolometer area of 35 µm × 35 µm, this corresponds to a detector membrane thickness of about 0.2 µm. Such a detector membrane would be too thin for efficient absorption of infrared radiation. For enhanced long-wavelength infrared absorption, a quarter wavelength resonant cavity is used (Figure 2.4b). The metal studs connecting the bolometer of a focal plane array to a CMOS (complementary metal oxide semiconductor) readout integrated circuit (ROIC) also adjust the cavity width. The ROIC is an integrated circuit multiplexer that couples to a focal plane array sensor that reads the individual electrical FPA outputs. The ROIC transforms the small detector signal to a relatively large measurable output voltage.

For an application of these detectors in thermal imaging systems, it has to be considered that bolometers are not perfect in DC operation. The previously discussed 2.54 nW change of incident radiation power for a long-wavelength camera system (7–14 μm) with a standard 24° objective and an object temperature change from 303 to 302 K will cause a radiation-induced bolometer temperature change (total conductance of $8.8 \cdot 10^{-8}$ W K^{-1} assumed) of only about 29 mK. However, the correct object temperature change of detector temperature of 1 K can only be measured by the bolometer if the intrinsic bolometer temperature is stable compared to the temperature change caused by the radiation power change. This means that if the detector temperature is changed by 29 mK due to other mechanisms, for example, a change in temperature of the whole detector assembly by 1 K also leads to an apparent 1 K object temperature change. Therefore in most cases, the bolometer array in a thermal imaging system is mounted on a Peltier element and its temperature is thermoelectrically stabilized. Additionally, a flag with known temperature is placed with a period of some minutes into the optical path of the camera for a recalibration of the detector array.

2.2.4
Photon Detectors

2.2.4.1 Principle of Operation and Responsivity

Photon detectors convert the absorbed electromagnetic radiation directly into a change of the electronic energy distribution in a semiconductor by the change of the free charge carrier concentration. This process is called *internal photoelectrical effect*. Solids as semiconductors exhibit a typical electronic band structure with energy bands or energy states ("allowed" electron energies) and energy gaps ("forbidden" electron energies, see Figure 2.5a,b). For the radiation detection, the quantum energy of the photons $E = h\nu$ must exceed an energy threshold ΔE, which is the excitation energy of an electronic transition between the valence (uppermost, almost with electrons fully occupied band) and the conduction band (lowermost, almost unoccupied band) or between an energy state within the energy gap (caused by a impurity energy level) and an energy band. Figure 2.5c depicts different excitation processes. For the intrinsic photoeffect, the interband excitation of an electron–hole pair at photon energies exceeding the energy gap of the semiconductor occurs. The extrinsic photoeffect describes the excitation of an electron or hole from an impurity level to the conduction or the valence band of the semiconductor, respectively. For this excitation, the photon energy must exceed the ionization energy of the impurity that is the energy difference between the impurity energy level and the corresponding band edge.

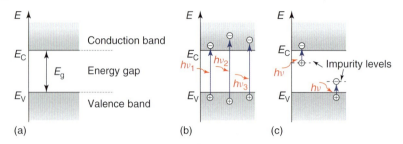

Figure 2.5 (a) Simplified energy-band structure of solids. E_V and E_C denote upper edge of valence band and lower edge of conduction band, respectively. (b) Intrinsic photoeffect: Photons of varying energy $h\nu$ (length of arrows) may excite electrons into the conduction band while leaving holes in the valence band. (c) Extrinsic photoeffect: impurities form localized electronic states within the energy gap. Photons need much lower energy for transitions between impurity levels and electronic bands.

Because of the energy threshold ΔE for photon detectors, the spectral sensitivity region is limited by a cutoff wavelength λ_{cutoff}:

$$\Delta E = h\nu_{\text{cutoff}} = \frac{hc}{\lambda_{\text{cutoff}}} \tag{2.22a}$$

$$\lambda_{\text{cutoff}} = \frac{hc}{\Delta E} \tag{2.22b}$$

The photon detector is only sensitive for $\nu \geq \nu_{\text{cutoff}}$, that is, $\lambda \leq \lambda_{\text{cutoff}}$. The responsivity and the specific detectivity are strongly wavelength dependent. This behavior is caused by the fact that the detector acts as a photon counter and the detector signal corresponds to the number of incident photons $Z_{\text{radiation}}$ per time, also denoted as *quantum flux*. The responsivity is defined by the ratio of the detector signal and the incident radiative power $\Phi_{\text{radiation}}$ (Table 2.1). If we assume a monochromatic radiation at a wavelength λ, the quantum flux Z can be calculated from the incident radiative power $\Phi_{\text{radiation}}$ and vice versa:

$$Z_{\text{radiation}} = \frac{\Phi_{\text{radiation}}}{h\nu} = \lambda \frac{\Phi_{\text{radiation}}}{hc} \quad \text{or} \quad \Phi_{\text{radiation}} = \frac{1}{\lambda} hc Z_{\text{radiation}} \tag{2.23}$$

The responsivity is defined by the ratio of the detector signal and the incident radiant power $\Phi_{\text{radiation}}$ (Table 2.1). We may write the signal current from the detector as

$$I_{\text{signal}} = e\eta Z_{\text{radiation}} = \frac{e\eta \lambda \Phi_{\text{radiation}}}{hc} \tag{2.24}$$

where η is the quantum efficiency, which means the number of free charge carriers per incident photon.

Therefore,

$$R_\lambda^I = \frac{I_{\text{signal}}}{\Phi_{\text{radiation}}} = \frac{e\eta\lambda}{hc}, \quad \frac{I_{\text{signal}}}{Z_{\text{radiation}}} = e\eta \tag{2.25}$$

The responsivity as well as the specific detectivity exhibit a spectral dependence proportional to the wavelength λ (Figure 2.6). The highest possible responsivity of

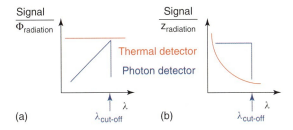

Figure 2.6 Wavelength dependence of the photon detector and thermal detector signals for constant incident radiant power (a) and quantum flux (b), respectively.

a photon detector for a quantum efficiency equal to unity ($\eta = 1$) depends only on the wavelength.

This results in theoretical triangular shape of the spectral dependent responsivity and detectivity curves for photon detectors. The deviation of the real spectral dependence (Figure 2.1) compared to the triangular-shaped curve results from the dependence of the quantum efficiency η near to the cutoff wavelength. For our simplified model, we have used a step function for the quantum efficiency η with $\eta = 0$ for $\lambda > \lambda_{\text{cutoff}}$ and $\eta = 1$ for $\lambda \leq \lambda_{\text{cutoff}}$. Deviations from the step function model $\eta(\lambda)$ are due to different parameters such as spectral dependence of absorption coefficient, nonequilibrium carrier lifetime, carrier diffusion length, detector geometries, and so on.

The quantum detector noise can be dominated by different noise sources. Two groups of noise can be distinguished: the dark current noise (Johnson–Nyquist noise, current noise, generation and recombination noise, etc. [1, 5]) and the radiation-induced noise. For an ideal detector, the noise is only due to the radiation-induced noise contributions. The signal detection can then be limited by the signal noise or by the background radiation.

2.2.4.2 D* for Signal Noise Limited Detection

We assume that the detector receives only signal radiation causing a signal current I_{signal} without any additional radiation from the background.

The shot noise current caused by the signal current (Eq. 2.24) is [5, 8]

$$I_N = \sqrt{2eI_{\text{signal}}\Delta f} = \sqrt{\frac{2e^2 \eta \lambda \Phi_{\text{signal}} \Delta f}{hc}} \qquad (2.26)$$

and the NEP$_{\text{SL}}$ for a signal noise limited detector becomes

$$\text{NEP}_{\text{SL}}(\lambda) = \frac{2hc\Delta f}{\eta \lambda} = \frac{2E_{\text{photon}}}{\eta}\Delta f \qquad (2.27)$$

The NEP$_{\text{SL}}$ depends only on the photon energy. For a wavelength $\lambda = 10\,\mu\text{m}$, a quantum efficiency $\eta = 1$ and a bandwidth of $\Delta f = 1$ Hz, we estimate an extremely low NEP of about $4 \cdot 10^{-20}$ W. In this case, a photon flux of only two photons per second will produce a signal-to-noise ratio equal to unity.

The specific spectral detectivity $D^*(\lambda)$ becomes dependent on the detector area:

$$D*(\lambda) = \frac{\eta\lambda}{2hc}\sqrt{\frac{A_D}{\Delta f}} \tag{2.28}$$

and for a detector area of 25 µm × 25 µm, we obtain $D^*(\lambda = 10\,\mu m, \eta = 1, \Delta f = 1\,Hz) = 6.3 \cdot 10^{16}\,cm\,Hz^{1/2}\,W^{-1}$. This value must be considered as an ideal theoretical result. It does not reflect the situation of a photon detector in a thermocamera, which also receives background radiation.

2.2.4.3 D* for Background Noise Limited Detection

Now we consider a more realistic situation of a photon detector in our camera. The detector will be exposed to signal as well as background radiation.

We can use Eq. (2.24) again for the calculation of the signal current. The noise current is modified by additional background radiation, generating an additional noisy detector current. Modifying Eq. (2.26) we find

$$I_N = \sqrt{2eI_{signal}\Delta f} = \sqrt{\frac{2e^2\eta\lambda\left(\Phi_{signal} + \Phi_{background}\right)\Delta f}{hc}} \tag{2.29}$$

We had discussed above the extreme case where background radiation was negligible. The other extreme would be if the photocurrent due to background radiation would be the dominant noise contribution.

Assuming that the signal radiation power can be neglected compared to the background radiation power and using the expressions for the signal current from Eq. (2.24) and the noise current from Eq. (2.29), we can write the specific spectral detectivity D^*_{BLIP} for a detector characterized by BLIP (background limited infrared photodetection) as

$$D^*_{BLIP}(\lambda) = \sqrt{\frac{\eta\lambda}{2hc\left(\frac{\Phi_{background}}{A_D}\right)}} = \sqrt{\frac{\eta\lambda}{2hc\,E_{background}}} \tag{2.30}$$

with the irradiance $E_{background}$ from the detector background. Note that Eq. (2.30) is valid for monochromatic background radiation with wavelength λ. It can be rewritten in terms of photon flux $Z_{background}$. $Z_{background}$ from a hemisphere is given by absorbed background photon flux:

$$Z_{background} = A_D \int_0^{\lambda_{cutoff}} M_\lambda\left(\lambda, T_{background}\right)\frac{\eta\lambda}{hc}d\lambda \tag{2.31}$$

Now the value of $D^*_{BLIP}(\lambda)$ can be derived for a detector in the presence of thermal background radiation (background temperature $T_{background}$) within a FOV characterized by the full cone angle θ of received background radiation (Figure 2.7). For hemispherical background, Eq. (2.31) gives the number of background photons. For a limited FOV of 2θ linear angle as seen from the detector, the result of Eq. (2.31) has to be multiplied by $1/\sin^2(\theta/2)$.

For the calculation of $D^*_{BLIP}(\lambda)$, we assume a black background $\varepsilon_{background} = 1$ with uniform temperature $T_{background}$, the triangular-shaped spectral responsivity

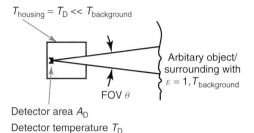

Figure 2.7 Detector aperture defined FOV of a cooled detector for thermal background radiation.

distribution of the photon detector with quantum efficiency $\eta = 1$ for $\lambda \leq \lambda_{\text{cutoff}}$, and 0 for $\lambda > \lambda_{\text{cutoff}}$ according to Eq. (2.25):

$$D^*_{\text{BLIP}}(\lambda) = \frac{\lambda}{hc} \sin^{-1}\left(\frac{\theta}{2}\right) \left\{ \int_0^{\lambda_{\text{cutoff}}} \frac{2\pi c}{\lambda^4 \left(e^{\frac{hc}{\lambda k T_{\text{background}}}} - 1\right)} d\lambda \right\}^{-1/2} \quad (2.32\text{a})$$

and can be approximated by

$$D^*_{\text{BLIP}}(\lambda) = \left\{ \sin\left(\frac{\theta}{2}\right) \left[\frac{4\pi (kT_{\text{background}})^5}{c^2 h^3}\right]^{1/2} \right. $$
$$\left. \cdot x_c \left(x_c^2 + 2x_c + 2\right)^{1/2} e^{-\frac{x_c}{2}} \right\}^{-1} \quad (2.32\text{b})$$

with dimensionless parameter x_c for $x_c \gg 1$:

$$x_c = \frac{hc}{\lambda k T_{\text{background}}} \quad (2.33)$$

Figure 2.8 depicts $D^*_{\text{BLIP}}(\lambda)$ according to Eq. (2.32b) calculated for a background temperature of 300 K and different aperture sizes. We assume that the housing that contains the detector and forms the aperture for the background radiation exhibits a temperature much lower than the background, for example, the temperature of liquid nitrogen $T_{\text{chamber}} = 77$ K. Therefore, the influence of the radiation from the inside of the detector housing on the D^* value can be neglected.

According to the results shown in Figure 2.8, the specific detectivity of a BLIP photon detector can be increased by decreasing the FOV for thermal background radiation. The minimum aperture size in a thermal imaging system is given by the FOV of the camera optics. In addition to this limit, one has to consider that we have disregarded all other detector noise processes (as dark current noise of the detector) for calculating D^*_{BLIP}. If the noise caused by background radiation is decreased by reducing the aperture size, the other noise processes become more important or even dominant. In this case, the aperture D^*-dependence saturates and the D^* value is not longer background radiation limited.

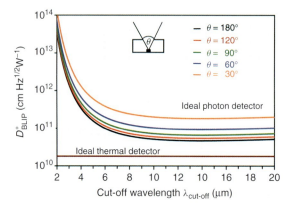

Figure 2.8 D^*_{BLIP} for an ideal photon detector depending on the cutoff wavelength and the field of view θ for background radiation with $T_{background} = 300$ K (ideal thermal detector for comparison).

From this discussion, we can draw another conclusion. If we want to limit the spectral range of a camera by using spectral filters, these filters should be cold. The background radiation emitted by a cold filter at a low temperature, such as 77 K, can be disregarded. The detector receives only the background radiation within the spectral range given by the filter transmittance. This results in a decrease in background radiation striking the detector and can cause a further increase in the specific detectivity D^*.

If we assume the detector D^*_{BLIP} at 2π FOV, the possible temperature resolution of thermal imaging systems in the LW and the MW region can be compared. As shown in Section 1.3.2.7, the radiation temperature contrast (change of excitance with object temperature change) can be calculated from the derivative of the spectral excitance $M_\lambda(\lambda, T)$ with respect to the object temperature T. This result can be related to the NEP by multiplying the derivative of $M_\lambda(\lambda, T)$ by $D^*(\lambda)$. Finally, we have to integrate this expression with respect to the wavelength region of the camera.

In this case (see also Eq. (2.3)),

$$\int_{\lambda_1}^{\lambda_2} D^*(\lambda) \frac{dM}{dT} \Delta T d\lambda \sim \int_{\lambda_1}^{\lambda_2} \frac{dM}{dT} \frac{\Delta T}{NEP} d\lambda \sim \frac{\text{signal change induced by } \Delta T}{\text{noise}}$$

(2.34)

The result gives a measure of the possible temperature resolution of the detector.

Figure 2.9 depicts the results of such a calculation for an MW and an LW system. All influences of camera optics have been neglected, that is, only the detector properties will affect the result. For the calculation, a triangular-shaped wavelength dependence of the responsivity has been assumed for the photon detector and a constant responsivity for the bolometer detector. The marked areas in Figures 2.9b,c represent the results of the integration and are a measure of

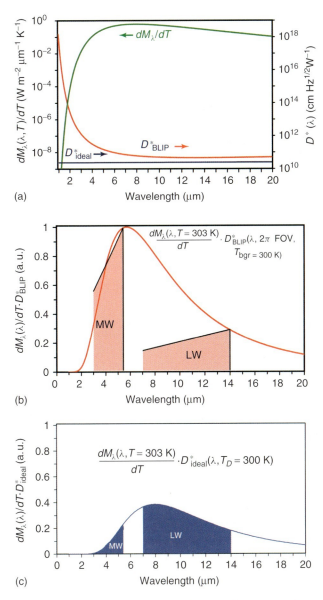

Figure 2.9 (a) Comparison of spectral dependent D^* for BLIP photon detector (FOV 2π with 300 K background) and ideal thermal detector with 300 K detector temperature with dM_λ/dT spectral dependence for an object temperature of $T_{object} = 300$ K. (b) Comparison of the temperature resolutions of a BLIP photon detectors in the LW (7 to 14 μm) and in the MW (3–5.4 μm, InSb detector at 77 K assumed) region. (The curve is normalized with the maximum value of its spectral dependence.) (c) Comparison of the temperature resolutions of an ideal thermal detector in the LW and in the MW region. (The curve is normalized with the maximum of the spectral dependent photon detector curve from (b).)

the signal-to-noise ratio for an infinitesimal temperature change of a 303 K object. The possible temperature resolutions are related to the reciprocal surfaces of the colored areas in the LW and MW region, depicted in Figure 2.9.

The ratio between the minimum resolvable temperatures in the LW and the SW regions for both types of detectors can be calculated using the ratio of the colored areas. For the photon detector operated at the BLIP limit, the ratio LW:MW is 1:0.8. Photon detectors with BLIP performance allow nearly the same temperature resolution of a thermocamera in the LW as well as in the MW spectral region. Because of the strong increase in the D^* value with decreasing wavelength and the slight increase in the temperature contrast dM_λ/dT from the LW to the MW spectral region, the MW detector surpasses the temperature resolution of the LW detector.

For the thermal detector with ideal performance, the ratio LW : MW is 1:8.8. Compared to the LW region, the temperature resolution of a thermal imaging system equipped with thermal detectors will drop by almost 1 order of magnitude if it were used in the MW region. Therefore, thermal detectors are suitable only for MW imaging systems for higher object temperatures.

Finally, we can compare the detector capabilities of the BLIP photon and the ideal thermal detectors for thermal imaging in the LW region. The ratio LW (thermal detector) : LW (photon detector) is 1:0.71. This demonstrates the capability of thermal sensors for thermal imaging in the LW region, although we have to consider that the performance of a photon detector can be further improved by decreasing the FOV for background radiation and the fact that the performance of available thermal detectors does not achieve the theoretical D^* limit.

2.2.4.4 Necessity to Cool Photon Detectors

Infrared photon detectors operating at BLIP conditions require cooling to low temperatures as can be seen by the following argument. The cutoff wavelength of the detector corresponds to the gap energy (Figure 2.5 and Eq. (2.22)). For a photon detector with a 10 μm cutoff wavelength, the energy gap is 124 meV. In general, the energy of photons is related to their wavelength via $E(eV) = 1.24/\lambda$ (μm). The optimum situation for detector operation is achieved if only incoming photons with $\lambda < \lambda_{cutoff}$ generate free charge carriers. From the discussion of noise in photon detectors, one expects that any additional thermal excitation process of charge carriers over the energy gap will increase the noise. At a temperature of $T = 300$ K, the thermal energy kT equals 25.9 meV. Using the energy gap of the 10 μm detector and this thermal energy, we can estimate the probability W of thermal carrier excitation using the Boltzmann equation (simplified model used):

$$W = \text{const.} \exp\left(-\frac{\Delta E}{kT}\right) \tag{2.35}$$

where const accounts for transition probabilities, density of states, and so on. For $\lambda = 10$ μm and $T = 300$ K, we obtain for the exponential part $8.3 \cdot 10^{-3}$. Because of this rather large factor for the excitation probability and the high density of states in the semiconductor material, the equilibrium concentration of thermally excited free carriers is much larger than the nonequilibrium carrier concentration caused by the radiation absorption. This causes a large noise and a reduced sensitivity of the photon detector.

If the temperature of the 10 μm detector is, however, reduced to the temperature of liquid nitrogen $T = 77$ K ($kT = 6.63$ meV), the Boltzmann factor for the probability of thermal carrier excitation is reduced by 6 orders of magnitude to the value of $7.6 \cdot 10^{-9}$. Cooling down to low temperatures very efficiently decreases the free charge carrier concentration and therefore the noise in the photon detector. The demand for cooling the photon detector to low temperatures increases with increasing cutoff wavelength due to the decreasing energy for carrier excitation. Almost all 7–14 μm photon detectors (for the LW cameras) operate at about 77 K. Detectors for the 3–5 μm MW region are thermoelectrically cooled to about 200 K or also cooled to 77 K. The detectors for the 0.9–1.7 μm SW region are thermoelectrically cooled.

2.2.5
Types of Photon Detectors

Photon detectors can be subdivided into different types depending on the operation principles. The most important photon detectors are classical semiconductor detectors, including photoconductors and photodiodes, and novel semiconductor detectors, including photoemissive Schottky barrier and band gap engineered quantum well infrared photodetectors (QWIPs).

2.2.5.1 Photoconductors
The photoconductor is a detector device composed of a single uniform semiconductor material.

Photoconductors monitor the incident radiation via the change of the semiconductor bulk resistivity. It is due to a creation of free charge carriers induced by the absorbed photons. Any change in photoconduction is detected by measuring the current–voltage characteristic.

The responsivity of a photoconductor can be derived from Eq. (2.25) but with the addition of a gain factor g. This gain factor represents the ratio of the carrier lifetime τ and the transit time for the carrier in the photoconductor. The latter is given by the ratio of the detector length and the carrier drift velocity [2]. If the carrier lifetime is greater than the transit time, the photocurrent is amplified ($g > 1$). This gain can be understood from the following argument: if charge carriers live longer than their transit time, they will leave the detector at one electrical contact. For charge conservation reasons, they will be replaced by new (secondary) carriers at the opposite contact. The process will repeat again and again until the time reaches the carrier lifetime. As a result, many charge carriers are due to a single excitation mechanism, that is, the process can be regarded as a photocurrent gain.

All D^* values calculated in Section 2.2.4 have to be divided by a factor of $\sqrt{2}$ for photoconductors due to additional recombination noise [3, 5].

The sensitivity of the photoconductor increases with the applied voltage or current. However, this goes along with the increased heat dissipation at the detector. As a consequence, large photoconductor arrays are difficult to cool. Moreover, the increase in operating current often leads to an increased $1/f$-noise behavior [5], which decreases the D^* value. The photocurrent gain can also be increased by increased carrier lifetime. Very sensitive photoconductors often exhibit much larger time constants than photodiodes. Regarding IR camera systems, long time constants are, however, not desirable. Detectors with larger time constants can only be operated at lower frame rates. As a result of these properties, photoconductors are not the photon detectors of choice for thermal imaging.

2.2.5.2 Photodiodes

Photodiodes consist of a p–n junction in a semiconductor. It is a photon-sensitive diode, producing a current or voltage output in response to incident optical radiation flux.

The cutoff wavelength is determined by the energy gap of the semiconductor used.

For focal plane arrays used in LW thermal imaging systems MCT (mercury–cadmium–telluride, $Hg_{1-x}Cd_xTe$) is the favorite semiconductor material. This II–VI-compound is an alloy composition of the wide energy gap semiconductor CdTe (energy gap 1.6 eV) and the semimetallic compound HgTe (energy gap -0.3 eV). The energy gap of $Hg_{1-x}Cd_xTe$ is controlled by the relative portions of the two binary semiconductors. The dependence of the energy gap on composition can be approximated by a linear dependence. For a composition of $x = 0.196$ at $T = 77$ K, a cutoff wavelength of 14 µm is observed [3].

Because of the strong dependence of the energy gap on composition caused by the strong variation of 1.9 eV from CdTe to HgTe and the low energy gap of 0.089 eV for a cutoff wavelength of 14 µm, a composition uncertainty of only 2% ($x = 0.196 \pm 0.004$) will cause a variation of the cutoff wavelength by $\Delta\lambda_{cutoff} = \pm 0.51$ µm, that is, $\lambda_{cutoff} = (14 \pm 0.51)$ µm. Therefore, MCT detectors pose problems in detector arrays. Small composition changes lead to changes of cutoff wavelengths and therefore also changes of NEP and D^* at fixed wavelength.

Therefore, the very complex technology and the demand for composition homogeneity of this material results in very high prices for two-dimensional detector arrays with a large number of detectors. Modern fabrication methods such as molecular beam epitaxy (MBE) and metalorganic chemical-vapor deposition (MOCVD) increase the manufacturability of this material and will hopefully lead to lower prices in the future.

MCT is also used for focal plane arrays in the MW spectral region (3–5 µm). However in this spectral region, it competes directly with indium antimonide (InSb). InSb is the most highly developed and widely used semiconductor material in the 1–5 µm spectral region and one of the most sensitive detector materials available in the MW spectral region. Compared to $Hg_{1-x}Cd_xTe$, the technology of InSb is less complicated [9] and enables large detector arrays of reasonable uniformity with more than 1 megapixel. The energy gap of this material is only temperature dependent and amounts to 0.23 eV at $T = 77$ K, causing a cutoff wavelength of about 5.5 µm. The excellent detector performance results in an operation that, similar to MCT, has D^*-value close to the BLIP limit (Figure 2.1). Many modern MW IR cameras use InSb detectors.

In recent years, progress for the SW spectral region (0.9–1.7 µm) was due to intensive technology development of the III–V compound semiconductors based on the binary semiconductor compounds GaAs and InAs. Today, it is possible to get high-sensitive and high-speed $In_{1-x}Ga_xAs$ photodiodes and photodiode arrays for the near infrared spectral region. $In_{1-x}Ga_xAs$ is an alloy consisting of GaAs (energy gap 1.43 eV corresponding to a cutoff wavelength of 0.87 µm) and InAs (energy gap 0.36 eV corresponding to a cutoff wavelength of 3.4 µm). The energy gap of the alloy can be tailored to the wavelength of interest by controlling the mixing of the two binary compounds.

The basic operational principle used in photodiode arrays of imaging systems is the external current flow to a readout circuit (short circuit current or reverse bias operation) caused by the incident infrared radiation. The incident photon with a quantum energy larger than the gap energy of the semiconductor is absorbed exciting an electron–hole pair. During their lifetime (average time from generation to recombination), the charge carriers move within the diode due to diffusion processes. The average distance is the diffusion length. If the electron–hole pair is generated inside the depletion layer or at a maximum distance of one diffusion length distant from the depletion layer, the electron–hole pair will be separated by the electric field of the p–n junction. The electron moves into the p-type region and the hole moves into the n-type region, causing a current to flow. This photogenerated current is given by Eq. (2.24) and resembles a reverse current in the photodiode. Mostly photodiodes are operated with reverse bias. The reverse bias widens the depletion layer that causes a decrease in the junction capacitance and allows a faster detector response due to the decreased time constant $\tau = RC$ of the photodiode. Typical photodiode time constants range from nanoseconds to microseconds, which allows frame rates of some 100 kHz. Additional limitations to the readout process are discussed in Section 2.4.

Using the reversed bias operation mode, the photodiode can also decrease the dark current noise [5]. The maximum increase in the specific detectivity D^* amounts to a factor of $\sqrt{2}$ if Johnson–Nyquist noise is the dominant noise process in the photodiode.

2.2.5.3 Schottky Barrier Detectors

A variety of metal films on semiconductors can form Schottky barriers [2]. For infrared applications, Schottky barriers on silicon are used [3]. The use of a silicon-based technology for the infrared detectors is advantageous for the construction of focal plane arrays since monolithic integration of the infrared sensor arrays with the readout circuit becomes possible. The complete technological process is similar to the well-developed standard silicon VLSI/ULSI process (very large scale integration/ultra large scale integration) for silicon-integrated circuits. This results in lower defect rates, excellent homogeneity of the detectors in the focal

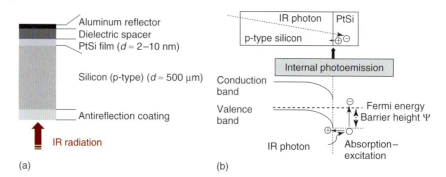

Figure 2.10 (a) Typical structure of a Schottky barrier infrared sensor and (b) the operation principle (for detailed description, see [3]).

plane array (response nonuniformity of less than 1% in large arrays), and cheaper manufacturing.

The most important metal used is platinum (Pt). Figure 2.10 depicts the typical structure of a Schottky barrier photodiode formed by Si/PtSi. These detectors act via the process of internal photoemission over the Schottky barrier height Ψ ($\Psi = 0.22$ eV for Pt/PtSi). The Si/PtSi Schottky barrier detector is cooled to $T = 77$ K. IR photons of photon $h\nu$ with energy $E_g^{Si} > h\nu \geq \Psi$ can pass the silicon ($E_g^{Si} = 1.1$ eV) but are absorbed in the silicide, thereby generating an electron–hole pair.

Photoexcited holes can diffuse to the Si/PtSi interface. The electrons accumulate on the silicide electrode and can be transferred to a readout circuit.

The low quantum efficiency of Schottky barrier detectors (typically 0.1–1%) is due to the fact that the absorption of the IR photons is weak (thin silicide layer) and only a small number of photoexcited holes can cross the barrier. Therefore, an additional $\lambda/4$ cavity with a dielectric layer and an aluminum reflector is used to enhance the photoemission process.

2.2.5.4 Quantum Well Infrared Photodetectors (QWIP)

The development of modern methods of crystal growth such as MBE and MOCVD allows the fabrication of semiconductor heterostructures. Different lattice matched semiconductor materials can be deposited in a perfect crystalline structure with atomic thickness resolution. The combination of different semiconductor materials with different band gaps in heterostructures causes a spatial variation of the conduction and valence band edge. Tailoring of the band edge distribution becomes possible by variation of semiconductor materials and the layer thickness. This method is called *bandgap engineering*. For the simplest case, an alternating sequence of abrupt changing materials such as GaAs and $Al_{1-x}Ga_xAs$ results in a cascaded band edge distribution (Figure 2.11). For small layer thickness in the nanometer range, drastic changes of the electronic and optical properties are observed due to quantum effects.

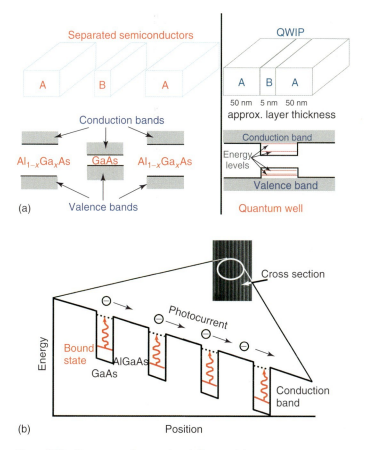

Figure 2.11 Structure and energy-band diagram (a) as well as excitation of electrons (b) in a biased GaAs/AlGaAs QWIP, after [10].

The band edge distribution forms alternating quantum wells and barriers for the electrons and holes. For an adequate layer composition and thickness in the nanometer range, quantized energy levels occur within the quantum wells (for a more detailed discussion of the charge carrier behavior and their wavefunctions within a box-shaped potential, see [11]). These energy levels are called *subbands*. An incident photon can excite an electron from the ground to the first excited state subband if the photon energy equals the energy gap of these two subbands (Figure 2.11). This energy gap is much smaller than the energy gaps of the two semiconductors used and corresponds to quantum energies of photons within the infrared spectral region. The narrow quantum well causes very low quantum absorption efficiency. To increase this efficiency, a multiquantum well structure with 50 or more quantum wells is used. The QWIPs responsivity spectra are much narrower and sharper than the spectra of intrinsic detectors due to their resonance intersubband absorption. The spectral dependence of the responsivity

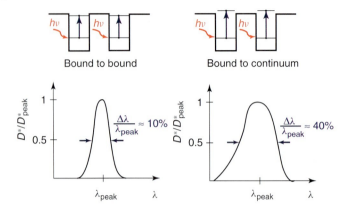

Figure 2.12 Typical spectral responsivities of bound excited state and continuum QWIPs.

of QWIPs can be tailored from narrow ($\Delta\lambda/\lambda \sim 10\%$) for bound excited states to wide ($\Delta\lambda/\lambda \sim 40\%$) for excited states in the continuum band just above the energy barrier [10, 12] (Figure 2.12).

The spectral bandwidth of QWIPs can be further increased by replacing a single quantum well staple with a superlattice structure consisting of staples of several quantum well types separated by very thin barriers [13].

In QWIPs, the photocurrent gain depends on the operating bias voltage and varies from about 10 to about 50%. This behavior is connected to the probability of electron capture into the quantum wells while traveling to the contact of the detector. Typical time constants of QWIPs are in the millisecond range, that is, higher than the time constants of photodiodes. This limits the maximum frame rate of a QWIP focal plane array camera.

For the operation of QWIPs, low detector temperatures are required, typically $T = 77$ K. With decreasing temperatures, the number of thermally excited charge carriers decreased strongly according to the Boltzmann distribution function. This results in an increasing responsivity. Furthermore, the noise current in the QWIP, which is limiting the NEP or its D^*-value, also strongly decreases with detector temperature. The dominant noise in QWIP devices is due to the shot noise resulting from the dark current in the detector. The dark current originates from three main processes: quantum well to quantum well tunneling, thermally assisted tunneling, and classical thermionic emission. The last mechanism decreases the responsivity and establishes the major source of the dark current at higher operation temperatures. The maximum D^* values for a given peak response wavelength (λ_{peak}) and detector temperature for GaAs/AlGaAs QWIPs were empirically fitted to measurement results [14]:

$$D^* = 1.1 \cdot 10^6 \exp\left(\frac{hc}{2\lambda_{peak} kT}\right) \text{ cm Hz}^{1/2}\text{W}^{-1} \qquad (2.36)$$

This equation was used to predict the limit of D^* of a QWIP as a function of peak wavelength and for $T = 77$ K (Figure 2.13).

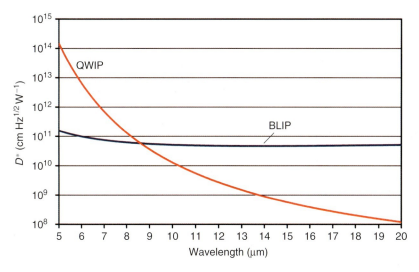

Figure 2.13 Predicted spectral dependent maximum D^*_{QWIP} at peak wavelength for QWIP detectors operated at 77 K [14] compared to D^*_{BLIP} for photon detectors at 2π FOV and 300 K background temperature.

A variety of III–V-semiconductor combinations are used for QWIPs, such as GaAs/AlGaAs, InGaAs/InP, InGaAs/InAlAs, and AlGaInAs/InP, covering the spectral range from 3 to 20 µm. QWIPs are the only genuine narrowband infrared detectors and are therefore ideally suited for narrowband applications (Chapter 7). In addition, they can be used as bi- or multispectral sensors. Such multispectral sensors can be assembled in a stack arrangement of QWIPs (independently contacted) with different peak wavelengths [15]. A modern approach to bispectral detectors are QWIPs operating in two different spectral regions with switchable sensitivity from the LW to the MW by increasing the bias voltage.

The QWIPs have made their way from the laboratory to commercial high-performance thermal imaging systems with large megapixel focal plane arrays [16]. QWIPs offer high thermal resolution (some millikelvin), spatial resolution, excellent homogeneity, low fixed pattern noise, low $1/f$-noise, high pixel functionality, and high yield at moderate cost.

2.3
Basic Measurement Process

The basic measurement process in radiation thermometry is described using the concept of the radiometric chain including all phenomena influencing the detection of the radiation emitted by an object at a certain temperature (Figure 2.14).

The radiometric chain starts with the emission of thermal radiation by the object at a temperature T_{object}. For the description of the basic measurement process, we

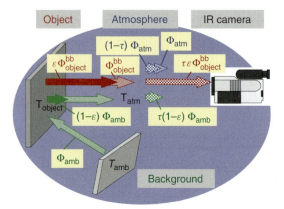

Figure 2.14 Radiometric chain – basic measurement process.

always use radiant power contributions Φ_i (i indicates the contribution mechanism) to the detector signal. These values are related to other radiometric quantities (Table 1.3), and they shall include the camera properties as the detector FOV, the spectral response of the detector, the transmittance of the camera optics, and so on.

To simplify matters, we limit the following discussion to opaque gray body objects (Section 1.4.2). The radiant power of the object Φ_{object} (T_{object}) compared to the radiant power of a blackbody Φ_{object}^{bb} (T_{object}) is given by $\Phi_{object}(T_{object}) = \varepsilon \Phi_{object}^{bb}(T_{object})$. The reflectivity of the opaque gray object is given by $r = 1 - \varepsilon$, according to Eq. (1.31). The object receives thermal radiation from its surrounding at ambient temperature T_{amb} and will reflect the radiant power $r \Phi_{amb}(T_{amb}) = (1-\varepsilon)\Phi_{amb}(T_{amb})$. The ambient temperature is often indicated as apparent reflected temperature. The radiant power emitted and reflected by the object toward the camera has to pass through the atmosphere. Because of the absorption and scattering processes in the atmosphere, the radiant power is attenuated. This can be described by multiplying the radiant power contribution from the object and the surroundings by the atmospheric transmittance τ_{atm}. If one assumes that the atmospheric transmittance is dominated by only absorption losses (scattering mechanism neglected), the atmosphere at a temperature T_{atm} will also emit a radiant power $(1 - \tau_{atm}) \Phi_{atm}(T_{atm})$.

Therefore, the camera detects a radiant power mixture with contributions from the object, the ambient, and the atmosphere. The total radiation power incident on the detector Φ_{det} can be written as

$$\Phi_{det} = \tau_{atm} \varepsilon \Phi_{object}^{bb}(T_{object}) + \tau_{atm}(1-\varepsilon)\Phi_{amb}(T_{amb}) + (1-\tau_{atm})\Phi_{atm}(T_{atm}) \quad (2.37a)$$

Strictly speaking, Eq. (2.37a) represents the spectral radiant power. For calculating the detector signal, this equation has to be integrated over the wavelength in order to account for the spectral dependence of detector responsivity and atmospheric transmission.

2.3 Basic Measurement Process

For blackbody radiation ($\varepsilon = 1$) and short measurement distances ($\tau_{atm} = 1$), the radiant power Φ_{det} equals the object radiant power. This behavior is used for the camera calibration process (Section 2.4.5).

The IR camera measures the radiance. Therefore from energy conservation, the signal "brightness" will be independent of the measurement range if $\tau_{atm} = 1$.

The radiant power emitted by the object can be calculated from

$$\Phi_{object}^{bb}(T_{object}) = \frac{\Phi_{det}}{\tau_{atm}\,\varepsilon} - \frac{(1-\varepsilon)}{\varepsilon}\Phi_{amb}(T_{amb}) - \frac{(1-\tau_{atm})}{\tau_{atm}\,\varepsilon}\Phi_{atm}(T_{atm}) \tag{2.37b}$$

The value Φ_{det} is determined from the measured sensor signal. According to Eq. (2.37b), the following parameters are necessary for the correct evaluation of the $\Phi_{object}^{bb}(T_{object})$:

- object emissivity ε;
- ambient temperature T_{amb};
- atmospheric temperature T_{atm}; and
- atmospheric transmittance τ_{atm}.

The values for object emissivity, the ambient, and the atmospheric temperature can be entered directly into the camera software. The atmospheric transmittance is calculated by the camera software using the LOWTRAN model (Section 1.3.2). For the calculation of the atmospheric transmittance, the atmospheric temperature, the relative humidity, and the measurement distance are the necessary input parameters.

The object temperature is determined from the object radiation power Φ_{object} (T_{object}) (Eq. (2.37b)) using the calibration curve of the camera (Section 2.5).

It is quite obvious that the described radiometric chain is very sensitive to any external influences on the radiation power detected by the camera. Any variation in the object emissivity within the measured area or a background with a spatial temperature variation requires a correction of the measured radiation power for all camera pixels. The necessary accuracy of the correction increases with decreasing object emissivity, that is, increasing reflectance contributions. Figure 2.15 illustrates the portion of object radiation (first term in Eq. (2.37a)) with regard to the total detected radiation as a function of emissivity for a MW and a LW camera with a constant spectral sensitivity in the wavelength range 3–5.5 or 7–14 µm, respectively.

On the one hand, object radiation contributions are lower for the LW than for the MW range; on the other hand, they are decreasing for decreasing object temperature.

The background is not only a source for reflected radiation from the object, but it also serves as a reference. The object is seen in front of a background of given temperature and whether the object can be distinguished from the background depends on the thermal contrast between object and background.

Figure 2.16 depicts the difference between object radiation and background radiation normalized to the object radiation. Obviously, if the object and background have

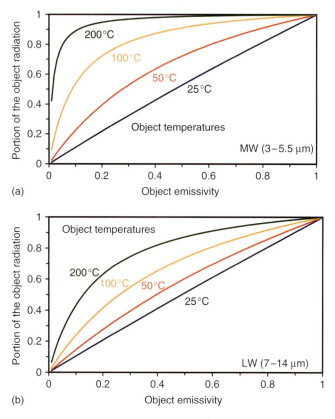

Figure 2.15 Portion of the object radiation with regard to total received by the camera for different object emissivities at different object temperatures and at an ambient temperature of 22 °C.

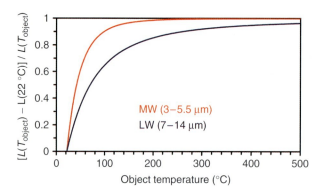

Figure 2.16 Difference of object radiance and background radiance normalized to the object radiance at a background temperature of 22 °C and different object temperatures.

identical temperatures, thermal imaging cannot identify any object, irrespective of emissivity. Differentiation starts if $T_{object} - T_{amb} >$ NETD.

Additional optics, filters, windows, and so on, can have a strong influence on the detected radiant power and its object temperature dependence. For a correct temperature measurement or signal correction, the complete radiant chain with all additional elements has to be analyzed carefully. This consideration results in a more complex relationship as shown in Eq. (2.37b), with all computable influences of the additional elements in the radiometric chain.

2.4
Complete Camera Systems

The main purpose of an infrared camera is to convert infrared radiation into a false color (including gray scale) visual image. This visual image should represent the two-dimensional distribution of the infrared radiation emitted by an object or a scene. For a temperature measuring system, the visual image displays the object temperatures. Therefore, the main components of an IR camera are the optics, the detector, the cooling or temperature stabilization of the detector, the electronics for signal and image processing, and the user interface with output ports, control ports, and the image display (Figure 2.17).

The understanding of the complete camera operation requires a more detailed discussion of these main camera components and their interaction in the measurement process.

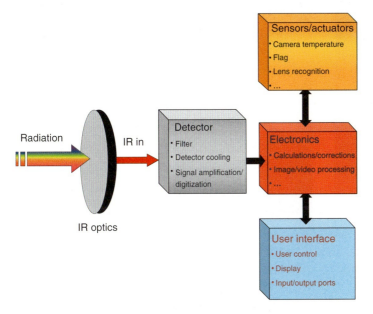

Figure 2.17 Block diagram with main IR camera components.

2.4.1
Camera Design – Image Formation

Two basic camera concepts for thermal imaging are scanning and staring systems (Figure 2.18).

In scanning systems, the image is generated as a function of time row by row similar to TV screens (Figure 2.18a); in staring systems, the image is projected simultaneously onto all pixels of the detector array (Figure 2.18b).

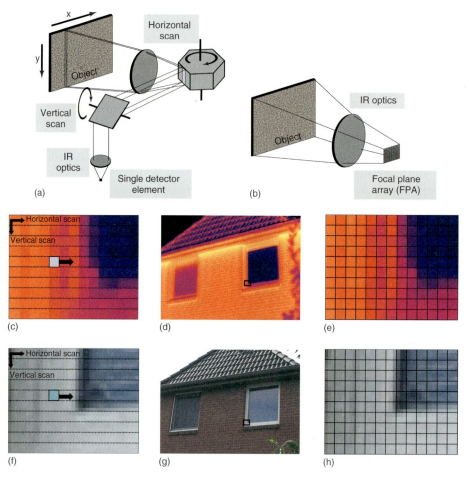

Figure 2.18 Basic principles of image formation for scanning (a) and FPA cameras (b) Vis image of a house (g) recorded with 2987 × 2177 pixels. The small black rectangle is shown enlarged at the sides (f,h). The left-hand side image illustrates the line after line scanning, whereas the right-hand side image refers to the pixel grid of an FPA. (c–e): IR image of an LWIR camera (320 × 240 pixel) and scan (c) as well as FPA mode (e) of image formation.

2.4.1.1 Scanning Systems

The heart of a scanning system mostly consists of a two-dimensional optical scanner with a single element infrared detector. Some systems also use a linear detector array or more complex detectors [17, 18]. In all cases, the image is built up sequentially. The detector's instantaneous field of view (IFOV) is moved across the FOV of the camera (Figure 2.18). A rotating mirror for horizontal scan allows the radiation measurement along a horizontal line of the scene. Mostly a polygon-shaped mirror is used to reach a vary high scanning efficiency [17]. A second mirror is used to switch the vertical position of the scanned line. For short image formation times at satisfactory pixel numbers, fast photon detectors and short signal integration times are needed. The following simplified estimate will show the limitations of a scanning imaging system. A scanning camera such as AGEMA ThermoVision 900 [19] with a single detector element at an image size of 272×136 pixels (136 lines and 272 pixel in one line) offers a full frame rate of 15 Hz. In a simplified consideration, the sequential readout process results in a time frame for one pixel of 1.8 µs. In the line scan mode at 272 pixels per line, a line frequency of 2500 Hz (or lines per second) is achieved corresponding to a time frame per pixel of 1.5 µs.

Assuming that the scanning system could also deal with larger pixel numbers, one can estimate that the line frequency and the frame rate would decrease for 320×240 pixels to about 2100 lines per second or 7.2 Hz full frame rate and for 640×480 pixels to 1050 lines per second or 1.8 Hz full frame rate.

In principle, faster frame rates seem possible by reducing the time frame for 1 pixel. However, it would not make sense to reduce this time interval, since the object signal would decrease and, hence, the signal-to-noise ratio would decrease.

Because of the continual improvements in focal plane detector technology and the limited number of pixels and limited frame rates, the scanning systems for thermal imaging became less important. Nevertheless, the scanning systems offer some advantages over starring systems as, for example, the radiometric accuracy caused by the fact that radiation from all image pixels is detected by only one detector element.

Today, IR line scanners are used for thermal imaging of moving objects or scenes such as band processes (Figure 2.19). For the imaging, only the line scan mode is necessary. The second dimension of the thermal image is built up by the object or scene movement. Line scanners allow very high scan rates up to 2500–3000 lines per second (for lower pixel numbers), variable pixel numbers in a measured line up to about 5000 (for lower scan rates), and a large FOV up to 140° [20–22]. Such line scanners often use one or two (at different temperatures) internal IR emitters with known emissivities and temperature for recalibration during every mirror revolution. For this purpose, the emitters are positioned outside the object FOV and the recalibration process is made in between the line scans (Figure 2.19).

2.4.1.2 Staring Systems – Focal Plane Arrays (FPA)

In focal plane arrays, the detectors are arranged in a matrix of columns and rows. Compared to scanning systems, focal plane arrays allow higher frame rates at

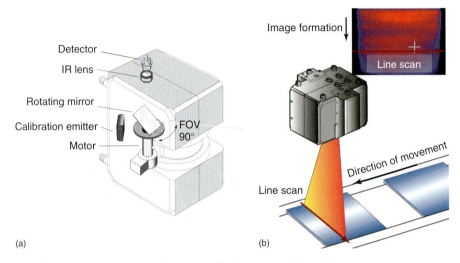

Figure 2.19 (a) Operational principle of a line scanner MP 150 (Raytek GmbH, [21]) and (b) image formation at a band process. Image courtesy: Raytek GmbH, Berlin, Germany.

higher pixel numbers. The main advantage of the staring system in addition to the absence of moving mechanical parts and the possibility to use thermal IR detectors is that the detector array covers the whole FOV simultaneously for the entire frame time [3]. This causes a reduced bandwidth for each detector resulting in a signal-to-noise ratio improvement. Theoretically the signal-to-noise ratio is increased by the square root of the image pixel format. Practically this gain cannot be obtained because the FPA consists of individual detector elements, causing a spatial or fixed pattern noise due to their nonuniform properties [23].

A focal plane array for IR imaging itself consists of two parts: the infrared sensor made from an infrared radiation sensitive material and the ROIC made from silicon. Nowadays, ROICs have two functions (Figure 2.20). First, they realize the simple signal readout and second they contribute to the signal processing with the signal amplification and integration or with multiplexing and analog-to-digital conversion.

One of the most important problems in technology results from the fact that in contrast to FPAs for the visible spectral range two different material classes have to be electrically integrated. This can be done by either monolithic or hybrid integration technology. For a hybrid system, both parts are made separately. Later they will be mechanically and electrically joined by special mounting processes such as flip-chip bonding [24]. For monolithic arrays, first the ROIC is made from silicon. After this process, the infrared sensors are built up on top of the ROIC by IR material thin film deposition, lithography, and etching. This is a very complex process because all technological processes and materials used for the fabrication of the IR sensor matrix must be compatible to the CMOS process used for the fabrication of the ROIC and must not change the ROIC properties. Because of

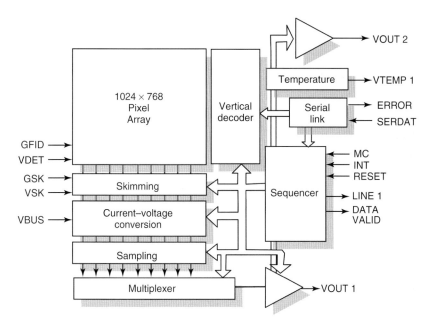

Figure 2.20 Readout integrated circuit (ROIC) for a 1024 × 768 pixels bolometer array. Image courtesy: ULIS, France.

lower fabrication costs, higher performance, durability, and reliability, there is a trend toward monolithic integration as, for example, for microbolometer FPAs [25]. The improvements in FPA technology lead to the following trends:

- The number of pixels in arrays is increasing. FPA cameras with up to 1024 × 1024 pixels are commercially available [19]. Figure 2.21 illustrates the different array sizes for a Vis 1 megapixel image. Figure 2.22 depicts infrared images of the same object scene for different pixel numbers.
- The pixel size is decreasing. Today, focal plane arrays with 17 μm × 17 μm microbolometer [26] pixels or 15 μm × 15 μm photon detector pixels are available. The reducing array size reduces the cost of the optics by reducing the size at a given F-number. At a given F-number, the cost for lens material will depend approximately on the square of the lens diameter.
- The NETD is decreasing. This requires a strong improvement of the detector performance (D^*) to overcompensate the effect that a simple decrease of the detector size will result in an increase of the NETD. Today IR cameras achieve NETDs of about 45 mK for microbolometer and about 10 mK for photon detectors at 30 °C object temperature.
- The trend of decreasing pixel size is accompanied by an increasing fill factor. The *fill factor* is defined as the ratio between the infrared-sensitive cell area and the whole cell area of an FPA (Figure 2.23). It is not possible to arrange the detectors fully contiguous in the FPA. The single detector elements must be electrically and thermally (for microbolometer arrays) separated from each other and be

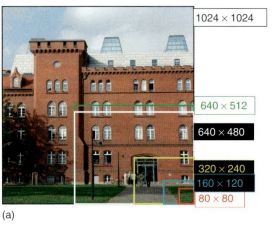

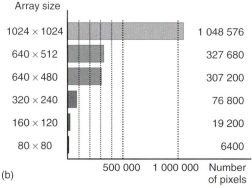

Figure 2.21 Illustration of the displayed image sections for different array sizes with the same optics (a) and the detectors at the same distance. Comparison of the total pixel numbers for different array sizes (b).

connected to the ROIC. Therefore, the infrared-sensitive area remains smaller than the cell area. This finite fill factor influences the camera performance such as, for example, the spatial resolution and MTF (Section 2.5.3). Today, microbolometer FPAs achieve a fill factor larger than 80% and photon detector FPA's larger than 90%. Figure 2.23 illustrates the increase of the fill factor. If a typical quadratic cell is assumed, a fill factor of 49% requires that the side length of the sensitive cell amounts 70% of the whole cell side length. If the fill factor is increased to 81%, this side length ratio has to be increased to 90%.

- The time constant for microbolometers has decreased to about 4 ms [26]. This allows to increase the time resolution of the camera and to use higher frame rates. As discussed in Section 2.2.3, the thermal properties of the bolometer structure have an influence on the time constant. In a complex manner, they affect all detector performance parameters. Therefore, the decrease in the bolometer time

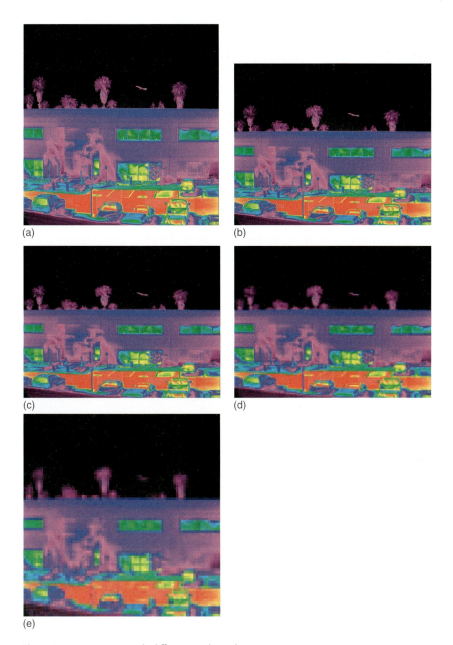

Figure 2.22 IR images with different pixel numbers generated from the same megapixel image. (a) 1024 × 1024 pixels; (b) 640 × 512 pixels; (c) 320 × 240 pixels; (d) 160 × 120 pixels; and (e) 80 × 80 pixels. Image courtesy: A. Richards, FLIR Systems.

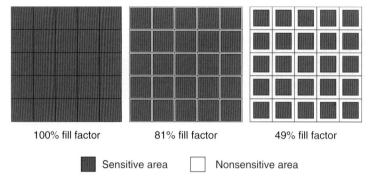

Figure 2.23 Fill factor of an FPA.

constant while simultaneously decreasing the detector area and NETD results in a very complex problem. Overall, these trends result in an increase of image spatial and thermal resolution, image information content, and time resolution of the imaging process.

2.4.1.3 Nonuniformity Correction (NUC)

Before IR detectors can be used for quantitatively measuring temperatures, two important image processing procedures must be performed. First, the differing gains and possible signal offsets of different individual detectors must be accounted for. Second, an absolute temperature calibration (Section 2.4.5) is required for quantitative analysis. Both procedures are interrelated, although in principle they can be distinguished from each other.

As mentioned above, FPAs are composed of many individual detector elements having different signal responsivities (gain or slope) and signal offsets. The same applies to scanning systems with detector lines. Figure 2.24a depicts this situation for some detector elements.

The spread in gain and offset results in a spread of detector signals of different pixels for the same incident radiant power (indicated by arrow). If this nonuniformity of the individual detectors becomes too large, the image becomes unrecognizable. Therefore, for all imaging systems consisting of more than one detector, the nonuniformity has to be corrected [27] (Figure 2.24). This is called *nonuniformity correction* (NUC).

For most commercial cameras, the NUC procedure is performed during the factory camera calibration process and the correction parameters are stored in the camera firmware. Only some R&D cameras allow the users to perform the calibration procedure and NUCs.

The correction procedure (Figure 2.24a) starts mostly with a signal offset correction to bring the response of each detector for a given object temperature range within the dynamic range of the electronics. Within the second step, the signal slope (detector signal dependence on the object temperature) is corrected. After this second step, all detectors should have the same signal slope and their response is

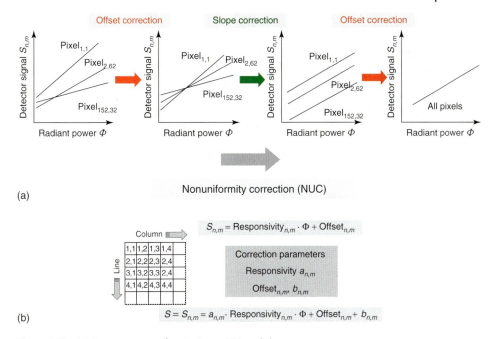

Figure 2.24 (a) Detector nonuniformity in an FPA and the nonuniformity correction to normalize the different FPA detector responsivities and offsets shown for three selected pixels. (b) Mathematical model of NUC corrections.

within the dynamic range of the electronics. Finally, an additional offset correction is necessary to get identical response curves for the different detector elements.

In practice, the NUC is performed by placing a rough (nonreflecting) gray body ($\varepsilon > 0.9$) of homogeneous temperature directly in front of the detector or camera lens. As a consequence, each pixel should yield the same signal. The deviations are corrected electronically. If this procedure is done for a single temperature (one-point NUC), the correction works only close to the used temperature. Usually two different temperatures are used (two-point NUC), which gives a reasonable correction between and in the vicinity of the chosen temperature. As a result, a detector array will yield a very uniform signal. Figure 2.25 depicts two IR images of the same scene, the only difference being that before the second was recorded a NUC was performed. The improvement in image quality via noise reduction is obvious.

However, a perfect detector uniformity in between the calibration points cannot be realized because of the possible slightly different nonlinear responsivity curves. This situation for a correction at two calibration points is illustrated in Figure 2.26. The remaining detector nonuniformity causes the so-called fixed pattern noise.

The NUC procedure forces the detector pixels to a uniform signal due to the uniform input of a homogeneous object. In principle, it is, however, also possible to perform NUCs based on arbitrary objects. Figure 2.27a depicts the IR image

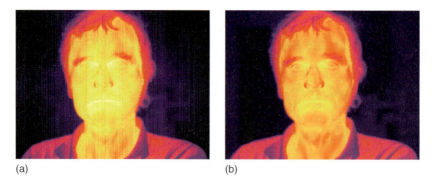

Figure 2.25 IR images of the same scene before (a) and after (b) NUC.

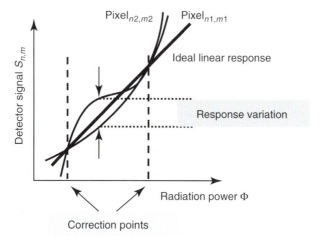

Figure 2.26 Two-point NUC as detectors derivate from linearity in their responsivity. The remaining response variation between the correction points due to the deviation from linearity is responsible for the fixed pattern noise. Deviations from linearity are strongly exaggerated to illustrate the effect.

of a person, which was recorded directly after using a homogeneous background for a NUC. Figure 2.27b depicts the IR image of the person, which was recorded directly after using the same scene for a NUC. The large radiance differences between background and object (person) were corrected to vanish via the NUC; therefore, the person is (nearly) invisible. In Figure 2.27c, the person has left and the camera only observes a homogeneously tempered background. Still, the camera now detects a feature where the person was standing, although the radiance of the wall is the same everywhere. Figure 2.27 thus illustrates that electronic correction of detector gain and offset can change apparent radiance distributions appreciably. Therefore, NUCs must be performed with great care.

2.4 Complete Camera Systems | 111

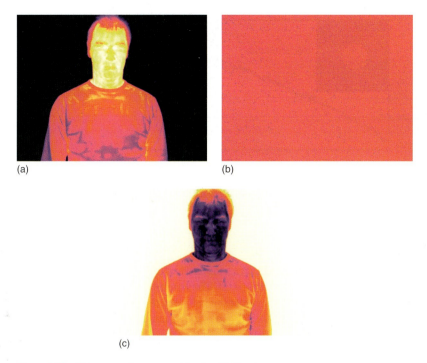

Figure 2.27 IR images of a person (a) after NUC using a homogeneous background (b) after NUC using the scene shown in (a) image (c) observing homogeneous background after the person has left with the NUC of the (b) image.

In principle, a NUC only requires that objects of homogeneous temperature be placed in front of the camera. The exact value of the temperature need not be known to get an IR image with low noise level. If, however, a quantitative analysis is needed, a temperature calibration must be performed. This is done by detecting objects of known fixed temperatures – usually stabilized blackbody scenes – with the camera for a number of different object temperatures. These temperature versus signal calibration curves are also stored in the firmware.

The quality of the NUC and the respective correctness of measured object temperatures relies on the stability of the detector operation. Thermal drifts will change the detector response curve of the pixels and a new NUC must be performed. This means that to ensure a proper correction, FPA cameras have to be recalibrated periodically. For a correction during the camera operation, mostly a tempered flag or autoshutter is used for a single point offset NUC. If this autoshutter operation is switched off, the periodical camera recalibration NUC will be disrupted. The missing internal correction will lead to a drift in the indicated temperature. Figure 2.28 depicts the situation for a LWIR thermal detector (bolometer) camera and a MWIR photon detector (PtSi) camera. For a period of about 2 h, the temperature of a 70 °C blackbody emitter (temperature-stabilized calibration source) was simultaneously measured with both cameras. The average

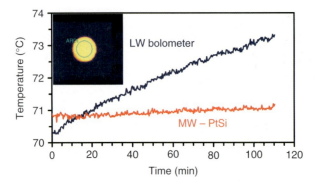

Figure 2.28 Results of a long-term temperature measurement at a stabilized 70 °C blackbody emitter with switched off operation of camera autoshutter.

temperature within the marked area was analyzed. At $t = 0$, the autoshutter operation of the cameras was switched off. The bolometer equipped LWIR camera shows a much stronger drift behavior of the indicated temperature (about 3 K) than the photon detector equipped MWIR camera (less than 0.5 K).

The physical reason for the drift follows from the discussion in Section 2.2.3. The bolometer detector signal is generated by the net radiation power transferred to the detector, which is given by the difference between the radiation power received from the object and the radiation power emitted by the detector itself. In addition, the thermal detector exchanges radiation with, for example, the camera housing and the optics. Therefore, a small change in the temperature balance inside the camera will also change the transferred net radiation power and will cause measurement errors.

In contrast, in photon detectors the signal is generated from the incident radiation alone. A temperature change in the camera will also cause a change in detector signal, but the deviations are much smaller than for the bolometer. Moreover, bolometer responsivity strongly depends on the detector temperature due to the high temperature coefficient (Section 2.2.3). Small temperature changes in the bolometer will cause strong signal changes, that is, wrong temperature measurements if automatic periodic recalibration is turned off. According to the quantitative estimate (Section 2.2.3), the error of the indicated temperature at the end of the long-term measurement of about 3 K after 2 h corresponds to a temperature change of about 100 mK in the bolometer. This example demonstrates the importance of a proper auto shutter operation for correct temperature calibration especially in bolometer FPA cameras.

Finally it should be stressed that the detector signal is linear only in a certain output range, for example, from 20 to 80% of the detector saturation signal. The NUC works well only within these limits. For very low signals as well as close to saturation (Figure 2.29c), an increased fixed pattern noise level will show up, that is, the NUC does not work anymore.

2.4 Complete Camera Systems | 113

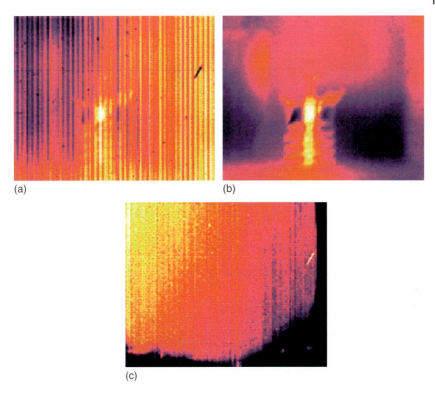

Figure 2.29 Illustration of NUC limitations due to detector signal saturation at high radiance levels. (a) Image of a cold light bulb without NUC, (b) image of the light bulb at room temperature after NUC with homogenous room temperature background, and (c) image immediately after bulb explosion due to high voltage pulse application (extremely high temperatures cause detector signals close to saturation level).

2.4.1.4 Bad Pixel Correction

An ideal detector array consists of 100% perfect detectors, which – after performing an NUC – will all give the same output signal when receiving the same radiant power.

Unfortunately, however, while manufacturing large detector arrays yields are less than 100%. This means that there are the so-called bad pixels consisting either of really nonfunctionary detectors or of detectors whose gain and offset are too far off the average such that they cannot be corrected for by an NUC. Very good manufacturers state that their detector arrays are only allowed less than 0.01% of bad pixels. For example, a 640 × 512 pixels camera would only be allowed to have at maximum 32 bad pixels in the whole array of 327 680 pixels. Probably all commercial cameras have bad pixels. They are usually corrected electronically by replacing the signal of the bad pixel by the weighted average of its neighbor pixels. Problems only arise if bad pixels form clusters. In this case, the respective

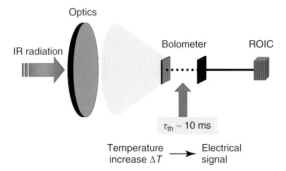

Figure 2.30 Signal generation process and signal readout in a bolometer camera (dotted line illustrates the time delay of bolometer temperature increase with τ_{th}).

array location is not suitable for accurate temperature measurements of very small objects (Section 2.5.5).

2.4.2
Photon Detector versus Bolometer Cameras

Today, the majority of IR cameras are equipped with microbolometer FPAs. These thermal imagers meet the demands of most practical applications and are much less expensive than photon detector FPA cameras. Microbolometers are characterized by relatively low sensitivities/detectivities, broad/flat spectral response curves, and slow response times of the order of 10 ms. Microbolometer FPAs are mostly temperature stabilized by a Peltier element. Because of their operation as thermal detectors, bolometers do not offer the possibility to fix the integration time (Figure 2.30). Rather, the "integration time" is given by the thermal time constant of the bolometer detectors. The frame rate of these cameras cannot be changed by the user. The image formation of microbolometer FPA cameras is characterized by sequential readout of the detector lines within the full frame time, also called *rolling readout*.

For more demanding applications requiring higher sensitivities or/and time resolutions such as in R&D, photon detector FPA cameras are used. For the MW and the LW spectral region, the FPA is cooled down to liquid nitrogen temperature $T = 77$ K. For the SW region, the detectors are cooled by multistage Peltier elements. The photon detector FPAs exhibit a much larger responsivity/detectivity and a spectral response varying with wavelengths (Figure 2.1). Photodiode arrays offer time constants down to about 1 μs. Therefore, photon detector cameras offer smaller NETD values and higher frame rates than microbolometer cameras.

The direct conversion of the photon flux into an electrical signal (photocurrent) and the small time constants allow a snapshot operation of photon detector FPA cameras. The photocurrent of each pixel is charging a capacitor during a user selected integration time (some microseconds to full frame time;

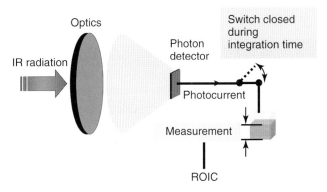

Figure 2.31 Signal generation process and signal readout in a photon detector camera.

Figure 2.31). The readout by the ROIC is mostly possible in integrate-while-read or integrate-then-read mode. Because of the fast direct signal generation by the photon detectors and the readout process, the camera measurement can be triggered.

The maximum achievable frame rate is limited by the speed of data transfer and storage. If we assume a detector with a time constant much below 1 μs and a full frame integration time of 1 μs, a frame rate of 1 MHz is theoretically possible. For the readout, mostly a dynamic range of 14 bits is used. This would result in a data rate of 14 Mbit/s per pixel. If the array size increases to a common size of 320 × 240 pixels (76 800 pixels) such a high frame rate would require a data rate of about 1 Tbit/s. Today, the practical limit of the maximum data rate given by the readout electronics is much below this value. For example, the MW camera FLIR SC 8000 equipped with an InSb FPA of 1024 × 1024 pixels allows a maximum full frame rate of 132 frames/s using a 16-channel readout [19]. This results in a practical overall data rate of about 2 Gbit/s. For higher frame rates, the windowed readout mode is used. This mode mostly allows a user selectable random size and location FPA windowing. The decreased pixel number allows a faster frame rate, for example, for the SC 8000 at 160 × 120 pixels 909 frames s^{-1}. For capturing high-speed events, the FLIR SC4000 and the SC 6000 high-speed cameras offer maximum frame rates of 48 kHz (2 × 64 pixels) and 36 kHz (4 × 64 pixels), respectively [19]. Figure 2.32 depicts the FPA window size of an FLIR SC6000 camera relative to the maximum frame rates.

With respect to the characteristics given for photon detector FPAs, thermocameras with a PtSi FPA form an exception. This is due to the low quantum efficiency. As discussed in Section 2.2.5, the sensitivity of PtSi detectors is limited and large integration times are necessary to achieve a sensitive detection. Therefore, cameras equipped with PtSi detectors commonly use a full frame time constant of 20 ms for signal integration.

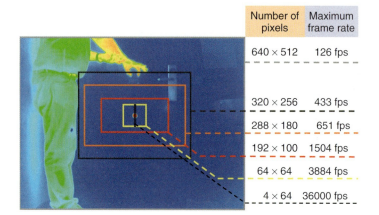

Figure 2.32 FPA window sizes of the FLIR SC6000 and maximum frame rates (random and user-defined windowing possible).

2.4.3
Detector Temperature Stabilization and Detector Cooling

For optimum operation of infrared detectors in thermal imaging systems, a stabilization of the detector temperature (thermal detectors) or a detector cooling (photon detectors) is necessary.

For the bolometer FPA and SW InGaAs FPA cameras, thermoelectric heaters/coolers are used. The thermoelectric cooling/heating uses the Peltier effect [28] to create a heat flux between the junctions of two different materials (Figure 2.33). A DC current is forced through thermoelectric materials. During operation, heat is removed from one type of junctions and creates the cold side of the element. At the other type of junctions heat is generated, forming the hot side of the element. If the current direction is changed, the heat flux direction will also be changed. This causes a cooling of the former hot side and a heating of the former cool side of the Peltier element. The cooling efficiency depends on the materials used. The best performance is achieved using semiconductor materials. The Peltier coolers are arranged in one- to more than three-stage arrangements (Figure 2.33). Peltier elements are used for both cooling InGaAs detector arrays in SW cameras and stabilizing the temperature of microbolometer arrays in LW cameras. Detectors in bolometer cameras are usually operated at temperatures of around 30 °C, which is mostly above ambient temperature. In principle, the temperature stabilization could also work by cooling the detector if ambient temperature would exceed the chosen detector temperature. Bolometer temperatures should be different from ambient temperatures in order to avoid a temperature change of the bolometer detector sensitivity with changing ambient temperature (camera temperature; see Section 2.2.3). A one-stage cooler offers maximum temperature differences of about 50–75 K between the hot and the cold side, which is sufficient for the FPA temperature stabilization.

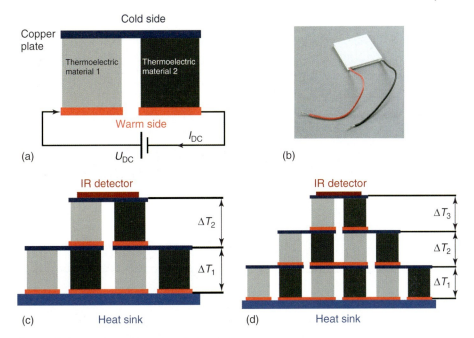

Figure 2.33 Operational principle of Peltier coolers (a), example of a Peltier cooler (b) two- and three-stage Peltier coolers (c and d).

In contrast, photon detectors require increased cooling with increasing wavelength (Section 2.2.4). MW and LW photon detectors mostly operate at about 77 K.

Multiple-stage Peltier coolers are used to increase the temperature difference. Unfortunately, the cooling efficiency decreases with decreasing temperature, which means that temperature differences decrease from stage to stage [28]. Therefore, the number of cooling stages is mostly limited to three. With a three-stage cooler, a maximum temperature difference of about $\Delta T = 120$–140 K can be reached, that is, $T = 150$–170 K. Therefore, the application of thermoelectric cooling for FPAs is restricted to the near infrared or SWIR spectral region.

In the beginning of IR camera technology for detector cooling in MW and LW cameras, liquid nitrogen at 77 K ($-196\,°\text{C}$) was used. Later, an electrical solution to cryogenic cooling using the Stirling process [29] was developed (Figure 2.34). The Stirling process removes heat from a cold finger with the FPA by a thermodynamic cycle. This heat is dissipated at the warm side. The efficiency of the Stirling process is relatively low but meets the requirements of the necessary IR detector cooling due to the low heat capacity and small heat losses. Stirling coolers allow detector temperatures of 77 K and below. The small cooler size and weight, electrical (e.g., electrical power consumption) and mechanical (e.g., mechanical vibrations) properties, reliability, and life time (more than 8000 h of operation guaranteed) of the Stirling coolers led to widespread use in thermal imaging systems.

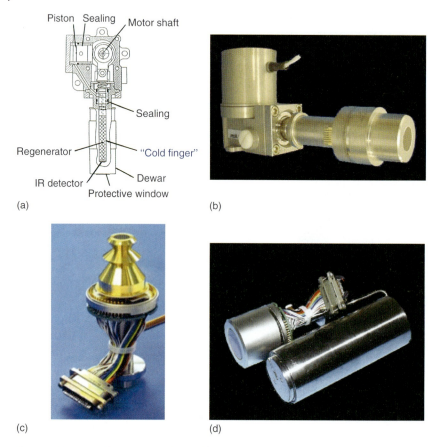

Figure 2.34 (a) Stirling cooler working with helium gas for IR detector cooling and (b–d) examples of cooled FPA assemblies. (Image courtesy: Infrared Training Center, FLIR.)

2.4.4
Optics and Filters

2.4.4.1 Spectral Response
Most thermocameras are characterized by a broad spectral response. Figure 2.35 depicts the spectral response curves of a number of IR cameras, normalized to their maximum value.

2.4.4.2 Chromatic Aberrations
Broadband detection is mostly accompanied by considerable chromatic aberrations of the lenses, which cannot be corrected completely. These and other aberrations affect the object–image transformation by the optics and limit the important image quality parameters, for example, spatial resolution or MTF (Section 2.5). Figure 2.36

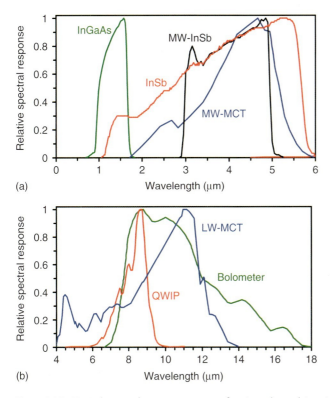

Figure 2.35 Typical spectral response curves of various thermal imaging systems.

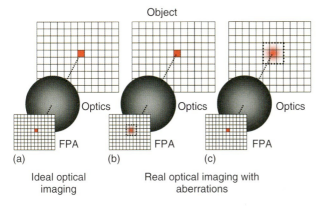

Figure 2.36 Comparison of ideal and real optical imaging. Ideal system (a) and realistic situation due to optics aberrations (b and c). For details see text.

depicts the aberration-induced blurring of an object field during the transformation into the image.

Ideal optical imaging means that the detector element receives radiation only from a well-shaped object field as depicted in Figure 2.36a. Aberrations will cause a spreading of the object radiance distribution, that is, a blurring effect: the radiation of the object field is also seen by the neighboring detector elements (Figure 2.36b). On the one hand, if this detector signal would be the result of optical image generation due to an ideal aberration-free lens, the blurring would be interpreted as being due to a large object size, that is, aberrations diminish the spatial resolution. On the other hand, the object radiance on the detector pixel decreases due to the blurring. This can also cause an incorrect temperature reading, since the chosen pixel receives less radiation from the respective object area plus additional radiance contributions (of differing amount) from neighboring pixels (Figure 2.36c).

2.4.4.3 Field of View (FOV)

The object field is transformed to an image within the FOV of the camera. The FOV is the angular extent of the observable object field (Figure 2.37). Sometimes the FOV is also given as the area seen by the camera.

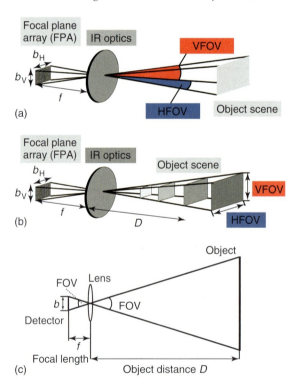

Figure 2.37 Calculation of the cameras angular field of view (a) and the area seen by the camera (b). Two-dimensional cross section (c).

2.4 Complete Camera Systems

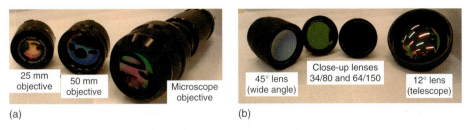

Figure 2.38 (a) Different lenses for the MWIR SC6000 camera and (b) the LWIR camera SC2000 (right).

The FOV can be easily constructed from geometrical optics. Rays passing through the center of a (thin) lens are not refracted. Therefore, rays that finally hit the four edges of the detector array limit the FOV.

The FOV depends on the camera lens and the FPA dimensions. For a camera lens with a focal length f and an FPA with the linear dimension b, the FOV can be calculated as follows:

$$\text{FOV} = 2\arctan\left(\frac{b}{2f}\right) \tag{2.38}$$

The object area with the length x seen by the camera at a given object distance D can be calculated from

$$x = 2D\tan\left(\frac{\text{FOV}}{2}\right) \tag{2.39}$$

The rectangular shape of the FPA causes different horizontal fields of view (HFOV) and vertical fields of view (VFOV).

Different camera optics allow a change of the camera FOV. Figure 2.38 depicts some camera lenses for the MW FLIR SC 6000 and the LW SC 2000 cameras with different focal lengths.

The respective fields of view are given in Table 2.2.

Table 2.2 Camera field of view for different lenses.

Camera Lens	FLIR SC 6000 (640 × 512 pixels, 25 μm pixel pitch) HFOV°/VFOV°	FLIR SC 2000 (320 × 240 pixels) HFOV° VFOV°
12°	–	12°/9°
Standard 24°	–	24°/18°
45°	–	45°/33.8°
Close up 34/80	–	24°/18°
Close up 64/150	–	24.1°/18.1°
25 mm (focal length)	35.5°/28.7°	–
50 mm (focal length)	18.2°/14.6°	–

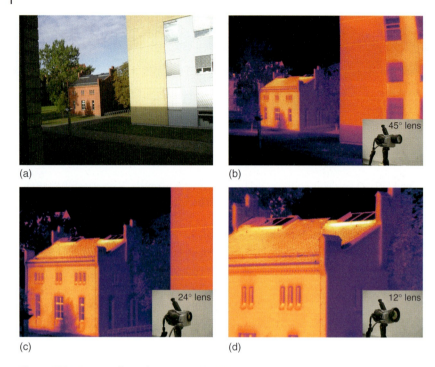

Figure 2.39 Images of an object scene (a), FLIR SC2000 LWIR images with different lenses: 45° wide angle (b); standard 24° lens of the camera (c); and 12° telescope lens (d). The angle refers to horizontal direction.

The label x/y for the close-up lenses is related to the FOV; however, the numbers also directly give the minimum object distance (y) and the respective horizontal object size (x) at this distance. If the horizontal number of pixels is known, one can immediately estimate the best possible spatial resolution. For example, an $x = 34$ mm with 320 pixels means that each pixel reflects an object size of $34\,\text{mm}/320 \approx 0.1$ mm.

Figure 2.39 depicts the IR images of an object scene for the different LW camera lenses. The smaller the FOV of the camera, the better the spatial resolution of the IR image. The knowledge of the camera FOV is necessary to estimate the smallest object that can be detected and to ensure a correct temperature measurement. From the FOV, IFOV can be determined by dividing the FOV by the number of FPA pixels in a line (horizontal or HIFOV) or in a column (vertical or VIFOV; see Section 2.5.3).

2.4.4.4 Extender Rings

Some infrared cameras offer the possibility to use extender rings in between the camera and the camera objective. Extender rings are known from macrophotography and offer an inexpensive and easy possibility to increase the magnification.

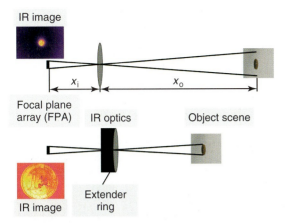

Figure 2.40 Increase in object magnification using extender rings by increasing the distance from the lens to the detector and decreasing the distance from the camera to the object.

The purpose of the extender rings is to move the objective lens farther from the FPA to magnify the image (Figure 2.40). Applying the extender rings decreases the distance between the camera and the object.

A simplified estimate is possible using the thin lens formula. It relates object distance x_o, image distance x_i, and lens focal length via

$$\frac{1}{f} = \frac{1}{x_o} + \frac{1}{x_i} \tag{2.40}$$

When observing objects in distances of meters and using lenses with focal lengths in the centimeter range $x_o \gg f$ and $x_i = f + z$ with $z \ll f$. Therefore,

$$\frac{1}{f} = \frac{1}{x_o} + \frac{1}{f+z} \tag{2.41}$$

which gives

$$x_o = \frac{f(f+z)}{z} \tag{2.42}$$

The ratio of the image size C_i and the object size C_o is given by

$$\frac{C_i}{C_o} = \frac{x_i}{x_o} = \frac{(f+z)z}{f(f+z)} = \frac{z}{f} \tag{2.43}$$

Assume, for example, a 50 mm lens that without extender ring has a $z = 5$ mm, corresponding to an object distance of $x_o = 0.55$ m and $C_i/C_o = 1/10$. Introducing an extender ring of $0.25'' \approx 6$ mm gives $z^* = 11$ mm, $x_o^* = (50 \cdot 61)/11 \approx 0.277$ m, and $C_i^*/C_o^* = 11/50 \approx 1/4.54$. Obviously, the image size has increased to

$$C_i^* = C_i \frac{z^*}{z} \approx 2.2 \cdot C_i \tag{2.44}$$

124 | *2 Basic Properties of IR Imaging Systems*

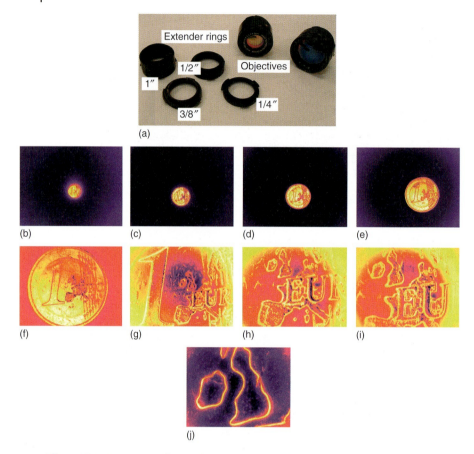

Figure 2.41 Image magnification by using a 50 mm lens and extender rings (b–i) and a microscope optics (j). Extender ring sizes, c: 0.25″, d: 0.375″, e: 0.5″, f: 1″, g: 2.125″, h: 3.125″, i: 4.125″.

Figure 2.41 depicts thermal images of a €1 coin for different extender rings, recorded at minimum possible object distances as well as the result using the microscope objective.

The use of extender rings is accompanied by some drawbacks such as decreasing FOV with corresponding loss of radiation with increasing extender ring thickness. This will cause large uncertainties in the object temperature determination. The strongest degradation of the image quality is observed at the image corner. A much better image quality is observed if close-up lenses or a microscope objective is used (Figure 2.41j).

2.4.4.5 Narcissus Effect

In IR imaging, an optical reflection effect occurs which is unknown from imaging in the visible spectral range. The so-called narcissus effect occurs if there is

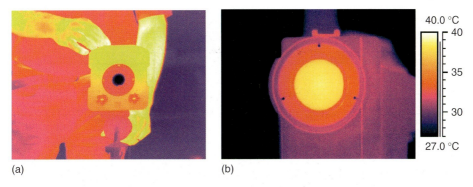

Figure 2.42 Narcissus effect looking on a polished room temperature metal plate for a MWIR camera (77 K cooled InSb FPA), (a) and with an LWIR camera (temperature-stabilized microbolometer FPA, (b).

a reflection from the camera lens, a window, or an object such that the camera detects the reflection of itself. The narcissus effect becomes pronounced if the reflecting object is in the focus of the camera. This effect results in a dark (cold) or a bright spot (warm) in the infrared image depending on the detector temperature. Figure 2.42 depicts the narcissus effect for a cooled InSb MW camera and a bolometer LW camera, which observe a polished steel plate at room temperature (about 22 °C). The LW camera measures a maximum temperature of 39 °C for object emissivity of 1 (perfect 100% mirror) on the lens area. The MW camera measures a temperature below −60 °C. One possibility to avoid this effect is to change the viewing angle to a value such that there is no incident reflected radiation from the camera detector area on the camera objective.

The Narcissus effect is particularly important if using filters, for example, for narrow band detection or other plane parallel partially transparent plates between camera and object.

Some R&D cameras such as an FLIR SC 6000 allow the users to perform NUCs. A one-point NUC can be performed by placing a homogeneous tempered blackbody in front of the camera objective. Figure 2.43 illustrates that such one-point NUC can be used to eliminate the Narcissus effect if a reflecting object, for example, a silicon wafer optically polished on both sides is placed in front of the camera. Silicon is transparent in the MW IR spectral region but reflects about 50% (Figure 1.49). Figure 2.43a depicts the object scene after a NUC with a homogeneous background. If the silicon wafer is placed in front of the camera, the Narcissus effect is clearly seen (Figure 2.43b). Now an additional NUC is performed using a homogeneous background but with the wafer in front of the camera. After this procedure, the Narcissus effect is eliminated (Figure 2.43c). Removing the silicon wafer will now lead to an inverse Narcissus effect as shown in Figure 2.43d.

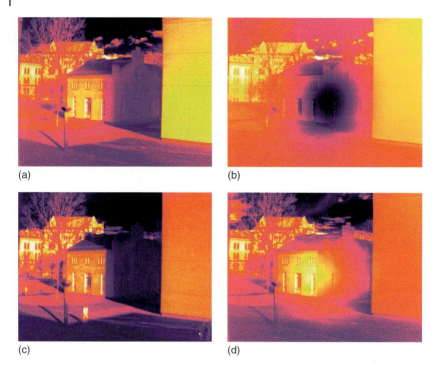

Figure 2.43 Elimination of Narcissus caused by a both side polished silicon wafer positioned in front of the camera objective performing a NUC of a FLIR SC 6000 camera. (a) Object image without the silicon wafer after NUC. (b) Object image with the silicon wafer in front of the camera objective showing the Narcissus in the center of the image. (c) Object image with the silicon wafer in front of the camera objective after performing an additional NUC (silicon wafer during the NUC in front of the camera objective). (d) Object image after removing the silicon wafer showing an inverse Narcissus in the center of the image.

2.4.4.6 Spectral Filters

The additional use of spectral filters can expand the applications of thermal imaging. As discussed in Chapter 1, the object optical IR properties, for example, emissivity, are important for thermal imaging. Not only materials such as glass or plastics but also combustion flames can be characterized as selective emitters and are characterized by a wavelength-dependent emissivity. An analysis of such selective emitters becomes possible at wavelengths with strong absorption in these materials. The camera can be spectrally adapted to these wavelengths by inserting appropriate filters. The spectral filters are described as short-pass (SP), long-pass (LP), bandpass (BP), and narrow bandpass (NBP) filters (Figure 2.44).

In addition to these spectral filters, neutral density (ND) filters are used to attenuate the IR radiation without a spectral filtering (wavelength-independent transmittance below 100%), for example, to prevent a detector saturation at higher signal levels. It is important that only interference or dichroic filters are used.

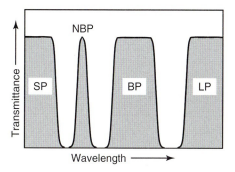

Figure 2.44 Spectral transmittance of different types of optical filters.

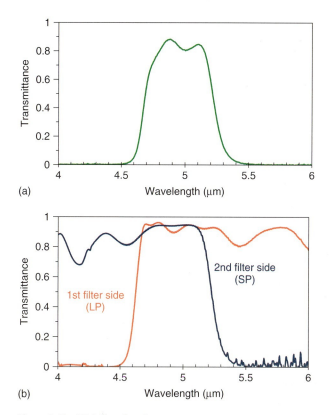

Figure 2.45 NBP filter for glass temperature measurement (a) spectral filter transmittance (b) separate spectral transmittance of the interference filter layers acting as LP and SP filters on the two sides of the sapphire substrate.

The spectral filtering is based on the reflectance and transmittance by interference effect, and it is not due to absorption by the filter. A filter absorption would cause a temperature increase in the filter. According to Kirchhoff's law, any absorbing filter would also emit very efficiently (Chapter 1). Detection of this additional radiation by the camera would lead to measurement errors. Sometimes the optical layers on both sides of the filter form the spectral dependent transmittance of the filter (Figure 2.45). The optical layers on one side of the filter substrate will only partly block the incident light (SP filter with cutoff wavelength λ_{cutoff} on one side of the substrate and LP filter with cut-on wavelength $\lambda_{cuton} > \lambda_{cutoff}$ on the other side of the substrate). The spectral transmittance of the BP filter is formed by the cut-on and cutoff behavior of the combined SP and LP filters. If the filter substrate material is not transparent in the complete IR spectral region, as for example, sapphire (used for MWIR filters), the object radiation must be incident on the filter side with the SP filter that blocks the longer wavelengths of substrate absorption (e.g., due to strong absorption behavior of sapphire at wavelengths above 5 μm the LWIR region must be blocked; see also Figure 1.64).

Table 2.3 gives a compilation of some standard filters used in thermal imaging.

For demanding applications with spectral adaptation, cold filters are used in front of the photon detector housing. Cold filters reduce the detector signal generated from the 300 K background radiation (e.g., inside camera parts) and cause an increased signal contrast.

Table 2.3 Some standard IR filters for thermography.

Wavelength (μm)	Type	Application
2.3	NBP	Measurements through glass
3.42	NBP	Measurements of plastics temperature (polyethylene)
3.6–4	BP	Reduction of atmospheric influences at longer measurement distances
3.6	LP	Reduction of influence of sun reflections
3.7–4	BP	Measurements through flames
3.9	BP	Measurements of high object temperatures (reduces incident radiation on the detector)
4.25	NBP	Measurements of flame temperatures
5.0	NBP	Measurement of glass temperature
7.5	LP	Measurements only in the LWIR spectral region
8.3	NBP	Measurements of Teflon temperatures

SP, short pass; NBP, narrow band pass; BP, band pass; LP, long pass.

2.4.5
Calibration

The infrared camera as a radiometer allows to measure some radiometric quantities such as radiance or radiant power. The purpose of calibration is to determine the accurate quantitative relations between camera output and incident radiation. For the calibration procedure, black bodies (emissivity close to unity) at different temperatures are used, since their radiometric quantities such as excitance M or radiance L and their spectral quantities are well defined (Section 1.3). Therefore, the calibration process gives a relation between camera signal and blackbody temperature. During camera calibration, the signal of each pixel depending on the blackbody temperature is determined. Therefore, the calibration process also forms a NUC of the detector array. After the camera calibration, all pixels will give the correct temperature information of the object. This is a NUC on the quantitative temperature scale, whereas the NUC discussed in Section 2.4.1.3 qualitatively adjusted all detector pixel responses as depending on incident radiance.

During calibration, the camera aperture is completely covered by the black body. The distance between the camera and the black body is small so that an atmospheric transmittance of unity can be assumed. This also means that atmospheric emittance is zero, that is, the incident radiance L and hence the camera output signal $S_{out}(T_{bb})$ for a given thermocamera with range $[\lambda_1, \lambda_2]$ depend only on the temperature of the blackbody (Eq. 2.45):

$$S_{out}(T_{bb}) = \text{const.} \int_{\lambda_1}^{\lambda_2} \text{Res}_\lambda L_\lambda (T_{bb}) \, d\lambda \qquad (2.45)$$

The output signal depends on the spectral camera response Res_λ determined by the spectral detector responsivity and the spectral dependence of the optics transmittance and a camera characteristic constant value including the properties of the camera optics such as the F-number of the camera lens. Changing the camera lens may change the F-number and the spectral transmittance of the optics. Therefore, the calibration has to be performed for each lens separately. If filters are used, the spectral camera response is changed and a recalibration of the camera with the filter is necessary. Mostly the calibration parameters for different configurations (e.g., different additional lenses of the camera) are stored in the camera firmware. The usage of additional lenses can be detected by the camera itself if the camera can detect a lens code. This can be done, for example, by a magnetic coding of the lens and a Hall-sensor reading at the camera. If the camera detects an additional lens, the corresponding calibration curve is uploaded to the firmware.

The relative output signal dependence on blackbody temperature $S_{out}(T_{bb})$ calculated for a MW camera (3–5.5 μm) and LW camera (8–14 μm) is shown in Figure 2.46. For the calculation, the spectral camera response Res_λ was assumed to be constant (boxcar assumption) within the given wavelength limits and zero outside.

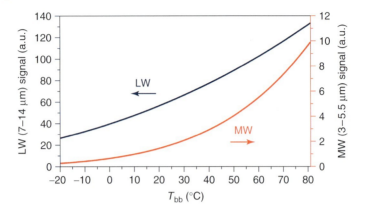

Figure 2.46 Blackbody temperature-dependent camera output signal generated by the incident blackbody object radiation for a PtSi MW (3.5–5 μm) and a LW camera (7–14 μm) with constant spectral response.

The calibration curve is then approximated by a mathematical fit function $S_{out}(T_{bb})$. A polynomial or a typical exponential fit function can be used:

$$S_{out}(T_{bb}) = \frac{R}{\exp\left(\frac{B}{T_{bb}}\right) - F} \tag{2.46}$$

Using a least square fit procedure, this function is adjusted to the calibration curve calculating the values R (response factor), B (spectral factor), and F (form factor) for the best fit. Most thermocameras offer overall temperature measurement ranges from below 0 °C up to 1000–2000 °C. With respect to the object temperature dependence of the radiance, the whole temperature range has to be divided into smaller temperature intervals to ensure an optimum temperature sensitivity and to avoid a saturation of the detector signal. Therefore, the overall temperature range of the camera is divided in different measurement subranges and the dynamic range of a specific measurement is limited to the used measurement subrange. If a scene with a larger temperature dynamic range is observed, it has to be decided as to what is the best measurement subrange for the analysis or one has to take some images with different measurement subranges. R&D cameras offer the possibility to capture large dynamic range scenes using superframing (Section 3.3). The measurement subrange is changed by changing the signal amplification and/or by using additional optical filter (ND or spectral filters) to attenuate the incident radiation on the detector to ensure detector operation within its linearity range.

The real camera sensitivities (Figure 2.47) differ from boxcar-shaped spectral response, which was assumed for Figure 2.46.

The spectral sensitivity of the camera has an effect on the characteristics of the calibration curve (Figure 2.48). Compared to the actual detector curve, the boxcar spectral sensitivity curve has a slightly larger slope, that is, should have a slightly better temperature sensitivity. The real bolometer and also QWIP camera

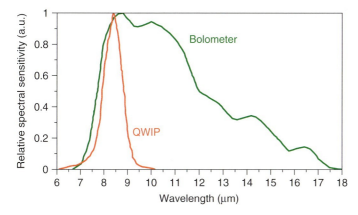

Figure 2.47 Typical spectral sensitivities for a LW bolometer and a LW QWIP camera.

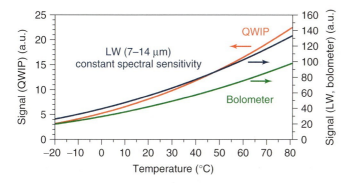

Figure 2.48 Blackbody temperature-dependent camera output signal for a LW bolometer camera and a QWIP camera generated by the incident blackbody object radiation (for spectral sensitivities according to Figure 2.47).

calibration curves can be fitted using Eq. (2.46). Because of the limited bandwidth of the QWIP spectral sensitivity, this camera will exhibit a much smaller signal for the same object temperature as a bolometer camera. However, the slope of the signal to object temperature dependence is larger for the QWIP camera. In connection with the large D^*-values of the QWIP sensors, this behavior will cause a better temperature resolution.

Besides spectral dependencies, the camera output signal is affected by the incident object radiation as well as by the additional radiation from camera parts such as the optics, the detector window, and the inner parts of the camera. The output signal of the camera, therefore, represents the sum of these two radiation fractions. The additional radiation depends on the camera temperature and causes a signal offset. The influence of this additional radiation can be corrected by a camera-temperature-dependent calibration. The quality of this correction determines the temperature measurement accuracy.

During the measurement process, the camera temperature is measured by temperature sensors inside the camera and the camera output signal is corrected using the data from the camera-temperature-dependent calibration. The additional use of a tempered flag inside the camera offers the possibility of a one-point camera recalibration. This corrects the output signal by an offset and can be used to perform an NUC of the detector array (Section 2.4.1). All these corrections will only operate if the camera state is characterized by a thermal equilibrium. A rapid change of the ambient temperature will cause a thermal shock behavior, which leads to incorrect temperature measurements. This thermal shock behavior results from the thermal nonequilibrium inside the camera connected with temperature gradients. Because of the thermal mass of the camera, it can take a long time to achieve the equilibrium state (Section 2.4.6).

Today, thermal imaging systems offer a dynamic range of 12–16 bits. This is the equivalent of 4096–65 536 signal levels and offers the possibility to expand the measurement range. If, for example, a camera is operated within the measurement range −20 to −80 °C, it is also possible to measure temperatures slightly below or above the measurement range limits. As an example, Figure 2.49 depicts the detector digitization for a 12-bit system. The temperature information of the whole measurement range is captured and available for the image analysis. For the representation and the analysis, it is possible to fix a level and a span (Figure 2.49). These two parameters only affect the image representation and correspond to brightness and contrast of the image, respectively. The temperatures within the span range can be depicted in levels of gray or false colors (palette). Mostly 8-bit scales with 256 colors are used with predefined color palettes.

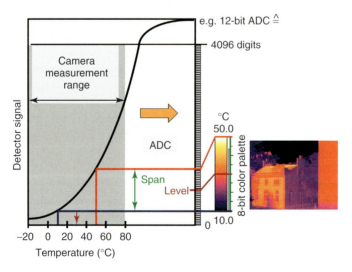

Figure 2.49 Digitization of the detector signal for a 12-bit system, level, and span.

2.4.6
Camera Operation

Accurate temperature measurement using thermal imaging requires a correct camera operation. If, for example, the camera is not yet in thermal equilibrium, it will not perform well in field measurements, although it would have passed all camera specifications during tests in the laboratory. In the latter, the cameras are started several hours before the tests such that they are in thermal equilibrium. As a consequence, all manually or automated performed camera correctives refer only to thermal equilibrium conditions. To enable accurate calibrations, most IR cameras have inside temperature sensors to monitor temperatures of camera parts. During the calibration process, the cameras are placed in an environmental chamber. The data collected from measurements at different camera ambient and object (blackbody) temperatures are stored in the firmware. These data are used to correct the camera temperature output. This correction, referred to as *ambient drift correction*, is hence correct for a similar thermal steady-state behavior of the camera as during calibration process.

To demonstrate the associated uncertainties in temperature measurements, two typical situations of camera operation with possibly transient thermal camera conditions are discussed. First, incorrect temperature readings may occur directly after switching on the camera and, second, measurements after rapid changes of camera ambient temperatures can also result in measurement errors.

2.4.6.1 Switch-on Behavior of Cameras

The apparent object temperature drifts due to the camera temperature drift after switch-on were measured using a cooled MWIR PtSi camera and an LWIR microbolometer camera. Figure 2.50 depicts measured blackbody temperatures after switch-on of the cameras, using stabilized blackbodies at 30 and 70 °C (long time temperature stabilization for more than 2 hours prior to the start of measurement with $\Delta T_{bb} \leq 0.1\,°C$). The measurements started immediately after the cameras were ready for image acquisition. This switch-on process takes about 5 min for the LWIR microblometer camera and 10 min for the cooled MWIR PtSi camera.

Both cameras show characteristic signal changes within the first 90–120 min after switch-on.

The LWIR microbolometer camera exhibits a more distinctive change in the temperature measured. Immediately after switch-on, the LWIR camera measures 2–3 K higher temperatures depending on the object temperature. For the MWIR camera, smaller deviations between 0.5 and 1 K were measured. For both cameras, the autoshutter operation (periodically measurement of tempered flag for one-point recalibration) was switched on. The observed steps of time-dependent temperature result from the autoshutter operation. The autoadjust procedure strongly influences the temperature measurement of the LW camera. The bolometer detectors signal is generated by the net radiation incident on the detector (Section 2.2.3). Furthermore, the bolometer sensitivity strongly depends on its temperature. Therefore, thermal

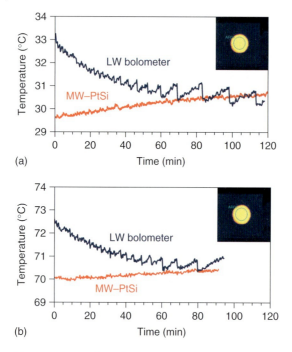

Figure 2.50 Results of the long-term temperature measurement at a blackbody emitter stabilized at 30 °C (a) and 70 °C (b) with an MWIR and an LWIR thermocamera after camera switch-on.

nonequilibrium conditions as well as bolometer temperature drift will affect the temperature measurement more strongly.

The time to reach thermal equilibrium within the cameras takes more than 90 min, but the temperature readings of both cameras achieve the given camera specification of a temperature accuracy of ±2 K after about 10 min.

2.4.6.2 Thermal Shock Behavior

The second situation to be discussed is known as *thermal shock behavior* of the cameras. During field measurements the ambient temperature often changes rapidly, for example, while taking the camera from outside into a heated building during winter or vice versa. This can cause a thermal shock for the thermocamera and can result in an incorrect temperature reading since the camera is not in thermal equilibrium, although it may have been switched on hours earlier. As discussed in Section 2.4.5, the accuracy of the radiometric temperature measurement depends on the accuracy of the object signal calculation. The measured detector signal has to be corrected for the additional radiation from, for example, the camera optics or other parts within the camera housing. If the ambient temperature is changed rapidly, the camera will exhibit transient temperature changes with temperature gradients inside the camera. This limits the accuracy of the detector

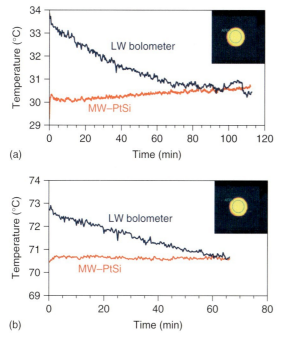

Figure 2.51 Results of the long-term temperature measurement at a blackbody emitter stabilized at 30 °C (a) and 70 °C (b) with an MWIR and an LWIR thermocamera after rapid change of camera ambient temperature from 13 to 23 °C and camera switch-on.

signal correction algorithms and causes measurement errors. Because of the large camera thermal mass, one has to expect large thermal time constants of minutes to hours.

To analyze this thermal shock behavior, the same two cameras of Figure 2.50 were tempered to 13 °C in an environmental chamber for 2 h in order to get steady-state temperature conditions of the camera. The ambient temperature was 23 °C and the relative humidity 50% (dew-point temperature above 13 °C). After 2 h, the cameras were taken out and switched on. Measurements started ($t = 0$) immediately after cameras were ready for recording temperatures (Figure 2.51). Similar to Figure 2.50, temperature-stabilized blackbodies were studied at 35 and 70 °C.

Within this experiment, both the switch-on as well as the thermal shock behavior are observed simultaneously. Similar results as for the pure switch-on measurements are observed. The LWIR bolometer camera is more sensitive to the thermal shock due to the 10 K increase in the ambient temperature with a temperature inaccuracy of 3–4 K at the beginning of the measurements. The MWIR camera is within the accuracy specification immediately after the start of the measurements. For the LWIR bolometer camera, it takes 20–30 min to reach the ±2 K accuracy specification of the camera.

From Figures 2.50 and 2.51, it can be concluded that steady-state camera temperature conditions are necessary for accurate temperature measurements. Because of the more sensitive signal correction process in bolometer cameras, their measurement accuracy is more sensitive to transient temperature changes due to either switch-on or thermal shock.

2.4.7
Camera Software – Software Tools

IR cameras without analysis software would just provide qualitative false color images of objects. However, whenever quantitative results are needed like temperatures, line plots, or reports, software tools are indispensable.

All manufacturers of thermal imaging systems provide a variety of software tools ranging from general-purpose software packages including, for example, thermal image analysis and generation of infrared inspection reports to sophisticated software packages which, for example, offer camera control functions, real-time image storage at selectable integration times, and frame rates or radiometric calculations.

The general-purpose software mostly contains the following key features:

- level and span adjustment;
- selectable color palette and isotherms;
- definition of spot analysis, lines, and areas with temperature measurements for maximum, minimum, and average temperatures;
- adjustment of object parameters (emissivity) and measurement parameters (e.g., humidity, object distance, ambient temperature);
- creation of professional customized reports with flexible design and layout (e.g., export to other software such as Microsoft Office, including visible images).

For a very detailed analysis of static or transient thermal processes sophisticated software especially designed for R&D is available. Beyond the features of the general-purpose software, such software usually includes more complex data storage, analysis, and camera operation tools:

- remote control of the camera from the PC (most or all camera parameters can be controlled by the PC);
- high-speed IR data acquisition, analysis, and storage;
- digital zoom of the infrared image, use of subframes;
- raw data acquisition and analysis, radiometric calculations;
- automatic temperature vs time and three-dimensional temperature profile plotting, additional graphic, and image processing tools;
- different data format export, automatic conversion to, for example, JPEG-, BMP-, AVI-, or MATLAB-format;
- data export to other common software applications such as Microsoft Office;
- thermal image subtraction;

- definition of different regions of interests (ROIs) with different shapes, separate emissivity, and so on.
- customized camera calibration and NUCs.

2.5
Camera Performance Characterization

The system performance evaluation of thermal imagers is well standardized and a lot of research has been done on characteristics and testing of imaging systems [17, 18, 30–32]. In general, it is, however, difficult for most practitioners to relate the relevance of the system performance parameters to their applications and the measurement results.

The performance of a thermal imaging system is described by a number of parameters such as thermal response, detector and electronic noise, geometric resolution, accuracy, spectral range, frame rate, integration time, and so on. These parameters can be divided into two groups: objective and subjective parameters (Table 2.4).

The temperature resolution given by NETD and the spatial resolution given by IFOV are important objective performance parameters. They both significantly affect the image quality.

An evaluation of the quality and performance limits of thermal imagers more oriented to practical applications requires a combination of these parameters. Additionally, the subjective factor (the ability to detect, recognize, and identify temperature differences while using the camera system) due to the human observer has to be taken into consideration. The minimum resolvable temperature difference (MRTD) and the minimum detectable temperature difference (MDTD) combine the objective and subjective parameters and are directly related to applications.

For any practical thermal imaging, the following question must eventually be answered:

Is the camera suited to my application on the basis of its performance parameters? The accurate knowledge of the most important performance limits of the thermal imaging system used and their relevance to the application is crucial for correct temperature measurements and interpretation of the results. Therefore, camera performance parameters and their influence on practical measurement results are discussed in more detail.

2.5.1
Temperature Accuracy

The specification of accuracy (or, more precisely, inaccuracy) gives the absolute value of the temperature measurement error for blackbody temperature measurements. For most thermocameras, the absolute temperature accuracy is specified to be ±2 K or ±2% of the temperature measured. The larger value is valid. The

Table 2.4 Camera performance parameters.

	Name	Definition, description	Unit	Significance
Objective parameters	Temperature accuracy	Absolute error of blackbody temperature measurement	K, °C, °F	Absolute measurement accuracy
	NETD (noise equivalent temperature difference)	Minimum temperature difference for SNR = 1	K, °C, °F	Temperature resolution
	FOV (field of view)	Angular extent of camera observable object field	degree, °	Detected object field
	IFOV (instantaneous field of view)	Angular extent of object field from which radiation is incident on the detector	mrad	Spatial resolution
	Frame rate	Frequency at which unique consecutive images are produced	1/s, Hz	Time resolution
	Integration time	Period of signal integration	s	Time resolution, Sensitivity
	SRF (slit response function)	Ratio of signal response for a slit size emitting object depending on slit width and infinite object width	–	Spatial resolution
	MTF (modulation transfer function)	Fourier transform of the camera response when viewing an ideal point source.	–	Spatial resolution
Subjective parameters	MRTD (minimum resolvable temperature difference)	Minimum temperature difference at which a 4 bar target can be resolved by a human observer dependent on the spatial frequency of the bar target.	K, °C, °F (dependent on spatial frequency)	Recognition of temperature difference
	MDTD (minimum detectable temperature difference)	Minimum temperature difference at which a circular or square object can be detected by a human observer dependent on the object size.	K, °C, °F (dependent on object size)	Detection of small size objects with low temperature difference

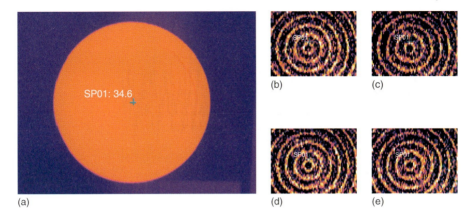

Figure 2.52 NETD measurement – experimental setup, blackbody image (a) and thermal fluctuations (b–d).

temperature measurement errors arise as a consequence of errors in radiometry connected with the calibration procedure, the long- and short-term variability of the camera sensitivity, the limited accuracy of object radiation calculation from the measured radiant power (Section 2.3), and so on. For the short-term reproducibility of a temperature measurement, a value of ±1% or ±1 K is typical.

2.5.2
Temperature Resolution – Noise Equivalent Temperature Difference (NETD)

The temperature resolution of a radiometric system is given by the NETD (see also Section 2.2.2), which quantifies the thermal sensitivity of the thermal imager. This parameter gives the minimum temperature difference between a blackbody object and a blackbody background at which the signal-to-noise ratio of the imager is equal to unity. The NETD is determined by the system noise and the signal transfer function [4, 33].

Experimentally the temperature-dependent NETD can be determined from the fluctuations of the measured temperature, analyzing the radiation from a heated and temperature-stabilized blackbody.

Figure 2.52a depicts a thermal image while the camera is directed to the blackbody source. A temperature-stabilized source was used with typical circular structures because of the geometry. For each blackbody temperature, a spot temperature measurement is recorded using thousand images (with 50 images/s, this takes approximately 20 s). Figure 2.52b–d gives a magnified view of the marked area at different times, demonstrating temperature fluctuations. The experiments were performed with a 3–5 μm PtSi MW camera (AGEMA THV550).

Figure 2.53a is a graph of the measured temperatures of a specific pixel as a function of time. It nicely illustrates the thermal noise of the imaging system since the thermal changes of the source happens at a much larger timescale.

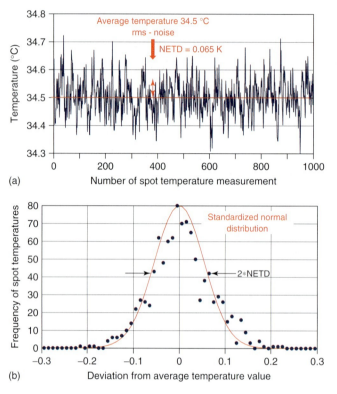

Figure 2.53 Results of an NETD measurement. (a) Time dependence of spot temperatures (marked in Figure 2.52) measured during 20 s with a frame rate of 50 Hz. (b) Frequency distribution of measured spot temperatures centered around the average object temperature.

The frequency distribution of the measured spot temperatures can be approximated by a standardized normal distribution, as expected for random noise processes (Figure 2.53b).

Using the measured temperature data, the rms (root-mean-square) value of these temperature fluctuations can be calculated which defines the NETD. For the example shown in Figure 2.53, the NETD equals 0.065 K.

The experimental NETD represents the half-width of the standardized normal distribution. This means that 68.3% of the measured temperatures are within the range $T = (34.50 + 0.065)\,°C$.

From this measurement, we can conclude that the deviation from the correct temperature value in a single measurement can be much higher than the NETD because the NETD represents the rms deviation. It can be seen that 95.4% of the temperatures measured are within the range $T = (34.50 + 0.13)\,°C$. This corresponds to the 2σ-value and 99.7% within $T = (34.50 + 0.192)\,°C$ (3σ-value).

Theoretically the *NETD* is defined as the ratio of the signal noise and the signal transfer function (differential ratio of signal changes dS to temperature changes dT;

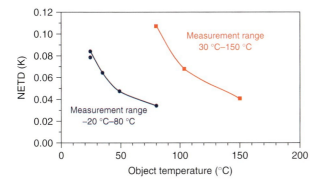

Figure 2.54 Temperature dependence of the NETD (rms – value of the temperature fluctuations measured) as a function of object temperature and measurement range.

dS/dT). The noise is given by the system (detector noise, amplifier noise, etc.) and independent on the object temperature measured. Because of the change of the signal transfer function with the object temperature T_{obj} the NETD is decreasing with increasing T_{obj} (Figure 2.54). This is caused by the temperature-dependent thermal derivative due to the spectral response of the imager. The measured object signal is strongly increasing with increasing temperature (stronger than a linear dependence). For the complete wavelength range ($\lambda = 0 \rightarrow \infty$), the temperature-dependent signal $S(T) \sim T^4$ (Section 1.3.2), that is, the NETD $\sim T^{-3}$. For limited spectral ranges, however, $S(T) = \int_{\lambda_1}^{\lambda_2} S_\lambda(T)d\lambda$ varies as shown in Figure 2.46. For the MW range, the signal change is governed by an exponential, leading to a strong decrease in NETD with increasing object temperature. Therefore, within a measurement range selected at the thermal camera (e.g., -20 to $80\,°C$), the NETD is decreasing with increasing temperature (Figure 2.54).

If the measurement range is changed to a higher temperature range as shown in Figure 2.54, the NETD is strongly increased due to the decreased sensitivity of the camera (signal decrease due to the insertion of a filter but unchanged noise resulting in a smaller signal-to-noise ratio or higher NETD). The lowest NETD values are always obtained at the upper temperature limits of the measurement ranges. Because of the temperature dependence of NETD, it only makes sense to give values at well-defined temperatures.

In Figure 2.55, some temperature measurement results using different measurement ranges of the camera are shown. For the measurements, an $80\,°C$ blackbody was used as the target and all other measurement conditions are the same. Obviously the temperature measurement within in the -20 to $80\,°C$ range exhibits the lowest noise (or NETD). The measurement within the highest temperature range is already outside of the useful part of the calibration curve.

Besides, Figure 2.55 also illustrates the changes of absolute temperature values. The two lower ranges are well within stated accuracy range in contrast to $100-250\,°C$ range result. This illustrates the fact that although cameras may still be used outside the specified ranges, the results must be interpreted with care.

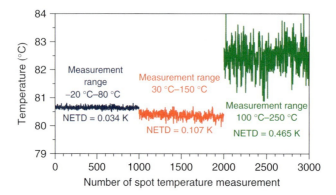

Figure 2.55 Temperature measurement at an 80 °C blackbody using different measurement ranges of the thermal camera; the NETD was calculated from the rms of the observed temperature fluctuations.

2.5.3
Spatial Resolution – IFOV and Slit Response Function (SRF)

The IFOV gives the angle over which one detector element of the focal plane array senses the object radiation [18] (Figure 2.56). Using the small angle approximation, the minimum object size whose image fits on a single detector element for a given distance can be calculated from object size = IFOV · distance (Table 2.5).

For example, a focal length of $f = 50$ mm and a single pixel size of 50 μm give an IFOV of 1 mrad.

At a distance $D = 5.5$ m, this IFOV refers to a minimum object size of 5.5 mm.

It is important to have in mind that the IFOV is only a geometric value calculated from the detector size and the focal length of the optics (Figure 2.57a). The IFOV can also be determined by dividing the FOV by the number of pixels, for example, the IFOV amounts to 1 mrad for 20° FOV and 320 pixels. The system resolution is additionally influenced by the diffraction of the optics. This is described by the *slit response function* (*SRF*), which is defined as a normalized dependence of the system response to a slit size object with a variable slit width. Experimental analysis of the SRF has been carried out. The test configuration is shown in Figure 2.57b. The angle Θ is the observable slit angle at the given slit width seen from the detector.

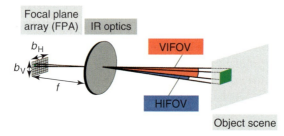

Figure 2.56 Angular IFOV for a detector pixel.

Table 2.5 Minimum object size corresponding to IFOV = 1 mrad at different object distances.

Distance (m)	Minimum object size (mm)
1	1
2	2
10	10
50	50

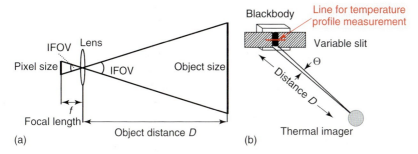

Figure 2.57 Spatial resolution (a) two-dimensional definition of IFOV; (b) experimental determination of the SRF.

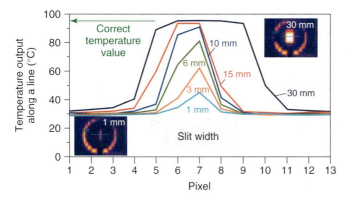

Figure 2.58 Measured temperature of the object viewed through a slit with variable width along a line for a measurement distance of 5.5 m between the thermal camera with 20° lens (FOV) and the blackbody with the variable slit.

The blackbody was heated up to a temperature of 95 °C at a distance of 5.5 m from the thermal camera. The object size was changed by a variable slit from 30 to 1 mm. The temperature was measured along a line perpendicular to the slit width (Figure 2.58). With decreasing object size, that is, slit width, the peak output becomes smaller.

The SRF represents a function of the measured object signal difference at a defined slit width normalized to that measured with a very wide slit (correct object signal difference of the slit size target). From the peak output for each slit width, the SRF can be calculated (Figure 2.59). This function provides both the imaging and the measurement resolution.

The *imaging resolution* is usually defined as that angular width of the object seen from the camera that gives a 50% response in the SRF. However, the absolute minimum size of the object for an accurate temperature measurement is twice or threefold the IFOV (in our case at a slit width of 12 or 18 mm at a measuring distance of 5.5 m using the THV 550) to reach 95 or 99% of the SRF, respectively.

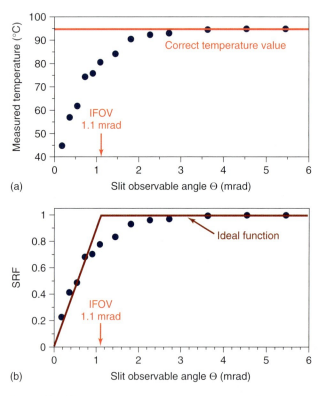

Figure 2.59 Measured peak output temperatures at different slit width (a) and slit response function SRF (b) as a function of the angular width of the object, seen from the camera (slit width 1, 2, 3, 4, 5, 6, 8, 10, 12.5, 15, 20, 25, 30 mm, respectively).

The deviation from the ideal SRF is caused by the aberrations of the camera optics as discussed in Section 2.4.4.

From these results, the minimum necessary object size for accurate temperature measurements can easily be calculated using the formula given above for the actual measurement distance, multiplying the object size with a factor of 2 or 3.

2.5.4
Image Quality: MTF, MRTD, and MDTD

The MRTD measures the compound ability of a thermal imaging system and an observer to recognize periodic bar targets within the image shown on a display. It is the minimum temperature difference between the test patterns and the blackbody background at which the observer can detect the pattern. This capability is governed by the thermal sensitivity (NETD) and the spatial resolution (IFOV) of the imaging system but strongly depends on other influencing variables such as the used palette, the ability of the observer to distinguish between different colors, and so on. ASTM standard MRTD and MDTD test methods have been described in [34, 35].

The MRTD is measured by determining the minimum temperature difference between the bars of a standard four-bar target (Figure 2.60) and the background required to resolve the thermal image of the bars by the observer. Different bar dimensions are used to do the analysis with different spatial resolutions. The regular four-bar target is characterized by a spatial frequency (cycles per length or cycles per milliradians).

The contrast in the thermal image measured for different spatial frequencies of the bar targets is decreasing with increasing frequencies or narrower line pairs due to the limited resolution of the camera (SRF as a result of given IFOV and limited optics quality; Figure 2.61). The black plate was heated in an environmental chamber to 60 °C and was then placed as the hot background behind the four-bar target structures. The function contrast measured at the four-bar target structures versus spatial frequency is called *modulation transfer function* (*MTF*). For standard determination of MTF, the measured object signals (raw signals) are used. A researcher software (as FLIR ThermaCAM Researcher) is needed to access the raw signals.

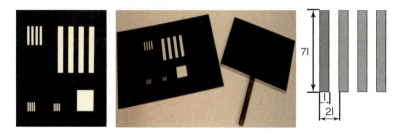

Figure 2.60 Test patterns (four-bar target and a black plate as the background).

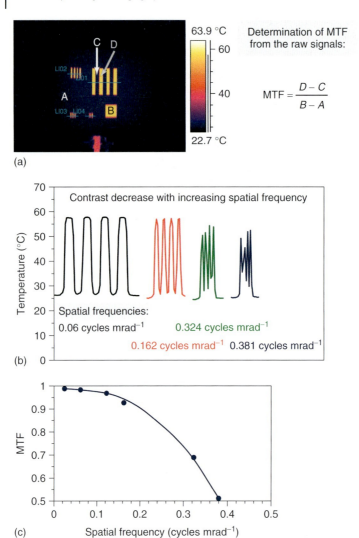

Figure 2.61 Temperature contrast measured at different spatial frequencies. (a) Thermal image of the test patterns during measurement (1.2 m measurement distance). (b) Temperature line profiles for different test patterns. (c) Spatial frequency-dependent MTF.

The MTF shown in Figure 2.61 was calculated from the ratio of the measured object signal differences (detected raw signals) and the true object signal difference of the four-bar target and the background.

For the MRTD measurements, the same equipment was used. The thermal images (50 frames/s) were stored during the cooling of the hot background plate down to room temperature. For the determination of target and background

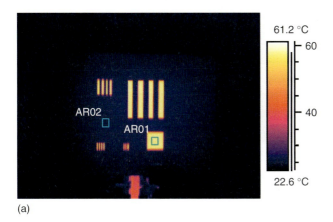
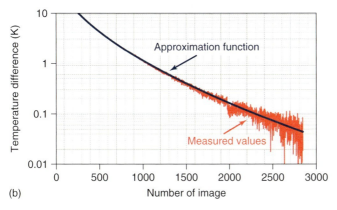

Figure 2.62 MRTD measurement – determination of temperature difference between bar target and background.

temperatures, the average temperature from the areas marked in Figure 2.62 was used. By analyzing the temperature difference between these two temperatures during the cooling of the background from 60 °C to room temperature, a functional fit for the time-dependent temperature difference can be found. With this fit, we can assign a temperature difference to every image during the cooling period. Such a procedure is necessary to determine the correct values for the very low temperature differences where the measured temperature differences are strongly influenced by temperature fluctuations due to the system noise (Figure 2.62).

Figure 2.63 depicts a series of images, observed during the cooling period while using the autoadjust scale and the iron palette. Images at different temperature differences between bar target and background are shown. With decreasing temperature differences, lower contrast details cannot be seen. First, the bar target corresponding to the largest spatial frequency cannot be observed any longer.

Figure 2.64 shows the images that correspond to the individual subjective MRTD of one of the authors for the targets with different spatial frequencies. Using these

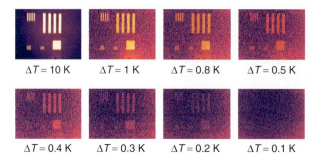

Figure 2.63 Images at different temperature differences between bar target and background for a PtSi MW camera.

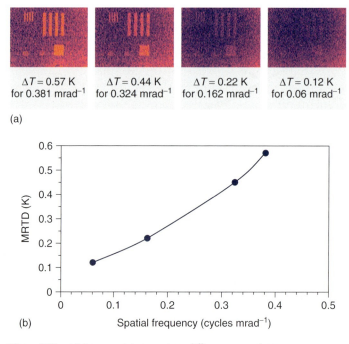

Figure 2.64 (a) Images at temperature differences equal to determined MRTD for different spatial frequencies and (b) determined spatial frequency dependence of MRTD.

results, one can get the spatial frequency (number of line pairs per milliradians or cycles per milliradians) dependent MRTD for a 24 °C temperature level. We note that the selection is much easier when analyzing the time sequences compared to presenting still images, since human perception is very sensitive to changes as a function of time.

Because the MRTD is determined not only by the objective parameters (NETD, IFOV) but also by a lot of subjective parameters (observers ability to distinguish

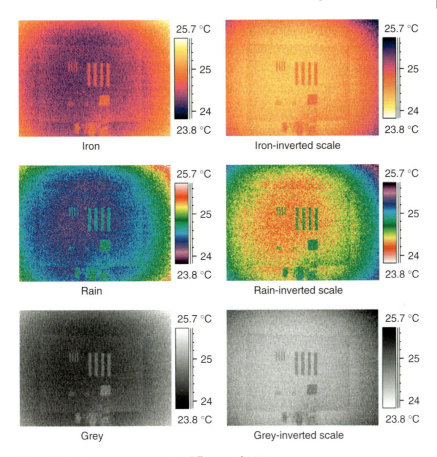

Figure 2.65 Image at a temperature difference of 0.5 K using different palettes and autoadjust scale.

between different colors, display quality, etc.), it is important to look for optimum camera parameters to get low MRTD values.

In Figure 2.65, the same image is shown using different palettes. The different contrast behavior within the images is clearly demonstrated.

The MRTD result is strongly influenced by the used level and span. Small temperature differences can only be detected if span and level are set to a maximum color contrast (Figure 2.66).

The MDTD measures the compound ability of the thermal imaging system and an observer to detect a small size object. It is the minimum temperature difference between a circular or square object and the background necessary to detect the object. It is measured as MRTD versus the inverse spatial size or the angular size of the object. Figure 2.67 depicts a series of images of a square object with an angular size of 23 mrad during cooling period.

The individual subjective MDTD for the square object with an inverse angular size of 0.043 mrad^{-1} was determined by one of the authors to be 0.05 K.

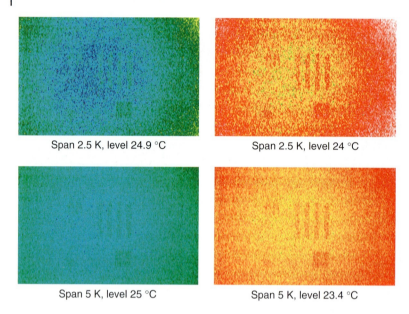

Figure 2.66 Image at a temperature difference of 0.2 K with different level and span adjustment for rain palette.

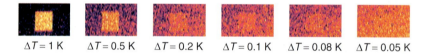

Figure 2.67 Images at different temperature differences between a square object of 23 mrad angular size and the background.

MRTD and MDTD are also very important when estimating the range for IR camera detection of objects (Section 10.5.2).

2.5.5
Time Resolution – Frame Rate and Integration Time

The accurate analysis of transient thermal processes requires a sufficient time resolution of thermal imaging compared to the characteristic thermal time constant of the process to be investigated. Most practitioners use bolometer cameras. As described in Section 2.4.1, bolometer cameras do not offer the possibility of a selectable integration or exposure time in contrast to photon detector.

In data sheets of imaging systems equipped with bolometer focal plane arrays, the time resolution is usually just characterized by the frame rate as the relevant camera parameter. Mostly this value is assumed to be connected with the time resolution for the imaging analysis. A simple experiment demonstrates that the frame rate alone will not give the time resolution of a camera. For this experiment,

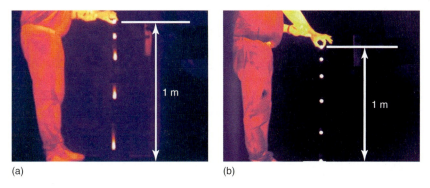

Figure 2.68 Thermograms at 50 Hz frame rates of a free falling ball (heated to a temperature of 70 °C, start at a height of 1 m) measured with an LW bolometer camera FLIR SC 2000 (a) and an MW InSb camera FLIR SC 6000 with 1 ms integration time (b).

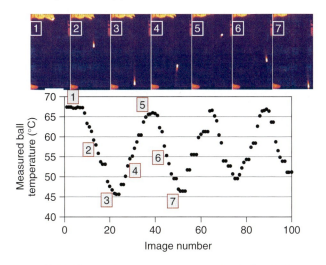

Figure 2.69 Measured temperature of free falling ball with LW bolometer camera.

a free falling rubber ball with a diameter of 3 cm was used. The rubber ball was heated to about 70 °C. Starting at a height of 1 m above the ground, the free falling ball was analyzed by thermal imaging with IR cameras for an object distance of about 4 m. Figure 2.68 depicts the results of the measurement using a bolometer camera FLIR SC2000 and an InSb camera FLIR SC6000 (photon detector camera with selectable integration time) both operating with a frame rate of 50 Hz.

With increasing velocity of the ball during the free fall motion, the FLIR SC2000 image of the ball becomes blurred due to the increasing speed of the ball.

The time dependence of the measured ball temperature during the movement of the ball can be analyzed. The ball with a temperature of about 70 °C always exhibits the maximum temperature in the infrared image. Figure 2.69 depicts the measured ball temperature during the movement of the ball. The measured temperature decreases with increasing ball velocity until the ball hits the ground. After bouncing from the ground, the velocity of the ball decreases and the determined temperature increases to about the actual ball temperature measured at the top, where the ball is at rest. This process of measured temperature change recurs during the periodic down and up movement of the bouncing ball. The correct ball temperature can only be determined when the ball is at rest. From these results, it can be concluded that the analysis of transient thermal processes or temperatures from moving objects for bolometer cameras requires a detailed investigation of the camera limits given by the thermal detector time constant. For photon detector cameras, the selectable integration time has to be adjusted to the process to be analyzed.

The large camera response time causes another result. The blurring of the image of the ball also displays that the infrared image will not reflect the real spatial temperature distribution. This can be discussed in more detail using the temperature distribution in the thermal image connected with the movement of the ball. Figure 2.70 depicts the results of a temperature profile measurement along the falling line of the ball, which for this image had already fallen about 92 cm. It is obvious that nonzero temperature differences in a 16 cm range of the measured line profile are observed, although the falling ball had only a diameter of 3 cm.

This behavior is caused by the time constant of the bolometer detectors in the FLIR SC2000 camera. Thermal detectors such as bolometers exhibit a time constant within the millisecond range (Section 2.2.3). The measured temperature line profile corresponds to the temperature rise and decay process of the bolometer sensors of the camera. It is not caused by the rolling readout process of the bolometer FPA. If the camera is rotated by 90°, the signal will be essentially the same.

From the falling distance s, the speed of the ball v can be calculated using $v = gt$ with gravitational acceleration $g = 9.81 \text{ m s}^{-2}$. Therefore, the falling distance shown in Figure 2.70b can be transformed into a timescale using $s = 1/2gt^2$ (Figure 2.70c).

At the lower end of the ball at the largest falling distance in the IR image, that is, for the largest falling time, the respective bolometer detector signal starts to rise due to IR radiation from the ball. Behind the actual position of the ball in the image, that is, at those detectors, which no longer receive IR radiation from the ball, there is a drop in the signal, characterized by the detector time constant. Therefore, the observed signal decay on the small time side of the signal directly reflects this time constant. Figure 2.70c depicts this detector signal rise and decay. For easier analysis, the signal was normalized to the maximum temperature difference ΔT_{max}.

One can estimate an exponential time dependence with a time constant τ of about 10 ms. The time resolution of the camera is limited by this time constant τ.

To get a correct temperature reading with an uncertainty of about 1 °C, one needs at least a 99% signal measurement (Section 2.4.4). This corresponds to a measurement time of $5\tau = 50$ ms. At a frame rate of 50 Hz, an image is recorded every 20 ms, but the detector responds more slowly to temperature changes. Therefore, not only the image of the falling is blurred but also the measured

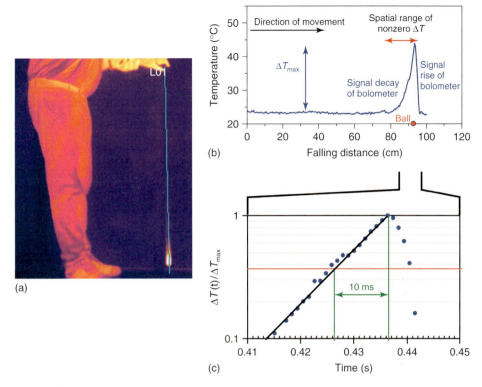

Figure 2.70 Temperature measurement at a free falling ball (heated up to 70 °C) with the LW bolometer camera FLIR SC2000. Thermal image of the falling ball at 93 cm falling height (a). Profile of the temperature distribution along the line in the image (b). Ball size has correct scaling. Time dependence of the temperature measured along the falling line with a time scale calculated from the falling distance s using $s = 1/2gt^2$ (c).

temperature of the falling ball will be incorrect (Figures 2.69 and 2.70). At a falling distance of 93 cm, the ball has a speed of 4.27 m s^{-1}. Therefore, a pixel receives the thermal radiation from the ball for only about 6 ms (1.3 mrad IFOV, 4 m measurement distance and a ball diameter of 3 cm), which is usually less than the time constant of the bolometer. As a consequence, the maximum signal is much lower than the correct 100% signal (Figure 2.71). Furthermore, the single bolometer signal itself from a 6 ms time window cannot be accurately analyzed, since it also depends on the quality of the optics. Because of the extended λ range, aberrations of the lenses may introduce additional uncertainties. Therefore, the signal rise of the bolometer close to the end of the ball should not be analyzed quantitatively. In contrast, the signal decay is only governed by the time constant of the detector.

For InSb photon detectors of the FLIR SC6000, the detector time constants are much lower (nanoseconds to microseconds) and therefore these detectors respond much faster. For the FLIR SC6000 variable, integration time settings from 9 μs to

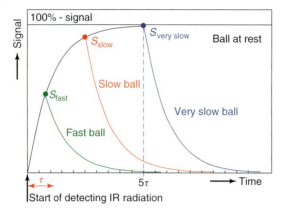

Figure 2.71 Signal rise and decay at the bolometer for different speed v of the ball (ball diameter d) compared to the bolometer time constant τ-fast ball $d/v < \tau$-slow ball $\tau < d/v < 5\tau$-very slow ball $d/v = 5\tau$.

full frame time are possible. For the experiment with the falling ball, (Figure 2.70) an integration time (Section 2.4.2) of 1 ms was used. No blurring of the image of the ball is observed and the temperature of the ball can be correctly determined over the whole falling distance.

References

1. Hudson, R.D. and Hudson, J.W. (eds) (1975) *Infrared Detectors*, John Wiley & Sons, Inc., Dowden, Hutchinson, and Ross.
2. Lutz, G. (1999) *Semiconductor Radiation Detectors*, Springer, Berlin, Heidelberg and New York.
3. Dereniak, E.L. and Boreman, G.D. (1996) *Infrared Detectors and Systems*, John Wiley & Sons, Inc.
4. Wolfe, W.L. and Zissis G.J. (eds) (1993) *The Infrared Handbook*, Infrared Information Analysis Center, Environmental Research Institute of Michigan.
5. Kingston, R.H. (1978) *Detection of optical and infrared radiation*, Springer Series in Optical Sciences, Springer-Verlag, Berlin, Heidelberg and New York.
6. Rogalski, A. and Chrzanowski, K. (2002) Infrared devices and techniques. *Opto-Electron. Rev.*, **10** (2), 111–136.
7. Li, B., Huang, S., and Zhang, X. (2004) Transient mechanical and electrical properties of uncooled resistive microbolometer focal plane arrays; Infrared detector materials and devices. Proceedings of SPIE, vol. 5564, pp. 123–132.
8. Bonani, F. and Ghigne, G. (2001) *Noise in Semiconductor Devices*, Sprinter, Berlin, Heidelberg and New York.
9. Capper, P. and Elliot, C.T. (eds) (2001) *Infrared Detectors and Emitters: Material and Devices*, Kluwer Academic Publishers.
10. Gunapala, S., Bandara, S., Bock, J., Ressler, M., Liu, J., Mumolo, J., Rafol, S., Werner, M., and Cardimona, D. (2002) GaAs/AlGaAs based Multi-Quantum well infrared detector arrays for Low-Background applications. Proceedings of SPIE, vol. 4823, pp. 80–87.
11. Hudson, R.D. (1969) *Infrared Systems Engineering*, John Wiley & Sons, Inc.

12. Ryzhii, V. (2003) *Intersubband Infrared Photodetectors*, Selected Topics in Electronics and Systems, vol. 27, World Scientific Publishing Co. Pvt. Ltd.
13. Bandara, S.V., Gunapala, S., Rafol, S., Ting, D., Liu, J., Mugolo, J., Trinh, T., Liu, A.W.K., and Fastenau, J.M. (2001) Quantum well infrared photodetectors for low background applications. *Infrared Phys. Technol.*, **42** (3-5), 237–242.
14. Rogalski, A. (1994) GaAs/AlGaAs quantum well infrared photoconductors versus HgCdTe photodiodes for Long-Wavelength infrared applications. Proceedings of SPIE, Vol. 2225, pp. 118–129.
15. Soibel, A., Bandara, S., Ting, D., Liu, J., Mumolo, J., Rafol, S., Johnson, W., Wilson, D., and Gunapala, S. (2009) A super-pixel QWIP focal plane array for imaging multiple waveband sensor. *Infrared Phys. Technol.*, **52**(6), pp. 403–407.
16. Simolon, B., Aziz, N., Cogan, S., Kurth, E., Lam, S., Petronio, S., Woolaway, J., Bandara, S., Gunapala, S., and Mumolo, J. (2009) High performance two-color one megapixel cmos roic for qwip detectors. *Infrared Phys. Technol.*, pp. **52**(6), pp. 391–394.
17. Holst, G.C. (1993) *Testing and Evaluation of Infrared Imaging Systems*, JCD Publishing Company.
18. Holst, G.C. (2000) *Common Sense Approach to Thermal Imaging*, SPIE Press, Bellingham.
19. http://www.flir.com (2010).
20. http://www.laser2000.de (2010).
21. http://www.raytek.com (2010).
22. http://www.ircon.com (2010).
23. Milton, A.F., Barone, F.R., and Kruer, M.R. (1985) Influence of nonuniformity on infrared focal plane array performance. *Opt. Eng.*, **24** (5), 885–862.
24. Lau, J.H. (ed.) (1995) *Flip Chip Technologies*, McGraw-Hill Inc.
25. Kruse, P.W. (2002) *Uncooled Thermal Imaging – Arrays, Systems and Applications*, SPIE Press, Bellingham and Washington, DC.
26. http://www.ulis-ir.com (2010).
27. (2009) *The Ultimate Infrared Handbook for R&D Professionals*, FLIR Systems Inc.
28. Rowe, D.M. (ed.) (1995) *CRC Handbook of Thermoelectrics*, CRC Press.
29. Organ, A.J. (2005) *Stirling and Pulse-tube Cryo-coolers*, Professional Engineering Publishing, Antony Rowe Ltd.
30. Chrzanowski, K. (2002) Evaluation of commercial thermal cameras in quality systems. *Opt. Eng.*, **41**(10), pp. 2556–2567.
31. Sousk, S., O'Shea, P., and Van Hodgkin (2004) Measurement of uncooled thermal imager noise. Infrared Imaging Systems: Design, Analysis, Modeling and Testing XV, SPIE vol. 5407, pp. 1–7.
32. Levesque, P., Brémond, P., Lasserre, J.-L., Paupert, A., and Balageas, D.L. (2005) Performance of FPA IR cameras and their improvement by time, space and data processing. *QUIRT-Quant. Infrared Thermography J.*, **2** (1), 97–111.
33. DeWitt, D.P. and Nutter, G.D. (1989) *Theory and Practice of Radiation Thermometry*, John Wiley & Sons, Inc.
34. ASTM E 1213-2002. *Standard Test Method for Minimum Resolvable Temperature Difference for Thermal Imaging Systems* 2002.
35. ASTM E 1311-2002. *Standard Test Method for Minimum Detectable Temperature Difference for Thermal Imaging Systems* 2002.

3
Advanced Methods in IR Imaging

3.1
Introduction

This chapter gives a brief overview of some advanced methods used in IR thermal imaging. Compared to the basic methods of recording and manipulation of IR images, for example, by adjustment of measurement parameters (Chapter 2), these methods contain additional features for data collection and data processing. Examples are the use of additional components such as narrowband spectral filters, application of heat sources for thermal stimulation of the investigated object, or various image-processing tools.

3.2
Spectrally Resolved Infrared Thermal Imaging

IR thermal imaging is a spectral measurement within limited bands (SW, MW, LW) of the electromagnetic spectrum (Chapter 1). The image results from the integration of all spectral signal contributions within the spectral band used, that is, 0.9–1.7 µm (SW), 3–5 µm (MW), or 7–14 µm (LW).

The use of these wide spectral bands increases the object radiance detected by the camera and enables the measurement of objects with low temperatures at room temperature or below with good signal-to-noise ratio (Chapter 2).

Reducing the spectral bandwidth is particularly useful when dealing with nongray, that is, spectrally selective emitters with $\varepsilon = \varepsilon(\lambda)$ (Section 1.4, Figure 1.29). For such objects, broadband detection has first to deal with the problem of a reasonable assumption for the average value of ε within the chosen IR band. Second, for some selective emitters such as gases, only a tiny part of the spectrum is used and broadband detection automatically leads to poor signal-to-background ratio (Section 7.3, Figure 7.7).

In contrast, spectrally resolved IR thermal imaging will have the advantage of first a better-defined emissivity due to the reduced bandwidth, which will allow to also quantitatively characterize spectrally selective emitters such as plastics (Section 9.6.5). Second, for gases, it can drastically increase the

Infrared Thermal Imaging. Michael Vollmer and Klaus-Peter Möllmann
Copyright © 2010 WILEY-VCH Verlag GmbH & Co. KGaA, Weinheim
ISBN: 978-3-527-40717-0

signal-to-background ratio by only detecting signal changes in the relevant part of the spectrum (Chapter 7). Narrow spectral band and dual-band measurements using pyrometry have already been in successful application for a long time. They allow, for example, radiation thermometry in steel and glass production or plastics processing [1], that is, in cases where product temperature is the most critical operating parameter during manufacturing.

The most advanced spectrally resolved thermal imaging techniques combine the spatially resolved measurements with an IR camera with IR spectroscopy, which ultimately allows measurements for each single pixel.

3.2.1
Using Filters

The use of spectral BP or NBP filters (Section 2.4.4) is the easiest way to perform spectrally resolved IR thermal imaging. A large variety of these NBP and BP filters covering the complete spectral region from SWIR to LWIR are available [2]. R&D cameras, especially, offer the possibility to mount an additional spectral filter between the camera lens and the detector (Figure 3.1). These filters mostly exhibit the same temperature as the optics, hence they are called *warm filters*. They should be made from nonabsorbing materials to prevent self-heating and radiation emission toward the detector. They block radiation transmission outside the spectral BP by complete reflection using interference from a thin layer of IR-transparent films (alternating layers) made from different materials with different refractive indices (Section 1.6).

For some applications such as GasFind cameras (Chapter 7) cold filters with temperatures close or equal to the detector temperature are used to increase the sensitivity by avoiding self-radiation of the filter and by blocking the radiation from warm parts of the camera. These cold filters form an integral part of the cold

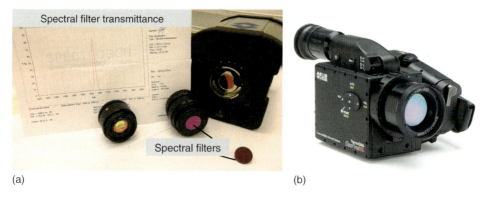

Figure 3.1 MWIR cameras for operation with spectral filters. (a) FLIR SC 6000 camera. The warm filters can be directly mounted between camera lens and window of the detector housing. (b) FLIR GasFindIR HSX camera with cold filter. (Image courtesy for (b): FLIR Infrared Training Center.)

detector engine of the camera as depicted in Figure 3.1b for a GasFind camera. Some research cameras that can house more than four cold filters are available with wheels. Cold filters may be made from partially absorbing materials in the used camera range.

3.2.1.1 Glass Filters

Narrow spectral band thermal imaging using filters can be applied to analyze the temperature and the temperature distribution of selective emitters, for example, glass. Thermal images obtained by using glass filters are depicted in Figure 3.2 using an MWIR camera (3–5 µm).

Glasses exhibit a transparent behavior from the Vis to the near infrared (NIR) region. At about 2-µm wavelength many glasses become semitransparent due to absorption, and opaque between 3 µm and at most about 5 µm (Figure 3.2, for details see [3]). Transmission as a function of wavelength varies with glass

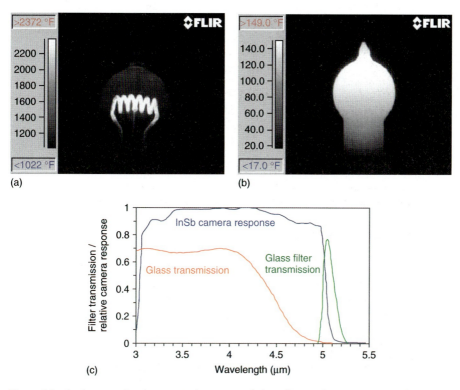

Figure 3.2 Application of a glass spectral filter to eliminate transparency for correct temperature measurement of the glass itself. (a) Thermal image in the MWIR region of an operated light bulb without additional spectral filtering, (b) thermal image in the MWIR region of the light bulb with an additional spectral glass filter, and (c) comparison of InSb MWIR camera spectral sensitivity distribution and transmission spectra of glass (exemplary) and the used glass spectral filter. (Image courtesy: FLIR Infrared Training Center.)

composition and depends on glass thickness (Section 1.5.3 and Figure 1.54). Therefore, measurement of glass temperatures by radiation thermometry requires the application of a spectral filter with narrowband transmission above 5 µm. However, the spectral response of detectors like InSb used in the MWIR, strongly decreases for $\lambda > 5$ µm. Furthermore, absorption due to water vapor strongly affects atmospheric transmission above 5 µm, where glass is opaque. Obviously, a compromise must be found. The spectral filters commonly used are transparent in the 5–5.2 µm region.

Figure 3.2a depicts an example of a light bulb that was analyzed during operation. In broadband mode (or using a filter at $\lambda < 4$ µm), the glass transmits a large part of the radiation. Therefore, the camera looks through it and the hot filament is clearly seen. The measured temperatures do not correctly resemble the filament temperature unless the proper attenuation due to glass transmission is taken into account.

With the narrow band glass filter, the camera only detects the radiation from the bulb because the glass is opaque in this spectral region. The temperature scale in Figure 3.2b refers to camera calibration without the filter. Therefore, correct temperature readings require a recalibration of the camera with the filter (Section 2.4.4).

3.2.1.2 Plastic Filters

Plastic films are typical examples for selective emitters. As for organic materials, most polymers exhibit the fundamental carbon–hydrogen (C–H) absorption band at 3.43 µm (see Figure 3.3 and the discussion of GasFind cameras in Chapter 7). This represents a region where plastic films can become opaque and reach high emissivity values depending on the film thickness. The plastic film absorption only covers a small part of the spectral response region of the camera. Therefore, without spectral filtering, the camera will look through the plastic film and its temperature cannot be measured (see also Section 9.6.5). Since absorption bandwidth varies with film thickness, measurements depend on the spectral filter width. If it is too large as shown for filter type 1 in Figure 3.3, the camera signal will strongly change with changing film thickness.

For the NBP filter, measurements are more accurate unless the film thickness is below a minimum thickness, that is, if it is too thin to be completely opaque at the respective wavelength.

For some plastics such as polyester and other ester-related polymers, there is also a strong carbon–oxygen (C–O) stretching band in the LWIR region that can be used for the radiation thermometry [1] with thermal detectors in pyrometers or cameras (Figure 3.4).

Another example for a thermal imaging application using spectral filtering is the detection of gases (Chapter 7).

3.2.1.3 Influence of Filters on Object Signal and NETD

The use of spectral filters will decrease the signal-to-noise ratio and consequently increase the NETD. Therefore, the application of NBP filters requires sufficiently high object temperatures. This situation is exemplarily analyzed for the use of NBP

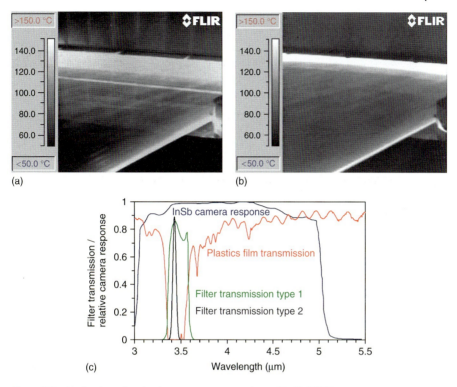

Figure 3.3 Application of a plastics spectral filter to eliminate transparency for correct temperature measurements with MW cameras. (a) Thermal image of a plastic film without additional spectral filtering, (b) thermal image of a plastic film with an additional spectral plastics filter, and (c) comparison of InSb MWIR camera spectral sensitivity distribution and transmission spectra of polyethylene plastic film and spectral filters made of two different plastics. (Image courtesy: FLIR Infrared Training Center.)

plastics filter such as the one of type 2 as shown in Figure 3.3 with a maximum transmittance of 0.89 at 3.43 μm center wavelength. Figure 3.5 depicts the portion of signal for an MWIR camera (constant sensitivity is assumed for the spectral range 3–5 μm) with spectral filter with respect to the signal without filter as a function of object temperature for different full width at half maximum (FWHM) of the filter. The signal depends on the FWHM of such a filter. For simplicity, theoretical signals were computed for top hat filter transmission curves.

For a 100-nm FWHM and a maximum transmittance of 0.89 the signal drops by a factor of about 100 at 20 °C object temperature and 50 at 120 °C. Consequently, the NETD is increased by this factor, that is, if the NETD without filter is 25 mK, it increases to >2.5 K at 20 °C and 1.25 K for 120 °C, respectively. Figure 3.6 depicts the object-temperature-dependent camera signal for the different filter configurations.

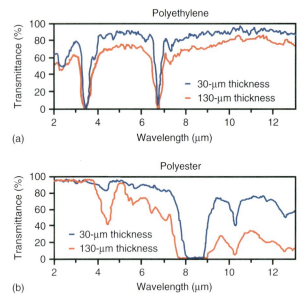

Figure 3.4 Spectral transmittance of polyethylene (a) and polyester (b) films with different thicknesses, showing strong absorption bands at around 6.7 μm (a) and 8.3 μm (b). (Image courtesy: Raytek GmbH Berlin.)

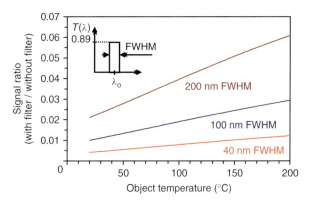

Figure 3.5 Signal ratio for an MWIR camera operated with and without narrow bandpass spectral filter at 3.43 μm with different FWHM (see inset).

Figure 3.6 can be used to determine exemplarily the minimum object temperature necessary for the same camera signal or NETD as for camera operation without the spectral filter. The camera signal at 30 °C object temperature is equivalent to the one with the 200-nm FWHM filter at 141 °C and for the 100-nm filter at 175 °C, respectively. If a 40-nm FWHM filter is used, the necessary object temperature is above 220 °C.

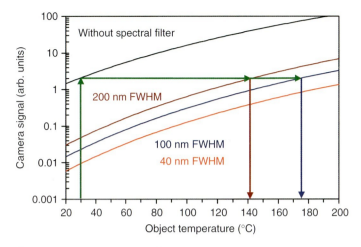

Figure 3.6 Object-temperature-dependent signal for an MWIR camera with different 3.43-μm spectral filters.

3.2.2
Two-Color or Ratio Thermography

A thermal image results from both the emitted radiance of an object and the reflected radiance from the object due to the surroundings. For large ε-values, reflection contributions of opaque objects are small. However, objects like metals with very low emissivities pose problems, since emitted radiance is low and reflected radiance is high. Therefore, many approaches have tried to develop an "emissivity-free" radiation thermometry [1]. One of the ideas is to measure the radiance within two or more narrow spectral bands and to determine the object temperature from the signal ratio [4–6] while at the same time eliminating the influence of the emissivity under certain conditions.

In the following, we assume a photon detector measurement at two monochromatic wavelengths. If spectral bands are used, the radiance values have to be replaced by the integral of the spectral radiance density (see below).

Neglecting a possible signal reduction due to atmospheric transmission along the path (short measurement path $\tau_{atm} = 1$) for the radiance L_{inc} incident on the detector at the wavelength λ, an object temperature T_{obj}, object emissivity ε_{obj}, and background temperature T_{bgr}, we assume (compare Eq. (2.37a))

$$L_{inc}(\lambda) = \underbrace{\varepsilon_{obj}(\lambda, T_{obj}) L_{BB}(\lambda, T_{obj})}_{\text{emitted object radiance}} + \underbrace{\left(1 - \varepsilon_{obj}(\lambda, T_{obj})\right) L_{BB}(\lambda, T_{bgr})}_{\text{reflected background radiance}} \quad (3.1)$$

If the radiance is measured at two different wavelengths ($\lambda_2 > \lambda_1$), the ratio can be determined as follows:

$$\frac{L_{inc}(\lambda_2)}{L_{inc}(\lambda_1)} = \frac{\varepsilon_{obj}(\lambda_2, T_{obj}) L_{BB}(\lambda_2, T_{obj}) + (1 - \varepsilon_{obj}(\lambda_2, T_{obj})) L_{BB}(\lambda_2, T_{bgr})}{\varepsilon_{obj}(\lambda_1, T_{obj}) L_{BB}(\lambda_1, T_{obj}) + (1 - \varepsilon_{obj}(\lambda_1, T_{obj})) L_{BB}(\lambda_1, T_{bgr})} \quad (3.2)$$

The emissivity can affect the ratio of Eq. (3.2) via two different effects. First, the absolute value of ε is important for the relative importance of the reflected background radiance. Second, wavelength-dependent variation of ε may lead to considerable differences between the radiance contribution at λ_1 and λ_2. We deal with both topics separately in this chapter.

The ratio (Eq. (3.2)) generally exhibits a complex relation between the signals at the two wavelengths and will not allow a simple object temperature determination. For certain conditions, however, the analysis can be simplified.

The simplest result is obtained if, first, the reflected background contributions can be neglected with respect to the object radiation, and second, the objects are true gray bodies, that is, $\varepsilon_{\text{obj}}(\lambda_1, T_{\text{obj}}) = \varepsilon_{\text{obj}}(\lambda_2, T_{\text{obj}})$. In this case, the ratio

$$\frac{L_{\text{inc}}(\lambda_2)}{L_{\text{inc}}(\lambda_1)} = \frac{L_{\text{BB}}(\lambda_2, T_{\text{obj}})}{L_{\text{BB}}(\lambda_1, T_{\text{obj}})} \tag{3.3}$$

is independent of emissivity and only depends on the chosen wavelengths and the object temperature.

In the following, we will discuss the three major sources of errors for the determination of object temperatures from Eq. (3.3).

First, there can be errors due to influence of reflected background radiation. Even if reflected radiation can be neglected, Eq. (3.3) is usually simplified for analysis of T_{obj} by replacing Planck's law with the so-called Wien approximation. This leads, second, to wavelength-dependent errors even for true gray bodies. Third, any unknown additional variation of emissivity between the two chosen wavelengths λ_1 and λ_2 can lead to very large measurement errors for the resulting object temperatures.

3.2.2.1 Neglecting Background Reflections

The transition from Eq. (3.2) to Eq. (3.3) is nontrivial. For small ε_{obj}-values, it follows from Eq. (3.2), that if the reflected radiation cannot be neglected even for objects with true gray-body behavior $\varepsilon_{\text{obj}}(\lambda_1, T_{\text{obj}}) = \varepsilon_{\text{obj}}(\lambda_2, T_{\text{obj}})$, the measured ratio is not independent of the emissivity. Figure 3.7 depicts the wavelength-dependent ratio of emitted object radiance and reflected background radiance for a background temperature of 27 °C at different object temperatures and emissivities. The ratio is strongly decreasing with increasing wavelength. Low object emissivities require a measurement at short wavelengths to obtain an insignificant error if the reflected background radiance is to be neglected. A useful measure is that $L_{\text{obj}}/L_{\text{bgr}} > 100$. In this case, typical errors by reducing Eq. (3.2) to Eq. (3.4)

$$\frac{L_{\text{inc}}(\lambda_2)}{L_{\text{inc}}(\lambda_1)} = \frac{\varepsilon_{\text{obj}}(\lambda_2, T_{\text{obj}}) L_{\text{BB}}(\lambda_2, T_{\text{obj}})}{\varepsilon_{\text{obj}}(\lambda_1, T_{\text{obj}}) L_{\text{BB}}(\lambda_1, T_{\text{obj}})} \tag{3.4}$$

will be of the order of a few percent only. The broken lines in Figure 3.7 refer to this criterion.

Consider, for example, an object emissivity of 0.02. At 200 °C object temperature, the measurement wavelengths need to be below 2 μm whereas higher temperatures of 500 or 1000 °C weaken the constraint to wavelengths below 3.5 and 4.5 μm,

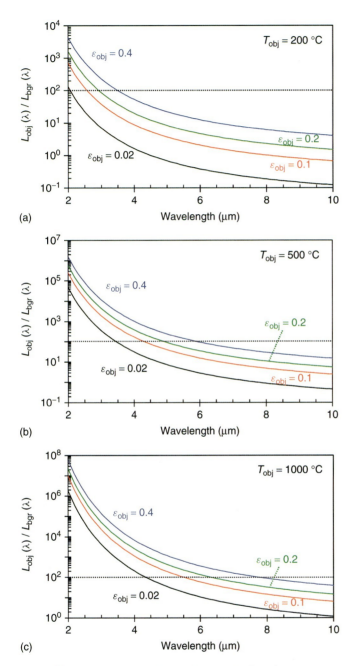

Figure 3.7 Ratio of emitted object radiance and reflected background radiance as a function of wavelength for different object emissivities at different object temperatures and a 27 °C background (300 K) (see text for details).

respectively. If an object with $\varepsilon = 0.02$ would be measured at $\lambda = 4.5\,\mu m$ for $T_{obj} = 200\,°C$, object radiance and reflected background radiance would be equal, resulting in large errors if the reflection contribution is neglected.

It is very important to note that even at very high object temperatures of about $1000\,°C$, the use of long wavelengths above $8\,\mu m$ would only allow to quantitatively measure with percentage accuracy for emissivities $\varepsilon_{obj} > 0.4$. The longer the operating wavelength, the higher is the emissivity needed for correct measurements. This explains why two-color measurements always use short wavelengths.

Once it is obvious that Eq. (3.4) can be used for measurements, one benefits from the following advantage for actual measurements: if the transmission between object and detector changes, for example, owing to transient contamination of the atmosphere or possible windows by dust, smoke, and so on, the ratio of Eq. (3.4) only changes if attenuation differs for the two chosen wavelengths. If they are close to each other and object signal changes are the same for both wavelengths, the ratio signal is not affected at all.

3.2.2.2 Approximations of Planck's Radiation Law

In the following, we will neglect any reflection contributions, that is, start with Eq. (3.4).

We first discuss the ratio of detected radiance due to object exitance at different wavelengths using the Planck distribution function (Eq. (1.15)):

$$\frac{\varepsilon_{obj}(\lambda_2, T_{obj})L_{BB}(\lambda_2, T_{obj})}{\varepsilon_{obj}(\lambda_1, T_{obj})L_{BB}(\lambda_1, T_{obj})} = \frac{\varepsilon_{obj}(\lambda_2, T_{obj})}{\varepsilon_{obj}(\lambda_1, T_{obj})} \frac{\lambda_1^5}{\lambda_2^5} \frac{e^{\frac{hc}{\lambda_1 k T_{obj}}} - 1}{e^{\frac{hc}{\lambda_2 k T_{obj}}} - 1} \quad (3.5)$$

The ratio of the Planck functions can be simplified for both the long and the short wavelength region.

For the long wavelength spectral region, the exponential can be linearized $e^x \approx 1 + x$ if $x = \frac{hc}{\lambda k T} \ll 1$, that is, $\lambda T \gg 14\,400\,\mu m\,K$. This spectral region of the Planck distribution function can be replaced by the Rayleigh–Jeans radiation law (Figure 3.8):

$$L_{BB}^{Rayleigh-Jeans}(\lambda, T_{obj}) = 2c\frac{kT_{obj}}{\lambda^4} \quad (3.6)$$

If the Planck law in Eq. (3.5) would be replaced by this Rayleigh–Jeans approximation, a ratio measurement to determine the object temperature in this spectral region would fail since the temperature in numerator and denominator will cancel each other in Eq. (3.5). Besides, the restriction $\lambda T \gg 14\,400\,\mu m\,K$ does require very large temperatures. Even for $\lambda = 10\,\mu m$, the restriction is $T \gg 1440\,K$.

The other extreme exists for short wavelengths. If $x > 5$, we find that $\frac{1}{e^x - 1} \approx e^{-x}$ (accuracy better then 1%). This condition is fulfilled for $\lambda T < 2897.8\,\mu m\,K \approx (1/5) \cdot 14\,400\,\mu m\,K$, where $\lambda T = 2897.8\,\mu m\,K$ resembles the maximum of the Planck radiance function (Eq. (1.16)). In this case, we find

$$L_{BB}^{Wien}(\lambda, T_{obj}) = \frac{2hc^2}{\lambda^5} e^{-\frac{hc}{\lambda k T_{obj}}} \quad (3.7)$$

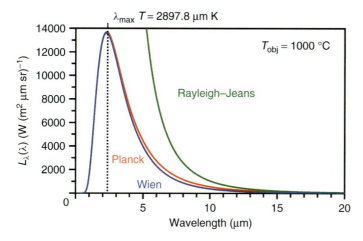

Figure 3.8 Comparison of the Planck distribution function, Wien approximation, and Rayleigh–Jeans approximation.

which holds rather well for all short wavelengths below the maximum, as can be seen in Figure 3.8.

In this case, the temperature dependence of the spectral radiance is exponential. An object temperature change will have a much stronger effect on the radiance than any emissivity changes. Therefore, the ratio method is applied in this spectral region.

Using this Wien approximation, one can easily determine the object temperature that is needed to ensure that background radiation may be neglected. For $\lambda T < 2897.8 \, \mu m \, K$ the ratio of emitted object radiance and reflected background radiance at a given wavelength λ can be written in the Wien approximation as

$$\frac{L_{obj}(\lambda, T_{obj})}{L_{bgr}(\lambda, T_{bgr})} = \frac{\varepsilon_{obj}}{(1 - \varepsilon_{obj})} e^{-\frac{hc}{\lambda k}\left(\frac{1}{T_{obj}} - \frac{1}{T_{bgr}}\right)} \quad (3.8)$$

Provided the Wien approximation holds, Eq. (3.8) can be rewritten to estimate the minimum object temperature that is needed to obtain a predefined ratio of object to reflected background radiation.

$$T_{obj}^{min} = \left(\frac{1}{T_{bgr}} - \frac{\lambda k}{hc} \ln\left[\frac{L_{obj}(\lambda, T_{obj})}{L_{bgr}(\lambda, T_{bgr})} \left(\frac{1 - \varepsilon_{obj}}{\varepsilon_{obj}}\right)\right]\right)^{-1} \quad (3.9)$$

For example, $T_{backgr} = 300 \, K$, $\varepsilon = 0.02$, $\lambda = 2 \, \mu m$ and a ratio of 100 yields $T_{obj}^{min} \approx 192 \, °C$, which correlates well with Figure 3.7.

A word of caution regarding the use of Eq. (3.9): if values are chosen for which the Wien approximation does not work, for example, ratio = 100, $\varepsilon = 0.2$, $T_{bgr} = 300 \, K$, $\lambda = 10 \, \mu m$, unreasonably high temperatures result.

3.2.2.3 T_{obj} – Error for True Gray Bodies within Wien Approximation

Using Eq. (3.7), Eq. (3.4) can be written as

$$\frac{L_{inc}(\lambda_2)}{L_{inc}(\lambda_1)} = \frac{\varepsilon_{obj}(\lambda_2, T_{obj})}{\varepsilon_{obj}(\lambda_1, T_{obj})} \frac{\lambda_1^5}{\lambda_2^5} e^{-\frac{hc}{kT_{obj}}\left(\frac{1}{\lambda_2} - \frac{1}{\lambda_1}\right)} \tag{3.10}$$

The object temperature can be calculated using the measured signal ratio by taking the natural logarithm of both sides of Eq. (3.10) and solving for T:

$$\frac{1}{T_{obj}} = \frac{k}{hc} \frac{\lambda_1 \lambda_2}{(\lambda_2 - \lambda_1)} \left[\ln\left(\frac{L_{inc}(\lambda_2)}{L_{inc}(\lambda_1)}\right) + 5 \ln\left(\frac{\lambda_2}{\lambda_1}\right) \right.$$
$$\left. + \ln\left(\frac{\varepsilon_{obj}(\lambda_1, T_{obj})}{\varepsilon_{obj}(\lambda_2, T_{obj})}\right) \right] \tag{3.11}$$

The last term represents the temperature error due to unequal emissivities at λ_1 and λ_2. Therefore, we can write the object temperature in terms of an apparent temperature $T_{app,mono}$.

$$\frac{1}{T_{obj}} = \frac{1}{T_{app,mono}} + \frac{k}{hc} \frac{\lambda_1 \lambda_2}{(\lambda_2 - \lambda_1)} \ln\left(\frac{\varepsilon_{obj}(\lambda_1, T_{obj})}{\varepsilon_{obj}(\lambda_2, T_{obj})}\right) \tag{3.12a}$$

with

$$\frac{1}{T_{app,mono}} = \frac{k}{hc} \frac{\lambda_1 \lambda_2}{(\lambda_2 - \lambda_1)} \left[\ln\left(\frac{L_{inc}(\lambda_2)}{L_{inc}(\lambda_1)}\right) + 5 \ln\left(\frac{\lambda_2}{\lambda_1}\right) \right] \tag{3.12b}$$

The apparent temperature for monochromatic wavelengths λ_1 and λ_2 results from Eq. (3.11) for true gray bodies, that is, $\varepsilon_{obj}(\lambda_1, T_{obj}) = \varepsilon_{obj}(\lambda_2, T_{obj})$. This assumption is usually made, that is, interpretation of measurements assume $\varepsilon_{obj}(\lambda_1, T_{obj}) = \varepsilon_{obj}(\lambda_2, T_{obj})$ and hence the ratio signal is analyzed for blackbody temperatures, giving $T_{app,mono}$. If, however, ε-differences occur, ($\varepsilon_{obj}(\lambda_1, T_{obj}) \neq \varepsilon_{obj}(\lambda_2, T_{obj})$), the real object temperature, which can be deduced from the ratio differs from the apparent temperature (see below).

Equation (3.12) contains two conflicting requirements for the selection of λ_1 and λ_2.

The wavelength-dependent factor decreases with increasing spectral separation $\Delta\lambda = (\lambda_2 - \lambda_1)$ of the radiance measurement but for large values of $\Delta\lambda$ the gray-body assumption will become less valid. The influence of these conflicting requirements on the measurement errors has to be analyzed in detail to find optimum measurement conditions [7].

The difference $\Delta T = T_{obj} - T_{app,mono}$ between real object and apparent object temperatures, evaluated for two monochromatic wavelengths (Eq. (3.12)) resembles the easiest analyzable case. Unfortunately, there are, however, additional temperature errors that occur owing to

- finite transmission of used filters, in particular, if they differ for the two wavelengths,
- finite spectral widths FWHM (Full Width at Half Maximum) of the filters,
- different spectral widths of the two filters.

3.2 Spectrally Resolved Infrared Thermal Imaging

For simplicity, let us assume two top hat filters with $\tau(\lambda_1) = \tau(\lambda_2) = 1$ and the same finite FWHM. This finite width will induce additional errors.

For finite spectral filter width the detector signals are determined by the integrals

$$L_{\text{inc}}^{\text{Planck}}(\lambda_1) = \frac{1}{\Delta\lambda} \int_{\lambda_1 - \frac{\Delta\lambda}{2}}^{\lambda_1 + \frac{\Delta\lambda}{2}} L_{\text{BB}}^{\text{Planck}}(\lambda, T_{\text{obj}}) \, d\lambda \tag{3.13a}$$

$$L_{\text{inc}}^{\text{Planck}}(\lambda_2) = \frac{1}{\Delta\lambda} \int_{\lambda_2 - \frac{\Delta\lambda}{2}}^{\lambda_2 + \frac{\Delta\lambda}{2}} L_{\text{BB}}^{\text{Planck}}(\lambda, T_{\text{obj}}) \, d\lambda \tag{3.13b}$$

The object temperature is calculated from the ratio of Eq. (3.13a,b) using Eq. (3.11).

For finite FWHM $\Delta\lambda$ the signal ratio differs from the signal ratio for monochromatic measurement by a factor $A_{\Delta\lambda}$:

$$\frac{L_{\text{inc}}^{\text{Planck}}(\lambda_2)}{L_{\text{inc}}^{\text{Planck}}(\lambda_1)} = A_{\Delta\lambda} \frac{L_{\text{BB}}^{\text{Wien}}(\lambda_2, T_{\text{obj}})}{L_{\text{BB}}^{\text{Wien}}(\lambda_1, T_{\text{obj}})} \tag{3.14}$$

Figure 3.9 depicts an enlarged schematic drawing of the Planck function, strongly exaggerating the deviations. The correction factor $A_{\Delta\lambda}$ depends on the chosen wavelengths as well as on the respective spectral widths of the filters and obviously leads to a modification of Eq. (3.12) for broadband detection according to

$$\frac{1}{T_{\text{obj}}} = \frac{1}{T_{\text{app},\Delta\lambda}} + \frac{k}{hc} \frac{\lambda_1 \lambda_2}{(\lambda_2 - \lambda_1)} \ln\left(\frac{\varepsilon_{\text{obj}}(\lambda_1, T_{\text{obj}})}{\varepsilon_{\text{obj}}(\lambda_2, T_{\text{obj}})}\right) \tag{3.15a}$$

with

$$\frac{1}{T_{\text{app},\Delta\lambda}} = \frac{1}{T_{\text{app,mono}}} + \frac{k}{hc} \frac{\lambda_1 \lambda_2}{(\lambda_2 - \lambda_1)} \ln A_{\Delta\lambda} \tag{3.15b}$$

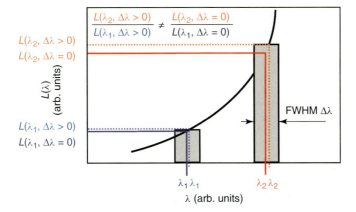

Figure 3.9 Schematic drawing of the Planck function illustrating the different radiance ratios for monochromatic measurements and measurements with finite filter FWHM $\Delta\lambda$. The finite filter width leads to apparent shifts of the center wavelengths, which are different for λ_1 and λ_2, giving rise to errors.

The deviation from $T_{app,mono}$ and $T_{app,\Delta\lambda}$ depends on the difference of the ratio of the radiances in the Wien approximation and the one using Planck's law.

Using the correction factor $A_{\Delta\lambda}$, the relative difference between the incident radiance ratio and the Wien approximation ratio is given by

$$\frac{\frac{L_{inc}^{Planck}(\lambda_2)}{L_{inc}^{Planck}(\lambda_1)} - \frac{L_{BB}^{Wien}(\lambda_2)}{L_{BB}^{Wien}(\lambda_1)}}{\frac{L_{inc}^{Planck}(\lambda_2)}{L_{inc}^{Planck}(\lambda_1)}} = \frac{A_{\Delta\lambda} - 1}{A_{\Delta\lambda}} \tag{3.16}$$

Figure 3.10a depicts this quantity, that is, the relative change of the radiance ratio due to the Wien approximation for the two center wavelengths of the filter transmission curves. For monochromatic wavelengths (FWHM = 0), deviations occur

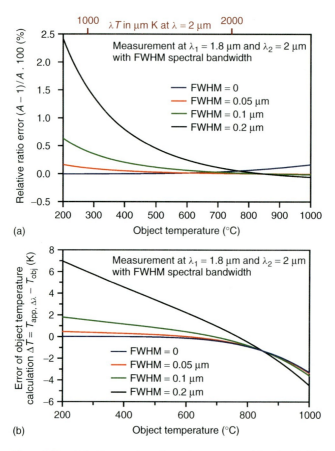

Figure 3.10 Object-temperature-dependent relative error of the 1.8- and 2-μm signal ratio $(A_{\Delta\lambda} - 1)/A_{\Delta\lambda} \cdot 100\%$ (according to Eq. (3.16)) applying the Wien approximation compared to the Planck function for different spectral bandwidth of the measurement spectral bands (a). The top scale refers to $\lambda = 2$ μm. Resulting error of the calculated object temperature using the signals at 1.8 and 2 μm and applying the Wien approximation (b).

for high temperatures (due to the limited applicability of the Wien approximation), whereas increasing filter width leads to errors in the percentage range for low temperatures.

These deviations have consequences for the deduced object temperatures according to Eq. (3.15a,b). Figure 3.10b depicts the difference of object and apparent temperatures. Largest temperature errors occur first for large filter widths at low object temperatures but also for very high object temperatures. Obviously, it is possible to perform measurements that are quite accurate (error < 2%) for gray bodies over an extended temperature range from 200 to about 800 °C and filter width below 0.1 µm.

The error discussion so far only referred to idealized conditions such as top hat filters with $\tau_{max} = 1$. In experimental studies, real filter transmission curves need to be used. Quite often, the filter curves differ in τ_{max} as well as FWHM for the two wavelengths, which may introduce additional errors. Real filter transmission curves $[\tau_1(\lambda), \tau_2(\lambda)]$ will modify Eqs. (3.13a,b) to

$$L_{inc}(\lambda_1) = \frac{1}{\Delta\lambda} \int_{\lambda_1 - \frac{\Delta\lambda}{2}}^{\lambda_1 + \frac{\Delta\lambda}{2}} \tau_1(\lambda) L_{BB}^{Planck}(\lambda, T_{obj}) d\lambda \qquad (3.17a)$$

$$L_{inc}(\lambda_2) = \frac{1}{\Delta\lambda} \int_{\lambda_2 - \frac{\Delta\lambda}{2}}^{\lambda_2 + \frac{\Delta\lambda}{2}} \tau_2(\lambda) L_{BB}^{Planck}(\lambda, T_{obj}) d\lambda \qquad (3.17b)$$

The different filter transmission behavior will therefore additionally influence the incident object radiances. The correction factor $A_{\Delta\lambda}$ from Eq. (3.14) will be modified to $A_{\Delta\lambda,\tau(\lambda)}$,

$$\frac{L_{inc}(\lambda_2)}{L_{inc}(\lambda_1)} = A_{\Delta\lambda,\tau(\lambda)} \frac{L_{BB}^{Planck}(\lambda_2, T_{obj})}{L_{BB}^{Planck}(\lambda_1, T_{obj})} \qquad (3.18)$$

and has to be determined from a blackbody calibration procedure.

3.2.2.4 Additional T_{obj}–Errors due to Nongray Objects

So far, only true gray objects were considered and resulting errors due to the Wien approximation, which is usually used to evaluate object temperatures. However, this most important assumption of ratio thermography, that is, $\varepsilon(\lambda_1, T_{obj}) = \varepsilon(\lambda_2, T_{obj})$ is often only fulfilled approximately. The consequences are obvious: if the object exhibits a nongray behavior the error of the temperature calculation will increase. The second term in Eq. (3.15a) represents the additional error due to the emissivity change. If we assume an emissivity difference of 1 or 5% for 2−1.8 µm the ratios of the measured signal will change by a factor of 1.01 or 1.05, respectively. Figure 3.11 depicts the resulting temperature measurement error depending on object temperature and filter FWHM.

This error was calculated in a manner similar to that in Figure 3.10, that is, it includes also the error due to the Wien approximation, the only difference being that the last term in Eq. (3.11) was zero in Figure 3.10 and nonzero in Figure 3.11.

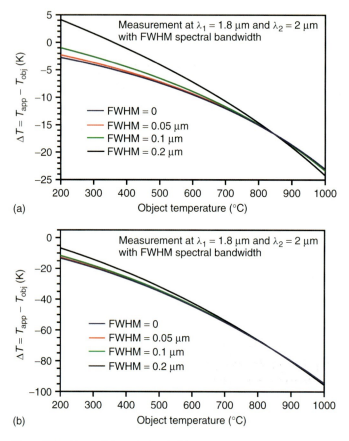

Figure 3.11 Error of the calculated object temperature using the signals at 1.8 and 2 μm as in Figure 3.10 with additional emissivity ratio $\varepsilon(2\ \mu m)/\varepsilon(1.8\ \mu m) = 1.01$ (a) and 1.05 (b).

The apparent object temperature is strongly affected by a spectral emissivity dependence. The error of the temperature calculation increases with object temperature. This behavior is caused by the decreasing slope of the Planck function for increasing object temperatures, that is, for increasing values of λT_{object}.

Figure 3.11 illustrates that the most important single error source in dual-wavelength pyrometry or thermography is the possibility of nongray bodies. Deviations of 5% can easily lead to temperature errors of the order of 10%. This stresses the fact, that whenever accurate object temperature measurements using dual-wavelength measurements are needed, one should fulfill the following requirements:

- Reflected background radiance should be at least smaller than 1% of the respective object radiance.

- Wavelengths should be in the SW spectral region, preferably ≤ 2 μm, in order to use the Wien approximation.
- Both filters should be narrow and have similar transmission and width.
- Wavelengths should be close to each other to reduce the probability of ε-changes of gray looking bodies.
- Never use the method for selective emitters such as plastics unless IR – spectra are well known and filters selected accordingly. Whenever IR – spectra are measured, they should refer to the same temperature region in order to avoid $\varepsilon(T)$ dependencies.

3.2.2.5 Ratio versus Single Band Radiation Thermometry

Ratio thermometry was motivated by eliminating unknown emissivities and still measuring object temperatures. The difficulties and restrictions of ratio thermometry, however, suggest the need to define the criteria when ratio thermometry should be used, that is, is advantageous to conventional thermometry if the emissivity is not accurately known.

We assume high object temperatures and short wavelength measurement, that is, that the Wien approximation can be used and background reflections can be neglected. The temperature measurement error for the ratio method has been discussed above. To estimate the measurement error for the single-band measurement, Eq. (3.7) can be used. We assume a known emissivity except for an uncertainty factor f_ε. The object emissivity can be written as $\varepsilon_{\text{obj}}(\lambda, T_{\text{obj}}) = f_\varepsilon \cdot \varepsilon_o(\lambda, T_{\text{obj}})$. Using Eq. (3.7), we can calculate the object temperature:

$$\frac{1}{T_{\text{obj}}} = \frac{1}{T_{\text{app,single}}} + \frac{k}{hc} \lambda \ln f_\varepsilon \quad (3.19)$$

with

$$\frac{1}{T_{\text{app,single}}} = \frac{\lambda k}{hc} \ln \left(\frac{\varepsilon_o(\lambda, T_{\text{obj}}) \, 2hc^2}{\lambda^5 L_{\text{incident}}(\lambda)} \right) \quad (3.20a)$$

The ratio measurement will have equal or better performance if

$$\lambda \left| \ln f_\varepsilon \right| \geq \frac{\lambda_1 \lambda_2}{(\lambda_2 - \lambda_1)} \left| \ln \left(\frac{\varepsilon_{\text{obj}}(\lambda_1, T_{\text{obj}})}{\varepsilon_{\text{obj}}(\lambda_2, T_{\text{obj}})} \right) \right| \quad (3.20b)$$

For our specific example of ratio measurement at $\lambda_1 = 1.8$ μm, $\lambda_2 = 2$ μm, and a single band measurement at $\lambda = \lambda_2 = 2$ μm we get

$$\left| \ln f_\varepsilon \right| \geq 9 \left| \ln \left(\frac{\varepsilon_{\text{obj}}(\lambda_1, T_{\text{obj}})}{\varepsilon_{\text{obj}}(\lambda_2, T_{\text{obj}})} \right) \right| \quad (3.21)$$

If this relation is fulfilled, the ratio measurement will be equal or more accurate than the single band measurement. Figure 3.12 depicts the limits for f_ε (single band temperature measurement at $\lambda = 2$ μm) dependent on the ratio of the object emissivities for the two-color measurement at $\lambda_1 = 1.8$ μm and $\lambda_2 = 2$ μm.

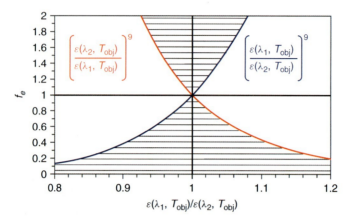

Figure 3.12 Limits of the emissivity uncertainty factor f_ε for one-color temperature measurement at $\lambda = 2\,\mu m$ dependent on the object emissivity ratio for a ratio temperature measurement at $\lambda_1 = 1.8\,\mu m$ and $\lambda_2 = 2\,\mu m$. The two-color temperature measurement exhibits a higher accuracy within the shaded area. Outside, the one-color measurement makes more sense.

3.2.2.6 Application of Two-Color Thermography

Figure 3.12 depicts an experimental demonstration of a two-color ratio measurement for thermal imaging. Two pieces of aluminum foil were fixed on a heating plate (Figure 3.13a). The surface temperature was also measured with a foil temperature sensor Pt100 to about 212 °C. For the radiometric measurement, an AGEMA THV900 camera (2–5.6 μm InSb, Stirling cooled) was used. Owing to the low emissivity of the aluminum, the thermal image (Figure 3.13b) exhibits a much lower apparent temperature of about 52 °C for the aluminum pieces. For the ratio thermography, two standard filters at 4.25 μm (for flame temperature measurements) and 3.9 μm (for measurements through flames) have been used (Section 2.4.4). Images of the heating plate with the aluminum pieces were taken in both spectral channels using the spectral filters. The resulting correct object temperature was calculated pixel by pixel using the Wien approximation as discussed above. Temperatures of about 200–240 °C are found for the aluminum pieces. This gives the corrected thermal image (Figure 3.13c). Owing to the low aluminum emissivity and the limited spectral range (large NETD values), the thermal image of the aluminum pieces exhibits a lot of noise. Although the result is not perfect, it shows that the ratio procedure using two wavelength measurement can help eliminate the influence of the object emissivity for the temperature measurement.

3.2.2.7 Extension of Ratio Method and Applications

An extension of dual-wavelength ratio radiation temperature measurement is the multiwavelength radiation thermometer. The temperature is determined from the ratios of three or more signals, taken at the same temperature but at different wavelengths. In this case, it is usually assumed that the problem of the wavelength dependence of emissivity can be solved by describing the logarithm of

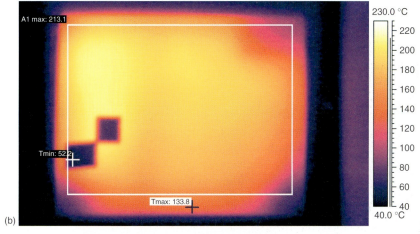

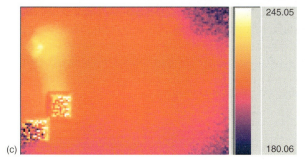

Figure 3.13 Experimental demonstration of two-color or ratio thermography. (a) Vis image of a heating plate with two aluminum pieces and the contact temperature sensor. (b) Thermal image of the heating plate at a temperature of about 212 °C in the MWIR range 2–5.6 µm. (c) Corrected thermal image calculated from thermal images taken at 4.25 µm at 3.9 µm applying the ratio procedure assuming the Wien approximation of Planck's radiation law. (Image courtesy: SIS Schönbach Infrarot Service.)

the emissivity as a series expansion in wavelength [8]. However, the sensitivity of the multiwavelengths radiation measurement to errors increases rapidly with the number of parameters in the emissivity model, that is, the number of wavelengths. So, in practice, the use of such thermometers cannot be regarded as a general solution to the emissivity problem, and very significant temperature errors can result even for materials that differ only slightly from gray-body conditions [9]. A critical analysis of the potential of multiwavelength radiation thermometry to derive true temperatures is given in [10]. To improve the accuracy of two-color and multicolor temperature measurements new methods using laser absorption [1] or relative laser reflectometry [11] have been developed.

In spite of the difficulties concerning the temperature accuracy, ratio radiation thermometry is being successfully applied in industrial processes for temperature measurement, for example, in steel production [12, 13] or in-process measurements of semiconductor manufacturing [14]. Also, dual-wavelength IR imaging is applied in automated on-line control [15].

Sometimes dual-band analysis combining measurements in the MWIR and LWIR region are recommended [16]. The LWIR operation can optimize the detection at lower object temperatures or in the presence of dust, fog, or smog. The MWIR can enhance the performance at higher object temperatures and higher humidity.

Correct temperature measurements can, however, only be successful if the emissivity exhibits a complete spectrally independent behavior and has the same value in both bands. Obviously, the large differences in wavelengths can more easily lead to changes in emissivity. Therefore, with respect to the real material properties, such dual-band measurements will be only successful for a few applications [17]. This technique is mostly used for qualitative analysis using differences in the radiative properties of the materials within the different spectral regions. The comparison and combination of thermal images taken in two different spectral bands can, for example, improve the defect analysis (such as detection of corrosion damage and delamination problems [18] or building material and conservation performance [19]).

3.2.3
Multi- and Hyperspectral Infrared Imaging

3.2.3.1 Principal Idea

Multispectral images capture image data at a variety of specific wavelength bands across the electromagnetic spectrum. A distinction is made between multispectral and hyperspectral imaging depending on the number of bands (up to 100 for multispectral and many more than 100 – usually a continuous spectrum – for hyperspectral imaging) used (Figure 3.14).

Multispectral imaging uses discrete spectral bands that are mostly separated, for example, by the use of spectral filters or detectors with different spectral sensitivity distributions.

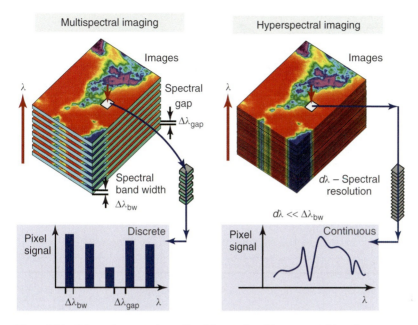

Figure 3.14 Schematic comparison of multispectral and hyperspectral imaging.

In contrast, hyperspectral imaging uses dispersive or interferometer instruments, which, in principle, allow detection of continuous spectra. Available technologies of IR spectrometry provide systems with excellent spectral properties. They essentially allow sampling the full continuum of a predefined wavelength region with excellent spectral resolution. The data are stored in so-called three-dimensional data cubes as depicted in Figure 3.14. One cube refers to a given time and consists of a series of separate IR images as a function of wavelength. Two dimensions are given by the spatial resolution of the IR camera (two-dimensional focal plane array sensor) and the third dimension is given by a wavelength or wavenumber scale generated by the spectrometer. In spectrometry, wavenumbers are often used instead of wavelength. The wavenumber is the reciprocal wavelength and is therefore proportional to photon energy.

Each separate image represents the spatial radiance distribution within a narrow spectral band, whose width is determined by the spectral resolution of the instrument.

Therefore, each single pixel contains an IR spectrum, that is, contains the same information as a point measurement with a nonimaging IR spectrometer.

In practice, two types of different spectrometer arrangements are used – the slit-based and the interferometer-based systems. Both systems use FPA detectors, however, the ways in which they use the detectors are different. In slit-based systems, single lines of the array are used (such detector linear arrays are used in line scanners), that is, one direction, for example, the horizontal, provides spectral information along a line of an object. Adjacent lines of the array receive signals of

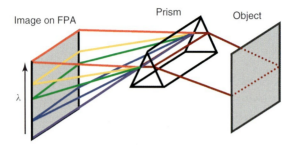

Figure 3.15 Principle setup of a slit-based system with a prism.

the same object line, but from a different spectral region. This is accomplished by using prismatic dispersion elements or a grating [20, 21]. The principal setup with a prism is shown in Figure 3.15.

The FPA detector therefore records spatial information along a line and spectral information in the pixel rows perpendicular to this line. Two-dimensional spatial images are recorded similarly by line scanners (Section 2.4), that is, spatial information as the second coordinate is recorded sequentially either due to a relative increment between imager and object or, if there is no possibility to move the object or the imager, by using an additional scan mirror to shift the image across the slit [20]. The number of spectral bands will be given by the number of pixels in one FPA row [20].

The second type of hyperspectral imagers uses the concept of interferometer-based spectrometers, which is a well known and extremely successful technique in IR spectrometry. The combination of IR imaging with the so-called FTIR (Fourier Transform IR Spectrometry) gives the most powerful quantitative measurement devices available in IR imaging with regard to spectral resolution [22, 23]. The IR hyperspectral imager depicted in Figure 3.16 uses a Michelson-type interferometer to spectrally modulate the object signal on a detector array with a 320×256 pixels format. The use of InSb and MCT focal plane arrays allows the operation within the extended MWIR (1.5–5 µm) and the LWIR (8–11 µm). The spectral resolution is defined by the spectrometer and can be varied within the range $0.25-150 \text{ cm}^{-1}$. The radiometric accuracy of the imager allows a temperature accuracy of better than 1 K after a calibration procedure.

3.2.3.2 Basics of FTIR Spectrometry

In the following, a brief, simplified description of the FTIR principle is given in order to get an understanding of hyperspectral IR imaging.

Most FTIRs are based on a Michelson interferometer as shown in Figure 3.17.

The interferometer consists of a radiation source, a beam splitter, a fixed and a movable mirror, as well as a detector. The movable mirror can be moved forward and backward very precisely. The incident radiation is split into two beams at the beam splitter. One beam is transmitted to the movable mirror and the other is reflected to the fixed mirror of the interferometer. After being reflected from the mirrors, these two beams hit the beam splitter for a second time. Again, one will be

Figure 3.16 (a) Hyperspectral imager "FIRST Hyper-Cam" and (b) scheme of setup for an environmental application. (Image courtesy: Telops Inc., Quebec, Canada; *www.telops.com*.)

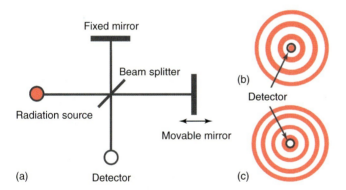

Figure 3.17 Michelson interferometer with a movable mirror to change the optical path difference of the two beams generated by the beam splitter (a). Radiance distribution within the plane of the axially arranged detector for constructive (b) and destructive (c) interference for monochromatic radiation.

partially transmitted and the other is reflected. As a result, two recombined beams are formed, one traveling toward the detector and the other to the radiation source. In the following, we only focus on the beam striking the detector. This beam is due to a superposition of the two light beams that have traveled along different optical paths. The path difference, which depends on the geometrical path difference induced by the movable mirror, leads to a phase difference of the two recombining

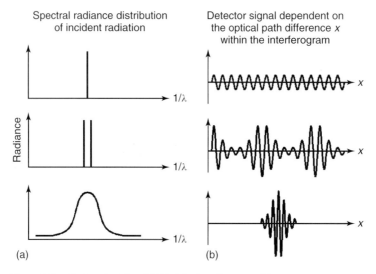

Figure 3.18 Basic principle of FTIR. The interferogram (b) generated from the detector signal during mirror movement depends on the spectral radiance distribution (a) of the incident radiation. For the interferograms the average detector signals (DC part) have been subtracted.

waves of IR radiation. For the two waves in phase, constructive interference results, whereas a phase shift by π (geometrically a shift by $\lambda/2$) leads to destructive interference. This means that the detector will either receive maximum signal (constructive interference) or minimum signal (destructive interference).

To simplify the problem, we assume monochromatic radiation at first. The superposition of the two beams will produce an interference pattern as shown in Figure 3.18 top.

Suppose the detector receives maximum signal. If the movable mirror is moved a distance $\lambda/4$, the reflected radiation has passed an additional distance $2 \cdot \lambda/4 = \lambda/2$, which means that it interferes destructively with the other light beam from the fixed mirror, giving rise to minimum signal. For a mirror distance of $\lambda/2$, the total shift is λ, that is, there will be a maximum signal again. Continuous movement leads, therefore, to a wavelike signal that can be described by a sine function.

This is registered dependent on the movable mirror position, and is called an *interferogram*, representing the raw FTIR data. If the radiation source emits additionally a second wavelength with a minor wavelength difference, we will get an interferogram that is the superposition of the two single interferograms (Figure 3.18 middle). One could proceed to add more wavelengths with successively more complex interferograms. Figure 3.18 bottom shows the result if the radiation source emits a broad continuous spectrum, for example, the one of thermal IR radiation of an object. The resulting interferogram will now be the superposition of the interference contributions of all wavelengths and looks significantly more complex than a single-wavelength interferogram. It contains the information about the source radiance at all wavelengths, the sinusoid frequency of each wavelength in the interferogram being inversely proportional to its wavelength.

The interferogram of a broadband radiation source is characterized by a big spike (centerburst) at zero optical path difference (Figure 3.18). At this point all waves are in phase. This causes a maximum signal contribution resulting in maximum detector signal. If the optical path difference is increased the detected radiance decreases and becomes an oscillating one.

The FTIR raw-data file consists of a large number of data pairs representing the detector signal as a function of different optical path differences during mirror movement.

If FTIR is applied to an imaging system, we need to multiply the number of data pairs by the number of detector pixels of the focal plane array. All wavelengths contributing to the interferograms are analyzed to extract the spectral information. Fourier transformation is applied in detail to the interferogram to get the spectral radiance of all wavelengths forming the observed interferogram. The extremely large number of data points requires a fast algorithm (Cooley–Tukey algorithm, discovered in 1965) and a high computational power [24]. Today, specially designed Fourier microprocessors are used most often to achieve a very high transformation speed. The Fourier transformation converts the interferogram into the spectrum (Figure 3.19). In spectral imaging, these spectra form the third dimension of the data cube.

3.2.3.3 Advantages of FTIR Spectrometers

The importance of interferometers for spectrometry is based on two fundamental advantages compared to dispersive instruments known as *Fellgett* and *Jaquinot* advantages [25]. These advantages, in addition to a few others, also enable the adaptation of an IR spectrometer to an IR camera.

The Fellgett advantage is also called the *multiplex advantage*. In a conventional dispersive spectrometer, the wavelengths are observed sequentially while scanning the grating. In an FTIR spectrometer, the detector signal at any given time already contains the contribution of all wavelengths. This causes a strongly increased signal-to-noise ratio. As a consequence, FTIR can provide a spectrum much faster than dispersive spectrometers can with the same signal-to-noise ratio.

For spectral imaging, this is a crucial point, in particular, if scenes are studied with rapid radiance changes as a function of observation time. The longer the time needed to record a spectrum for a given object point, the larger can be the effect of transient changes. Unfortunately, any time dependence on the radiance received from the source during the recording time of the interferogram (i.e., while the movable mirror is still moving) will lead to mathematical artifacts in the subsequent Fourier transformation. This can cause complete misinterpretation of the spectra (Figure 3.20).

Therefore, fast mirror movement in the interferometer and high frame rates in the spectral imagers are necessary to prevent these problems. This requires low integration times or, in other words, very fast photon detectors. Lowering

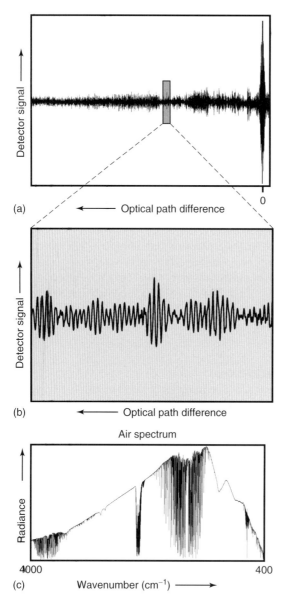

Figure 3.19 Fourier transformation to convert the interferogram (a) into the wavenumber-dependent spectrum (c). For the example, a spectrum of air was chosen. The feature in the middle of the spectrum corresponds to CO_2, the absorption close to the maximum of the radiance curve reflects water vapor. Obviously, it is impossible to guess, just from looking at the interferogram, as to which absorption feature will be present. Besides, the complete interferogram (a), an expanded view to the left-hand side of the centerburst is shown (b) to better resolve the rapid oscillations in the interferogram.

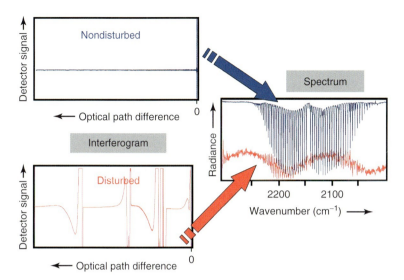

Figure 3.20 Influence of radiance variations during the mirror scan and interferogram data acquisition on the resulting spectrum of a CO gas sample after Fourier transformation. Disturbance was artificially introduced by briefly blocking the signal three times during data recording. The spectrum calculated from the disturbed interferogram does not allow to detect the characteristic CO spectral absorption lines.

the integration times will, however, decrease the detector signal, which means that the signal-to-noise ratio of the radiation detection during acquisition of the interferogram becomes important.

In order to quantitatively characterize the signal-to-noise ratio in thermal imagers with spectrometers, beyond the quantities NEP and NETD, the noise-equivalent spectral radiance (NESR) is introduced [26]. The spectrally dependent NESR determines the minimum incident spectral radiance necessary to reach a signal-to-noise ratio equal to unity at a given spectral resolution and wavelength. The NESR is given in terms of nW $(cm^2 \; sr \; cm^{-1})^{-1}$, depending on the wavelength and the spectral resolution for a single scan (single interferogram measurement).

The Jaquinot or throughput advantage of FTIR spectrometers over dispersive spectrometers is caused by the fact that no slits are used to define the spectral resolution. This causes a much higher detector signal and therefore higher signal-to-noise ratio at each point of the interferogram. However, in reality, there are also some slitlike limits in interferometers. The spectral resolution is affected by the level of collimation of the beams in the two paths of the interferometer. This defines a useful input aperture of the spectrometer especially if a high level of spectral resolution should be reached.

One additional advantage of FTIR spectrometry is the so-called Connes advantage. Usually, a frequency-stabilized laser operating in the visible spectral range, for example, a He–Ne laser, is used as an internal wavelength standard. The corresponding FTIR instruments are self-calibrated spectrometers because the moving mirror position is measured very accurately from the laser interferogram. The

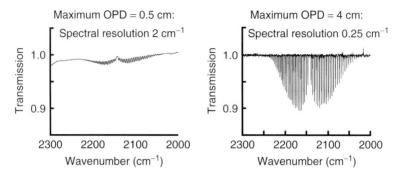

Figure 3.21 Influence of the maximum optical path difference in the interferometer on the spectral resolution of FTIR spectrometry using the example of the rovibrational MWIR absorption spectrum of the CO molecule

laser wavelength or frequency stability determines the wavelength accuracy of the FTIR.

Another advantage of FTIR spectrometers is the achievable spectral resolution. As a first approximation, the spectral resolution of a FTIR is given by the reciprocal of the maximum achievable optical path difference in the Michelson interferometer. For example, a spectral resolution of $2\,\text{cm}^{-1}$ requires an optical path difference of at least 0.5 cm, whereas the typical best resolution of $0.25\,\text{cm}^{-1}$ requires path differences of at least 4 cm (which, of course, results in a longer time for recording the spectrum). Figure 3.21 depicts an example of analyzing a CO sample with poor ($2\,\text{cm}^{-1}$) and good ($0.25\,\text{cm}^{-1}$) spectral resolution. The center of the absorption band at a wavenumber of about $2150\,\text{cm}^{-1}$ corresponds to a wavelength $\lambda = 1/(2150\,\text{cm}^{-1}) = 4.65\,\mu\text{m}$.

3.2.3.4 Example of a Hyperspectral Imaging Instrument

Imaging spectrometers have recently become commercially available for general scientific use.

Both spatial and spectral information of an object scene can be captured simultaneously by IR hyperspectral imagers. Today this technology is applied to military, airborne (combustion processes), research, and environmental (chemical agent detection and investigation) applications [26–28]. As a unique investigation tool for identification of chemical substances or constituents of gas mixtures and solids, IR hyperspectral imaging will most probably find many additional application fields. Figure 3.22 depicts the detection and identification of a cloud consisting of a SF_6/NH_3 gas mixture with a LW (8–11 μm) hyperspectral camera. The temperature distribution of the object scene is shown in a gray scale. The SF_6 and NH_3 identified from their spectral absorption signature (see Chapter 7 and spectra in Appendix 7.A) are depicted in different colors. The data were captured at an image size of 320×128 pixels and a spectral resolution of $4\,\text{cm}^{-1}$ [27]. A complete data cube was

Figure 3.22 Detection and identification of constituents in a gas mixture using hyperspectral imaging. Different spectral channels for the absorption features of NH_3 (yellow) and SF_6 (violet) easily allow to distinguish several gas species simultaneously. In addition, the temperature of the nongaseous opaque object is indicated by the gray scale. (Image courtesy: Telops Inc., Quebec, Canada; *www.telops.com*.)

measured within 5 s. This even allows to detect the cloud displacement in nearly real time (frame rate, 0.2 Hz).

3.3
Superframing

The lower sensitivities of bolometer cameras and QWIP cameras with regard to systems with photon detectors have an important consequence in relation to the dynamic range that can be recorded for given integration times. As discussed in Section 2.4.2, bolometer cameras have fixed integration times that are related to the frame rates, whereas photon detectors allow selection of much lower integration time from about 10 μs to several milliseconds.

Figure 3.23 depicts two typical signal responses for a low- and a high-sensitivity detector with well-defined spectral bands versus object temperature for a defined integration time. The signal is usually represented by a 12–14 bit number, that is, a count rate. This signal is almost linearly proportional to the in-band radiance, that is, the radiance detected within the detector spectral band. On the other hand, the in-band radiance depends nonlinearly on object temperature (integration of the Planck function across a chosen wavelength band of detector). Therefore, the detector signal depends nonlinearly on object temperature as is shown in the schematic plot of Figure 3.23. For very low temperatures, the signal is below the noise level. For ranges between about 10–20% and 80–90% of the maximum signal, there is a well-defined relation between object temperature and detector

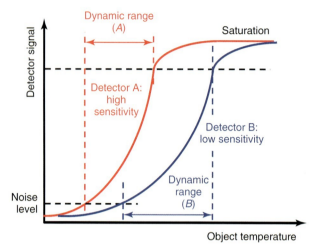

Figure 3.23 Schematic plot of the object-temperature-dependent detector signal for low- and high-sensitivity detectors, respectively.

signal. This region may be used for quantitative analysis and defines the dynamic range of the detector for the given integration time. For higher temperatures, the detectors will be saturated.

Both detectors in Figure 3.23 are chosen to start with their response at about the same object temperature level. Obviously, the lower sensitivity detector has a larger dynamic range, that is, a larger object temperature range can be simultaneously recorded. Photon detectors usually have smaller dynamic ranges. This is a well-known fact for users of MW and LW systems: the temperature ranges of bolometer cameras are usually larger than the one from photon detector cameras (Section 2.4.2). In order to deal with the obvious disadvantage of smaller temperature ranges for photon detector cameras, a method to extend the dynamic range, called *superframing*, has been introduced [29].

3.3.1
Method

Superframing uses the fact that photon detector response curves depend on integration times. First, they are shifted to a higher temperature, and, second, the slopes of the response curves decrease for smaller integration times. Figure 3.24 schematically depicts three response curves of the same detector for three different integration times.

The response for the longest integration time (red curve) is steepest and only has a small dynamic range. However, the signals close to the saturation limit can be easily recorded well within the dynamic range of the detector response with a slightly smaller integration time (black curve). This response curve has a dynamic range that partially overlaps with the one of longest integration time.

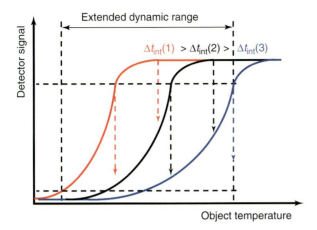

Figure 3.24 Response curves of a detector for different integration times.

For sufficiently large object temperatures, the detector will again be saturated for the chosen integration time but it is also possible to record the object scene with a still smaller integration time (blue curve). Each integration time will have its own dynamic range and within this range, a quantitative determination of object temperature is possible, once the detector is calibrated for the respective integration time. Overall, the recording of the same object scene with a certain number of differing integration times (e.g., up to four) will increase the overall dynamic range. It is then possible to combine the temperature information from all integration time measurements to receive one single result with a very much higher dynamic range.

Such a procedure is useful whenever an object scene contains sources with very large temperature differences, the span of which is beyond the temperature region within the dynamic range for the chosen integration time. Besides using IR images of the same object scene for different integration times (typical use for photon detector cameras), it is also possible to use superframing for any camera by combining IR images of the same scene recorded for different temperature ranges, for example, by inserting a filter.

If integration time is changed using, for example, four preset times, images as a function of time are recorded with a fourth of the possible image frame rate with single integration time. As an example, a camera operating at maximum frame rate of 120 Hz and using four subframes would achieve a maximum superframe rate of 30 Hz due to the four subcycles recorded for each full frame of the object scene.

Figure 3.25 illustrates how a superframe image is practically formed from a series of IR images.

The four subframes show images of the same scene, here the wheel of a car with hot brake disc behind. Subframe 1 is almost entirely saturated with exception of background around the wheel and a few parts between brake disc and outer wheel. The subsequent subframes (recorded with either shorter integration time or additional filters) allow to observe more and more details of the hot parts of the

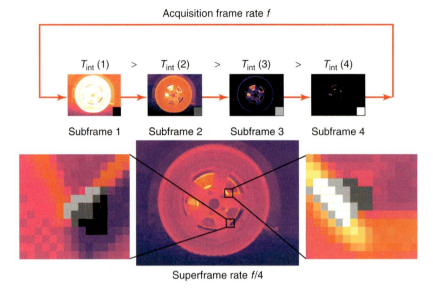

Figure 3.25 Scheme for generation of a superframe image from four subframes (see text for details). The method is also known as *dynamic range extension*.

scene, and, finally, subframe 4 only detects the hottest parts of the scene without saturation. The superframe image, shown below is a superposition of the subframes generated postrecording using the following algorithm: the procedure starts with the subframe representing the lowest measurement range. If a pixel in the first subframe is saturated, the corresponding pixel of the next subframe is analyzed. If its pixel signal is within the dynamic range of the detector, its signal representing radiance or via the calibration curve the temperature is used for building up the superframe. If the pixel is saturated in the second subframe as well, the algorithm analyzes the third subframe, and so on. This procedure is repeated for all cycles. The algorithm results in one superframe for the chosen pixel per cycle. The procedure is performed for all pixels and the resulting superframe image will have high contrast and wide temperature range offering an increased dynamic range that corresponds to a signal depth of 18–22 bit compared to the typical 14 bit of single images. In Figure 3.25, two enlarged image sections of $15 \times 13 = 195$ pixels are shown, which directly illustrate how various subframes contribute to the resulting superframe image. In each small image, a smaller subset of about 30–35 pixels was selected and replaced by either black (subframe 1), dark gray (subframe2), light gray (subframe 3), or white (subframe 4) colored areas. Consider, for example, the image section to the right. No part in it is saturated: the yellow colors resembling high temperatures were taken from subframe 4, indicated by the white pixels, the medium temperatures, represented by orange stem from subframe 3 and the lower temperatures indicated by pink were taken from subframe 2. Similarly, the left-hand image section uses only pixels from subframes 1–3. Superframing operates well if the object scene does not change during the recording of the subframes, that is, during one recording cycle. Therefore, only high frame rate IR cameras can be used for the superframing of temporarily dynamic scenes with large brightness dynamic range.

3.3.2
Example for High-Speed Imaging and Selected Integration Times

Figure 3.26 depicts FLIR SC 6000 thermal images of a propeller-driven aircraft "Beachcraft King Air" for two different integration times.

The image in Figure 3.26a was captured with a long integration time of 2 ms according to a temperature range of about 20–80 °C. A signal saturation for the hot parts of the aircraft, especially the exhaust fumes, occurs. If the integration time is reduced to 30 µs the range of the temperature measurement is changed to 80–300 °C. The image in Figure 3.26b only depicts the temperatures of the hot parts of the aircraft and some of the hot exhaust gases.

Figure 3.27 depicts the result of the superframing procedure using four subframes, the two shown in Figure 3.26 as well as those with integration times of 500 and 125 µs. This superframe presents an increased dynamic range of 20–200 °C without any loss of temperature resolution compared to each individual subframe.

3.3.3
Cameras with Fixed Integration Time

As mentioned above the superframing algorithm can also be used for standard IR cameras, for example, equipped with bolometer FPA by using filters to change the temperature range. For quantitative measurements, the object scene must, however, be characterized by steady-state conditions during the subframe recording. Figure 3.28 depicts a stationary object scene with temperatures in the range of 16–270 °C. The measurement ranges of a LWIR bolometer camera FLIR SC 2000 used to grab these images are 40–120 °C (Figure 3.28a) and 80–500 °C (Figure 3.28b). The hot parts of the image (a halogen lamp with 270 °C maximum temperature, a blackbody emitter at 260 °C and a lamp with 150 °C maximum temperature) are above the measurement range in Figure 3.28a and the cold objects are below the measurement range in Figure 3.28b.

A superframe is generated by using the described process. In Figure 3.28a, all temperatures above 120 °C have been replaced by the temperatures of the other image. Figure 3.29 depicts the superframe condensed from the possible extended dynamic range of 40–500 °C to a temperature range 16–270 °C, which is adapted to the object scene. This image has the same temperature resolution as in the subranges.

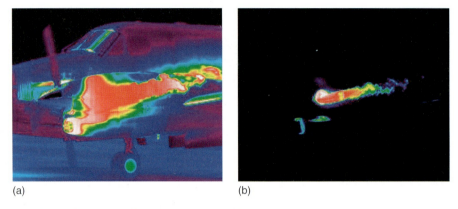

Figure 3.26 Thermal images of a Beachcraft King Air grabbed with a FLIR MW SC 6000 camera at 2 ms (a) and 30 μs (b) integration time. (Image courtesy: FLIR Infrared Training Center.)

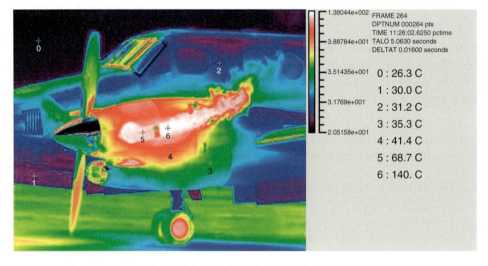

Figure 3.27 Superframe of four subframe images with 2-ms, 500-μs, 125-μs, and 30-μs integration time. (Image courtesy: FLIR Infrared Training Center.)

3.4
Processing of IR Images

IR images represent a spatial radiance distribution incident on the detector array. The first image processing is done inside the camera firmware by calculating the temperature of the object scene from the detector signals using the camera calibration parameters and various user-defined parameters, for example, emissivity, ambient temperature, humidity, measurement distance. The result is a false

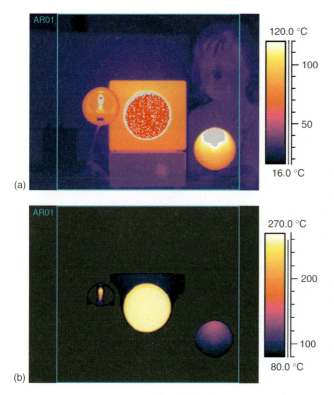

Figure 3.28 Thermal images of a high dynamic scene. (a) Image captured using the measurement range 40–120 °C. Temperatures above the upper limit of the measurement range (saturated pixels) are shown in red. (b) Image captured using measurement range 80–500 °C.

color image of the temperatures, which has to be interpreted by the observer. Further, postrecording image processing can, on the one hand, help improve the pictorial information for human interpretation of IR images. On the other hand, it can prepare images for the measurements of features. More advanced procedures of image processing will use the image data for autonomous machine perception.

3.4.1
Basic Methods of Image Processing

Some basic methods of image processing, for example, image fusion, image building, or image subtraction can be applied using the typical R&D software in IR cameras. Moreover, numerous programs (e.g., Adobe Photoshop) can also be used for postprocessing of images such as digital detail enhancement.

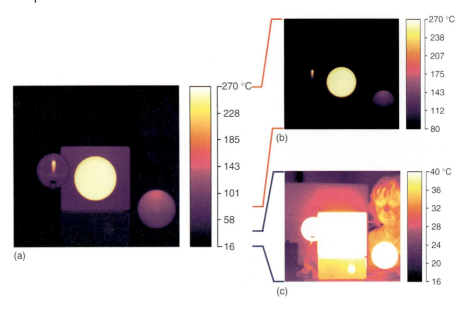

Figure 3.29 (a) Superframe generated by image postprocessing from the marked regions of the subframes depicted in Figure 3.28. (b,c) Superframe depicted with two different level and span parameters to demonstrate the increased signal dynamic range.

3.4.1.1 Image Fusion

In general, image fusion is a process of combining relevant information from different images into a single image [30]. In thermography, image fusion refers to the combination of IR images with Vis images from megapixel cameras. The IR image can be either superimposed on the Vis image (picture-in-picture overlay) (Figure 3.30) or merged with the Vis image (see below). This significantly enhances the information content within a single image. For instance, the temperature distribution depicted by the IR image will sometimes not reflect the structural properties of the investigated object. Superimposing the IR image on the visual image can also solve the problem of identifying observed temperature anomalies with features of the object. Overall, superimposition of IR images on visible images simplifies the interpretation considerably.

Some IR cameras offer image fusion not only by superimposing IR images on visible ones but also by merging visual and IR images. This merging function operates in real time [31] and offers the possibility to highlight object areas in Vis images that are above or below a predefined temperature threshold, or areas within a predefined temperature interval. Such a thermal image fusion can also be done in a postrecording mode using image-processing software. Figure 3.31 depicts some examples for the merging of visual and IR images.

(a) (b)

Figure 3.30 (a) FLIR P660 camera with 640 × 480 pixels IR resolution and 3.2 Megapixel digital camera (above IR lens) allowing IR image overlay on Vis images. (b) Example overlay image of an indoor inspection of a roof area. (Image courtesy: FLIR Infrared Training Center.)

3.4.1.2 Image Building

A typical practical problem of an IR inspection is that the object scene is sometimes so large that the field of view of the available camera optics can only capture part of it. This can, for example, be due to constraints regarding possible object distances. Alternatively, one may be able to get complete images of extended objects, however, with the disadvantage of low spatial resolution due to the required large object distance. Both these problems can be solved with image-building software tools. This postrecording image-processing software has been developed, in particular, at the request of the building inspection industry and is an image-assembling software that allows to combine and align different images to create a high-resolution overview image. Obviously, the method only works for stationary conditions, since multiple repositioning of the camera can lead to quite a large time span for the recording of a single large image.

We now discuss this application in more detail. The camera FOV is determined by the f-number of the camera optics (many standard lenses having 20–24° FOV). The spatial resolution can be simply calculated by dividing the FOV by the linear number of FPA pixels and is represented by the IFOV (Section 2.5.3). If the object scene to be analyzed is too large, for example, an elongated building as shown in Figure 3.32, compared to the camera FOV, the use of an additional wide-angle lens is possible. It increases the IFOV by a factor while at the same time reducing the spatial resolution by the same factor. The resolution reduction can cause measurement errors and misinterpretations for small object sizes (Section 2.4.4). This problem is avoided by capturing a number of thermal images which cover different parts of the object scene. All these frames offer spatial resolution according to the detector IFOV. The thermal images depicted in Figure 3.32 were recorded using a camera with 1.3 mrad IFOV. Considering the object distance of about 40 m, the images will exhibit a spatial resolution of about 5 cm. The combined image exhibits the same

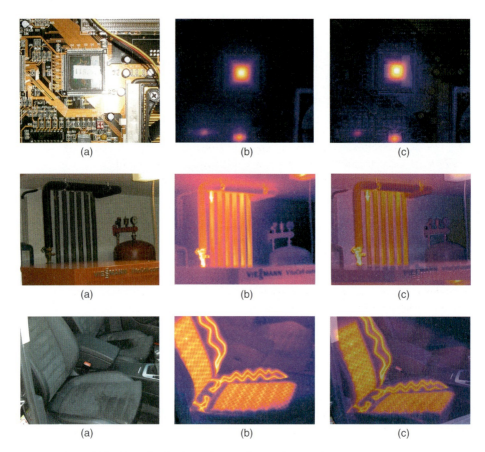

Figure 3.31 Examples for thermal image fusion. (a) Visual image; (b) IR image; and (c) fusion image. Top: Electronic board under load, middle: inspection of pipes of a heating system, and bottom: seat heating of a car.

Figure 3.32 High-resolution overview image of a building with 932 × 230 pixels combined from three separate thermal images with 320 × 240 pixels per image.

spatial resolution. If wide-angle camera lens had been used to capture the same horizontal object size, the FOV would have increased by a factor of 3. The IFOV would have been about 4 mrad and the image would offer only a spatial resolution of 15 cm. This difference in spatial resolution is shown in Figure 3.33. It depicts an example of the same object scene that was first recorded with a wide-angle lens with 45° FOV (Figure 3.33a). The same scene is shown in Figure 3.33b as a combined image of six single IR images recorded with a telelens with 12° FOV. The difference in spatial resolution is obvious. If a quantitative analysis is requested, the high-resolution combined image would be much better, whereas the wide-angle image could cause wrong interpretations due to the low spatial resolution.

IR image-building software usually offers fully radiometric composite images that can also be analyzed with the typical analysis software used for single images [31].

Image building is suited for large static object scenes and can be used for a large variety of applications, for example, inspection of buildings, electrical systems at transformer stations, panels, or electronic boards.

As an additional example Figure 3.34 depicts a composite image of an electronic board. For the analysis, a 34/80 close-up lens of a 320 × 240 pixels LWIR camera was used to obtain a high spatial resolution of about 100 μm. The image field of a single image equals to 34 mm × 25 mm at an object distance of 80 mm.

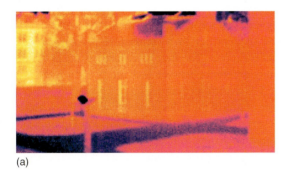

(a)

(b)

Figure 3.33 Thermal image captured with a 45° lens (a) compared to the combination of six individual images using a 12° lens and image building software (b).

Figure 3.34 Composite IR image (864 × 602 pixels) of an electronic board with load. The composite image has been assembled from 3 × 3 single images. Object size of a single image of 34 mm × 25 mm (320 × 240 pixels).

3.4.1.3 Image Subtraction

Image subtraction is a method of image processing that can, first, help to identify small temperature changes, second, suppress the influence of thermal reflections and third, if applied consecutively, it can visualize time derivatives, for example, those due to transient phenomena.

Image subtraction can be done in two ways: by subtracting a reference image from each image of a recorded sequence or by a consecutive process, that is, subtracting each image from its precursor. The first procedure results in a new sequence showing the temperature difference between each image of the sequence and the reference image. Figure 3.35 depicts an example for this procedure using a plate with holes. The temperature of one of the holes is slightly increased by about 0.6 °C using a microheater. It is very difficult to find this hot spot in in Figure 3.35a (Result) due to the small temperature change compared to the temperature span in the IR image. Figure 3.35b (Baseline) shows the temperature distribution before the hole is heated up. The comparison of the baseline image and the result image does not really help find the position of the hot spot but the subtracted image

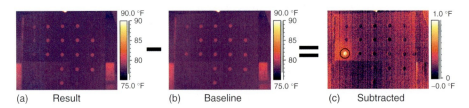

Figure 3.35 Thermal images of a plate with holes. After grabbing the baseline image (b), one of the holes is heated by a microheater. The result image (a) shows the resulting temperature distribution. The baseline image is subtracted from the result image. The resulting subtracted image (c) clearly exhibits the temperature difference. (Image courtesy: FLIR Infrared Training Center.)

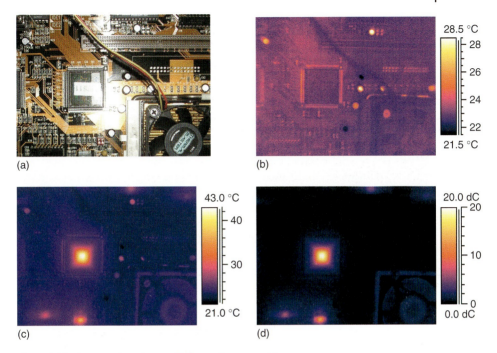

Figure 3.36 Image subtraction applied to a thermographic analysis of an electronic board. (a) Visual image, (b) infrared image at ambient temperature without applied voltage (reference image), (c) infrared image of the electronic board with applied voltage (source image), and (d) subtracted image (source image−reference image).

(Figure 3.35c) clearly depicts its position. The temperature is changed by about 1 °F (5/9 °C). This example demonstrates that image subtraction is suitable to detect very small temperature changes in object scenes.

Image subtraction can also be used in applications where reflections of ambient temperature should be suppressed. Figure 3.36 depicts an electronic board. From the visual as well as from the IR image, high-reflecting electronic devices are clearly seen. If the board is operated, these reflections will add to the emission of the current-induced self-heated electronic devices. Therefore, the simple IR image does not allow to correctly estimate the temperature change under load.

The suppression of background radiation reflections can be understood from Eq. (2.36). On neglecting atmospheric influences, the detected radiance will consist of the emitted object radiation and the reflected radiation of the background:

$$L_{\text{det}}(x, y) = \varepsilon(x, y) L_{\text{object}}^{\text{bb}}(T_{\text{object}}) + [1 - \varepsilon(x, y)] L_{\text{amb}}(T_{\text{amb}}) \quad (3.22)$$

The $\varepsilon(x, y)$ equals the object emissivity at the position (x, y). If we calculate the difference of detected signals detected for two different object temperatures but the same ambient temperature, the signal is no longer influenced by the reflection:

$$\Delta L_{\text{det}}(x,y) = \varepsilon(x,y)\left\{\left[L^{\text{bb}}_{\text{object}}(T_{2,\text{object}}) - L^{\text{bb}}_{\text{object}}(T_{1,\text{object}})\right]\right\} \qquad (3.23)$$

The subtracted image will only depend on the temperature difference. However, the correct emissivity must be applied to measure the correct temperature difference.

From Eqs. (3.22) and (3.23), it is clear that image alignment is critical for image subtraction. A small misalignment between the source and the reference image will cause different positions (x, y) with different emissivity values and the subtraction algorithm will fail. This is illustrated in Figure 3.37 for two similar IR images for the electronic board with no load. With proper alignment, the subtracted image resembles the image noise, whereas slight misalignments result in presence of additional features although no temperature change occurred during recording.

3.4.1.4 Consecutive Image Subtraction: Time Derivatives

In the second method of image subtraction, consecutive image subtraction is performed. In this case, the difference signal for a pixel (x, y) of the nth image in the sequence $\Delta S(x, y, n)$ is computed via

$$\Delta S(x,y,n) = S(x,y,n) - S(x,y,n-1) \qquad (3.24)$$

Since subsequent images n and $n-1$ are all separated by a time interval Δt, defined by the frame rate of the recording, the consecutive image subtraction represents a calculation of the first derivative of the image in time domains. If this algorithm is applied to the self-heating process of the electronic board components shown in Figure 3.38, the differences of the time-dependent temperature increase become obvious. Figure 3.38 depicts a series of images that represent consecutive subtraction results during the self-heating process if a voltage is applied. From the consecutive image subtraction, the different time constants during the self-heating process are obvious. The component at the bottom of the board exhibits a faster temperature increase than the device in the middle. It reaches stationary conditions much faster during board operation. Therefore, it has already disappeared in the final subtracted images. The temperature of the device in the middle increases on a much longer time scale. At the end of the self-heating process when stationary conditions are reached, all components will disappear in the consecutive subtracted images of the board.

If the consecutive subtraction algorithm is applied a second time to an already consecutive subtracted image sequence, the second derivative of the temperature in time domains will be calculated. It can provide additional information about the time-dependent temperature behavior of the investigated object and is used, for example, in pulse thermography ([32] and Section 3.4).

Consecutive image subtraction can be also used to extract features of moving objects in real time if the object scene and the background are either more or less static [33] or featureless. Figure 3.39 depicts an example of such a system: a water beaker was heated from below resulting in convection patterns at the upper surface (Section 5.3.3). Convection currents have only small temperature

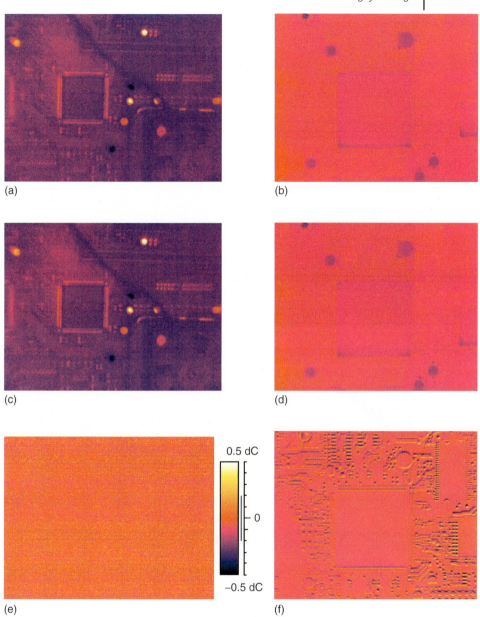

Figure 3.37 Influence of image misalignment on the result of the image subtraction. Source (top) and reference (middle) images for electronic boards with no load. Left: Correct alignment of source (a) and reference image (c) will result in a correct subtracted image (e). Right: Misalignment between the source (b) and reference image (d) will result in an incorrect subtracted image (f).

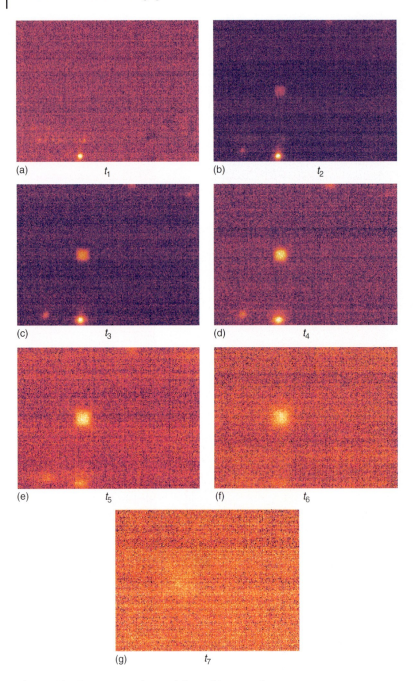

Figure 3.38 Consecutive subtracted thermal images of an electronic board during self-heating due to applied voltage ($t_1 < t_2 < \ldots < t_7$). Transient phenomena with different timescales are observed.

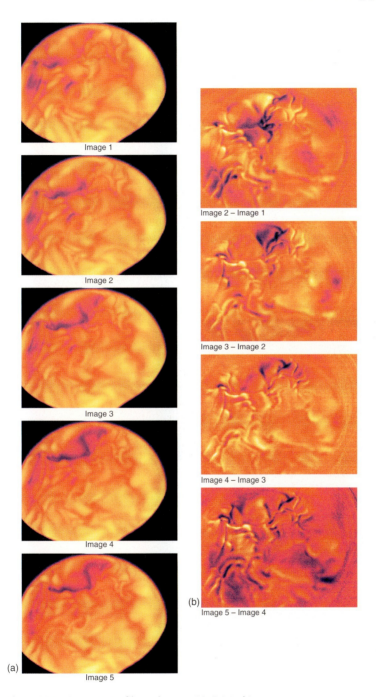

Figure 3.39 Convection of heated water. (a) Original infrared images grabbed every 0.2 s. (b) Images from consecutive subtraction.

differences compared to the rest of the water and features change rapidly. Whereas comparison of two consecutive images with regard to changes is difficult, the subtracted images easily show only the locations where changes occur because of convection currents.

3.4.1.5 Image Derivative in the Spatial Domain

For still, that is, static images, it is sometimes desirable to increase image contrast, for example, by enhancing thermal contours of objects. The corresponding image-processing algorithm calculates spatial derivatives. This can be done by exporting the radiance or temperature data from the IR image into a spreadsheet program. The calculation of the derivative is done in x- or y-coordinate direction by consecutive subtraction of columns or lines, respectively. This procedure will increase the image contrast for image regions with nonzero temperature gradients $\frac{dT}{dx} \neq 0$ or $\frac{dT}{dy} \neq 0$. In Figure 3.40, the original IR image is depicted in comparison to the calculated image derivative in x-coordinate, that is, the horizontal direction. Obviously, the image derivative exhibits enhanced contours of vertical and inclined features caused by the temperature gradients.

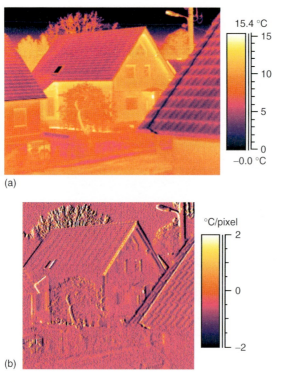

(a)

(b)

Figure 3.40 Comparison of an original infrared image (a) and its image derivative (b) in the spatial domain along the horizontal coordinate.

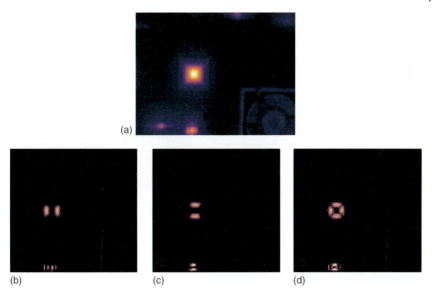

Figure 3.41 An IR image of an electronic board (a) processed to yield spatial derivatives along x (horizontal) coordinate (b) and y (vertical) coordinate (c) and the combination of the two derivative images (d).

However, horizontal features, for example, the roof edge of the house, cannot be seen, as expected. For them to show up the derivative in y-direction must be computed.

Figure 3.41 depicts an example where the derivation algorithm was applied both to x- and y-coordinate directions for an electronic board under load. The derivative images for x and y can be combined to get improved information about the temperature gradients in both directions. The combined image, however, still gives incorrect results at the corners of the self-heated processor because temperature gradients not correctly aligned toward the coordinate axes are shown incorrectly.

Therefore, an improved procedure for calculating the temperature gradients within the thermal image should be applied. Figure 3.42a depicts the mathematical procedure. For each pixel, the signal difference to all eight neighboring pixels is calculated. The maximum difference is used as the new pixel signal. This algorithm is applied to all pixels of the image and results in the derivative image shown in Figure 3.42b. Obviously, the result of this algorithm reflects the contours of temperature changes of objects more accurately.

3.4.1.6 Digital Detail Enhancement

An IR camera signal contains usually a 12–14 bit information, which, in false color, resembles 4096–16 384 different hues. Perception problems occur because of the limitations of human visual recognition. We discuss the gray-scale images first.

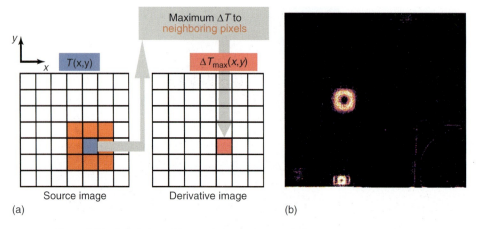

Figure 3.42 Calculation of image derivative in spatial domain by using the maximum temperature difference between neighboring pixels of the image.

A human observer can distinguish only approximately 128 levels of gray equal to 7 bits in an image. If automatic gain control with linear gradation is applied, the human observer cannot detect low-contrast objects within the high dynamic object scene, see, for example, image, Figure 3.43a. The idea is to change the linear image gradation curve to a nonlinear one that is characterized by small changes if large signal changes occur and large changes in regions with small signal changes. This procedure has been applied to the image, Figure 3.43a, resulting in the changed image, Figure 3.43b. Apparently, this nonlinear algorithm enables the observer to detect also the low-contrast object details.

Figure 3.43 also compares gray scale with color representations of the same image with the same content of information. Obviously, color scales are better suited for image enhancement procedures. Nevertheless, gray-scale images are still often used in security and surveillance applications (Section 10.5), probably because they allow a direct comparison to low-light level visual images that are anyhow gray-scale images. For all other purposes, color palettes are superior for image enhancement with regard to visual detection of features by human observers.

One supplier of IR cameras, FLIR Systems, has developed an advanced nonlinear image-processing algorithm that allows to find low-temperature contrast objects in a high dynamic range object scene [31]. The detailed image is enhanced so that it matches the total dynamic range of the original image. As a result of the algorithm, low- as well as high-contrast objects can be seen simultaneously by the human observer. This algorithm is especially important for a number of security and surveillance thermal imaging applications.

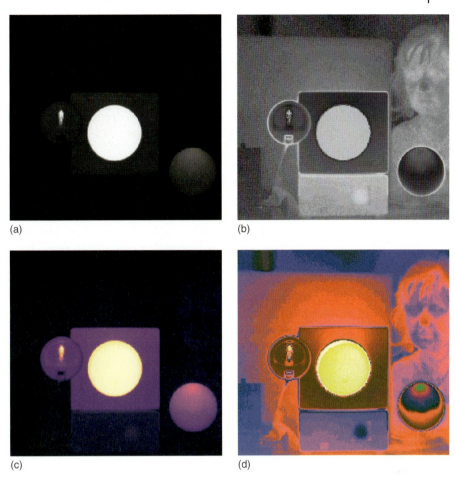

Figure 3.43 Detail enhancement by an application of nonlinear gradation to a high dynamic range infrared image. (a,b) 8-bit gray-level representation of high dynamic range infrared image from Figure 3.29a: (a) with linear image gradation curve; (b) application of nonlinear image gradation curve. (c,d) The same image, represented with an 8-bit color scale (iron) representation: (c) with linear image gradation curve; (d) application of nonlinear image gradation curve.

3.4.2
Advanced Methods of Image Processing

Thermal imaging systems with the goal of process visualization and computer-based process observation have become more and more important in many industrial fields. Experts guess that so far only 20% of all possible applications of industrial image processing are realized. As a consequence, developments in the field of industrial image processing offer enormous potential for fast and sustained growth.

The enhanced use of image processing became possible owing to the fast development of first, thermal imaging systems and, second, high-speed multicore processors. Twenty years ago, image-processing algorithms for a large number of pixels would have lasted several hours or days, whereas, today, a full image inspection consisting of the whole visualization and classification process is finished within a few seconds. This means nearly real-time visualization, which offers the possibility of fast controlling and/or adjusting on-line production, for example, using robotics. The advantages are obvious, in particular, if the visualization is applied in real-time quality control of industrial products.

The most important applications of industrial image processing are the inspection of single work pieces, band processes, recognition of characters, patterns, and codes, and the two- or three-dimensional survey of objects [34–39]. The variety of applications emphasizes the universal character of such image-processing software solutions. In industrial production processes, reliable and fast tools to minimize errors and prevent production breaks are essential to ensure high productivity and quality levels. This is why more and more industrial branches invest in automatic process visualization systems. Another advantage of such automated image-processing systems is the possibility to collect important process parameters in databases for subsequent offline process analysis studies.

The following section gives only a brief general overview of some procedures applied in processing of IR images for automation processes in industry. They are considered to be advanced in this section with regard to IR imaging, since they are usually not included in typical IR image-processing software packages from the manufacturers of IR cameras.

In IR imaging, the basis of every processing model is a thermal image or a series of thermal images. Compared to the object scene, any thermal image already suffers from a loss of information due to the unavoidable transformation of real-world contours into pixels (problem of discretization). A constrained number of pixels build a matrix of signal values, which is the basis of the following mathematical processing steps. The goal is to extract relevant information like identification of defects or given patterns, such that reliable analysis and classification tools can be applied.

As an example, Figure 3.44 depicts typical defects in a glass sheet, which are not detectable in the visible spectrum but clearly seen in the thermal image. The defects are visible in the IR image since any defect in a glass sheet will exhibit a slightly different cooling behavior than the rest of the bulk material, that is, give rise to temperature differences. In addition, cracks in glass will cause slightly higher emissivities due to the cavity effect (Section 1.4.4). Identifying these defects is crucial since they could easily result in failure of the product, for example, breaking of the glass before the guaranteed lifetime is reached. Therefore, in the case of Figure 3.44, the IR images are recorded immediately after the last thermal process during manufacturing. Obviously, if a large number of glass sheets is produced, image-processing algorithms can help analyze images, identify, and classify such defects in real time. In case of defect detection, there will be a notification of

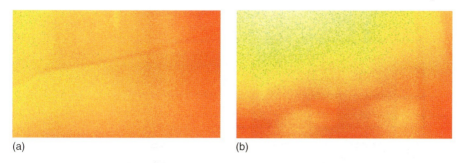

Figure 3.44 Two possible defects visualized by thermal images: (a) a long break in a glass sheet and (b) a couple of hot spots in a glass sheet (Image courtesy: Raytek GmbH Berlin.)

the process operator. Such an automated procedure, hence, constitutes a quality control step.

Advanced image processing for automation or other purposes can be divided into three major parts: first, preprocessing of images (noise reduction); second, segmentation (locate relevant information); and third, feature extraction and reduction. In some cases, feature extraction uses specific pattern recognition tools, which produce output false color images with especially marked defect locations. These can form the basis for quality control criteria.

3.4.2.1 Preprocessing

Perhaps the most important part in any image processing for pattern recognition or image classification is preprocessing of the raw data. On the one hand, it is necessary to reduce noise and eliminate pixel artifacts in the images by applying low-pass filters; on the other hand, phenomena like distortion, twisting, scaling, or translation can occur with regard to the object geometry. This requires the input images to be transformed geometrically.

Noise Reduction Noise in IR images is given by random changes of pixel signals caused by single pixel noise as well as fixed pattern noise (Chapter 2). The filters used to reduce noise are characterized as low-pass filters. The frequency information (low pass means low frequency) refers to the spatial frequency of the system. Definitions are similar to the treatment of diffraction in optics. Fixed pattern noise is characterized by a spatial periodicity referring to the pixel size, which can be regarded as analogous to wavelength λ. Then, $2\pi/\lambda$ is referred to as *wave vector* or *spatial frequency*. A small value λ implies high spatial frequency and vice versa. Obviously, the pixel size is the smallest possible spatial periodicity in IR images, that is, it resembles the highest possible frequency f_{max}. Therefore, fixed pattern noise (as well as single pixel errors) refers to high spatial frequencies. Consequently, it may be reduced by applying low-pass filters, which suppress high frequencies, while at the same time preserving edges and contours. The easiest low-pass filters are just averaging pixels with their neighbor pixels, for example,

208 | 3 Advanced Methods in IR Imaging

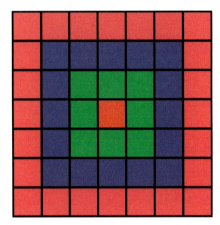

Figure 3.45 Section of pixel array of an IR image. The new value of the signal of the center pixel (red) can be calculated by either averaging its signal values with the 8 nearest neighbors (green, operator: 3 × 3 matrix), with 24 neighbors (blue, operator: 5 × 5 matrix), or 48 neighbors (pink, operator: 7 × 7 matrix).

each pixel may be computed as the average of its value with the one of the 8 nearest neighbors (using neighbors with 1 pixel size distance) or its 26 nearest neighbors (using neighbors with up to 2 pixel size distances) asf. Since the corresponding total filter widths are then either 3 pixels (8 nearest neighbors) or 5 pixels (24 neighbors), the respective frequencies are only one-third or one-fifth of the highest frequency f_{max}.

The application of such averaging filters is mathematically described by matrices. Taking into account only nearest neighbors will lead to (3 × 3) matrices, including second-nearest neighbors to (5 × 5) matrices (Figure 3.45). Owing to limitations of computational power (easy but CPU-intensive calculations), the maximum filter size is usually described by (7 × 7) matrices. For simplicity, we present the principal idea for these and all of the following filters by using (3 × 3) matrices, only.

The averaged pixel signal $\hat{S}(x, y)$ is mathematically evaluated by computing the trace of the matrix product of the averaging operator (3 × 3) matrix **A** and the (3 × 3) signal matrix centered around the pixel at row x and column y (Eq. (3.25)):

$$\hat{S}(x,y) = Tr\left\{\begin{pmatrix} a_{11} & a_{12} & a_{13} \\ a_{21} & a_{22} & a_{23} \\ a_{31} & a_{32} & a_{33} \end{pmatrix} \begin{pmatrix} S_{x-1,y-1} & S_{x-1,y} & S_{x-1,y+1} \\ S_{x,y-1} & S_{x,y} & S_{x,y+1} \\ S_{x+1,y-1} & S_{x+1,y} & S_{x+1,y+1} \end{pmatrix}\right\}$$

$$= a_{11}S_{x-1,y-1} + a_{12}S_{x,y-1} + a_{13}S_{x+1,y-1}$$
$$+ a_{21}S_{x-1,y} + a_{22}S_{x,y} + a_{23}S_{x+1,y}$$
$$+ a_{31}S_{x-1,y+1} + a_{32}S_{x,y+1} + a_{33}S_{x+1,y+1}$$

(3.25)

The colored bars denote the scalar product of the respective row of **A** with the column of the signal matrix. Since the trace is computed, only the indicated three diagonal components of the matrix are needed. From Eq. (3.25) it is obvious that the resulting average value depends critically on the choice of the a_{ij} values of matrix **A**.

Let us come back to noise in IR images. The problem in averaging is the need to preserve edges and contours such that no relevant information in the image gets lost. Several filters are available, the most often used being simple box filters, Gauss filters, or median filters [35]. Regarding preserving edges, the Gauss filter works best, the median filter is second, while less satisfying results are obtained for the simple averaging box filter.

The simple box averaging filter $\mathbf{A}_{box}$ and the Gauss filter $\mathbf{A}_{Gauss}$ are described by the matrices

$$\mathbf{A}_{box} = \frac{1}{9} \begin{pmatrix} 1 & 1 & 1 \\ 1 & 1 & 1 \\ 1 & 1 & 1 \end{pmatrix} \qquad \mathbf{A}_{Gauss} = \frac{1}{16} \begin{pmatrix} 1 & 2 & 1 \\ 2 & 4 & 2 \\ 1 & 2 & 1 \end{pmatrix} \qquad (3.26)$$

According to Eq. (3.25), it is quite easy to interpret the averaging procedure from the form of the matrices. The box filter applied to the signal matrix just computes an average where each of the 9 pixels contributes with the same weight. In contrast, the center pixel has the largest weight for the Gauss filter with the horizontal and vertical nearest neighbor pixels contributing twice as much as the more distant diagonal neighbors.

The median filter for a (3×3) signal matrix centered around pixel (x,y) resembles a procedure where the nine signal values of the pixels are ordered and the middle value, for a (3×3) matrix the fifth value, is defined as result. Because of this procedure, the median value is less sensitive to single value deviations, that is, single pixel errors, than the other filters.

Figure 3.46 (a) depicts the thermal image of a glass sheet with a long break. The images after a Gaussian (b) and a median (c) filtering (both of size 7×7) are shown for comparison.

3.4.2.2 Geometrical Transformations

For comparing a processed object with a reference object (e.g., manufactured products), both have to have identical geometrical properties (position and orientation). But before that, it is appropriate to start with the challenging and exciting part of

Figure 3.46 Comparison of different low-pass filters. (a) Original image (raw-data image). (b) Image after a Gaussian filtering. (c) Image after a median filtering. Both the Gaussian and the median filtering preserve the edge. (Image courtesy: Raytek GmbH, Berlin, Germany.)

defect detection and image classification. Mathematically, the possible geometric variations are described by projective and affine transformations.

Affine transformations consist of a multiplication of the original Cartesian coordinates with a (2 × 2) matrix and a vector addition to represent a translation. They can be described by

$$\begin{pmatrix} x' \\ y' \end{pmatrix} = \begin{pmatrix} a_{11} & a_{12} \\ a_{21} & a_{22} \end{pmatrix} \begin{pmatrix} x \\ y \end{pmatrix} + \begin{pmatrix} t_x \\ t_y \end{pmatrix} \qquad (3.27)$$

and all affine transformations are special cases of that general equation.

The three basic affine transformations are

1. Rotation by α : $\begin{pmatrix} a_{11} & a_{12} \\ a_{21} & a_{22} \end{pmatrix} = \begin{pmatrix} \cos\alpha & -\sin\alpha \\ \sin\alpha & \cos\alpha \end{pmatrix}$

    and $\begin{pmatrix} t_x \\ t_y \end{pmatrix} = \begin{pmatrix} 0 \\ 0 \end{pmatrix}$ \qquad (3.28)

2. Translation by t : $\begin{pmatrix} a_{11} & a_{12} \\ a_{21} & a_{22} \end{pmatrix} = \begin{pmatrix} 1 & 0 \\ 0 & 1 \end{pmatrix}$

    and $\begin{pmatrix} t_x \\ t_y \end{pmatrix} \ne \begin{pmatrix} 0 \\ 0 \end{pmatrix}$ \qquad (3.29)

3. Scaling by s_x, s_y : $\begin{pmatrix} a_{11} & a_{12} \\ a_{21} & a_{22} \end{pmatrix} = \begin{pmatrix} s_x & 0 \\ 0 & s_y \end{pmatrix}$

    and $\begin{pmatrix} t_x \\ t_y \end{pmatrix} = \begin{pmatrix} 0 \\ 0 \end{pmatrix}$ \qquad (3.30)

It is possible to simplify this representation by including the translation within a single (3 × 3) matrix operator:

$$\begin{pmatrix} x' \\ y' \\ 1 \end{pmatrix} = \begin{pmatrix} a_{11} & a_{12} & t_x \\ a_{21} & a_{22} & t_y \\ 0 & 0 & 1 \end{pmatrix} \begin{pmatrix} x \\ y \\ 1 \end{pmatrix} \qquad (3.31)$$

Any change in position or scaling can be constructed by combining these basic affine transformations and applying them to every single pixel in the image.

Problems that can emerge are obvious. An input image with integer pixel coordinates is transformed by an affine transformation into a result image. After transformation back to the input image noninteger coordinates result. Therefore, a loss of information may be caused by the constrained resolution of pixels and the affine transformations. For example, Figure 3.47 depicts the discretization and an affine transformation of an ellipse. In the transformed output image, the original ellipse is scaled, rotated, and translated.

The transformation can therefore result in a new pixel center position between four adjacent pixel centers, (Figure 3.48).

The resulting signal strengths for each pixel are calculated by working in the subpixel area. A simple nearest neighbor algorithm can be applied to calculate

3.4 Processing of IR Images | 211

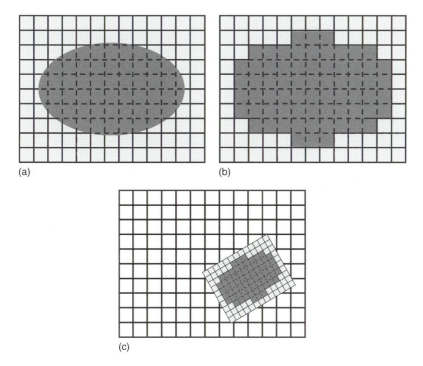

Figure 3.47 Problems caused by an affine transformation. (a) Ellipse with smooth contours. (b) Discretized and scaled ellipse. (c) Transformed ellipse (rotation, translation, scaling).

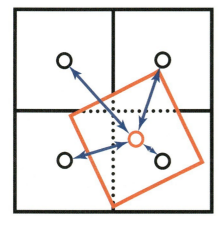

Figure 3.48 Geometry for calculating the signal strength of transformed pixels. The lower right pixel is the nearest neighbor, whose signal will dominate.

the closest pixel. The signal of the nearest neighbor pixel will dominate the signal strength of the transformed pixel. A better interpolation results if the new signal is calculated from signals of all neighboring pixels weighted according to their center distances to the new pixel center (Figure 3.48).

3.4.2.3 Segmentation

Typically, the first step after preprocessing is segmentation of the input data, that is, searching for ROI (region of interest). The main goal is to determine the zones in the image at which the relevant information is located. There are several possibilities of image segmentation like region-growing segmentation, histogram-based segmentation, and edge detection. The decision on which type of segmentation procedure has to be applied depends on the concrete goals of the image-processing procedure. In some cases, only a segmentation between foreground and background using a histogram method is necessary; in others, one may need to locate certain objects by edge detection algorithms.

All pixels that are classified as belonging to a ROI have one or more similar properties. In the thermal image depicted in Figure 3.49, there are, for example, 4 ROIs (one rectangle, three circles). These objects can be segmented very easily from the background using the corresponding histogram and defining a temperature threshold. Alternatively, ROIs may be defined by using edge detection and a fill algorithm.

3.4.2.4 Feature Extraction and Reduction

After the segmentation, that is, defining ROIs in the image, the procedure of feature extraction can start. Pixels in the signal matrix that represent noncontinuous jumps represent existing structures or edges in the image. To emphasize these discontinuities with sizes of the order of 1 pixel a high-pass filtering is needed by applying several local pixel operators.

Most of the different methods used may be grouped into two categories: gradient and Laplacian. The gradient methods detect edges by looking for extreme values (maxima and minima) of the first derivative of the image. In principle, the Laplacian

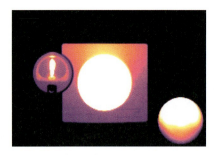 

Figure 3.49 Segmentation of a thermal image: determining regions of interest (ROIs). (a) Original thermal image and (b) colored ROI and gray background.

method, which searches for positions where the second derivative of the image becomes zero, is equivalent.

The most commonly used local operators of the gradient method are the simple difference operator (Figure 3.42), Sobel masks (Eq. (3.32)), or Roberts' cross operators (Eq. (3.33)):

$$\mathbf{So}^y = \begin{pmatrix} -1 & -2 & -1 \\ 0 & 0 & 0 \\ +1 & +2 & +1 \end{pmatrix} \quad \mathbf{So}^x = \begin{pmatrix} -1 & 0 & +1 \\ -2 & 0 & +2 \\ -1 & 0 & +1 \end{pmatrix} \quad (3.32)$$

$$\mathbf{R}_1 = \begin{pmatrix} -2 & -1 & 0 \\ -1 & 0 & +1 \\ 0 & +1 & +2 \end{pmatrix}, \quad \mathbf{R}_2 = \begin{pmatrix} 0 & -1 & -2 \\ +1 & 0 & -1 \\ +2 & +1 & 0 \end{pmatrix} \quad (3.33)$$

These matrices are applied to pixel areas exactly as discussed before. In contrast to the averaging matrices, which only contained positive numbers, the gradient operators must contain negative numbers, since they resemble derivatives, that is, the difference signals of neighboring pixels is calculated. From the structure of the matrices, it is obvious, that the Sobel mask computes the horizontal/vertical spatial derivatives with more emphasis on the direct nearest neighbors in these directions. In contrast, the Roberts' crosses resemble spatial derivatives in the directions of the two diagonals with more emphasis on the diagonal nearest neighbors.

As an example, the magnitude of the gradient for a pixel at row i and column j is calculated for the Sobel mask according to

$$So_{i,j} = \sqrt{\left(So^x_{i,j}\right)^2 + \left(So^y_{i,j}\right)^2} \quad (3.34)$$

where $So^x_{i,j}$ and $So^y_{i,j}$ represent the results when applying the Sobel mask to the pixel area centered around image point (i, j). Figure 3.50 depicts the edge detection in an IR image using the Sobel masks according to Eq. (3.32) for all pixels of the original image.

Depending on the directional preference of the local operator, certain edges are preferentially extracted while others may not be found. This is, of course, undesirable in some cases. This is why the use of isotropic operators can be advantageous. Examples for such isotropic (symmetric in all directions) operators are Laplace filters, which are approximations of the second-order derivatives (Eq. (3.35)).

$$\mathbf{L}_1 = \begin{pmatrix} -1 & -1 & -1 \\ -1 & 8 & -1 \\ -1 & -1 & -1 \end{pmatrix} \quad \mathbf{L}_2 = \begin{pmatrix} 1 & 0 & 1 \\ 0 & -4 & 0 \\ 1 & 0 & 1 \end{pmatrix}$$

$$\mathbf{L}_3 = \begin{pmatrix} -1 & -2 & -1 \\ -2 & 12 & -2 \\ -1 & -2 & -1 \end{pmatrix}$$

Laplace filters (3.35)

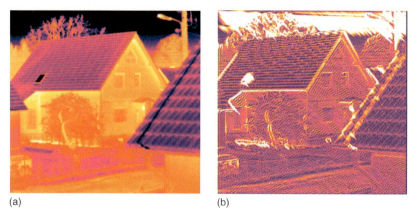

Figure 3.50 Application of the Sobel masks for edge detection. (a) original image and (b) gradient image of $S_{i,j}$.

The structure of the three chosen Laplacian operators leads to second derivatives. This can be seen, from the calculation, which leads to the eight differences from the center pixel to all of its neighbors. If grouped in four pairs along the four major directions (vertical, horizontal, and two diagonals) the two differences for each direction have different signs, which means that again their difference is computed, namely, the second difference or derivative. The three chosen Lapalacian operators L_1, L_2, and L_3 differ in the following properties. L_1 resembles just the average of the second derivatives in all four directions. L_2 only calculates the second derivatives in the diagonal direction and L_3, again, calculates the second spatial derivatives in all four directions, however, with twice as much weight of the horizontal and vertical nearest neighbors.

Both the first- and second-order derivative operators are vulnerable toward 1-pixel disturbances such that a previous low-pass filtering is mandatory.

The most important goal of image processing is to find optimal edge detection algorithms. The most popular edge detection procedure is the so-called Canny algorithm, which claims to combine good detection, good localization, and minimal response. This means, that as many existing edges as possible should be marked exactly at the correct position and independent of noise. In addition, no false edges should be detected due to noise or pixel errors.

"Optimal operators" such as the Canny or the Marr–Hildreth operators have a multistep architecture [36]. In the Canny procedure, the image is first convoluted with a Gaussian matrix and next, Sobel masks are used to extract the existing edges. To eliminate nonedge points, a nonmaxima-suppression algorithm gets started as the third step. In this step, the gradient values and the gradient directions are evaluated such that pixels that are not local maxima in a 3 × 3 neighborhood and whose gradient direction is not the same as the one of the maximum get classified as nonedge points. The advantage of this elimination procedure is good localization and ideal edges with a width of only 1 pixel. In the fourth step, a so-called hysteresis threshold algorithm can extract these edges directly via a contour-following algorithm. Therefore, an upper threshold for the gradient matrix of real edge points is set. These pixels build the starting points of the contour-following algorithm. All points along the contour above a defined lower threshold (mostly between 2 or 3 times smaller than the upper threshold) are the connecting pixels, which are used to define a complete contour.

3.4.2.5 Pattern Recognition

In some applications, the acquired images contain specific patterns or geometrically parameterizable objects like, for example, straight lines, parabola, circles, and so on. In these cases, a coordinate transformation from the original x-y-space to a so-called Hough space can be applied to extract these patterns. All pixels in the gradient matrix (result of feature extraction) with intensities above a certain threshold are used for the transformation. The simplest case is to search for collinear points or nearly straight lines.

Every straight line in Cartesian coordinates can be identified as one point in the corresponding Hough space. Figure 3.51 depicts the transformation procedure for the Hough space associated with linear features. A straight line $y = mx + b$ is plotted in Cartesian coordinates, that is, a rectangular (x,y) coordinate system. Within the Hough transformation, such a line is characterized by two other parameters (d, α) where $d = x_0 \cos(\alpha) + y_0 \sin(\alpha)$. The parameter d represents the shortest distance between the line and the origin whose orientation is necessarily perpendicular to the chosen line. The angle $\alpha (-90° < \alpha < 90°)$ is the angle between the x-axis and this shortest distance line. Angles lying without the range $-90° < \alpha < 90°$ are replaced by $\alpha \pm 180°$.

In order to understand how the Hough transformation is applied to image analysis, a rule must be defined that states how extracted image features are treated. This rule and the respective procedure are described for a simple example.

Each already extracted object image consists of a large number of pixels, that is, image points that were found via feature extraction. First, we consider one of these, a single point P, in the Cartesian object space. This point is transformed into a whole series of points in Hough space by first drawing an arbitrary line through P. This line refers to a well-defined point in Hough space. Second, we let the chosen line start to rotate with the rotation axis through P. This obviously defines a large number of lines in Cartesian space and hence also a large number

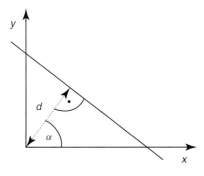

Figure 3.51 Principle of Hough transformation from coordinates (x, y) to coordinates (d, α). The shortest connection from the origin to the line hits the line at point (x_0, y_0).

of transformed points in Hough space. This means, that each image pixel refers to a very large number of points in Hough space. Third, a second pixel point Q in the image is chosen and again, all possible different lines through Q are transformed into Hough space. This procedure is repeated for each pixel, which was previously defined as ROI in feature extraction. If the number of image points is large an even much larger number of points result in Hough space.

The usefulness of Hough space becomes obvious if we consider, for example, a line defect found by the feature extraction algorithm. Arbitrarily distributed image points in the IR image would result in a distribution of points in Hough space that is quite homogeneous. If, however, pixels along a line in the image are transformed, each of the line pixels will also contribute to the same spot in Hough space, since each pixel will contribute a line that is parallel to the image line. As a result, this line will lead to an accumulation point in Hough space.

Consequently, if all image pixels of extracted features are used for the transformation, one only needs to look for accumulation points. These define the line feature unequivocally according to the definition of Hough space (Figure 3.51). Therefore, a simple algorithm is applied to the problem: after all points of extracted features are transformed, the maxima, that is, accumulation points in Hough space are rated via a threshold. Positively rated points are transformed back to visualize the problem areas of the extracted features.

Figure 3.52 depicts an example to visualize the procedure with three straight lines that were defined as extracted ROI. All image points along these lines were used to plot points in Hough space. Obviously, three accumulation points were found, which, as expected, resemble the three chosen lines.

Another advantage of such a representation is the independence of the algorithm toward noncontinuous structures that are mostly found in the acquired images. If, for example, only fragments of a line are detected in the IR image, they still lead to an accumulation point on Hough space.

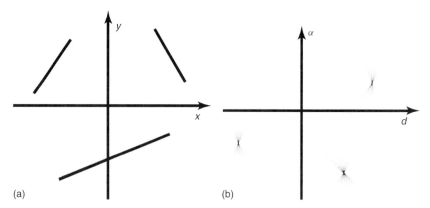

Figure 3.52 Hough transformation for three straight lines. (a) Straight lines in Cartesian coordinates and (b) the corresponding accumulation points in the Hough space.

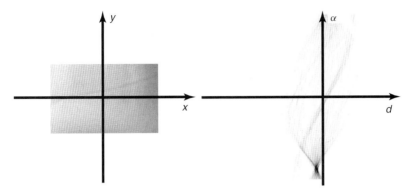

Figure 3.53 Hough transformation of a break in a glass sheet.

Figure 3.53 depicts an application of the Hough transformation to the break in a glass sheet shown in Figure 3.44a.

The speed of calculation decreases with the dimension of the Hough space. The easiest two-dimensional Hough space refers to linear features. If a search for objects that can be represented by circles is performed one needs a three-dimensional Hough space (the two center coordinates + the radius) and for an ellipse search one needs a five-dimensional Hough space. In practice, these higher-dimensional Hough calculations are not used because of a lack of speed as a consequence thereof. Several other more advanced methods, such as the generalized Hough transformation [40] or template matrices [41], exist, which can treat other structures and problems as well.

3.5
Active Thermal Imaging

In most applications of IR imaging, the investigation is done passively, that is, the camera observes a scene and detects the thermal radiation emitted from objects. Because of the thermal radiation from the surroundings, a thermal contrast in an image is only observed if temperature differences to ambient temperature exist. The method is called *passive* since the existing temperature distribution of the object scene is analyzed without imposing an additional heat flow to the object. This means that the observed temperature patterns are solely due to existing temperature differences. These often lead to nearly dynamic equilibrium conditions, that is, object surface temperatures do not change as a function of time (at least not very much). Typical examples are building inspections, where temperature differences are due to heating from inside, while the outer walls are cooled by the surrounding air. Such passively recorded images may visualize inner structural details of a building. A typical example are half-timbered houses (Section 6.2.1). The surface temperature differences measured by thermal imaging are caused by a spatial variation of the heat flow due to different thermal properties of the materials or constructions (Chapter 4).

Unfortunately, there are often situations where no natural temperature differences are present or – if they are present – they may not be sufficiently strong, or envelopes of objects may be too thick for an identification of structural elements below the surface. If one still wants to have a look beneath the surface, that is, detect structural details below the surface, active methods are needed.

They are based on the heating (in principle, cooling is also possible) of the surfaces to be investigated. These processes provide a nonstationary temperature gradient inside the structure under test, which affects the observable surface temperature distributions.

Therefore, active thermography can also be considered as non-steady-state, nonequilibrium, or dynamic thermography. The surface temperatures of the sample are monitored as a function of time and the transient heat flow generated through the sample subsurface will cause transient anomalies in the surface temperature distribution. Figure 3.54 gives a brief overview of active thermal imaging techniques. A sample is heated by either absorption of radiation (as shown), by electrical heating (current flowing through specimen), eddy currents, or, for example, ultrasound. The energy transfer to the sample can be either continuous, in a modulated continuous form (e.g., harmonic), or via pulses (rectangular or other shaped pulses). Detection of the heat transfer within the sample – which is affected by internal defects/structures – is done via measuring surface temperatures either in reflection or in transmission as a function of time. Neglecting the trivial cases of continuous heating or step heating, the most important methods of active thermal imaging are pulse thermography and lock-in thermography. They are applied widely in many different fields such as inspection of aircraft, solar panels, or objects made from composite materials with carbon or glass fiber structures and many more.

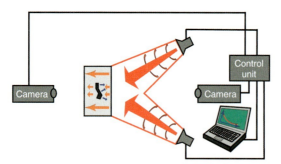

Figure 3.54 Schematic setup for active thermography techniques. A sample with internal structure is heated, for example, by absorbing visible light. Surface temperatures are measured using IR cameras as a function of time. They are affected by structures/intrusions/anomalies within the sample.

The idea of nondestructive testing using transient temperature behavior has a long history and predates the development of the first thermal imagers. In the 1960s, a lot of laboratory work was done in detecting transient surface temperature of objects. Temperature-sensitive colors, point-measuring sensors, and other instrumentation were used. All these methods were, however, severely limited by low temperature sensitivity and slow response times. For more detailed historical perspectives see [42]. The late 1970s mark the beginning of successful developments of both lock-in and pulse thermography methods [43, 44]. With increasing use of thermal imaging systems, and new heating methods, as well as electronic and computer technology during the last decades, these two methods became well established in the general field of NonDestructive Testing (NDT) [45].

This field is so wide that here, only a very brief overview on the fundamentals of transient heat transfer and active thermographic methods can be given. For more detailed information about these methods, one should refer to the literature [46, 47]. A large number of papers on the application of active thermography in NDT are available at the NDT database [48].

Active thermography methods are based on transient heat transfer in solids. The aim of all methods is to detect, locate, and characterize buried material discontinuities in solids [45, 49–54]. The correct quantitative description of 3D transient heat transfer in such systems is a very complex problem. The simulation of the transient heat transfer in active thermography requires the prediction of 3D temperature distribution using the 3D heat conduction differential equation for anisotropic thermal material behavior (e.g., for composites). A simplification of the heat transfer using 1D or 2D models will only give rough estimates or qualitative information of the temperature distribution [45, 55] and will make the interpretation of the measurement result difficult. Nowadays, the Finite Element Methods (FEMs) are the most promising tools for 3D heat transfer modeling if the experimental conditions are considered correctly [55]. A description of these methods is beyond the scope of this section. Here, we discuss first the relevant material parameters, which help to understand whether a material discontinuity may be detectable with active techniques. Second, the simplest method, pulse thermography, is discussed briefly along with some examples that demonstrate the sensitivity of the method. Third, the basics of lock-in thermography are presented with some applications to aircraft inspection. Finally, pulse phase thermography, which is simple to perform but complex concerning data analysis is briefly discussed.

3.5.1
Transient Heat Transfer – Thermal Wave Description

The transient heat transport in a solid body is characterized by two dynamic quantities. These are the thermal diffusivity a_{diff} and the thermal effusivity e_{eff}. Whereas the first describes how fast thermal energy diffuses through a material, the second represents some kind of thermal inertia. (In many publications, effusivity is simply denoted by e; here we rather use e_{eff} in order to distinguish it from the number $e = 2.718\ldots$.)

As also discussed in Section 4.4, the thermal diffusivity a_{diff} is the ratio of the thermal conductivity λ to the volumetric heat capacity (i.e., the product of specific heat capacity and density):

$$a_{\text{diff}} = \frac{\lambda}{\rho c_p} \tag{3.36}$$

The thermal diffusivity is a measure of the thermal energy diffusion rate through the material. The diffusion rate will increase with the ability to conduct heat and decrease with the amount of thermal energy needed to increase the temperature. Large values of diffusivity mean that objects respond fast to changes of the thermal conditions. Therefore, this quantity governs the timescale of heat transfer into materials. If a material has voids or pores in its structure, then the thermal conductivity and density decrease, which means the thermal diffusivity changes. As a result, the heat transfer within the material is affected, leading to observable changes of surface temperatures in the vicinity of the defects.

The *thermal effusivity* e_{eff} is defined as the square root of the product of the thermal conductivity and the volumetric heat capacity:

$$e_{\text{eff}} = \sqrt{\lambda \rho c_p} \tag{3.37}$$

This quantity is often referred to as *thermal inertia*. If the response of a heat input is analyzed, the effusivity will represent the ability of the material to increase its temperature according to $T \sim 1/e_{\text{eff}}$. Therefore, this quantity will govern how much the temperature of an object changes owing to input of thermal energy. Again, if a defect below a surface has a different temperature than its surroundings, this will affect the observed surface temperatures.

The effusivity also has another effect on heat transfer within a material. Considering a thermal contact between two materials with different effusivities $e_{\text{eff},1}$ and $e_{\text{eff},2}$ one often characterizes the thermal behavior with the thermal mismatch factor Γ:

$$\Gamma = \frac{e_{\text{eff},1} - e_{\text{eff},2}}{e_{\text{eff},1} + e_{\text{eff},2}} \tag{3.38}$$

$\Gamma = 0$, that is, equal effusivities, implies that there is no thermal mismatch (the interface of the two materials cannot be detected by a temperature measurement at the surface). For a perfect thermally conducting first material, $\Gamma = 1$ and for a perfect thermally insulating material, $\Gamma = -1$.

If the transient thermal behavior of a composite material is analyzed, the thermal mismatch factor will describe the change in thermal transit time compared to a homogeneous material. The effusivity, in this respect, behaves similar to, for example, the index of refraction in optics when describing the reflection of optical waves being incident on the interface between two media. An optical interface cannot be detected if the two indices of refraction are identical. In this case, the wave just passes the interface undisturbed without changing its speed. In an even more general scheme, any wave is characterized by a wave resistance or impedance, which depends on the material properties. If the wave hits an interface to another material, reflections can only be observed if there is a change in impedance.

As a matter of fact, one also often uses the concept of thermal waves in active thermography, in particular, if periodical heating processes are described, for example, in lock-in thermography [45].

It can be shown that the Fourier equation, assuming harmonic object heating at the surface provides a harmonic temperature field within the object with the same frequency but different amplitude and phase. Therefore, the concept of thermal waves, which can be described by the theory of wave physics, can be introduced. The main characteristic of thermal waves is the strong decay as a function of depth in the object. This decay can be characterized by the thermal diffusion length, which resembles a thermal penetration depth:

$$\mu = \sqrt{\frac{a_{\text{diff}}}{\pi f}} \tag{3.39}$$

The thermal diffusion length depends on the thermal diffusivity a_{diff} and the frequency f of the thermal wave (or heat stimulation). This expresses the fact that low-frequency thermal waves will penetrate deeper into a material than the high-frequency waves. The penetration depth will increase with increasing diffusivity. μ can give a first idea of possible depth ranges. Consider, for example, the outer skin of aircraft constructions, which is made of a few millimeters aluminum and below some carbon fiber composite materials. Aluminum has a diffusivity of around 10^{-4} m^2 s^{-1}, which, for a frequency of 10 Hz would result in a μ value of about 1.8 mm. Since μ corresponds to a depth where the signal has decreased exponentially to a value $e^{-1} = 0.36...$, it seems reasonable to assume that thermal signatures may be detectable in depths of several millimeters.

As discussed above, using the comparison to light waves, thermal waves are reflected at boundaries of materials only if the materials have different effusivities. If we look at the boundary between a solid and a gas, the effusivity of the solid is much higher than the effusivity of the gas. Applying Eq. (3.38) we will get $\Gamma = 1$ and the solid behaves as a thermal wave mirror for the gas to solid interface [45]. As a result, any reflection from internal boundaries will lead to changes of the thermal wave propagation in the object, which will, in turn, cause changes of observable surface temperature distributions.

In summary, thermal waves exhibit all wave phenomena such as reflection, refraction, and interference. The propagation of thermal waves in inhomogeneous solids will be affected by these effects and will result in a well-defined time-dependent surface temperature distribution, that can be analyzed by transient temperature measurements.

3.5.2
Pulse Thermography

Pulse thermography is one of the most popular methods in active thermography because it is extremely easy to carry out. The setup just needs the possibility of pulsed heating, for example, a system of flash lamps and time-dependent recording with an IR camera. The data analysis consists of analyzing IR images of the object as a function of time and thereby detecting surface temperature changes induced by subsurface structures. The only severe drawback is the limited spatial resolution with regard to depth on the one hand, and lateral resolution on the other.

The experimental setup of pulse thermography was already shown in Figure 3.54. Short pulses are applied with duration from a few milliseconds for high-conductivity material inspection such as metals to a few seconds for low-conductivity specimens such as plastics and graphite epoxy laminates [45]. Although appreciable heating powers are applied, the brief duration of the heating usually only leads to temperature rises of a few degrees above initial component temperature, which prevents any thermally induced damage to the component.

Figures 3.55 and 3.56 depict two different situations for possible pulse thermography inspections. Figure 3.55 refers to an object material that has three holes of different diameters and depths drilled from the bottom of the sample. The sample is heated with a pulse from the bottom. Thermal diffusion in the material can proceed faster in the region of the deepest holes, since heat must then only diffuse through a thin layer of object material before reaching the top surface. Therefore, the observable top surface temperatures are increased with respect to neighboring parts of the surface. This temperature difference will evolve as a function of time, the absolute values depending on thermal diffusivity and effusivity of the material. Figure 3.55 shows the situation schematically. Owing to the thermal diffusion, a lateral spreading takes place, and the hole with smaller diameter will have a less pronounced thermal signature than the hole with a larger diameter and of the same depth. Similarly, when comparing the two holes with equal diameters but of different depths, the temperature difference is smaller for the hole of lower depth. The various holes of different depth and diameter will also show up with best thermal contrast at different times of such an experiment.

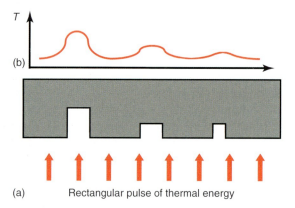

Figure 3.55 (a) Scheme of a pulse thermography experiment with a test structure composed of a material containing holes of given diameter and depth. The structure is illuminated from the bottom and temperatures are analyzed on the opposite side (b). Owing to better heat transfer through the thinner regions of the object, temperature anomalies appear, which depend on the ratio of thickness below the surface to the diameter.

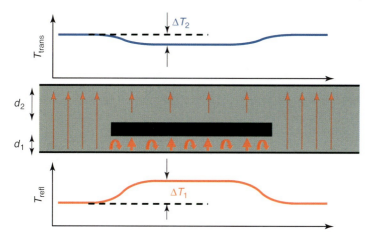

Figure 3.56 Scheme of a pulse thermography experiment with a test structure composed of a material with embedded second material of different thermal properties. Observable surface temperature differences depend on the thermal material properties of object and intrusion.

Figure 3.56 depicts a second example. A solid material 1 characterized by thermal diffusivity $a_{\text{diff},1}$, conductivity λ_1, and effusivity $e_{\text{eff},1}$, and so on has a subsurface structure of material 2, characterized by different values of thermal diffusivity $a_{\text{diff},2}$, conductivity λ_2, and effusivity $e_{\text{eff},2}$. Heat flow is induced by a thermal rectangular pulse from the bottom. If the effusivity $e_{\text{eff},2}$ of the substructure is much lower than the one of material 1, the heat flow will try to at least partially bypass the obstacle, that is, it will be lower behind the substructure. At the same time, blocking of heat flow by the obstacle (visualized by the curved arrows) leads to lower heat flow from the surface to the inside of the material. As a consequence, the surface temperatures at the bottom of the thermal energy input will decrease more slowly in the region of the intrusion, that is, the temperature profile across the structure will lead to a temperature rise above the region of intrusion. In contrast, the corresponding heat flow is lower on the opposite side, giving rise to a lower temperature with regard to adjacent parts of the object. Depending on the distance of the obstacle from the two surfaces, the temperature profiles will differ: the one at greater distance will be shallower owing to lateral spreading by thermal diffusion within the material. In case the intrusion is of a material with much higher effusivity than the surrounding material 1, the respective temperature profiles from front and back would be inverted. If the distances d_1 and d_2 from the intrusion to object surfaces are different, the temperature profile for the larger distance will, on the one hand, be more smeared out laterally due to the diffusion and, on the other hand, its thermal contrast will be lower.

A more quantitative analysis reveals [45] that the time to observe maximum thermal contrast of an intrusion in depth z is proportional to z^2, that is,

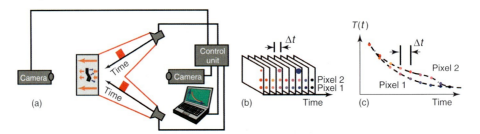

Figure 3.57 Principle and experimental arrangement of pulse thermography (a) and data acquisition/analysis (b,c). Images are recorded at time intervals Δt (b) and temperature plots of single pixels as function of time (c) show thermal contrast due to anomalies

$t_{obs} \sim z^2/a_{diff}$, while, at the same time, the thermal contrast C_{th} decreases because of lateral spreading of the thermal energy ($C_{th} \sim 1/z^3$). Deeper discontinuities will be observed later and with a reduced contrast. As a result, deep embedded structures usually appear very shallow with very poor thermal contrast. Overall, the size and depth of observable discontinuities are restricted. In isotropic media, the smallest detectable discontinuity should have a diameter of at least 2 times its depth below the surface. For anisotropic media, this restriction can even amount to a factor of 10. The criterion for still observing a structure is that the corresponding laterally induced observable surface temperature change ΔT is larger than the NETD of the camera.

Figure 3.57a depicts a typical experimental arrangement for observation in reflection or in transmission. Which approach is chosen depends on whether both sides are accessible for recording and, if this is the case, where the substructure is located. It can be better detected from the side where the depth below the surface is smaller.

A square pulse of thermal energy, in this case, visible light from flash lamps, is incident on the sample. For example, total thermal energies of 10–15 kJ are typical for pulse widths of 0.2 s. Figure 3.57b schematically depicts a time series of IR images recorded from the side of the illumination. The frames are separated by time intervals Δt. In each frame, the temperature decrease of two selected pixels are indicated (different colors refer to different temperatures). The time dependence of the temperatures of these two pixels is schematically plotted in Figure 3.57c. The temperature at the surface drops after the initial pulse, since the thermal energy diffuses into the material. In the example, the two chosen pixels differ in their transient behavior, one resembling the undisturbed material, the other a position with a subsurface structure that blocks the heat flow, thus leading to a higher surface temperature than the other pixel. The lateral spreading of the thermal energy pulse is visualized by the top spot in the IR frame sequence (b). It can be thought of as resembling a point source heating at the chosen given pixel. The corresponding spatially localized initial high temperature spreads out owing to thermal diffusion.

Figure 3.58 depicts a measurement result, recorded with a test structure similar to Figure 3.55. A solid laminate composite test object with three rows of flat bottom holes of different diameters was investigated. The diameter of the holes increased from top to bottom and the depth of the holes was increasing from left to right. After pulsed excitation, IR images were recorded from the back side as a function of time. As expected, the deep holes were already clearly visible within short times after excitation. With increasing time (from (e) to (h)) the holes with smaller depths also appear but the structures are more blurred because of the lateral heat

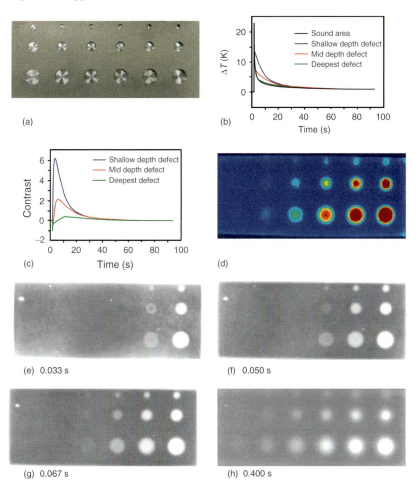

(e) 0.033 s

(f) 0.050 s

(g) 0.067 s

(h) 0.400 s

Figure 3.58 Flat bottom holes of different depth in solid laminate composite. (a) Vis image from backside. (b) Temperature evolution (thermal pulse at $t = 0$). (c) Temperature contrasts between the hole area and the homogeneous part of the sample. (d) False color representation of the IR image at $t = 0.4$ s. (e) Infrared image at $t = 0.033$ s. (f) Infrared image at $t = 0.050$ s. (g) Infrared image at $t = 0.067$ s. (h) Infrared image at $t = 0.400$ s. (Image courtesy: National Research Council Canada, Institute for Aerospace Research (NRC-IAR).)

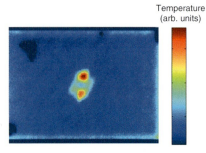

Figure 3.59 Impact damage of solid laminate. (Image courtesy: National Research Council Canada, Institute for Aerospace Research (NRC-IAR).)

diffusion. Comparing the holes with different diameters, but same depths, it is also obvious – as expected – that the smaller the hole, the more it gets spread out laterally. This can be seen best, for example, for the third structure from the right in the top row. At short times, it is not visible. Then it starts to show up with good contrast and rather sharp edge before beginning to get blurred edges and lower thermal contrast for long times.

Figure 3.59 depicts an example of a practical application. Here, pulse thermography was used to detect an impact damage in solid laminate.

3.5.3
Lock-in Thermography

Lock-in thermography [45, 46] is very similar to pulse thermography in terms of the setup, however, the pulsed thermal excitation is replaced by a sinusoidal input of thermal energy (Figure 3.60). This is most easily realized by illuminating an extended sample by a modulated lamp. This periodic, to be specific harmonic, heating input leads to a similar transient harmonic variation of the surface temperature of the object. As mentioned in Section 2.4.1, harmonic heating at the surface also leads to harmonic temperature variations at given depths within the object, although with a strongly attenuated amplitude as a function of depth, described by the thermal diffusion length μ (Eq. (3.39)).

Let us assume thermal excitation by using light absorption from a modulated light source. The harmonically varying input energy is partially absorbed leading to a strong temperature increase of the surface during the maximum of the input energy. While the excitation energy drops and passes through a minimum, the surface temperature drops too, because the initially absorbed energy diffuses into the inside of the object. After another half period of excitation, the surface is again heated by maximum input of thermal energy and so forth. As a result, the harmonic input will lead to a harmonically varying surface temperature distribution. Since the diffusion of thermal energy into the solid is taking place at the same time, one can also observe harmonic variation of temperature within certain depths of the object as a function of observation time. The observed surface temperature variation may be shifted by a phase angle with respect to the exciting light source owing to thermal inertia of the object. If the object is homogeneous, the phase shift will be similar for all observed surface pixels, which were recorded with the IR camera. If, however,

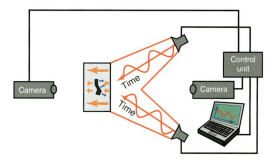

Figure 3.60 Setup for lock-in thermography: a sample is heated harmonically (sinusoidal excitation) as a function of time, for example, by modulated illumination, eddy currents, ultrasound, microwaves, and so on. The camera analyzes the spatial temperature distribution as a function of time by recording thermal images.

there is an intrusion or a thermal anomaly below the surface of the object, the thermal behavior changes. Either lower or higher heat flow due to the intrusion will manifest itself in a change of the phase of the observable surface temperature above the intrusion. As a consequence, any defect or structure within an object may lead to observable changes, here, predominantly an additional phase shift of the surface temperature with respect to the excitation. The analysis of the recorded images is therefore straightforward. One must record surface temperatures as a function of time for each pixel (x, y) and evaluate changes of the phase shift compared to a reference signal. Usually, it is sufficient to study a harmonic function by recording at least four data points per complete cycle (Figure 3.61), that is, the frame rate must be chosen according to the predefined frequency of excitation.

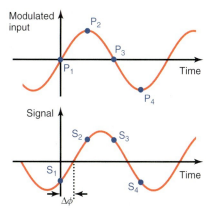

Figure 3.61 Measurement principle: the signal (surface temperature) varies harmonically but is shifted with respect to the excitation due to the modulated input.

Consider four well-defined points $P_1(t_1)$, $P_2(t_2)$, $P_3(t_3)$, and $P_4(t_4)$ of the modulated excitation where t_1 corresponds to a zero crossing of the excitation signal. These points define the times at which the temperature signal is analyzed, that is, the surface temperatures $T_1(t_1)$, $T_2(t_2)$, $T_3(t_3)$, and $T_4(t_4)$ within one cycle. Usually, the signal is shifted by a phase angle difference $\Delta\phi$ with respect to excitation.

If the temperature signal is given by an average value and a modulation, that is, $T(t) = T_{av} + S(t)$ one may construct two different result signals, an amplitude signal $A(x, y)$ and a phase signal $\phi(x, y)$ for each pixel (x, y) according to

$$A(x, y) = \sqrt{\left[S_1(x, y) - S_3(x, y)\right]^2 + \left[S_2(x, y) - S_4(x, y)\right]^2} \tag{3.40}$$

$$\phi(x, y) = \arctan\left[\frac{S_1(x, y) - S_3(x, y)}{S_2(x, y) - S_4(x, y)}\right] \tag{3.41}$$

The average temperatures cancel each other because of the differences, that is, these signal functions only depend on the harmonically induced temperature differences. If signals remain constant, the amplitude signal would be zero. Maximum amplitude signal results for a harmonic variation of temperature, that is, $S(t) = \Delta T_{max} \sin(\omega t + \phi)$ where the excitation varies like $\sin(\omega t)$. In this case, the amplitude signal would give $2\Delta T_{max}$, that is, the maximum temperature variation of the harmonic signal. The phase from Eq. (3.41), on the other hand, directly reflects the phase change between excitation and harmonic signal, as can be seen by calculating the signal differences and assuming the various signals each to be separated by $90°$ from the next.

The amplitude signal does depend on the sum of the squared temperature differences, whereas the phase signal only refers to the ratio of the respective temperature differences. Any local variations in illumination, absorption, or emissivity (i.e., higher or lower temperature differences between different pixels) will cancel out in the phase but show up in the amplitude signal. Therefore, one usually uses the phase signal, which is more sensitive to buried defects than the amplitude signal. It can investigate a depth range of about twice the thermal diffusion length μ (Eq. (3.39)). Evaluating the phase signal for all pixels allows to produce phase IR images, that is, the phase is correlated with a color or a gray scale (see below for examples). Images therefore reflect phase changes induced by subsurface structures or defects. The probed depth range depends on frequency according to Eq. (3.39). Therefore, the input energy source must be used at an optimal frequency, which depends on both the thermophysical characteristics of the object as well as its thickness.

The definition of phase and amplitude also stress the fact, that the recording frequency must be large enough such that at least four data points may be recorded during one cycle of excitation. If the four chosen points are distributed over several cycles, the results would no longer be unequivocal. Therefore, the method resembles a phase-sensitive detection if the signal is locked to the excitation frequency. This is similar to the lock-in technique used in other measurement science applications and, hence, explains the name.

Before presenting practical examples, some advantages and disadvantages of lock-in thermography must be mentioned. Lock-in thermography detects defects in depth ranges, which are related to the excitation frequency according to Eq. (3.39). Therefore, an inspection covering large range of depths requires longer times than other approaches such as pulsed thermography, since many different frequencies must be applied to the sample and each time, dynamic equilibrium must be established before starting the measurement. Compared to pulse thermography, the method is, however, more sensitive and may be used for larger depths. Furthermore, the energy required to perform lock-in thermography is generally smaller than in other active techniques, which may be important if the inspected part may suffer damage from high energy input.

3.5.4
Nondestructive Testing of Metals and Composite Structures

The main body of a passenger aircraft, also known as *fuselage*, is built as skeleton frame structure with a skin affixed to the frame elements. For reasons of weight, the skin only consists of a thin metal layer, usually aluminum, with a thickness (e.g., for Boeing 737) of the order of 1 mm. The stability is achieved by attaching a composite material, usually a carbon fiber reinforced material, to a substantial portion of the interior surface of the skin layer. The fuselage must remain intact for many flights, that is, for many pressurizing and depressurizing cycles. Aging aircraft may suffer from material fatigue, for example, cracks in the skin or loose connections between skin and supporting frame, and so on, which can lead to failure. Therefore, the aircraft fuselage is regularly scanned for defects.

Figure 3.62 depicts an example of such an aircraft testing [56]. Phase image lock-in thermography was performed on the outer skin of a Boeing 737 aircraft. Obviously the lock-in technique can easily look through the 1-mm aluminum skin and detect the supporting subsurface structure. One of the advantages of lock-in compared to conventional NDT techniques like ultrasound is that during a single measurement, large areas of typically 1 m^2 can be imaged. This considerably reduces the inspection times by a factor of 10 or so when compared to conventional methods. It was estimated that a lock-in inspection of the complete fuselage of a Boeing 737 can be done within about 100 h. The method is, for example, approved by the United States Federal Aviation Administration (FAA) and used by Boeing, Airbus or, for example, the Lufthansa airline.

The thermal input, needed to perform the measurement, only leads to maximum temperatures of about 40 °C, that is, thermally induced damages are nonexisting. The technique can detect all key defects such as delaminations, cracks, loose rivets, or water intrusions for metals (such as aluminum), carbon, or glass fiber reinforced plastics, as well as honeycomb structures.

As an example of a defect, Figure 3.63 depicts an example of delamination in a composite panel honeycomb structure from an airbus aircraft.

Test structures similar to those used for the pulse thermography (Figure 3.58) are depicted in Figure 3.64. A carbon fiber sample was prepared with flat bottom

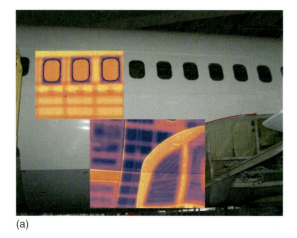

(a)

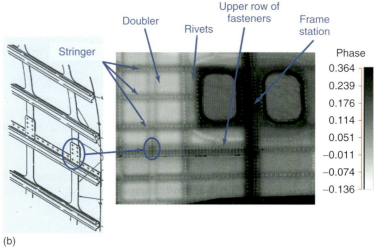

(b)

Figure 3.62 Lock-in thermography testing of a Boeing 737 aircraft. The images show selected sections as overlay of thermal phase images and visible images (a). The subsurface structures are easily detected and investigated with regard to anomalies, which would indicate defects. Often results are shown as gray-scale images (b). (Image courtesy: MoviTHERM, www.movietherm.com.)

holes and a larger milled out area (to the left). No structure can be seen visually from the other side, whereas the lock-in phase image easily detects all subsurface structures.

3.5.4.1 Solar Cell Inspection with Lock-in Thermography

Photovoltaic power generation is one of the most rapidly developing fields of renewable energies. Major challenges of the industry exist in the quality control of solar cells. In particular, defects in the semiconducting materials can reduce efficiencies of the cells. Therefore, reliable test procedures of the modules are

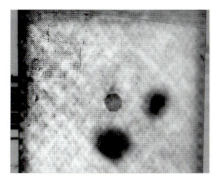

Figure 3.63 Lock-in phase image of a honeycomb structure from an aircraft showing delaminations. (Image courtesy: MoviTHERM, *www.movietherm.com*.)

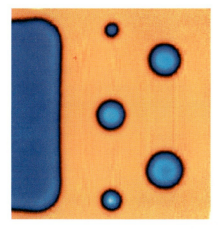

Figure 3.64 Lock-in thermography phase image of a test structure of a carbon fiber sample with flat bottom holes and a large milled out area. (Image courtesy: MoviTHERM, *www.movietherm.com*.)

needed, which are adjusted to the typical areas of modules in the square meter range.

Current solar cell tests, that is, photovoltaic cell tests, are based on three types of measurements: spectroscopy, electrical (contact) measurements, and IR imaging. Measured electrical parameters include the short-circuit current, open-circuit voltage, fill factor, ideality factor, series resistance, shunt resistance at 0.0 V, and reverse voltage breakdown. On the one hand, electrical contact measurements are straightforward and collect a complete set of parameters. Therefore, electrical cell tests can be quite time consuming. On the other hand, conventional IR imaging allows a faster detection of the major shunts by either applying a reverse bias voltage or by observing the cell temperature for normal operating conditions. The sensitivity and thermal resolution of standard thermography is, however, limited

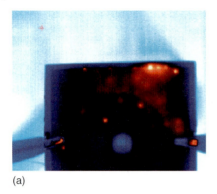 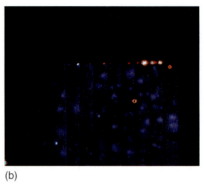

(a) (b)

Figure 3.65 Standard thermal image (a) recorded with an InSb camera of a 60 mm × 60 mm silicon solar cell. The lock-in image of the same cell (b) is the result of 800 individual images, recorded using a 10-Hz electrical AC input and an acquisition time of 20 s. Both images show shunt defects (orange areas) under steady-state reverse bias conditions. (Image courtesy: MoviTHERM, *www.movietherm.com*.)

by the detector NETD (Section 2.4.2), which amounts to about 20 mK for cooled InSb detectors and up to 80 mK for uncooled microbolometers. Therefore, refined methods such as lock-in thermography are needed to detect a wider variety of defects below this limit with temperature resolutions in the microkelvin range, for example, shunt conditions below a solar cell's metallization layer. Figure 3.65 depicts a standard (a) and lock-in (b) thermal image of the same solar cell. The triangular shapes on the left and right side (best seen in Figure 3.65a) are alligator clips that apply a bias voltage. The circular blue area of the conventional IR image is a reflection of the cold detector of the InSb camera.

A direct comparison of the two images reveals that, whereas the standard image, recorded with a cooled InSb camera shows thermal reflections from the detector as well as from the surroundings on the clips, there are no reflections visible on the lock-in image. The InSb camera detects some hot spots (shunts) as blurred, that is, strongly diffused orange regions in the upper right half, whereas the lock-in image shows a very sharp image with well-localized shunts. The shunt detection in both images is limited by the sensitivity of the detector, which is 20 mK for the InSb camera, but only 0.02 mK for the lock-in image.

Therefore, only severely shunted areas become visible as bright orange and localized spots in the InSb image. The darker orange regions are a result of weaker shunt defects. Locating the origins of these weaker shunts is extremely difficult, if not impossible, owing to the thermal diffusion (spreading of thermal energy over time) as well as the weak thermal radiation of the defect itself.

Overall, image quality and possibilities to extract quantitative information are strongly improved in the lock-in image. These sharp lock-in thermography images provide additional information, such as nonuniform heating of the cell, as revealed by lighter and darker blue areas. Therefore the lock-in technique also needs much less thermal energy input to then solar cells during the test procedure.

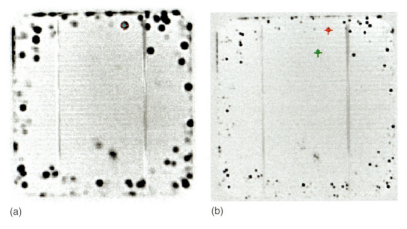

Figure 3.66 Phase images of shunt defects from lock-in measurements system at 10 Hz (a) and 200 Hz (b) sine wave stimulation. (Image courtesy: MoviTHERM, www.movietherm.com.)

Another example, which illustrates the effect of modulation frequency on the image is depicted in Figure 3.66, which shows the results of lock-in thermography testing for shunt defects on cells that were electrically excited with sine waves of different frequencies. Obviously – as expected – the image with lower frequency, which refers to a larger depth within the cell, is more blurred.

As a final example, Figure 3.67 depicts three images, which show the differences between a standard thermal image, on the one hand, and either amplitude or phase lock-in thermal images, on the other. Obviously, the phase image has the best quality, clearly detecting a shunt with high spatial resolution.

It should be noted, however, that which lock-in image is to be used may depend on the specific requirements and experimental conditions [46].

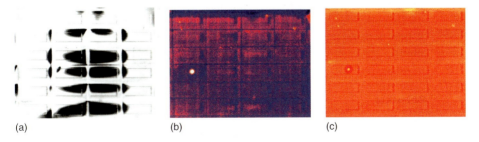

Figure 3.67 Comparison of different thermal images of an array of 5 mm × 10 mm solar cells. (a) standard image recorded with a cooled InSb camera, again showing the reflection of the cold detector, that is, the narcissus effect (Section 2.4.4.5). (b and c) Amplitude and phase lock-in images recorded using sinusoidal illumination with a diode array operating at $\lambda = 850$ nm and a frequency of 25 Hz. Shunts are clearly detected in both lock-in images. (Image courtesy: PVflex Solar GmbH, Fürstenwalde, Germany.)

3.5.5
Pulsed Phase Thermography

Pulse thermography is easy to perform, however, it suffers from limited thermal contrast and limited depth range. Lock-in thermography is better suited to locate depth ranges of defects by varying the frequency of excitation, however, this advantage is correlated with longer measurement times.

In order to overcome the latter drawback while, at the same time, having a simple setup, pulsed phase thermography was introduced as a link between pulsed thermography and lock-in thermography [45].

Pulsed phase thermography uses the concept of Fourier that any pulse of arbitrary shape can be described as a superposition of harmonic waves of different frequencies. Therefore, a square pulse, such as used in pulse thermography can also be considered to be composed of a multitude of different harmonic waves of different frequencies and associated amplitudes. The frequency distribution is unequivocally related to the shape of the pulse in the time regime, for example, a square pulse gives rise to a frequency spectrum defined by a $\mathrm{sinc}(x) = \sin(x)/x$ function. The fundamental idea behind pulsed phase thermography is the following. A square pulse is applied to a sample. This leads to heat flows into the sample, which may be changed owing to the presence of defect/substructures, exactly as explained for pulse thermography. Again, IR images of the changing surface temperatures are recorded as a function of time. The difference between pulsed phase and pulse thermography is just due to the data analysis procedure, that is, signal processing. In pulsed phase thermography, a fast Fourier transform is computed from the signals: since the input includes contributions of many frequencies simultaneously, the signal is thereby deconvoluted into the responses due to the individual frequencies. This means that results are obtained for a large spectrum of frequencies, thereby simultaneously probing different depth ranges of a sample.

From theory, the typical maximum frequencies associated with a pulse of width Δt can be computed. One finds, for example, that for a 10-ms pulse, the highest frequency that still has more than 90% of the maximum amplitude is 25 Hz; for a 0.1-s pulse; this maximum frequency is 2.5 Hz asf. Images must be recorded at least with twice this the maximum frequency in order to avoid any aliasing effects.

References

1. De Witt, D.P. and Nutter, G.D. (1989) *Theory and Practice of Radiation Thermometry*, 2007 Proceedings, John Wiley & Sons Inc., pp. 129–139.
2. www.spectrogon.com. (2010).
3. www.schott.com. (2010).
4. Hunter, G.B., Allemand, C.D., and Eagar, T.W. (1984) An improved method for multi-wavelength pyrometry. Thermosense VII, Proceedings of SPIE, vol. 520, pp. 40–46.
5. Hunter, G.B., Allemand, C.D., and Eagar, T.W. (1986) Prototype device for multiwavelength pyrometry. *Opt. Eng.*, **25** (11), 1223–1231.
6. Inagaki, T. and Okamoto, Y. (1994) Temperature measurement and radiometer pseudo gray-body

approximation. Thermosense XVI, Proceedings of SPIE, vol. 2245 (34), pp. 231–240.
7. Thevenet, J., Siroux, M., and Desmet, B. (2008) Brake disc surface temperature measurement using a fiber optic two-color pyrometer. 9th International Conference on Quantitative InfraRed Thermography, July 2–5, 2008, Krakow.
8. Coates, P.B. (1988) The least-squares approach to multi-wavelength pyrometry. *High Temp.–High Press.*, **20**, 433–441.
9. Saunders, P. (2000) Reflection errors and uncertainties for dual and multiwavelength pyrometers. *High Temp.–High Press.*, **32**, 239–249.
10. Neuer, G., Fiessler, L., Groll, M., and Schreiber, E. (1992) in *Temperature: its Measurement and Control in Science and Industry*, vol. 4 (ed. J.F. Schooley), American Institute of Physics, New York, pp. 787–789.
11. Svet, D.Ya. (2003) in *Temperature: its Measurement and Control in Science and Industry*, vol. 7 (ed. D.C. Ripple), American Institute of Physics, New York, pp. 681–686.
12. Peacock, G.R. (2003) in *Temperature: its Measurement and Control in Science and Industry*, vol. 7 (ed. D.C. Ripple), American Institute of Physics, New York, pp. 789–793.
13. Peacock, G.R. (2003) in *Temperature: its Measurement and Control in Science and Industry*, vol. 7 (ed. D.C. Ripple), American Institute of Physics, New York, pp. 813–817.
14. Gibson, G.C., DeWitt, D.P., and Sorrell, F.Y. (1992) In-process temperature measurement of silicon wafers. *Temp.: Meas. Control Sci. Ind.*, **6** (Part 2), 1123–1127.
15. Kourous, H., Shabestari, B.N., Luster, S., and Sacha, J. (1998) On-line industrial thermography of Die Casting tooling using dual-wavelength IR imaging. Thermosense XX, Proceedings of SPIE, vol. 3361, pp. 218–227.
16. Holst, G.C. (2000) *Common Sense Approach to Thermal Imaging*, SPIE Press, Bellingham.
17. Williams, G.M. and Barter, A. (2006) Dual-Band MWIR/LWIR Radiometer for absolute temperature measurements. Thermosense XXVIII, Proceedings of SPIE, vol. 6205.
18. Khan, M.S., Washer, G.A., and Chase, S.B. (1998) Evaluation of dual-band infrared thermography system for bridge deck delamination surveys. Proceedings of SPIE, vol. 3400, pp. 224–235.
19. Moropoulou, A., Avdelidis, N.P., Koui, M., and Kanellopoulos, N.K. (2000) Dual band infrared thermography as a NDT tool for the characterization of the building materials and conservation performance in historic structures. *MRS Fall Meeting: Nondestructive Methods for Materials Characterization, November 29–30 1999, Boston*, vol. 591, Publication Materials Research Society, Pittsburgh, pp. 169–174.
20. Beecken, B.P., LeVan, P.D., and Todt, B.J. (2007) Demonstration of a dualband IR imaging spectrometer. Proceedings of SPIE, vol. 6660, Infrared Systems and Photoelectronic Technology II, pp. 666004.1–666004.11.
21. Beecken, B.P., LeVan, P.D., Lindh, C.W., and Johnson, R.S. (2008) Progress on characterization of a dualband IR imaging spectrometer. Proceedings of SPIE, vol. 6940, Infrared Technology and Applications XXXIV, pp. 69401R-69401R-9.
22. Farley, V., Belzile, C., Chamberland, M., Legault, J.-F., and Schwantes, K.R. (2004) Development and testing of a hyper-spectral imaging instrument for field spectroscopy. Proceedings of SPIE, vol. 5546, pp. 29–36.
23. FIRST Hyper-Cam Datasheet, Telops Inc., Quebec. www.telops.com. (2010).
24. Kauppinen, J. and Partanen, J. (2001) *Fourier Transformations in Spectroscopy*, Wiley-VCH Verlag GmbH, Berlin.
25. Bell, R.J. (1972) *Introductory Fourier Transform Spectroscopy*, Academic Press, New York and London.
26. Moore, E.A., Gross, K.C., Bowen, S.J., Perram, G.P., Chamberland, M., Farley, V., Gagnon, J.-P., Lagueux, P., and Villemaire, A. (2009) Characterizing and overcoming spectral artifacts in imaging Fourier transform spectroscopy of turbulent exhaust plumes. Proceedings of SPIE, vol. 7304, pp. 730416–730416-12.
27. Farley, V., Vallières, A., Villemaire, A., Chamberland, M., Lagueux, P.,

and Giroux, J. (2007) Chemical agent detection and identification with a hyperspectral imaging infrared sensor. Proceedings of SPIE, vol. 6739, p. 673918.
28. Harig, R., Gerhard, J., Braun, R., Dyer, C., Truscott, B., and Mosley, R. (2006) Remote detection of gases and liquids by imaging Fourier transform spectrometry using a focal plane array detector: first results. Proceedings of SPIE, vol. 6378.
29. Richards, A. and Cromwell, B. (2004) Superframing: scene dynamic range extension of infrared cameras. Proceedings of SPIE vol. 5612, pp. 199–205.
30. Smith, M.I. and Heather, J.P. (2005) Review of image fusion technology in 2005. Thermosense XXVII, Proceedings of SPIE, vol. 5782.
31. www.flir.com. (2010).
32. Avdelidis, N.P., Almond, D.P., Ibarra-Castanedo, C., Bendada, A., Kenny, S., and Maldague, X. (2006) Structural integrity assessment of materials by thermography. Conference Damage in Composite Materials CDCM, Stuttgart, www.ndt.net. (2010).
33. Mat Desa, S. and Salih, Q.A. (2004) Image subtraction for real time moving object extraction. International Conference on Computer Graphics, Imaging and Extraction CGIV'04, pp. 41–45.
34. Jaehne, B. (2004) *Practical Handbook on Image Processing for Scientific and technical Applications*, CRC Press LLC.
35. Steger, C., Ullrich, M., and Wiedemann, C. (2008) *Machine Vision Applications and Algorithms*, Wiley-VCH Verlag GmbH.
36. Gonzalez, R.C. and Woods, R.E. (2008) *Digital Image Processing*, 3rd edn, Prentice Hall.
37. Möllmann, S. and Gärtner, R. (2009) New Trends in Process Visualization with fast line scanning and thermal imaging systems. Proceedings of the Conference Temperatur, PTB Berlin, ISBN 3-9810021-9-9.
38. Parker, J.R. (1997) *Algorithms for Image Processing and Computer Vision*, John Wiley & Sons, Inc.
39. Russ, J.C. (2007) *The Image Processing Handbook*, 5th edn, CRC Press Taylor and Francis Group, LLC.
40. Ballard, D.H. (1981) Generalizing the hough transform to detect arbitrary shapes. *Pattern Recognit.*, **13** (2), 111–122.
41. Brunelli, R. (2009) *Template Matching Techniques in Computer Vision: Theory and Practice*, John Wiley & Sons, Inc.
42. Shepard, S.M. (2007) Thermography of composites. *Mater. Eval.*, **65** (7), 690–696.
43. Milne, J.M. and Reynolds, W.N. (1985) The Non-destructive evaluation of composites and other materials by thermal pulse. Proceedings of SPIE, vol. 520, p. 119.
44. Kuo, P.K., Ahmed, T., Huijia, J., and Thomas, R.L. (1988) Phase – Locked image acquisition in thermography. Proceedings of SPIE, vol. 1004, pp. 41–45.
45. Moore, P.O. (ed.) (2001) *Nondestructive Testing Handbook*, 3rd edn, American Society for Nondestructive Testing, Columbus.
46. Breitenstein, O. and Langenkamp, M. (2003) *Lock-in Thermography*, Springer-Verlag, Berlin and Heidelberg.
47. Maldague, X.P.V. (2001) *Theory and Practice of Infrared Technology for Nondestructive Testing*, JohnWiley & Sons, Inc.
48. www.ndt.net. (2010).
49. Wu, D., Salerno, A., Schönbach, B., Hallin, H., and Busse, G. (1997) Phase sensitive modulation thermography and its application for NDE. Proceedings of SPIE, vol. 3056, pp. 176–183.
50. Sagakami, T. and Kubo, S. (2002) Application of pulse heating thermography and Lock-In thermography to quantitative non-destructive evaluations. *Infrared Phys. Technol.*, **43**, 211–218.
51. Grinzato, E., Bison, P.G., Marinetti, S., and Vavilov, V. (2007) Hidden corrosion detection in thick metallic components by transient IR thermography. *Infrared Phys. Technol.*, **49**, 234–238.
52. Maldague, X., Benitez, H.D., Ibarra Castenado, C., Benada, A., Laiza, H., and Caicedo, E. (2008) Definition of

a new thermal contrast and pulse correction for defect quantification in pulsed thermography. *Infrared Phys. Technol.*, **51**, 160–167.
53. Genest, M. and Fahr, A. (2009) Pulsed thermography for nondestructive evaluation (NDE) of aerospace materials. Inframation 2009, Proceedings, vol. 10, pp. 59–65.
54. Tarin, M. (2009) Solar panel inspection using Lock-in-Thermography. Inframation 2009, Proceedings vol. 10, pp. 225–237.
55. Weiser, M., Arndt, R., Röllig, M., and Erdmann, B. (2008) Development and test of numerical model for pulse thermography in civil engineering. ZIB-Report 08-45, Konrad-Zuse Zentrum für Informationstechnik, Berlin, December 2008.
56. Tarin, M. and Kasper, A. (2008) Fuselage inspection of boeing-737 using lock-in thermography. Proceedings of SPIE, vol. 6939, pp. 1–10.

4
Some Basic Concepts of Heat Transfer

4.1
Introduction

In this section, we discuss the relevance of measured surface temperatures of objects like buildings, electrical components under load, animals or humans, aircraft engines, and many more due to their emitted IR radiation using IR cameras. Taking nice IR images is just the start of analyzing the data and drawing conclusions concerning thermal properties of the objects under study. The main objectives are, for example, to gain insight into heat loss problems due to thermal insulation properties of buildings or leakages in industrial installations, or to understand heat sources which may lead to failure of electrical components and many more. For each case, one must usually start with the measured surface temperature to therefrom extract useful information. In general, the physics problem behind can be formulated as follows. If the surface temperature of an opaque object is given, what can we learn about the temperature distribution within the object and about the associated heat flows through the surfaces of the object?

Disregarding emissivity for the moment, it is common to all possible investigated objects that they must have a different temperature than their surroundings. This means that either the objects have energy sources or sinks within them, or they were heated or cooled before the observation started. In addition, during the observation, the temperature of the objects may change, that is, thermal equilibrium conditions are usually never fulfilled. In order to interpret the results of surface temperature measurements, one must therefore know about all processes which may lead to temperature changes of objects. In this chapter, we briefly discuss the three basic heat transfer modes. More information can be found in textbooks (e.g., [1, 2]). Then, the meaning of the measured surface temperatures with regard to the measurement conditions is discussed for a number of examples, that is, the problem of how the surface temperature can be used to extract meaningful information.

4.2
The Basic Heat Transfer Modes: Conduction, Convection, and Radiation

Temperature differences in any situation result from energy flows into a system (heating by electrical power, contact to thermal bath, absorption of radiation, e.g.,

Infrared Thermal Imaging. Michael Vollmer and Klaus-Peter Möllmann
Copyright © 2010 WILEY-VCH Verlag GmbH & Co. KGaA, Weinheim
ISBN: 978-3-527-40717-0

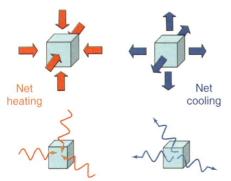

Figure 4.1 Energy flows (red) into a system, here a cube, lead to temperature rise. Energy flows (blue) from a system (cube) decrease the temperature. These energy flows are due to conduction, convection, and radiation acting via the surface of the system. For small absorption constants of the system material, radiation may also act within the volume.

microwaves, sun radiation, etc.) and energy flows from a system to the surrounding (Figure 4.1). The former leads to heating, whereas the latter results in cooling of an object. In thermodynamics, any kind of energy flow (also called *heat transfer*), which is due to a temperature difference between a system and its surroundings is usually called *heat flow*. In physics, one usually distinguishes three kinds of heat flow: conduction, convection, and radiation. As a matter of fact, the underlying physical processes for conduction and convection are very similar therefore, the distinction is rather artificial.

4.2.1
Conduction

Conduction refers to the heat flow in a solid or fluid (liquid or gas) which is at rest (Figure 4.2). Conduction of heat within an object, for example, a wall, is usually assumed to be proportional to the temperature difference $T_1 - T_2$ on the two sides of the object (thickness $s = s_1 - s_2$) as well as the surface area A of the object, that is,

$$\dot{Q}_{cond} = -\lambda \cdot A \cdot \frac{dT}{ds} \tag{4.1a}$$

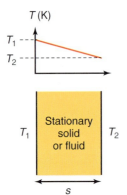

Figure 4.2 Conduction of heat takes place within a solid or stationary fluid.

For the specific example of a one-dimensional wall, steady-state conduction leads to a temperature which varies linearly with distance.

$$\dot{Q}_{\text{cond}} = -\lambda \cdot A \cdot \frac{dT}{ds} \approx \frac{\lambda}{s} \cdot A \cdot (T_1 - T_2) = \alpha_{\text{cond}} \cdot A \cdot (T_1 - T_2) \tag{4.1b}$$

In the following, we mostly refer to this simplified heat conduction equation. The heat transfer coefficient is defined as $\alpha_{\text{cond}} = \lambda/s$ where λ is the thermal conductivity of the wall material and s is the wall thickness. The heat transfer coefficient α_{cond} describes heat transfer in watts per unit area and kelvin, that is, it is given in W (m² K)⁻¹. Hence, the heat flux through the wall $\dot{Q}_{\text{cond}}$ in watts gives the energy flow per second through the wall of surface area A if the temperature difference between the inner and outer surface is given. Table 4.1 gives typical values of thermal conductivities of materials as well as their respective heat transfer coefficients for $s = 10$ cm "wall thickness."

Values of the heat transfer coefficient of conduction depend on the geometry of an object. To give a numerical example, we consider walls with thickness $s = 10$, 20, or 30 cm made of stones with $\lambda = 1$ W (m K)⁻¹. We find $\alpha_{\text{cond},10\text{ cm}} = 10$ W (m² K)⁻¹, $\alpha_{\text{cond},20\text{ cm}} = 5$ W (m² K)⁻¹, and $\alpha_{\text{cond},30\text{ cm}} = 3.3$ W (m² K)⁻¹.

The microscopic processes responsible for conduction of heat are most easily explained for gases by using the concept that in a gas, a large number of atoms/molecules move with different velocities and in different directions. The temperature of a gas determines the average kinetic energy and – related to it – the average velocity of the gas molecules; these averages are mean values of the corresponding energy and velocity distributions. Within a gas, there are a large number of collisions between molecules which permanently exchange energy and momentum. Individual molecules may change their energy and velocity; overall, the average value of a gas however stays the same if the gas is characterized by a well-defined temperature.

Now consider a gas which is in thermal contact with an object of given high temperature. It will correspondingly have high average kinetic energies and high average velocities in the vicinity of this object. In a certain distance to this object, the same gas may also be in contact with another object at lower temperature. The neighboring gas molecules have lower average kinetic energies and average velocities in its neighborhood. In the space between the two objects the gas molecules will suffer many collisions. These collisions lead to an energy transport from high kinetic energies to low kinetic energies. Obviously, conduction of heat just describes this energy transfer via molecular or atomic collision in the gas. For liquids the situation is quite similar, the main difference being that molecular distances are much shorter. In solids, there are no free-moving atoms or molecules; however, the atoms in a crystal lattice can vibrate. In the language of solid-state physics, these vibrations are called *phonons*. For solids that are in thermal equilibrium, there is a well-defined phonon distribution, giving rise to the corresponding average energy. If the solid is placed between two objects of different temperatures, the phonon distributions and hence the corresponding average energies differ. Similar to the collisions of gas molecules,

Table 4.1 Some approximate values for thermal conductivity of materials at $T = 20\,°C$ and the corresponding heat transfer coefficients for $s = 10$ cm. Values may vary depending on purity/composition.

Material	λ in W (m K)$^{-1}$	$\alpha_{cond} = \frac{\lambda}{s}$ in W (m^2 K)$^{-1}$	Material	λ in W (m K)$^{-1}$	$\alpha_{cond} = \frac{\lambda}{s}$ in W (m^2 K)$^{-1}$	Material	λ in W (m K)$^{-1}$	$\alpha_{cond} = \frac{\lambda}{s}$ in W (m^2 K)$^{-1}$
Aluminum (99%)	220	2200	Concrete stones	0.5–2 0.5–1.2	5–20 5–12	Water	0.6	6
Copper	390	3900	Dry wood	0.1–0.2	1–2	Oils	0.14–0.18	1.4–1.8
Silver	410	4100	Foams, Styrofoam	0.02–0.05	0.2–0.5	Air	0.026	0.26
Steel	15–44	150–440	Glass	0.8–1.4	8–14	CO_2	0.016	0.16

the phonon distributions change as a function of position between the two different temperatures. In conductors, there is, in addition, a contribution of free electrons to heat conduction which may even dominate over the phonon contribution. In principle, the corresponding energy transfer is similar to the one of gas molecules, described above.

4.2.2 Convection

Convection refers to the heat flow between a solid and a fluid in motion (Figure 4.3).

The energy flow $\dot{Q}_{cond}$ per second from the surface of an object with temperature T_1 into a fluid of temperature T_2 due to convection is usually assumed to follow a law similar to the one of conduction.

$$\dot{Q}_{conv} = \alpha_{conv} \cdot A \cdot (T_1 - T_2) \tag{4.2}$$

The heat transfer coefficient for convection depends on the nature of the motion of the fluid. In free convection, the current of the fluid is due to temperature and, hence, density differences in the fluid; in forced convection, the current of the fluid is due to external forces/pressure. Typical values for free convective heat transfer coefficients of gases above solids are cited to range between 2 and 25 W $(m^2\ K)^{-1}$, the exact value depending on flow conditions, wind speed, and moisture of the surface; for liquids they can be in the range 50–1000 W $(m^2\ K)^{-1}$. Figure 4.4

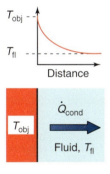

Figure 4.3 Convection of heat occurs between a surface and a moving fluid.

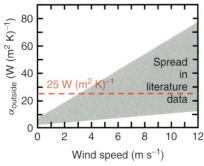

Figure 4.4 A large range of heat transfer coefficients for outside walls of buildings are reported as a function of wind speed. For many estimates of total heat transfer through walls (Section 4.3.3), a value of 25 W $(m^2\ K)^{-1}$ is used.

depicts as an example the range of values for convection heat transfer coefficients for outside walls of buildings reported in the literature.

The values for forced convections can be higher by an order of magnitude. A typical thermography application where forced convection plays a role is an outdoor building inspection when there is a finite wind speed, where α_{conv} depends on wind speed [3]. In many heat transfer estimates, the convective heat transfer coefficient is also assumed to include the radiative heat losses of the corresponding surfaces. The reason for this and its limitations are discussed in Section 4.5.

In order to understand the microscopic processes responsible for convection of heat, one needs a microscopic model for the system. Owing to the movement of the fluid across the surface, a boundary layer of a certain thickness is established. As a consequence, there is a velocity profile of the fluid due to the molecular forces between fluid molecules and surface molecules with zero velocity at the surface and the bulk fluid velocity at the distance of the boundary layer. Convection comprises two different mechanisms. First, convective heat transfer is due to the conduction of heat between the object surface and the fluid layer very close to the surface, that is, it is due to molecular motion and collision processes. The second convective contribution is due to the bulk motion of the fluid. The bulk motion of the fluid close to the boundary layer sweeps away the heat which was transferred via conduction from the surface. Obviously, a detailed microscopic modeling is rather tedious and involves many phenomena of fluid dynamics.

4.2.3
Radiation

The emission of thermal radiation was already treated in Chapter 1. In any realistic situation in IR imaging, an object of temperature T_{obj} is surrounded by other objects of background temperatures T_{surr}. For simplicity, we assume an object which is completely surrounded by an enclosure of constant temperature (Figure 4.5). Whenever objects with different temperatures are present, one needs to compute the corresponding view factors (Section 1.3.1.5 and 6.4.4) to find the net radiation transfer.

The object surface of T_{obj} emits radiation according to the radiation laws (Section 1.3.2). The total emitted power is given by the Stefan–Boltzmann law (Eq. (1.19)) corrected for the emissivity of the object. In the following, we assume gray objects. In addition, radiation from the surroundings is incident onto the object. This finally leads to a net energy transfer from the object with surface area A to the surroundings

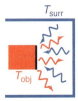

Figure 4.5 Whenever an object is placed in an environment of different temperature, there will be a net energy transfer via thermal radiation due to emission as well as absorption of radiation by the object.

$$\dot{Q}_{rad} = \varepsilon \cdot \sigma \cdot A \cdot (T_{obj}^4 - T_{surr}^4) \quad (4.3)$$

where $\sigma = 5.67 \times 10^{-8}$ W $(m^2\ K^4)^{-1}$.

This energy transfer equation differs from the ones for conduction and convection which were of the form $\dot{Q} = \alpha \cdot A \cdot (T_1 - T_2)$ due to the nonlinear dependence on temperature. Since any quantitative analysis concerning heat transfer is much easier for linear temperature differences, it is customary to approximate the radiative contribution also with a linear equation. This makes sense if temperature differences are small ($T_{obj} \approx T_{surr}$), since in this case

$$\begin{aligned}(T_{obj}^4 - T_{surr}^4) &= [(T_{obj} + T_{surr}) \cdot (T_{obj}^2 + T_{surr}^2)] \cdot (T_{obj} - T_{surr}) \\ &= k_{appr}(T) \cdot (T_{obj} - T_{surr}) \quad (4.4)\end{aligned}$$

where $k_{appr}(T) \approx 4 T_{surr}^3$. Using $\alpha_R = \varepsilon \cdot \sigma \cdot k_{appr}$, Eq. (4.3) can then be rewritten as

$$\dot{Q}_{rad} = \alpha_{rad} \cdot A \cdot (T_{obj} - T_{surr}) \quad (4.5)$$

which is of the same type as the heat transfer equations for conduction and convection. A discussion of applicability of this approximation is given below in Section 4.5.

4.2.4
Convection Including Latent Heats

Usually microscopic models for convective heat transfer treat the energy transfer as due to combined effects of conduction of heat within the boundary layer between a solid and the fluid, and effects due to fluid motion. In this case, internal thermal energy of the fluid is transferred to a solid or vice versa.

However, in addition, there are many convection processes where additional latent heats are exchanged. Such latent heats are associated with phase changes between the liquid and vapor states of the fluid. The most important and often used fluid in technology and nature is water. The corresponding heats for phase changes are called *heat of condensation* or *heat of vaporization* (when vaporization occurs at the liquid–gas interface, it is termed *boiling*). They add up to the transferred heat and give rise to strongly increased coefficients for the convective heat transfer, which can reach values between 2500 and 100 000 W $(m^2\ K)^{-1}$.

Microscopically, heat transfer due to evaporation and condensation can be understood as follows:

1) If a solid is covered with liquid droplets or a liquid film and it is surrounded by a gaseous fluid with low vapor pressure, the liquid droplets or the film can evaporate. In order to do so, the heat of vaporization is needed, which is extracted from the solid, thereby transporting energy from the solid to the gaseous fluid (evaporative cooling of the solid).
2) If heat is transferred from a solid to a liquid around it, and the surface temperature exceeds the saturation temperature, vapor bubbles start to grow on the surface. Finally upon growing they can detach from the surface, thereby

Table 4.2 Heat of vaporization for saturated water at various temperatures.

Temperature in K (°C/°F)	Heat of vaporization/ condensation in kJ kg^{-1}
273.15 (0/32)	2501
283.15 (10/50)	2477
293.15 (20/68)	2453
303.15 (30/86)	2430
373.15 (100/212)	2257

transporting the energy which was used for their generation from the solid into the liquid (cooling of the solid).

3) Condensation occurs when the temperature of a vapor is reduced below its saturation temperature. In industrial applications (and also in nature around us), this usually results from contact of the vapor with a cold surface. Upon condensation, the latent heat of condensation is released, that is, the condensing vapor transports energy from the vapor to the solid (heating of solid). Depending on whether droplets or a thin film is produced upon condensation, different microscopic theories are used for the estimation of the associated heat transfer coefficients.

For water, the latent heats are temperature dependent (Table 4.2).

Owing to the large value of the latent heats for water, the associated heat transfer coefficients can have drastic influences on temperature distributions of objects. In outdoor building thermography, the surface temperature distributions of walls or enclosures of objects strongly depend on rain and wind. The phenomenon is also well known from personal experiences in summer. Getting out of a swimming pool, lake, or the sea, one is usually still covered with many droplets of water. Sometimes one may start to shiver during a mild breeze while experiencing evaporative cooling.

One numerical example illustrates the cooling potential of water. One gram, that is, 1 cm^3, of water will transfer an amount of about 2450 J at a temperature of 20 °C. If this 1 cm^3 of water would be spread as a thin film on the metal cube with side length 60 mm in the example of Section 4.4.3, the film would have a thickness of about 0.28 mm. It would strongly increase the heat transfer coefficient, in particular if a fan would be used for artificial wind speed. Total vaporization of this minute amount of water would lead to a temperature drop of about 4.7 K, provided that the system would be thermally isolated from the surroundings. Experimental investigations of this effect are discussed in Chapter 6. We finally note that there can be additional heat sources present; for example, reaction enthalpies while studying chemical reactions (Section 8.3). Whenever such special problems are investigated, the corresponding energies must be taken into account.

4.3
Selected Examples for Heat Transfer Problems

4.3.1
Overview

Obviously, any object which is studied by thermography can be characterized by heat transfer due to both conduction within a wall/an enclosure as well as convection at the inner (if applicable) and outer surfaces of the object. Typical examples are hot liquids transported within tubes, electrical wires heated from within due to current, or buildings heated from inside. In thermography, one measures the surface temperatures of the objects with the goal of learning something about the object. Two extreme situations are that the relevant information is contained in the inhomogeneities of the thermal radiation, indicating thermal leaks in a qualitative analysis, and that one is interested in the absolute temperature values for quantitative analysis of a problem. We discuss the general problem of heat transfer for the example of building walls.

Figure 4.6 depicts five standard situations which resemble typical measurement situations including these two limiting cases. The conditions for these situations are summarized in Table 4.3.

The least interesting situation for IR imaging is thermal equilibrium (a), that is, any initial temperature differences have vanished since there are no heat sources (e.g., heater) or sinks (e.g., cooling system) available. If every parameter of the camera is properly adjusted, the IR image will just be a homogeneous area with no thermal signatures, independent of any possible emissivity contrast.

In cases (b) and (c), time-independent heat sources are present. In the case of building thermography, the heat source can be the heating system inside leading to constant inner temperature and the heat sink is the outside of the house at much lower constant temperature. For industrial applications, one may think of pipes transporting hot liquids. In this case, the hot liquid is filled in at constant flux, thus providing the time-independent heat source. Again, the heat sink is the outside of the pipe at much lower but constant temperature. Since we assume time-independent behavior, that is, that the inside and outside temperatures do not change with time, a stationary temperature distribution within a wall is reached. After initializing the corresponding measurement setup or process, it may take quite a while to reach this dynamic thermal equilibrium. Depending on the material properties of the wall/tube, the temperature may vary with distance as is shown for the wall (b) or stays more or less constant within the tube (c). The relevant parameter is the Biot number which is introduced below. Obviously, the case of large Biot numbers is usually encountered for buildings. In this case, there is a strong variation of temperature with distance. In contrast, the measured surface temperature is close to the inner temperature for small Biot numbers. The stationary situations (b) and (c) are the easiest for quantitative analysis.

Unfortunately, the standard application in IR imaging deals with time-dependent heat sources or sinks. For building inspections, this is due to a number of factors

Table 4.3 Parameters/conditions referring to standard situations in thermography.

(a)	Thermal equilibrium	No heat sources or sinks, after long time	IR images homogeneous, no thermal contrast
(b)	Temperature difference due to time-independent heat sources/sinks	Large Biot number: variations of T within wall/solid T profiles independent of time	Typical building inspections measure $T_{S2} < T_1$ (inside) useful information in spatial variations of T_{S2}
(c)		Small Biot number: nearly no variations of T within wall/solid T profiles independent of time	For example, pipe system with hot fluid measure $T_{S2} \approx T_1$ (inside) useful information also in spatial variations of T_{S2}
(d)	Temperature difference due to time-dependent heat sources/sinks	Large Biot number: variations of T within wall/solid	Typical building inspections measure $T_{S2} = T_{S2}(t)$
(e)		Small Biot number: nearly no variations of T within wall/solid	For example, pipe system with hot fluid measure $T_{S2} = T_{S2}(t)$

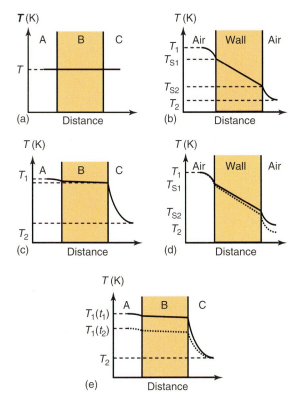

Figure 4.6 (a–e) Schemes for standard situations, encountered in thermography. They can be distinguished according to differences in temperatures T_1 and T_2 of two fluids (e.g., air on both sides of a wall or fluids A and C), to material properties of a solid material in between (e.g., a wall or solid B), and to time dependence of heat sources or sinks (see text and Table 4.3).

like the night setback of the heating cycle, potential solar load effects of the outside wall during the daytime, potential night sky radiant cooling, and changes of the ambient outside temperature during day and night (cloudy vs cloudless sky, etc.). In industrial applications with pipe systems, there can be transient changes both of the source (e.g., variations of initial temperature of the fluid) or of the sink (change of outside temperature). As a consequence, the spatial temperature profiles do additionally depend on time. This is illustrated for the case of large Biot number and lowering of outside temperature in (d) and for small Biot number and lowering of inside temperature in (e).

There are a number of questions related to this transient behavior. How much useful information can still be guessed, is quantitative analysis still possible or only a qualitative one, and – if transient effects should be avoided – how long are the corresponding time constants for these changes? As shown by many examples, the answers to the first two questions are easy: one can still extract a lot of useful information, in particular from the spatial variations across the investigated object surface. In most cases, comparison to earlier studies on the same object or similar objects also allows some semiquantitative analysis (e.g., criterion for potential failure of component). In the few cases where a quantitative analysis of absolute temperature is needed, one must however study the thermal time constants of the objects (see below).

In the following we present a few selected examples in detail.

4.3.2
Conduction within Solids: the Biot Number

Consider a solid, which is between two fluids of different but constant temperatures T_1 and T_2, as shown in Figure 4.6b,c. Assuming steady-state conditions, the heat flows due to conduction, convection, and radiation will lead to a spatial temperature distribution within the object. It is possible to get some idea on the temperature within the solid by using the so-called Biot number Bi.

$$Bi = \frac{\alpha_{conv}}{\alpha_{cond}} = \alpha \cdot s/\lambda \tag{4.6}$$

where α is the heat transfer coefficient from the object surface to the surroundings, s is the dimension of the object, and λ is the thermal conductivity of the object material (in this book we only discuss the Biot number and below, the Fourier number, although there are many other dimensionless quantities defined and used to describe the properties of heat and mass transfer depending on flow conditions [1, 2]. Such quantities as the Nusselt number, the Reynolds number, or the Prandtl number are particularly important for forced convection, where the heat transfer coefficient depends not only on geometry but also on whether the flow is laminar or turbulent).

The Biot number is a dimensionless quantity, usually describing the ratio of two adjacent heat transfer rates. In the present case, it describes the ratio of the outer heat flow from the surface to the surrounding, characterized by the convective heat transfer coefficient λ at the surface, and the inner heat flow within the object characterized by the conductive heat transfer coefficient $\alpha_{cond} = \lambda/s$. For $Bi \gg 1$, the outer heat flow is much larger than the inner heat flow. Obviously, this results in a strong spatial variation of internal temperature within the object. This is typical for walls of buildings (Figure 4.6b). If however, $Bi \ll 1$, the internal heat flow is much larger than the heat loss from the surface. Therefore, there will be temperature equilibrium within the object, that is, a homogeneous temperature distribution within the solid, and a large temperature drop at the boundary of the object with the surrounding fluid [1]. This is, for example, typical for metal tubes transporting hot liquids (Figure 4.6c).

As an example of time-dependent effects, we now discuss the situation of the cooling of objects. Consider, for example, an initially hot object of temperature T_{obj} which is in contact with surroundings of lower temperature. There shall be no further energy input into the object.

Figure 4.7 gives a schematic representation of the temperature distributions within initially hot one-dimensional objects as a function of time for different Biot

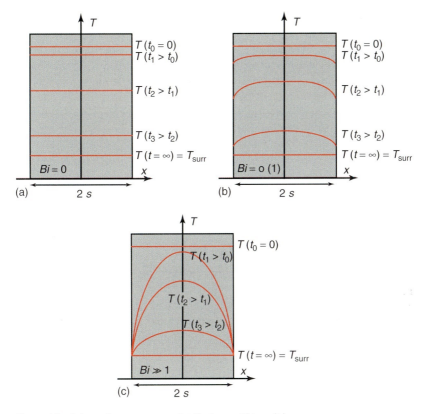

Figure 4.7 Schematic temperature distributions within solid objects upon cooling as a function of increasing Biot number ($Bi = 0$, $Bi = O(1)$, i.e., of the order of unity, $Bi \gg 1$). For finite object temperature, there is an additional temperature drop at the boundary to the surrounding fluid, which was omitted here.

numbers. For clarity, the temperature drops outside of the boundaries (similar to those in Figure 4.6) are omitted here. For Biot numbers less than 0.1, the temperature differences between the exact solution and the one assuming equilibrium within the object only lead to a maximum of 2% deviation [2]; hence, whenever $Bi < 0.1$ one may use this approximation [1], that is, assume constant temperature throughout the solid.

For larger Biot numbers, the conduction heat transfer within the solid proceeds more slowly than the convective heat transfer from the surface boundary. Therefore, the outside parts of the solid cool faster than the inside and a spatial temperature profile results. In this case, the surface temperature does not resemble a useful measure for the inside temperature or even an average temperature of the solid. For very large Biot numbers, the convective heat transfer dominates, the surface temperature drops very rapidly and then stays low, while the internal temperature drops only very slowly.

Figure 4.7 also gives an indication of how cooling curves of initially hot objects will be changing for larger Biot numbers. If the object were still to be characterized by some kind of mean temperature of the object, it is clear that this mean temperature would be larger than the surface temperature. Hence, compared to the case where the mean temperature equals the surface temperature, the heat losses would be smaller. Therefore, cooling would take longer. This can also be guessed from a simpler argument: the internal heat flow within the object is smaller than the heat flow due to convection from the boundary and therefore restricts the cooling time.

Whenever we deal with situations like those in Figure 4.6, we can also evaluate Biot numbers. A typical wall may have 24-cm thickness with stones of $\lambda = 0.5$ W (m K)$^{-1}$. In this case, the conductive heat transfer $\lambda/s \approx 2$ W(m^2 K) is of the order or less than the typical value for convective heat transfer of outside walls of 2–25 W(m^2 K). Hence, the corresponding Biot number, $Bi = 1-12.5$, is equal to or larger than unity, and we expect a large temperature drop with the building wall. In contrast, a stainless steel metal tube with thickness 2 cm and $\lambda = 15$ W (m K)$^{-1}$ gives $(\lambda/s) \approx 750$ W(m^2 K). In this case, the Biot number is given by $Bi \ll 1$, that is, that there is no temperature gradient within the tube.

Table 4.4 gives a summary of Biot numbers for some objects used in experiments (see below) plus several others for comparison. For small objects of metals or small objects filled with water, the condition $Bi < 0.1$ is usually fulfilled. However, in realistic building materials like brick or concrete of larger dimensions, this condition is no longer valid. The consequences for thermal time constants of objects are also discussed below.

4.3.3
Steady-State Heat Transfer through One-Dimensional Walls and U-Value

Figure 4.8 depicts a one-dimensional wall of an object. The fluid at the left side of the wall (e.g., the inside air) is at a high temperature T_1; the one at the right (e.g., the outside air) is at lower temperature T_2. The heat transfer from left to right is described by heat transfer equations. As discussed above, typical Biot numbers are larger than unity and one expects appreciable temperature drops within the wall. A typical qualitative result is shown in the figure.

Imagine this would be a building. There is a temperature drop from the inside air temperature T_1 to the inner wall surface temperature T_{S1}. Within the wall, the temperature drops to the outer wall surface temperature T_{S2} which is still above the outside air temperature T_2. Within a one-dimensional wall, there is a linear temperature drop. If the wall is composed of two or more different materials, there will be intermediate boundary temperatures T_B. One must keep in mind that in thermography one usually measures the surface temperatures and not the fluid (air) temperatures.

The dimensions x and y of the temperature boundary layers δ_{th}, that is, the distances from the walls where the corresponding fluid temperatures are reached, vary with the flow conditions [1, 2]. An order of magnitude estimate for the thermal

Table 4.4 Some material properties of objects and the corresponding Biot numbers. In massive building materials like brick the assumption $Bi \ll 1$ does no longer hold.

Object	Material(s)	α_{conv} in W (m² K)⁻¹	s in m	λ in W (m K)⁻¹	$\alpha_{cond} = \lambda/s$ in W (m² K)⁻¹	Biot number
Metal cubes	Aluminum, paint	2–25	$20 - 60 \times 10^{-3}$	220	11 000–3670	<0.01
Soft drink can (0.5 l)	Aluminum inside water	2	$\leq 1 \times 10^{-3}$ 3.3×10^{-2} radius	220 0.6	>220 000 18.2	$\ll 1 \approx 0.1$
Bottle (0.5 l) in fridge	Glass inside water	2	$\approx 3 \times 10^{-3}$ 3.3×10^{-2} radius	≈ 1 0.6	333 18.2	≈ 0.006 ≈ 0.1
Brick (for comparison)	Stone	2–25	12×10^{-2}	≈ 0.6	5	0.4–1
Pipe	Stainless steel	2–25	2×10^{-2}	15	750	0.033
Concrete (for comparison)	Composite with stones	2–25	20×10^{-2}	≈ 1	5	0.4–1

boundary layer thickness δ_{th} is

$$\delta_{th} \approx \lambda/\alpha \tag{4.7}$$

where λ is the thermal conductivity of the fluid and α is the heat transfer coefficient for convection at the boundary [2]. Using $\alpha_{inside} = 8$ W (m² K)⁻¹ and $\alpha_{outside} = 2$–25 W (m² K)⁻¹ for air as well as $\lambda_{air} \approx 0.026$ W (m K)⁻¹, we find $\delta_{inside} \approx 3$ mm and $\delta_{outside} \approx 13$ to 1 mm, that is, the thicknesses of thermal boundary layers for building walls are usually much smaller than other characteristic dimensions.

Note on the use of $\alpha_{outside}$ values: in many cases, $\alpha_{outside}$ ranges between 2 and 25 W (m² K)⁻¹. In most calculations concerning heat losses of building envelopes, the upper limit of $\alpha_{outside} = 25$ W (m² K)⁻¹ is chosen (usually assumed for wind speeds below 5 m s⁻¹) in order to get the worst case scenario, that is, the largest possible heat transfer rates. This limit will be too low for larger wind speeds. In the following, we also give results for the lower limit. For building inspections one may think of the lower limit for "no wind" situations. The upper limit value may also depend on moisture on the surface, leading to latent heat effects (Section 4.2.4).

The analysis of total heat transfer through the wall can be simplified using an analogy. The mathematical nature of Eqs. (4.1) and (4.2) is very similar to the one of Ohm's law for electrical circuits suggesting an analogy between heat transfer and charge transfer. In an electrical circuit, the driving force for charge transfer dQ/dt (i.e., current I) is the potential difference, that is, the voltage U, whereas in

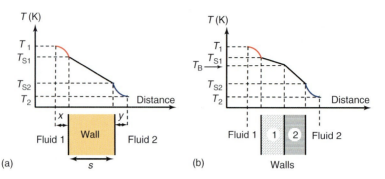

Figure 4.8 Schematic dependence of temperature changes due to heat transfer through a one-dimensional wall, embedded in fluids of different temperatures (a) or a double layer wall (b).

a thermal circuit the driving force for heat transfer is the temperature difference ΔT:

$$I = \dot{Q} = \frac{U}{R} : \text{(charge)} \Leftrightarrow \dot{Q} = \alpha \cdot A \cdot (T_1 - T_2) : \text{(heat)} \tag{4.8}$$

This analogy suggests to define thermal resistances $R_{th} = \frac{1}{\alpha \cdot A}$. Table 4.5 gives a comparison of the corresponding quantities. For conduction, the thermal resistance is $R_{th,cond} = \frac{s}{\lambda \cdot A}$ and for convection it is defined as $R_{th,conv} = \frac{1}{\alpha_{conv} \cdot A}$.

Using this analogy and the concept of thermal resistance, it is now easy to understand the limiting factors of heat transfer. A building wall constitutes a series connection of the thermal resistances of convection and conduction similar to an electrical series connection (Figure 4.9).

For the building wall of Figure 4.8, the thermal resistance R_{inside} of the inner convection leads to the temperature drop from the inside air temperature T_1 to the inner wall surface temperature T_{S1}. Within a one-dimensional wall, there is a linear temperature drop from T_{S1} to T_{S2} due to the thermal resistance R_{wall}, and the thermal resistance of the outer convection $R_{outside}$ leads to a drop from the outer wall surface temperature T_{S2} to the outside air temperature T_2.

Obviously, the total heat transfer for the wall problem can be written as

$$\dot{Q} = \frac{\Delta T_i}{R_i} = \frac{T_1 - T_2}{R_{total}} \tag{4.9}$$

Table 4.5 Equivalent quantities in electrical and thermal circuits. Please note that Q has different meanings for the two cases.

Quantities	Electrical circuit	Thermal circuit
Driving force	Potential difference U (voltage) in V	Temperature difference ΔT in K
Resistance	Electric resistance R in Ω	Thermal resistance R_{th} in K/W
Transfer quantity	Charge transfer $I = dQ/dt$ in C s^{-1} = A	Heat transfer $\dot{Q}$ in J s^{-1} = W

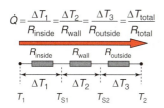

$$\dot Q = \frac{\Delta T_1}{R_{\text{inside}}} = \frac{\Delta T_2}{R_{\text{wall}}} = \frac{\Delta T_3}{R_{\text{outside}}} = \frac{\Delta T_{\text{total}}}{R_{\text{total}}}$$

Figure 4.9 Equivalent electrical circuit for heat transfer through wall with convection on both sides.

where ΔT_i denote the temperature drops at resistances R_i. The total thermal resistance is the sum of the individual resistances

$$R_{\text{total}} = \frac{1}{\alpha_{\text{conv,ins}} \cdot A} + \frac{1}{\alpha_{\text{cond}} \cdot A} + \frac{1}{\alpha_{\text{conv,out}} \cdot A} \tag{4.10}$$

As a consequence, the largest individual resistance dominates the total heat transfer.

For composite walls, it is often convenient to introduce an overall heat transfer coefficient U of a wall [1] (in Europe, it was previously also denoted as k-value; in the United States, the reciprocal $1/U$ is also called R-value [4]) by Eq. (4.11):

$$\dot Q = U \cdot A \cdot \Delta T \tag{4.11}$$

U is given in W (m² K)$^{-1}$ and describes the amount of energy per second, which is transmitted through a surface of 1 m² through a wall if the temperatures on both sides of the wall differ by 1 K. Adapting the actual surface areas and temperature differences then give the total heat transfer according to Eq. (4.11) (for unit change to BTU/(h ft² F), see [4]). Comparing Eq. (4.11) with (4.9) gives

$$U = \frac{1}{R_{\text{total}} \cdot A} = \frac{1}{\dfrac{1}{\alpha_{\text{conv,ins}}} + \sum \dfrac{s_i}{\lambda_i} + \dfrac{1}{\alpha_{\text{conv,out}}}} \tag{4.12}$$

Eqs (4.9) or (4.11) and (4.12) allow us to measure total heat transfer rates if thermal resistances, that is, heat transfer coefficients, are known or vice versa if the temperatures are measured.

These quantities shall be illustrated with a specific example of a composite wall (Figure 4.10).

A brick wall with thickness 24 cm ($\lambda_{\text{br}} = 0.5$ W (m K)$^{-1}$) has at the inside a 12-mm thick layer of plaster ($\lambda_{\text{pl}} = 0.7$ W (m K)$^{-1}$) and at the outside a 60-mm thick layer of Styrofoam ($\lambda_{\text{st}} = 0.04$ W (m K)$^{-1}$). The Styrofoam is additionally covered with a thin layer of special plaster on the outside, whose thermal resistance is neglected for simplicity. The heat transfer coefficients for convection are assumed to be $\alpha_{\text{inside}} = 8$ W (m² K)$^{-1}$ and $\alpha_{\text{outside}} = 25$ W (m² K)$^{-1}$. Therefrom, the U-value is found to be $U = 0.46$ W (m² K)$^{-1}$. It would decrease to about 0.32 W (m² K)$^{-1}$ if the Styrofoam layer would increase in thickness from 60 to 100 mm.

The total heat flow per area $\dot Q/A = U \cdot \Delta T$ is then about 14.8 W m^{-2} for 60-mm Styrofoam and 10.1 W m^{-2} for 100-mm Styrofoam. We only calculate temperatures

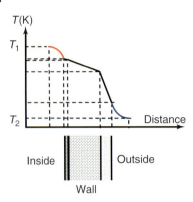

Figure 4.10 Example of composite stone wall of a building for winter conditions (for details, see text).

for the 60-mm Styrofoam example. The inside temperatures follow from inverting Eq. (4.9) to give

$$\Delta T_i = \frac{\dot{Q}}{A} \cdot \frac{1}{\alpha_i} \tag{4.13}$$

with coefficients α_i from Eqs. (4.1) and (4.2). Starting from inside, we obtain the following temperatures:

Air in room inside	$T_1 = 20\,°\mathrm{C}$
Inside wall	$T_{S1} = 18.2\,°\mathrm{C}$
Boundary plaster–brick	$T_{B1} = 17.9\,°\mathrm{C}$
Boundary brick–Styrofoam	$T_{B2} = 10.8\,°\mathrm{C}$
Outside wall	$T_{S2} = -11.4\,°\mathrm{C}$
Air outside	$T_{S1} = -12\,°\mathrm{C}$

that is, the freezing point temperature lies in the Styrofoam layer, as is desired. The necessity of building insulation becomes evident if we consider a wall system where the Styrofoam layer is missing and the brick stone thickness is larger by 60 mm such that the total thickness is the same. In this case, $U = 1.28\,\mathrm{W\,(m^2\,K)^{-1}}$ and the total heat flow per area would increase to 40.9 W m^{-2}, that is, about a factor of 2.8 larger than that with insulation. Besides much larger energy costs for heating, the freezing point lies within the bricks, which may cause structural problems if moisture enters the wall.

Regarding the emerging energy crisis and consequent legislative measures like more tight energy conservation regulations for buildings, a typical future application of thermography may be verification of building insulation by measuring U-values of building envelopes (Chapter 6). Table 4.6 gives some typical values for U for various materials or constructions.

Table 4.6 Typical *U*-values for certain building materials.

Material	Thickness (cm)	U in W $(m^2\ K)^{-1}$
Concrete wall, no thermal insulation	25	≈3.3
Brick wall	25	≈1.5
Brick wall plus thermal insulation	24 + 6	≈0.46
Massive wooden walls	25	≈0.5
Wooden or plastic entrance doors of houses	–	3–4
Windows (Section 4.3.4):		
Single pane	–	≈6
Double pane	–	≈3
For passive houses	–	≤ 1

4.3.4
Heat Transfer through Windows

Windows are present in many thermography applications, in particular in building inspections, and they are usually eye catchers in IR images. Owing to their importance, we present typical heat transfer characteristics. Problems in correctly interpreting surface temperatures are discussed in Chapter 6.

Consider a window with a glass size of $1.2 \times 1.2\ m^2$ and glass width of 4 mm and thermal conductivity $\lambda = 1\ W\ (m\ K)^{-1}$. The glass is surrounded by either a metal frame or a wooden frame of width 3 cm and thickness 5 mm ($\lambda = 220\ W\ (m\ K)^{-1}$ and $\lambda = 0.15\ W\ (m\ K)^{-1}$, respectively). The problem consists in calculating the U-value, the heat flux through pane and frame as well as surface temperatures for a single pane window and a double pane window when the inside air temperature is $20\ °C$ and outside air temperature is $-10\ °C$. In the latter case, the two panes shall be separated by 10 mm of pure air with $\lambda = 0.026\ W\ (m\ K)^{-1}$. The inner and outer heat transfer coefficients for convection shall be assumed as $\alpha_{in} = 8\ W\ (m^2\ K)^{-1}$ and $\alpha_{out} = 25\ W\ (m^2\ K)^{-1}$. Figure 4.11 depicts a cross-sectional view of the two pane window.

First, Eq. (4.12) is used to estimate U of the single pane window (glass alone): $U_{single} = 5.92\ W\ (m^2\ K)^{-1}$.

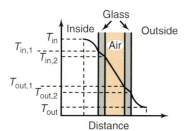

Figure 4.11 Cross-sectional view of double pane window.

For the double pane window (glass only), $U_{double} = 1.79$ W $(m^2\ K)^{-1}$. Therefrom the total heat transfer through the glass is found to be $\dot{Q}_{single} = 255.6$ and $\dot{Q}_{double} = 77.5$ W. Using Eq. (4.13), the various glass surface temperatures are evaluated. For the single pane, one finds for the inside and outside glass surfaces $T_{in} = -2.2\ °C$ and $T_{out} = -2.9\ °C$. Obviously, a single pane window is more or less isothermal. The very low inside glass surface temperature even allows generation of frost flowers. The double pane window has the huge advantage of the air-filled gap with low thermal conductivity.

The surface temperatures for the windows 1 (facing inside) and 2 (facing outside) are found to be $T_{in,1} = 13.3\ °C$, $T_{out,1} = 13.1\ °C$ and $T_{in,2} = -7.6\ °C$ as well as $T_{out,2} = -7.8\ °C$. Obviously, the window pane facing the inside has now much larger temperatures, which should be above typical dew point temperatures (Section 4.3.6). The inside of the outer window is however very cold; therefore, care must be taken so that the gas filling does not contain any water vapor, in order to avoid condensation. Very often gas fillings with noble gases are used.

Finally, the heat transfer through the frame shall be estimated. The total frame area is about $0.144\ m^2$ and Eq. (4.12) gives for the frame alone $U_{Al\text{-frame}} = 6.06$ W $(m^2\ K)^{-1}$ and $U_{wood\text{-frame}} = 5.04$ W $(m^2\ K)^{-1}$. Obviously, the heat transfer coefficients at the boundaries dominate the behavior. The heat flux through the frame alone is found to be $\dot{Q}_{Al} = 26.2$ W and $\dot{Q}_{wood} = 21.8$ W. Adding up the total heat flux through window and frame, we can define U-values for the whole window. For the single pane window, $U_{single,Al} = 5.93$ W $(m^2\ K)^{-1}$ and $U_{single,wood} = 5.84$ W $(m^2\ K)^{-1}$, and for the double pane window, $U_{double,Al} = 2.18$ W $(m^2\ K)^{-1}$ and $U_{double,wood} = 2.09$ W $(m^2\ K)^{-1}$.

4.3.5
Steady-State Heat Transfer in Two- and Three-Dimensional Problems: Thermal Bridges

In realistic applications, any object which is investigated with thermography is usually three-dimensional. The heat transfer in the corresponding geometries is more complex than the one-dimensional case. In particular, rectangular structures lead to a new phenomenon called *thermal bridge*. Consider, for example, the walls of a house. Figure 4.12 depicts a cross section of a corner segment of an outside wall. For the sake of simplicity a wall made of a single material is shown. Figure 4.12a shows schematically several isotherms (i.e., lines of constant temperature) within the wall for an inside temperature of $20\ °C$ and the outside temperature (in winter) of $-15\ °C$. Following the electrical analogy, isotherms in thermal physics correspond to equipotential lines in electrical situations. The current flows along electric fields, that is, the gradient of the electric potential. In the thermal situation, the heat flows along the gradients of the temperature distributions. This is indicated by the broken arrows.

In the planar sections of the wall (bottom and right hand side), the isotherms are parallel to the wall surfaces and one-dimensional calculations may be used. In this case, the heat which is transported to the outside wall area A comes from a similarly

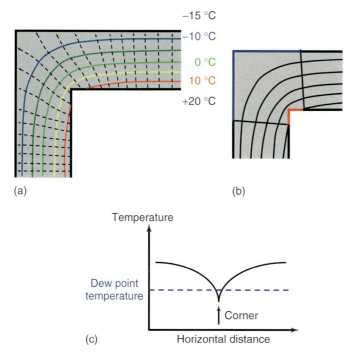

Figure 4.12 (a) Schematic illustration of a geometrical thermal bridge in a corner of a house during winter time. The contours of constant temperature are curved and the heat flux is perpendicular to them (broken lines with arrows). (b) The blue area of the outside wall, through which heat is transferred to the outside air, is much larger than the corresponding red area of the inside wall; the corner temperature drops. (c) If the corner temperature drops below the dew point temperature, mold may grow.

large surface area at the inside of the wall. In the corner section, however, the heat flow follows curved trajectories. As a result, there is a much larger outside wall area A_{out} (blue line in Figure 4.12b) than the corresponding inner wall area (red line Figure 4.12b), from which the heat is transported. As a consequence, the inner wall temperature must decrease in a corner (Figure 4.12c). For buildings this is very important: one must make sure that the corner temperature does not drop below the dew point temperature, that is, the temperature where condensation starts at the wall. If unnoticed, mold starts to build up. This can happen in situations with relative humidity around 80%.

Figure 4.12 is a schematic representation of the very general phenomenon called *thermal bridge*, which is observed in any building thermography. Thermal bridges can be due to geometry or due to neighboring materials with different thermal properties. Since thermal bridges lead to temperature differences, they are naturally present in IR images and the corresponding temperature change need not be due to a bad insulation. Thermal bridges are particularly important since temperatures may drop below the so-called dew point.

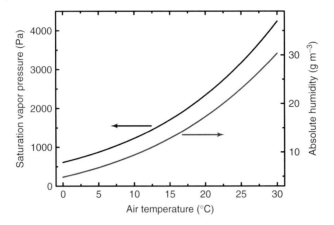

Figure 4.13 Vapor pressure and humidity as function of air temperature. Maximum absolute humidity (blue curve) corresponds to 100% relative humidity.

4.3.6
Dew Point Temperatures

In many applications of IR imaging, in particular from building thermography and outdoor inspections of industrial facilities, it is important to know whether the corresponding surfaces are dry or wet (of course for otherwise dry weather conditions), since wet surfaces under wind load suffer evaporative cooling which changes surface temperatures. Surfaces can become wet upon condensation of water vapor from the surrounding air. This happens whenever the air temperature drops below the corresponding dew point temperature.

This is due to the fact that air at a given temperature is only able to accommodate a certain percentage of water vapor. It is characterized either by the relative humidity or by the absolute humidity. The absolute humidity is a measure of the density of water vapor in air. Figure 4.13 depicts the maximum absolute humidity (related to the saturation vapor pressure) as a function of air temperature. This is the maximum amount of water which can be accommodated by air at this temperature. The relative humidity is a measure of the actual percentage of water vapor in the air, with regard to the maximum possible water vapor content. Typical indoor situations refer to 50% relative humidity. Whenever air contains more water vapor than the maximum possible amount, the water vapor will condense on surrounding surfaces as droplets of a thin film. This usually happens during cold nights in spring or fall. Air of a given water vapor content, say 50% relative humidity, starts in the evening with a high temperature of, say, 20 °C. This means that it contains $0.50 \times 17.3 \text{ g m}^{-3} = 9.65 \text{ g m}^{-3}$ water vapor. During clear night conditions, the atmospheric temperature may drop to less than 5 °C.

However, the relative humidity of the air increases during the cooling since colder air cannot accommodate as much water as warm air. In the present case,

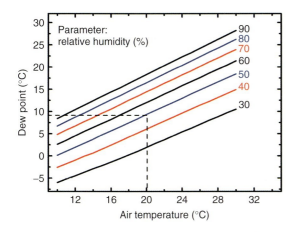

Figure 4.14 Dew point temperatures as a function of air temperature and relative humidity. For typical in-house temperatures of 20 °C and air with 50% relative humidity, the dew point is at 9.7 °C, that is, water will condense at any surface colder than 9.7 °C.

the air reaches 100% relative humidity at a temperature of 9.7 °C, which is the corresponding dew point temperature. Since the air cools down further, the excessive water vapor in the air starts to condense on colder surfaces of leaves of trees, grass, and so on. If this happens indoors, water vapor condenses at inside walls, which may lead to mold. Therefore, building inspections must look quantitatively at low temperatures, for example, from thermal bridges in corners, and so on, and investigate whether dew point temperatures are reached.

Figure 4.14 depicts plots of dew point temperatures as a function of air temperature, the parameter being relative humidity.

4.4
Transient Effects: Heating and Cooling of Objects

So far, steady-state conditions were assumed, that is, situations as shown in Figure 4.6b,c. In many cases, however, IR imaging is done while an object is either heated or is cooling down. Imagine, for example, a building wall which is exposed to the sun. It will absorb visible radiation and heat up, disguising any thermal signature of insulation problems of the wall itself. Similarly, electrical equipment under load may change temperature upon load changes during an investigation. In the following, the simplest theoretical description of temperature changes of solid or liquid objects due to heat sources is discussed. For example, only opaque objects will be assumed and the analysis must be modified for semitransparent objects [5]. We also later on only compute Biot numbers for quasi-steady-state conditions and neglect the transient case, which would require a more complex analysis with the need to introduce the Fourier number (e.g., [6]).

4.4.1
Heat Capacity and Thermal Diffusivity

Whenever an amount ΔQ of heat is transferred to or from an object, the object temperature changes according to

$$mc\Delta T_{\text{obj}} = \Delta Q \tag{4.14}$$

Here, m denotes the mass of the object and c the material-dependent specific heat (which, to first order, is assumed to be independent of T). Of course, a certain time is needed to reach the new thermal equilibrium within the object. The specific heat determines how much energy is needed for a given temperature change. Combined with the mass, $m \cdot c$ describes the energy storage ability of an object in Joules per Kelvin. In order to compare materials, one also uses the volumetric heat capacity $\rho \cdot c$ given in J (m^3 K)$^{-1}$ where ρ denotes the density of the material. This quantity describes the energy storage ability of an object per volume. Table 4.7 summarizes these quantities for a number of materials.

Knowing the energy storage capacity of an object and its thermal conductivity, it is possible to define the thermal diffusivity a_{diff} given in square meter per second (Section 3.5; we note that sometimes the diffusivity is denoted by α, which is also used for the heat transfer coefficients and for the absorptivity of radiation)

$$a_{\text{diff}} = \frac{\lambda}{\rho \cdot c} \tag{4.15}$$

Diffusivity is defined as the ratio of thermal conductivity and energy storage ability. The meaning of diffusivity becomes clear when considering a specific example. Consider a cube or sphere of a certain material at a given temperature which is immersed in a liquid of much higher temperature. The larger the thermal conductivity and the smaller the storage capacity of energy within the object, the faster is the thermal equilibrium established. Therefore, large values of a_{diff} mean that objects/materials will respond very quickly to thermal changes, establishing a new thermal equilibrium, whereas small values reflect materials where the corresponding processes take longer. Table 4.8 compares the thermal diffusivity for a number of materials.

As shown in Table 4.8, metals have large diffusivities, that is, they do redistribute heat much more quickly than other solids or liquids.

4.4.2
Short Survey of Quantitative Treatments of Time-Dependent Problems

The spatial and time-dependent distribution of energy in an object due to temperature differences is described by the heat diffusion equation [1]. The general form includes energy source or sink terms and the possibility of anisotropic thermal properties. Here we only consider the simplest case of one-dimensional heat transfer without any source or sink terms. In this case, the change of temperature with time at a given location within the object is connected to the spatial changes

Table 4.7 Specific heat, density, and volumetric heat capacity of some materials at 20 °C. For gases, the specific heat refers to constant pressure.

Material	Specific heat c in J (kg K)$^{-1}$	Density ρ in g cm^{-3}	$\rho \cdot c$ in J (m^3 K)$^{-1}$	Material	Specific heat in J (kg K)$^{-1}$	Density in g cm^{-3}	$\rho \cdot c$ in J (m^3 K)$^{-1}$
Aluminum	896	2.7	2.42×10^6	Foams Styrofoam	1300–1500	0.02–0.05	$2.6–7.5 \times 10^4$
Copper	383	8.94	3.42×10^6	Glass	500–800	2.5–4.0	$1.25–3.2 \times 10^6$
Silver	237	10.5	2.42×10^6	Water	4182	1.0	4.18×10^6
Steel	420–500	6.3–8.1	$2.6–4 \times 10^6$	Oils	1450–2000	0.8–1.0	$1.2–2 \times 10^6$
Concrete	840	0.5–5	$4 \times 10^5 – 4 \times 10^6$	Air	1005	1.29×10^{-3}	1.3×10^3
Stones	700–800	2.4–3	$1.7–2.4 \times 10^6$				
Dry wood	1500	0.4–0.8	$6 \times 10^5 – 1.2 \times 10^6$	CO_2	837	1.98×10^{-3}	1.7×10^3

Table 4.8 Thermal diffusivity for certain materials. For those with varying composition (steel, concrete, stones, wood, glass, oil) either specific materials or reasonable average values are given.

Material (metals)	a_{diff} (m² s⁻¹)	Material (other solids)	a_{diff} (m² s⁻¹)	Material (liquids, gases)	a_{diff} (m² s⁻¹)
Aluminum (99%)	90×10^{-6}	Concrete, Stones	0.66×10^{-6}	Water	0.14×10^{-6}
Copper	110×10^{-6}	Dry wood	0.17×10^{-6}	Synthetic oil	0.11×10^{-6}
Silver	170×10^{-6}	Foams, Styrofoam	0.7×10^{-6}	Air	20×10^{-6}
Stainless steel	4×10^{-6}	Quartz glass	0.85×10^{-6}	CO_2	9.4×10^{-6}

of temperature (Eq. (4.16)).

$$\frac{\partial T(x, t)}{\partial t} = a \cdot \frac{\partial^2 T(x, t)}{\partial x^2} \tag{4.16}$$

Here, a is the thermal diffusivity, introduced above (Eq. (4.15)). Equation (4.16) determines the temperature distribution completely, provided boundary conditions (e.g., initial temperature distribution, initial heat flux, etc.) are given. Unfortunately, Eq. (4.16) can only be solved analytically for a few special cases; mostly it is solved numerically. Here we sketch the solution for temperature distributions of the simple geometries of plane, cylinder, and sphere upon a sudden change of temperature of the surrounding (think of putting an object in a fluid of different temperature).

The geometries are approximated by one-dimensional objects, that is, we treat a plate of thickness 2s in x-direction, but of infinite dimension in y and z directions. An infinitely long cylinder with radial coordinate x and a sphere, defined by its radius x, are treated in a similar fashion. The solution $T(x, t)$ is usually expressed in terms of dimensionless temperature, dimensionless time (Fourier number), and Biot number (Table 4.9). $T(x, t)$ is then written as a so-called Fourier series expansion. Its numerical results are often depicted in terms of temperature at object surface ($T_{surface}$), temperature at object center (T_{center}), and temperature averaged over the object ($T_{average}$) as a function of Biot number with the Fourier number being a parameter.

As an example, Figure 4.15 depicts results for the dimensionless surface temperature $T_{surface}$ of a sphere (similar results can be found for the other geometries [7]).

Such figures can be used to graphically solve problems of transient phenomena. As an example, let us assume a sphere of diameter 2R which is initially at temperature T_0. At time $t = 0$ it is put into a fluid (the fluid, e.g., water, is assumed not to change its temperature upon heating up the sphere) at temperature T_∞. The problem is to use Figure 4.15 to find the surface temperature as a function of time (and from similar plots the temperature at the center and the average temperature of the sphere). The solution is simple. One needs to calculate the thermal diffusivity,

4.4 Transient Effects: Heating and Cooling of Objects

Table 4.9 Dimensionless quantities commonly used to represent solutions for $T(x,t)$.

Quantity	Dimensionless temperature Θ	Dimensionless time (Fourier number)	Biot number
Definition	$\Theta = \dfrac{T - T_\infty}{T_0 - T_\infty}$	$Fo = \dfrac{a \cdot t}{s^2} = \dfrac{\lambda \cdot t}{\rho \cdot c \cdot s^2}$	$Bi = \dfrac{\alpha_{conv}}{\lambda/s}$
Meaning	Fraction of realized temperature change	Ratio of heat transfer to change of stored thermal energy of object	Ratio of heat transfer at boundary to that within object

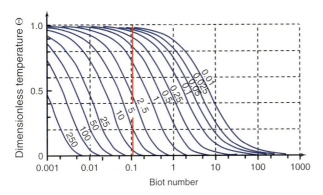

Figure 4.15 Schematic plots for dimensionless surface temperature of a sphere as a function of Biot number with the Fourier number as parameter (for details concerning the example (red line) see text, after [7]).

Biot number, and Fourier number as a function of time. For boiling water, one must assume a meaningful value for the convective heat transfer coefficient. The solution is then found as follows: one draws a vertical line at the position of the Biot number, which is relevant for the problem (e.g., red line in Figure 4.15). The intersections with the curves belonging to different Fourier numbers give the dimensionless temperatures for that Fourier number, which actually is a measure of time. Schematic results for the surface, center, and average temperatures of the sphere problem with small Biot number are given in Figure 4.16 for heating from 20 to 100 °C or cooling from 100 to 20 °C. They can be directly related to the temperature distribution within the sphere, as shown in Figure 4.16c (similar to plots in Figure 4.7).

Figures 4.15 and 4.16 illustrate the problems associated with the interpretation of surface temperature changes of spatial temperature distributions within objects. Figure 4.16, however, also shows that for small Biot numbers (as in this case), $T_{surface}$ still resembles a more or less reasonable approximation for the average temperature $T_{average}$.

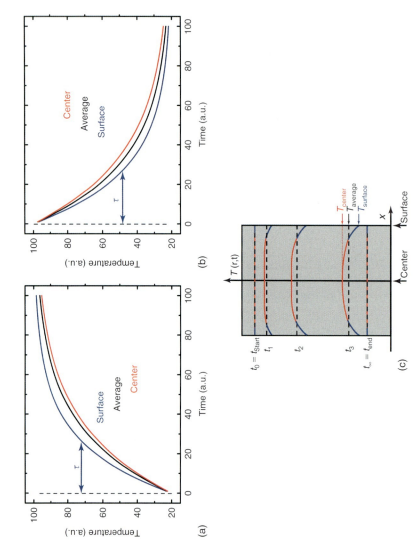

Figure 4.16 Schematic representation of surface, center, and average temperature of a sphere at small Biot number, heated (a) or cooled (b) due to an abrupt temperature, after dropping it into a fluid. The corresponding temperature distributions within the sphere are depicted in (c). From such curves one may determine corresponding time constants as indicated for the surface temperature plots.

4.4.3
Typical Time Constants for Transient Thermal Phenomena

In thermography, it is convenient to describe and interpret a thermal steady-state solution. Unfortunately, one often encounters situations where one observes a heating or cooling of objects due to thermal contact with another object, due to heat sources or sinks within the object, and so on. Whenever this happens, interpretation of IR images can become complex. In this case, it is desirable to at least get a best guess for the respective time constants which are associated with temperature changes. This will be illustrated for cooling processes which are analyzed using IR imaging.

4.4.3.1 Cooling Cube Experiment

Aluminum metal cubes of 20-, 30-, 40-, and 60-mm side length were heated up in a conventional oven on a metal grid. The cubes were covered with a high temperature stable paint to provide high emissivity (in detail, three sides were covered with the paint of $\varepsilon \approx 0.85$, whereas the other three sides were left as polished metal of $\varepsilon \approx 0.05$). Ample time for the heating was given such that all cubes were in thermal equilibrium within the oven at a temperature of 180 °C. The IR imaging experiment started when opening the oven and placing the metal grid with the cubes onto some thermal insulation on a table. Figure 4.17 depicts two snapshots of the cooling process.

Obviously, the smallest cubes cool best as can also be seen in the temperature profiles as a function of time (Figure 4.18). The temperature as a function of cooling time can be fitted with a simple exponential decrease, as can be seen in the expanded plot in Figure 4.18. Since we are dealing with small Biot numbers this can be explained by simple theoretical assumptions.

4.4.3.2 Theoretical Modeling of Cooling of Solid Cubes

For the metal cubes, the Biot numbers follow from the values of heat conductivity $\lambda \approx 220$ W (m K)$^{-1}$, size s between 20 and 60 mm, and typical values for heat transfer coefficients for free convection (solids to gases) in the range 2–25 W (m^2 K)$^{-1}$. We find λ/s between 11 000 and 3667 W (m^2 K)$^{-1}$ giving Bi $\ll$ 1, that is, we can expect a temperature equilibrium within any of the metal cubes. Supposing an initial temperature T_{init} of the cubes, energy conservation requires that any heat loss will lead to a decrease of the thermal energy of the cube, that is,

$$mc\frac{dT_{\text{obj}}}{dt} = -\dot{Q}_{\text{cond}} - \dot{Q}_{\text{conv}} - \dot{Q}_{\text{rad}} \qquad (4.17)$$

where m is the mass of the cube, c is the specific heat (here assumed to be independent of T), and dT_{obj}/dt denotes the decrease in (uniform) temperature of the objects (here cubes) due to the losses.

Using Eqs. (4.1–4.3) for the heat losses would lead to a nonlinear differential equation which is difficult to solve analytically. However, if the radiative cooling contribution can be used in the linearized form (Eq. (4.5)), Eq. (4.17) turns into a

268 | *4 Some Basic Concepts of Heat Transfer*

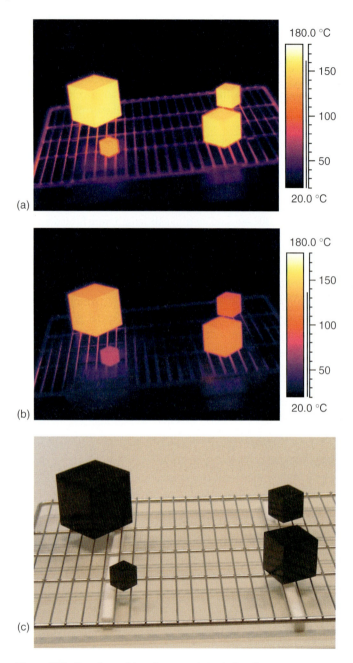

Figure 4.17 Two thermal imaging snapshots during the cooling of paint-covered aluminum cubes of different sizes and visible image, showing the cubes on a grid and thermal insulation on the lab table.

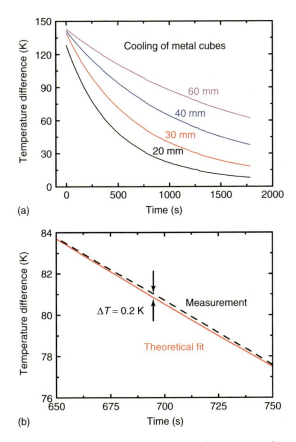

Figure 4.18 Temperature as a function of cooling time for the aluminum cubes of various sizes (a) and example for a simple exponential fit for the 40-mm cubes. (b) The fit is so close to the experimental data that only an expanded portion can illustrate the differences to the experiment.

conventional linear differential equation

$$mc\frac{dT}{dt} = -\alpha_{\text{total}} \cdot A \cdot (T_{\text{obj}} - T_{\text{surr}}) \quad \text{where}$$

$$\alpha_{\text{total}} = \alpha_C + \alpha_R = \alpha_C + \varepsilon \cdot \sigma \cdot k_{\text{appr}} \tag{4.18}$$

Here α_C accounts for the sum of conduction and convection. α_{total} additionally includes the linearized radiative heat transfer. The solution for $t_0 = 0$ for our metal cubes can be written as

$$T_{\text{obj}}(t) = T_{\text{surr}} + (T_0 - T_{\text{surr}}) \cdot e^{-t/\tau} \quad \text{with time constant}$$

$$\tau = \frac{\rho c (V/A)}{\alpha_{\text{total}}} \tag{4.19}$$

Here, ρ is the density of the object material, c is the specific heat, and V/A is proportional to x, the length of the cube. Equation (4.19) predicts that the difference between initial temperature T_0 and surrounding air temperature T_{surr} drops exponentially. The characteristic time constant τ describes the time after which the temperature difference has dropped to $1/e \approx 0.368$ of its original value. In this type of problem, it depends on the ratio of size/α_{total}. As expected, the larger the size and the smaller the effective total heat transfer coefficient, the longer it takes to cool the cube.

4.4.3.3 Time Constants for Different Objects

For the practitioner, it is important to get a feeling for the time constants τ involved in the heating and cooling of objects. In the following, several examples are given for typical values/ranges of τ. Rather than analyzing every situation in detail, order of magnitude estimates are given using Eq. (4.19), assuming a constant $(\alpha_{con} + \alpha_{rad}) = 15$ W (m² K)$^{-1}$ for the sum of convective and radiative heat transfer. Table 4.10 gives a summary. Numbers are rounded to give an indication of the order of magnitude. Equation (4.19) is based on the validity of $Bi < 0.1$. In all cases where Bi is larger, the time constants represent lower limits, that is, real times will be even longer. It has to be kept in mind that within one time constant, a temperature difference ΔT only decreases to $(1/e) \cdot \Delta T$, that is, thermal equilibrium requires at least five time constants or more.

Obviously, time constants for large buildings can be in the range of many hours. This is crucial in outdoor building thermography, in particular if solar load or night sky radiant cooling are effective as time-dependent additional heat sources or sinks.

Table 4.10 Typical time constants for heating and cooling of objects.

Object	Dimensions	Material constants	Comments	Time constant τ (s)
Aluminum metal cube	20–60 mm	$\rho = 2700$ kg m^{-3} $c = 900$ J (kg K)$^{-1}$	–	≈ 500 ≈ 1500
Halogen light bulb	Thickness 1 mm	$\rho = 2500$ kg m^{-3} $c = 667$ J (kg K)$^{-1}$	–	≈ 100
Liquid container for 0.5-l water	Cylinder, for example, $R =$ 3.35 cm $h =$ 14.2 cm	$\rho = 1000$ kg m^{-3} $c = 4185$ J (kg K)$^{-1}$ for water	Container material can add to heat capacity and τ	Order of 4000
Single brick stone	24 × 12 × 8 cm³	$\rho = 1500$ kg m^{-3} $c = 850$ J (kg K)$^{-1}$	Typical for buildings	≈ 1700
Concrete	0.3 × 0.3 × 1 m³	$\rho = 2000$ kg m^{-3} $c = 850$ J (kg K)$^{-1}$	Typical for building foundation	$\approx 10\,000$
Stone wall	0.24 × 3 × 10 m³	$\rho = 2000$ kg m^{-3} $c = 850$ J (kg K)$^{-1}$	Typical for stone buildings	$\approx 14\,000$

Estimating time constants is easy, if the temperature varies exponentially with time as given in Eq. (4.19). The validity of the underlying assumption that the total heat transfer rates depend linearly on the temperature difference (sometimes this dependence is denoted as Newton's law) is, however, not straightforward. As a matter of fact, it is quite surprising that the linearization of Eqs. (4.4) and (4.5) seems to work over the extended temperature range of around $\Delta T = 100$ K as indicated in Figure 4.18. As shown below, this unexpected behavior can be explained with the result that whenever temperature differences are below 100 K, a simple exponential function is sufficient to explain heating and cooling of objects. In this case, time constants as derived above may be used as estimates for the transient thermal behavior of the objects.

4.5
Some Thoughts on the Validity of Newton's Law

The cooling of objects with small Biot numbers according to Eq. (4.18) does lead to an exponentially decreasing temperature difference with time (Eq. (4.19)). We briefly summarize the assumptions which led to Eq. (4.19), sometimes referred to as *Newton's law of cooling* [8]:

1) The object is characterized by a single temperature ($Bi \ll 1$).
2) For small temperature differences $\Delta T (\Delta T \ll T_{obj}, T_{surr}$ with absolute temperatures in K), radiative heat transfer may be approximated by its linearized form where the heat transfer coefficient is constant (does not depend on temperature). In the following, we discuss in detail how small ΔT may be.
3) The convective heat transfer coefficient is assumed to stay constant during the cooling process.
4) The temperature of the surrounding stays constant during the cooling process; this means that the surroundings must be a very large thermal reservoir.
5) The only internal energy source of the object is the stored thermal energy.

It is quite easy to experimentally fulfill requirements (5) and (4) and (1). The convective heat transfer (3) assumption is more critical. In experiments, it can be kept constant using steady airflow around objects, that is, for forced convections. If experiments use free convections, α_c may depend on the temperature difference. In the following theoretical analysis, we however focus on the influence of the linearization of radiative heat transfer. In particular we discuss the question of whether the linearization of Eq. (4.18) does also work over extended temperature ranges.

In a more accurate description of such a cooling process, the radiation contribution must be treated in its original nonlinear form.

$$mc \frac{dT}{dt} = -\alpha_{con} \cdot A \cdot (T_{obj} - T_{surr}) + \varepsilon \cdot \sigma \cdot A \cdot (T_{obj}^4 - T_{surr}^4) \qquad (4.20)$$

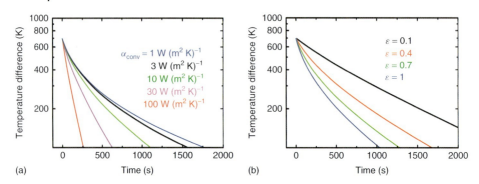

Figure 4.19 Numerical results of Eq. (4.20) for Al metal cubes. (a) Variation of convective heat transfer coefficient α_{conv} from 1 to 100 W (m² K)$^{-1}$ for fixed size ($s = 40$ mm) and $\varepsilon = 0.9$. (b) Variation of emissivity for fixed $\alpha_{conv} = 10$ W (m² K)$^{-1}$ and cube size $s = 40$ mm.

4.5.1
Theoretical Cooling Curves

Equation (4.20) was numerically solved for a specific example of painted aluminum cubes of 40 mm size since such cubes were used in one of the experiments. They may serve as a theoretical model system with simple geometry. The cooling depends on three parameters: first the cube size (in general, this relates to the energy storage capability of the object), second the convective heat transfer coefficient, and third the emissivity of the object, that is, the contribution of radiative heat transfer. The variation of cube size directly relates to the fact that the time constant for cooling is linearly proportional to size.

Figure 4.19 depicts the results while varying convection and emissivity independently in semilogarithmic plots. Newton's law would be represented by straight lines.

Obviously, one observes curved cooling plots, that is, deviations from the simple exponential behavior (straight line) for all investigated sizes. The variation of convective heat transfer coefficient (a) does have a strong impact on linearity of the plot. The larger α_{conv}, the more the plot follows a straight line. Similarly, the variation of emissivity (b) shows that very small emissivities (e.g., $\varepsilon < 0.2$), that is, small contributions of radiative heat transfer, clearly favor Newton's law, that is, exponential cooling. In practice, ε cannot be varied in such wide ranges. Polished metal cubes will have ε values of the order of 0.1, whereas those painted with high-emissivity paint will have emissivities around 0.9. For experiments, intermediate values for the total radiative heat transfer may be realized by only painting a few sides of the cubes and leaving the others polished.

The numerical results of Figure 4.19 were, of course, expected as they are a direct consequence of Eq. (4.20), which explains the conditions for having either linear or nonlinear behavior. If the convection term is large and the radiation contribution

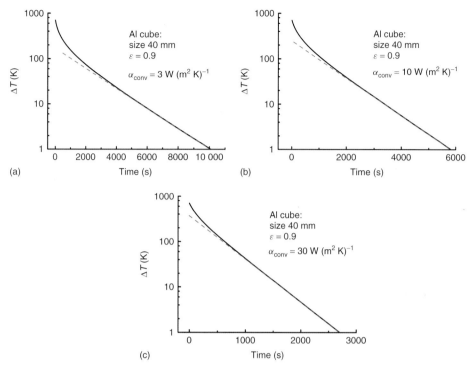

Figure 4.20 Theoretical cooling of Al metal cubes with size 40 mm and $\varepsilon = 0.9$ for different convective heat transfer coefficients (a–c). Newton's law would be a straight line such as the red broken lines, which closely describe the low temperature data, but show deviations for larger temperatures.

small, linear plots are expected and vice versa, whereas changing the size (ratio of mass and area) only has an effect on the timescale of the cooling process.

The degree of deviations from a straight line is depicted for three different convective heat transfer coefficients (3, 10, and 30 W (m² K)⁻¹) for 40-mm size cubes and $\varepsilon = 0.9$ in Figure 4.20. The initial temperature differences were assumed to be 700 K with regard to ambient temperature.

It is quite obvious that all plots show deviations from straight lines (red broken lines), which nicely fit the low temperature data. The larger the convective heat transfer, the smaller the deviations. For low convective losses of only 3 W (m² K)⁻¹, deviations can be expected for temperature differences as small as 40 K. In contrast for very high convective losses of 30 W (m² K)⁻¹, simple exponential cooling seems to work quite well for $\Delta T < 100$ K.

The results demonstrate that there is no general number for ΔT describing the range of validity of Newton's law. Rather, the corresponding temperature range depends on the experimental conditions and how close one looks for deviations. Plotting data only for small temperature differences can lead to the impression that

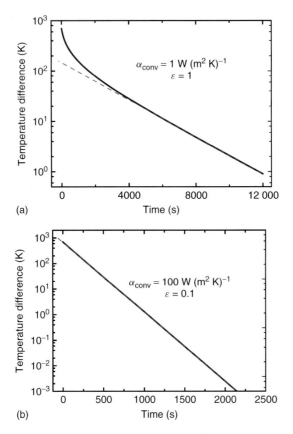

Figure 4.21 Extreme cases of cooling of Al metal cubes (s = 40 mm): small convection with large radiation heat transfer (a) and large convection with small radiative heat transfer (b).

the straight line fits work quite well since deviations are not as pronounced as for the high temperature range.

We finally consider the two extreme cases for cooling of objects, one where radiation dominates and another where convection dominates the cooling process. Results are depicted in Figure 4.21. The smallest imaginable realistic value for convective heat transfer is in the range of 1 W (m² K)$^{-1}$, the largest corresponding radiative heat transfer occurs for black bodies, that is, setting $\varepsilon = 1.0$ as may be realized by metal cubes covered with high-emissivity paint. In this case, the cooling curve starts to deviate from Newton's law for $\Delta T \approx 30$ K. In contrast, polished metal cubes with low emissivity reduce radiation losses. The convective losses may simultaneously be enhanced by directing fans with high air speed onto the objects. In this case, very high values of up to 100 W (m² K)$^{-1}$ seem possible. In this case, no deviation from the straight line plot is observable, that is, Newton's law would hold for the whole temperature range of $\Delta T = 500$ K.

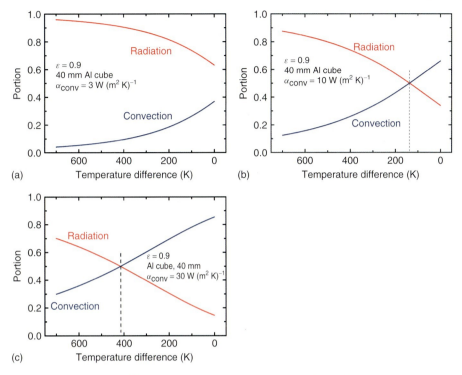

Figure 4.22 Relative contributions of convection and radiative cooling for Al cubes of 40 mm size with $\varepsilon = 0.9$ as a function of temperature. The initial temperature was 993 K, that is, $\Delta T = 700$ K. For small convective heat transfer coefficients, radiative cooling is dominant throughout the cooling process, whereas for larger convection there will be a change of the dominant cooling contribution at a certain temperature.

4.5.2
Relative Contributions of Radiation and Convection

In order to further understand the relevance of the nonlinearities during the cooling of objects, we now consider the relative contributions of radiation and convection heat transfer. Quick estimates for special cases are possible using Eqs. (4.2) and (4.3). Even close to room temperature, radiative losses are surprisingly large. This must be noted since many argue that radiation losses can be neglected close to room temperature which is incorrect. Let us assume a background temperature of $\approx 20\,°C$, that is, $T_{surr} = 293$ K. At 300 K, a blackbody ($\varepsilon = 1$) will then emit about 41 W m^{-2} which is of the same order of magnitude as typical convection losses of 63 W m^{-2} (for $\alpha_{con} = 9$ W (m^2 K)$^{-1}$ or 14 W m^{-2} (for $\alpha_{con} = 2$ W (m^2 K)$^{-1}$).

A small radiative or a large convective contribution reduce the nonlinear effects in Eq. (4.20). This is shown in Figure 4.22, which depicts the relative contributions of convective and radiative heat transfer for the 40-mm Al cubes by assuming $\varepsilon = 0.9$ and three different values, $\alpha_{conv1} = 3$ W (m^2 K)$^{-1}$, $\alpha_{conv2} = 10$ W (m^2 K)$^{-1}$, and $\alpha_{conv1} = 30$ W (m^2 K)$^{-1}$, for convective heat transfer.

For very small convective heat transfer of 3 W (m² K)⁻¹, radiation dominates the total energy loss of the object from the beginning to the end. This easily explains why one necessarily expects strong deviations from Newton's law for small temperature differences in this case. For larger convection coefficients like 10 or 30 W (m² K)⁻¹, there is a cooling time, that is, transition temperature difference, where the dominant cooling changes from radiation to convection. For 10 W (m² K)⁻¹, this happens at $\Delta T = 137$ K (after $\approx$840 s in a $T(t)$ plot), and for 30 W (m² K)⁻¹ it happens at $\Delta T = 415$ K (after 112 s). Qualitatively, it makes sense that the higher this transition temperature difference, the larger the range of validity of Newton's law of cooling. For the convective heat transfer of 100 W (m² K)⁻¹ and $\varepsilon = 0.1$ (Figure 4.21), convection would dominate radiation right from the beginning, which easily explains the linear plot.

From the theoretical analysis, it is clear that radiative cooling should lead to deviations from Newton's law of cooling above critical temperature differences, which, in some of the discussed cases, were below 100 K. This raises the question, why many experiments reported the applicability of Newton's law for a temperature difference of up to 100 K. The answer is simple [8]: if one waits long enough, any cooling process can probably be described by a simple exponential function.

4.5.3
Experiments: Heating and Cooling of Light Bulbs

The Al metal cubes could only be heated to about 180 °C. At these temperatures, the curvature of the cooling curves is not very pronounced, though definitely observable [8]. Therefore in order to study higher temperatures experimentally, we used light bulbs. Several light bulbs of different power consumption and size were tested. Experiments were performed with small halogen light bulbs (near cylinder diameter 11 mm, height 17 mm (Figure 4.23)). Figure 4.24 shows an IR image of the halogen bulb while being hot, as well as measured surface temperatures.

Figure 4.23 Examples of investigated light bulbs. Samples were placed in front of a room temperature cork board.

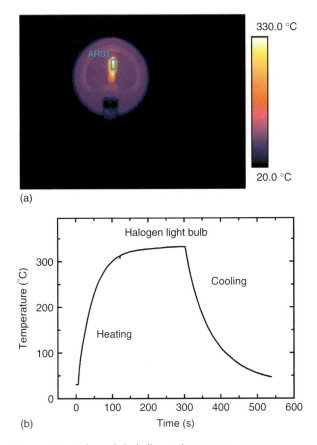

Figure 4.24 Halogen light bulbs reach maximum temperatures > 330 °C with small relative temperature variations (a). The heating and cooling of the halogen light bulb covers a temperature difference of more than 300 K (b).

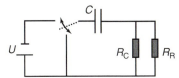

Figure 4.25 Equivalent electrical circuit representing heating and cooling of an object. Its heat capacity resembles the capacity. The thermal losses via convection and radiation are represented by two resistances in parallel.

For the following analysis, the maximum temperatures in a small area around the top of the light bulb were used (a test using average rather than maximum temperatures within the indicated area showed that the general form of normalized $T(t)$ curves as in Figure 4.24 changed very little, that is, any conclusions drawn from time constants, derived from these data, do hold).

For the measurement in Figure 4.24 the halogen light bulb was powered up until an equilibrium temperature was reached, then the power was turned off and the cooling curve was recorded. Compared to the cooling of the metal cubes with times of the order of 1000 s, the cooling of the light bulb occurs much faster. This is mostly due to the small amount of stored thermal energy in the mass of the light bulb. Hence this illustrates the characteristic fact that small systems with small stored energy $c \cdot m \cdot \Delta T$ have much faster time constants to reach their equilibrium. The smaller the system, the smaller the corresponding time constants (Table 4.10, see also chapter 8).

Similar to Section 4.3.3, the heating and cooling of objects can be described by equivalent electrical circuits. Figure 4.25 depicts an RC-circuit which can be connected to a voltage supply. The capacitor resembles the light bulb and the resistances correspond to the losses due to convection and radiation. The charging

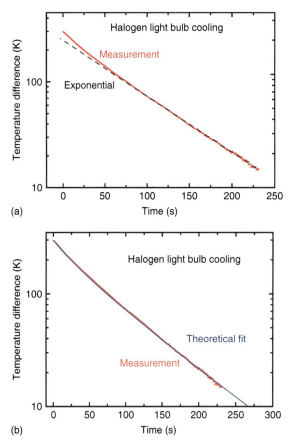

Figure 4.26 Cooling of the halogen light bulb. Again, deviations occur for $\Delta T > 100$ K as indicated by the straight line (a). The data can be fitted by a double exponential fit (b), where the time constant of the initial part is 16.7 s (amplitude 37.7 K) and that for the part with the slower decrease is 77.3 s (amplitude 259.3 K). For the smallest temperatures, the signal is very noisy. This is due to the fact that the temperature range of the camera was fixed at 80–500 K during the measurement; hence data below 80 K must be considered with care.

of the capacitor corresponds to the heating of the light bulb. The discharging corresponds to the cooling of the light bulb. Obviously, the voltage across the capacitor shows the same behavior as the temperature of the light bulb.

Experimental results for the cooling of the halogen light bulb (Figure 4.26) is nicely represented by an exponential for small ΔT values, but deviates for larger ΔT. This is exactly as expected theoretically.

References

1. Incropera, F.P. and DeWitt, D.P. (1996) *Fundamentals of Heat and Mass Transfer*, 4th edn, John Wiley & Sons, Inc., New York.
2. Baehr, H.D. and Karl, S. (2006) *Heat and Mass Transfer*, 2nd revised edn, Springer, Berlin and New York.
3. Möllmann, K.-P., Pinno, F., and Vollmer, M. (2007) Influence of wind effects on thermal imaging – is the wind chill effect relevant? Inframation 2007, Proceedings, ITC 121 A May 24, 2007.
4. Madding, R. (2008) Finding R-values of stud frame constructed houses with IR thermography. Inframation 2008, Proceedings vol. 9, pp. 261–277.
5. Sazhin, S.S. (2006) Advanced models of fuel droplet heating and evaporation. *Prog. Energy Combust. Sci.*, **32**, 162–214.
6. Sazhin, S.S., Krutitskii, P.A., Martynov, S.B., Mason, D., Heikal, M.R., and Sazhina, E.M. (2007) Transient heating of a semitransparent spherical body. *Int. J. Therm. Sci.*, **46**, 444–457.
7. VDI Wärmeatlas (in German) (2006) *Verein Deutscher Ingenieure, VDI-Gesellschaft Verfahrenstechnik und Chemieingenieurwesen (GVC)*, 10th edn, Springer, Berlin.
8. Vollmer, M. (2009) Newton's law of cooling revisited. *Eur. J. Phys.*, **30**, 1063–1084.

5
Basic Applications for Teaching: Direct Visualization of Physics Phenomena

5.1
Introduction

Infrared thermal imaging allows quantitative and qualitative imaging of a multitude of phenomena and processes in physics, technology, and industry. During the last decade, thermography has also started to become popular for physics teaching at universities since it allows visualization of phenomena dealing with minute energy transfer, for example, in processes involving friction, which cannot be easily demonstrated with other methods [1–4]. Therefore, in this chapter, the focus is on selected applications of qualitative IR imaging of phenomena for physics education. The examples are intended to inspire more experiments by demonstrating how IR imaging can be used in teaching physics and in visualizing fundamental principles and processes. Unfortunately, seemingly simple phenomena very often involve complex explanations. Therefore, despite the simplicity of the phenomena, a complete quantitative analysis is far beyond the scope of this chapter. The topics are arbitrarily divided into the classical categories of physics i.e., mechanics, thermal physics, electromagnetism, and optics, followed by radiation physics as an example for using thermography in "modern physics." Of course, many other applications, which are treated in later chapters can and should also be used for physics teaching such as, for example, thermal reflections (Section 9.2), detection of gases (Chapter 7), building insulation (Chapter 6), heat sources in electrical components (Section 9.7), and so on. More details to the physics of the phenomena can be found in nearly every textbook on introductory physics (e.g., [5, 6]).

5.2
Mechanics: Transformation of Mechanical Energy into Heat

A very important field for IR imaging in physics education concerns the visualization of mechanical phenomena involving friction. The most important everyday phenomenon concerns our ability to move around. Walking, riding bicycles, motorbikes, or cars is only possible due to frictional forces between the shoes/tires on one hand and the floor/the street on the other hand. Whenever there is a force acting

Infrared Thermal Imaging. Michael Vollmer and Klaus-Peter Möllmann
Copyright © 2010 WILEY-VCH Verlag GmbH & Co. KGaA, Weinheim
ISBN: 978-3-527-40717-0

along a given direction for a given distance, work is done, which is finally converted into thermal energy (often briefly but not very precisely denoted as heat). For sliding friction, this will ultimately lead to a temperature rise of the two areas that are in contact. In contrast, static friction which is the physical basis for walking or driving on vehicles with wheels will not convert work into thermal energy. In these cases, a closer look will show that inelastic deformations of the two touching objects will, however, also produce heat, which can be made visible with IR imaging.

5.2.1
Sliding Friction and Weight

Whenever two dry unlubricated solid surfaces slide over each other, there are frictional forces, which can be expressed by the empirical law $F_{\text{friction}} = \mu \cdot F_{\text{normal}}$, where $\mu < 1$ is the coefficient of friction and F_{normal} is the normal force with which each surface presses onto the other (e.g., [5, 6]). One distinguishes coefficients for static friction μ_{static} (no movement yet) and for kinetic friction μ_{kinetic}, that is, after a sliding of the two surfaces has been realized. Some typical μ-values for sliding friction, for example, for wood on wood or a car tire on the pavement of a street are in the range of 0.5. If some object is, for example, sliding across the floor, work must be done against frictional forces. Imagine that after a while the objects have a constant sliding velocity. In this case, the work is only used to overcome the kinetic frictional forces. It is ultimately converted into thermal energy, that is, the temperature of the two sliding surfaces will rise.

In order to analyze these effects of frictional energy transfer in more detail, two different weights of 1 and 5 kg respectively, were placed on small wooden plates and drawn simultaneously with constant speed across the floor (Figure 5.1). The heavier weight led to a much larger warming of the floor as expected, since the normal force was increased by a factor of 5. The plate surfaces were also heated up (not shown here). This experiment qualitatively demonstrates the effects of

Figure 5.1 Two weights of 1 kg (right) and 5 kg (left) were placed on wooden plates and simultaneously drawn across the floor. The temperature rise of the floor is easily observed.

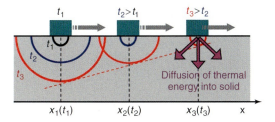

Figure 5.2 Transient thermal phenomena in sliding friction experiments arise from the fact that thermal energy diffuses into the bulk material, giving rise to spatial and time dependence of observable surface temperatures. The distance of energy diffusion into the bulk material as a function of time is indicated by the semicircles.

frictional forces in mechanics. A quantitative analysis would be quite complex. On one hand, it would require exact measurements of the frictional forces, and, on the other hand, the corresponding mechanical work would be split up into heating of both surfaces, depending on their thermal material properties. Finally, the diffusion of thermal energy from the directly heated contact surfaces would lead to transient effects, which means that a realistic modeling would require to record time sequences of this problem.

Figure 5.2 schematically illustrates (for the lower surface only) the transient effects of thermal energy diffusion into the bulk of the solid. It depicts a small object (blue–green) which is moving with constant speed across a solid surface. Three snapshots are shown at times t_1, t_2, and t_3 at which the object was at location x_1, x_2, and x_3 respectively. Owing to the work done by the object, the temperature at the contact spots rises to a maximum and then drops as a function of time due to first a lateral diffusion and second a diffusion of the energy into the bulk material. This transient behavior is characteristic for sliding friction phenomena.

Obviously, very simple looking basic physics phenomena become very difficult in a realistic quantitative analysis. This, however, is beyond the scope of this chapter, which only presents qualitative visualizations of physics phenomena.

5.2.2
Sliding Friction during Braking of Bicycles and Motorbikes

Very similar to the rising temperature of sliding planar surfaces as in the example above, the surfaces of bicycle, motor cycle, or automobile tires heat up during braking with blocked tires. The temperature of the contact spot of the tire with the pavement rises very quickly since the kinetic energy of the vehicle will be transferred into thermal energy. Figures 5.3 and 5.4 depict the temperature rise of the floor as well as of the tire for a bicycle after using the back pedal brake and for a motorbike on the road. For a motorbike braking from an initial velocity of 30 km h^{-1}, temperature rises can easily amount up to more than 100 K for the tire. The temperature rise of the floor again depends on the floor material, its

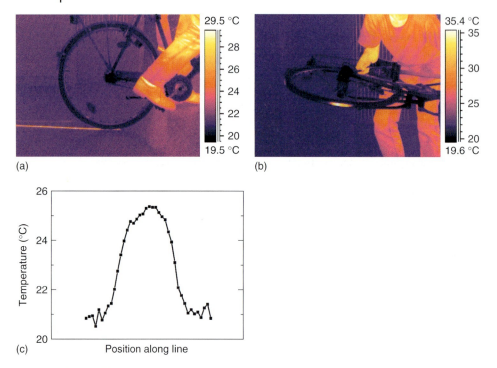

Figure 5.3 Sliding friction causes temperature rises during braking of a bicycle tire with a blocked wheel, using the back pedal brake (a,b). The heat transfer into adjacent locations on the floor is illustrated as temperature profile along a line perpendicular (c) to the trace on the floor, recorded several seconds after breaking.

thermal conductivity, heat capacity, and so on. It is usually smaller than the tire since the thermal energy is spread over a much larger area during the braking procedure. Figure 5.3c shows the temperature across the braking trace on the floor. It may be easily observed as a function of time, illustrating the transient thermal effects.

A very similar sliding friction phenomenon involves the use of the rim brakes. The contact between the friction pads (usually made of some kind of rubber) and the metal rim of the rotating wheel again uses sliding friction forces to transfer kinetic energy into thermal energy. Therefore, the rim itself as well as the friction pads can become very warm. Figure 5.5 depicts the wheel before and after a braking maneuver.

Braking maneuvers with blocked tires are not healthy for the tires. The hot spots on the tire go along with more material ablation at this location. This means that the lifetime of a tire will decrease for repeated braking maneuvers of this type

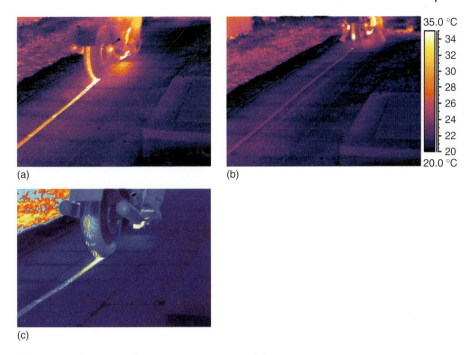

Figure 5.4 (a,b) Low-resolution LW camera image of the braking of a motorbike with blocked tires. The tire had a temperature reaching up to 100 °C immediately after stopping. (c) High-resolution image recorded with high-speed camera and smaller integration time.

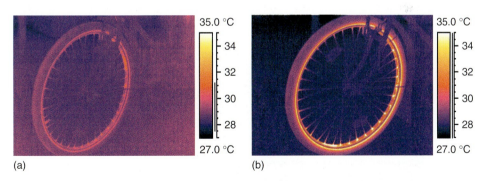

Figure 5.5 Sliding friction causes temperature rises during braking of a bicycle tire with a blocked wheel using the back pedal brake (a) and the rim brake (b).

(the same happens for "jackrabbit starts" where the accelerator pedal is pushed hard such that the wheels will skid leaving a similar black rubber trail behind as for braking with a blocked tire). In addition, the braking itself is not as efficient, since the coefficient of sliding friction is lower than the one for static friction. If sliding between tire and pavement is avoided, the slightly larger coefficient of static friction applies. In addition, during sliding, it is not possible to maneuver the car. For these reasons, modern cars are equipped with systems to prevent sliding friction during braking.

5.2.3
Sliding Friction: the Finger or Hammer Pencil

A very simple, but impressive demonstration of sliding friction and the corresponding temperature rises of surfaces is the use of a hammer or just a finger to write texts or equations on any convenient surface, for example, even the floor. The surfaces need not be too rough and their thermal conductivity should be not too large (in metals the thermal energy diffuses away very quickly, linoleum floors are excellent). Depending on finger speed and contact pressure, it is easily possible to achieve temperature differences of several Kelvin. Figure 5.6 depicts an example.

5.2.4
Inelastic Collisions: Tennis

Collisions are different mechanical phenomena that also involve energy transfer. One may think, for example, of two billiard balls, colliding with each other. Usually, one distinguishes elastic and inelastic collisions. Elastic collisions are those where the total kinetic energy of the objects before the collision is exactly equal to the total kinetic energy after the collision. Elastic collisions are idealized phenomena that are usually demonstrated in physics using apparatus to reduce any residual

Figure 5.6 Finger writing: work against sliding frictional forces lead to temperature increase of the surface.

friction effects, for example, by using an air rail system. In practice, most collisions in everyday life are inelastic, that is, part of the kinetic energy of a moving object is transferred into thermal energy during the collision process. Think, for example, of any ball (tennis, soccer, volley ball, basket ball, rubber ball, etc.) which falls from a certain height to the floor. It will collide with the floor leading to a rebound. From energy conservation, an elastic collision of the ball with the floor would give the ball enough energy to reach its original height, from which it was dropped. However, no real ball will reach the original height from which it was dropped, that is, part of the initial kinetic energy is lost. For new tennis balls (mass ≈ 57 g, diameter ≈ 6.5 cm), it is required that if dropped from a height of 2.54 m, they must at least reach a rebound height of 1.35 m. This corresponds to a loss of kinetic energy of about 0.67 J, that is, about 47% of the initial kinetic energy is lost. Even super balls lose about 20% of their kinetic energy upon bouncing from a floor [7].

Microscopically, the ball as well as the surface deform upon impact of the ball. Consider, for example, the ball. If the deformation changes its shape from the ideal initial spherical shape to a distorted shape while touching the floor, it stores potential energy. Such deformations are, however, never totally elastic, that is, reversible, because during deformations part of the energy is transferred into thermal energy. This means that whenever we observe falling objects, colliding with surfaces, we expect temperature increases of the surface spot of the falling object as well as of the surface spot on the floor, where it hits. Figure 5.7 shows an example for an inelastic collision of a tennis ball with a floor.

In this experiment, where the ball was hit with a racket by an amateur, a temperature rise of the ball of about 5 K was observed with a decay time of several seconds. Similar to the friction experiment, a quantitative analysis is more difficult. The tennis ball experiment is explained in more detail in Section 10.3 when discussing high-speed thermography.

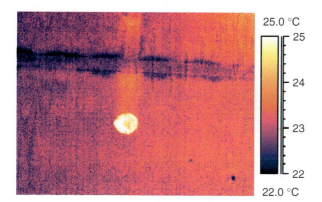

Figure 5.7 A tennis ball was hitting a carpet, resembling the court. The image was taken just after it touched the ground. The ball was also heated up during the collision, but – due to its fast movement – only left the vertical trace.

5.2.5
Inelastic Collisions: the Human Balance

If two objects stick together after a collision, it is termed as being *completely inelastic*. Imagine, for example, a piece of putty, falling to the floor. It will not rebound at all, that is, it loses all of its kinetic energy upon impact. The amount of energy transferred into thermal energy is larger than for inelastic collisions and therefore the corresponding temperature changes may be more easily observable. For the observation, it is however necessary to remove the object and turn it around after it has come to rest in order to measure the surface temperatures of the two contact areas.

Figure 5.8 shows an example of two persons of different mass $m_1 \approx 80$ kg and $m_2 \approx 120$ kg jumping down from a table to the floor. Both wear the same type of shoes. After landing, they quickly step aside and the contact spots on the floor are examined with IR imaging. Quite obviously, the heavier jumper gives rise to a higher surface temperature of the floor. One could easily argue that this was of course expected, since the heavier jumper started with a higher initial potential energy. This is correct, but again, a more thorough discussion shows that any kind of quantitative explanation will need much more information, for example, the contact area while hitting the floor maybe of different materials and hence different heat transfer properties of the soles of the shoes, and so on.

To elaborate, the two jumpers have different initial potential energies $m_i g h_i$ in the gravitational field of the earth, where h_i is the height difference between

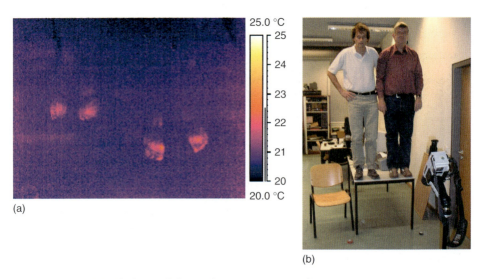

Figure 5.8 The human balance: The temperature rise due to an inelastic collision of people jumping from a table onto the floor (a) can be used to compare masses (weight forces) of the jumpers (b).

floor and center of gravity of the persons. In the following, we assume people of about the same height, $h_1 = h_2 = h$. Just before reaching the floor the potential energy has been completely transferred into kinetic energy $(1/2)\, m_i v_i^2$, where v denotes the velocity. During the completely inelastic collision, a first part of the kinetic energy is transferred into deformations of the floor and the shoes, a second part (i.e., the rest) into deformations within the body (muscles, knee joints, etc.). Ultimately, both parts will end up in thermal energy. Of course, it is only possible to measure the first part, that is, the temperature rise of the floor and the shoes with IR imaging. Unfortunately, again, the amount of energy dissipation within the body will depend on the jumper, that is, on the fact, whether and how muscles are stretched and it is not easily possible to guess the ratio of both contributions. Therefore, we assumed above that both jumpers try to jump in as similar a manner as possible, with muscles stretched. In this case, one expects that a similar ratio of energy would be dissipated into the shoe–floor contact area. It will then divide up into heating of the shoe and of the floor.

5.2.6
Temperature Rise of Floor and Feet while Walking

Walking on ice is very difficult, whereas walking on a dry street is easy. The difference between these two situations is that the frictional forces are much lower for the contact between shoes and ice as compared to that between shoes and pavement of a street. Obviously, friction is necessary for walking. However, although the phenomenon is one of the most natural for us, the details can become very complex. First, static friction is usually involved (the undesired sliding friction does apply to the ice, though). Usually there is no sliding between shoe and floor during walking. The shoe just touches the floor and then lifts up again. Since there is no distance traveled along the direction of an acting force, no mechanical work is done that could be converted into thermal energy.

The energetics of human walking and running has been studied in detail [8–10]. It involves work for accelerating and decelerating the legs plus the gravitational work associated with lifting the trunk at each step. The total power expended during walking finally leads to heating up of the body, sweating, and, to a small extent, also to a heating up of the two contact areas (this contribution has not been studied in detail so far; probably its portion of the total expenditure is at most in the percent-range, most likely in the single percent range). Microscopically, one may understand the mechanism for the heating from the inelastic collision experiment (Section 5.2.5). During each step, the shoe experiences something like an inelastic collision with the floor. Therefore, part of the original kinetic energy of the leg is transferred into deformation energy of the soles of the shoe as well as of the floor (and perhaps a small amount into the body of the walker). These deformation energies end up as thermal energy, that is, a temperature rise of the shoe and floor. How the total energy is split up depends again on the thermal properties of the two materials in contact.

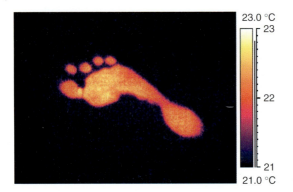

Figure 5.9 Energy dissipation due to walking can be visualized with IR imaging.

Figure 5.9 depicts an example of a person walking at constant speed with bare feet on a linoleum floor.

Since the foot hits the floor only with part of its surface, one easily sees the walking profile and the toes. The corresponding temperature rise is smaller for the shoes although still detectable, in particular when running, that is, hitting the floor with larger velocity (compare Figure 5.8).

5.2.7
Temperature Rise of Tires during Normal Driving of a Vehicle

Similar to walking, the driving of any vehicle with wheels is based on static frictional forces [7]. When a vehicle moves forward, its wheels rotate such that the bottom surface does not slide on the ground. Instead, a portion of the surface of each wheel touches the ground where it briefly experiences static friction. Then it moves up with a new portion of the wheel surface taking its place. This touch and release procedure involves only static friction; therefore, similar to the walking discussed above, this mechanism alone is not able to convert mechanical energy via work into thermal energy.

However, rolling of wheels on a surface involves more. The corresponding technical term for the resistance to motion is *rolling resistance* or *rolling friction*. Whenever a wheel or a tire rolls on a flat surface, it deforms the object as well as the surface. At the contact point/area, there are static frictional forces present. Sliding friction does not contribute since each contact spot on the tire is lifted up upon rolling. The deformation of the surface leads to reaction forces that have a component opposed to the direction of motion. As a matter of fact, the deformations of the surface lead to the seemingly paradox situation that any horizontally driving vehicle must drive upward (out of the hole due to the deformation).

As in the case of static and sliding friction, the frictional force is described as $F_{\text{friction}} = \mu_{\text{roll}} \cdot F_{\text{normal}}$ with μ_{roll} being the rolling friction coefficient. This coefficient is much smaller than typical static or dynamic friction coefficients, for

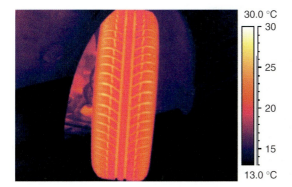

Figure 5.10 Homogeneous heating of tires during normal driving conditions. This is also a test of the quality of the profile of the tire; without profile, the heating would be more homogeneous across the tire surface.

railroad steel wheels on steel rails it is less than 0.001 and for car tires on asphalt about 0.03. Ideally, the deformations should be elastic, in which case, no thermal energy would be generated. In reality, part of the deformation is inelastic and the contact areas should warm up.

As a consequence, the tire of any transport vehicle will have elevated temperatures upon driving.

High-quality tires should have a homogeneously heated surface, provided that no braking with blocked tires or wheel spinning during accelerating contributes. Figure 5.10 depicts an example for a car. As expected, no hot spots are visible; however, the profile of the rather new tire is clearly visible.

This technique – investigation of tire surfaces after driving – is commonly used to analyze the quality of new tires, in particular of car tires for Formula One races.

5.3
Thermal Physics Phenomena

Although nearly all applications of IR imaging involve thermal phenomena, for example, by transferring mechanical or electrical energy into thermal energy and corresponding heating up of the surface of objects, there are some purely thermal physics phenomena which can be visualized using thermography. These include characteristic properties of heater systems, material properties like thermal conductivity, and also convection in liquids. IR imaging can be used to study the effect of phase transitions like in evaporative cooling or consequences of adiabatic processes like temperature differences due to adiabatic cooling. Finally, IR imaging offers possibilities for quantitative analysis of the heating and cooling of many objects.

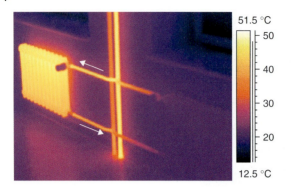

Figure 5.11 Radiator heater and water pipes with hot water (inflow) and slightly colder water (outflow).

5.3.1
Conventional Hot-Water-Filled Heaters

There are many indoor heating systems (e.g., wood, coal or oil stoves, or furnaces which use warm air heating or water-filled radiators) that can be analyzed with thermography. For physics teaching, it is very convenient to use hot water heaters whose hot water supply comes from a furnace. The hot water is usually driven by a pump and flows in pipes from room to room and, in each room, it also flows through the radiators, which transfer the heat to the room via convection and radiation.

Since the water is losing thermal energy to the heaters, it should be possible to detect a temperature difference between the inlet and outlet pipes of a radiator.

Figure 5.11 depicts a set of two radiator heaters in a lecture room close to the window as well as the vertical pipe system of inflow and outflow water. One of the heaters is turned on, the other is off. The image immediately visualizes the hot inlet water pipe and the slightly colder outlet water pipe. The water enters the radiator from top and slightly colder water flows out at the bottom as expected.

If IR image examples like these are shown when first introducing thermography, the confidence in the method increases. Probably many other well-known everyday life objects whose surface temperatures can also be measured separately with thermocouples can similarly build up confidence in this measurement technique.

5.3.2
Thermal Conductivities

In Section 4.2.1, conduction of heat was introduced as representing heat flow within a solid or fluid at rest, due to a temperature difference between its ends. The simplest theoretical system like a one-dimensional wall, which laterally extends indefinitely, can obviously not be measured with thermography. One needs to measure the surface temperature of objects.

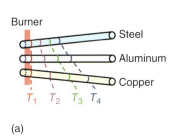 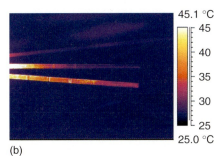

Figure 5.12 Typical setup of experiment to demonstrate thermal conductivity differences of materials (a). After heating has started the diffusion of heat into the rods proceeds with different velocities as indicated by the locations of the same temperature along the rods. (b) IR image of such a setup of steel, aluminum and copper rods, heated from one end (not seen in image) with a Bunsen burner.

A typical setup to demonstrate differences in thermal conductivity of solid materials is the following (Figure 5.12). Thin rods of different materials are horizontally fixed in such a way that one of their ends is free and the other end is heated (e.g., by flames of a Bunsen burner). Along the length of the rods, small pieces of wax can be attached at regular intervals (not shown in figure). The wax starts to melt at a certain temperature. Therefore, melting indicates that the critical temperature has been reached. The experiment is done by recording the times at which wax at given locations starts to melt. This allows visualization of the heat diffusion within the rods as a function of time. In particular, at a given distance from the heating location, the wax will start to melt earlier for rods with higher thermal conductivity. This experiment, nicely, but only qualitatively, demonstrates thermal conductivity. Unfortunately, it is not thermal conductivity alone that determines the outcome of such an experiment. First, for any quantitative analysis of thermal conductivity, a well-defined temperature difference is needed. Since, however, the rod ends are usually not fixed in a heat bath, but just end in air at room temperature, the whole rod will start to warm up and the end temperature will also increase with time. Second, convective heat losses due to the surface area of the rods and, third, the radiation losses will also contribute. As a consequence, it may well be that if these additional losses dominate, the temperature profile would not allow any precise conclusion concerning the thermal conductivity.

Figure 5.12 also depicts an IR image of such an experiment. Three rods of the same diameter, made of steel (top), aluminum (middle), and copper (bottom) were heated at one end using a flat flame Bunsen burner. The thermal conductivities (Table 4.1) increase from top to bottom. As expected, the Cu rod with the largest thermal conductivity is heated up much more quickly than steel and Al. In order to avoid saturation of the IR detectors close to the heating zone, it is best to first heat and then record images directly after turning off the heater.

A more well-defined setup for thermal conductivity measurements uses a fixed temperature difference between the two ends of a material, here water. A particularly

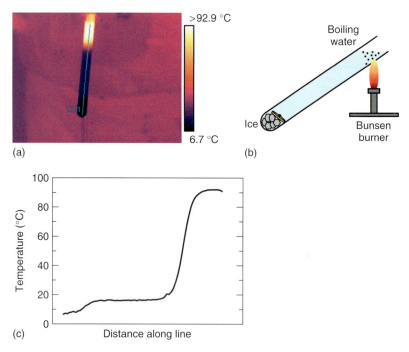

Figure 5.13 (a) IR image and setup for demonstration of thermal conduction in water. (b) Line plot of temperature reveals ice water at the bottom while the water at the top, recorded directly after turning off the burner, is still around 90 °C.

simple experiment is depicted in Figure 5.13. An ice cube is crushed and some pieces are put into a test tube. The tube is filled more or less completely with precooled water around 0 °C. Ice would normally float; therefore, we use some metal weight on top of the ice to keep the pieces at the bottom of the test tube. The ice within water reaches the lower temperature of 0 °C at the bottom of the test tube. The upper end of the test tube is then heated using a Bunsen burner until the water in the upper few centimeters of the tube starts to boil defining the upper temperature of 100 °C. The temperatures of 0 and 100 °C will be maintained as long as there is ice in the test tube (latent heat of melting) and as long as the upper parts are still covered by boiling water (latent heat of evaporation). Along the test tube, a temperature profile develops, which is mainly governed by the thermal conductivity of the water (the glass has a somewhat larger thermal conductivity). Furthermore, it is very thin such that – to first order – the Biot number can still be assumed to be small compared to unity. This means that the glass surface temperature should more or less resemble the water temperature inside the tube. Figure 5.13 depicts the experimental result from IR imaging. The line plot along the test tube shows that both glass and water are poor thermal conductors. The water at the bottom is still around 0 °C, whereas the water at the top has just boiled.

Of course, waiting for a long time will also lead to heat conduction, which will eventually equalize the temperatures.

5.3.3
Convections

In Chapter 4, convection was introduced as representing heat flow between a fluid and a solid. It is composed of both heat transfer due to conduction in the boundary layer around the solid and heat transfer due to bulk motion of the fluid that is outside the boundary layer. Both processes are difficult to visualize with IR imaging if the fluids are gases, unless the gases have strong absorption features in the thermal IR region (Chapter 7). Convections due to liquid fluids are easier to observe. Figure 5.14 depicts an ice cube, which is floating in a glass beaker filled with water at room temperature. If the ice cube and water were just at rest at the beginning, natural convection would start to build up, that is, water close to the ice cube would get cooler, thereby transferring part of its thermal energy to the cube, which would start to melt at the surface. The colder water has a higher density and will start to sink down in the beaker, thereby transporting warmer water to the surface. These slow convection currents would not be observable with IR imaging since water is not transparent in the thermal IR region (Section 1.5, Figure 1.55). However, convective bulk motion of water can be made visible by observing from above. The ice cube floats and therefore we can study the water surface convection currents. Since there is no natural lateral force driving such currents, the ice cube is given a little bit of an initial spin. Owing to this initial rotary movement some volume elements of the water, which have touched the cube and had already cooled down move, that is, flow away from the boundary layer. Therefore new water volume elements can come close to the cube transporting thermal energy from the water at room temperature to the ice cube at around freezing temperature. This leads to a melting at the ice cube surface. In Figure 5.14, water with the green shades is about 6–7 K cooler than the average water temperature. Figure 5.14 also

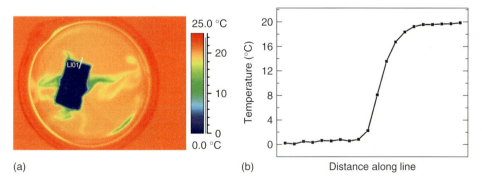

Figure 5.14 (a) Top view of water convection around a slowly rotating ice cube and (b) temperature profile along the white line. For details, see the text.

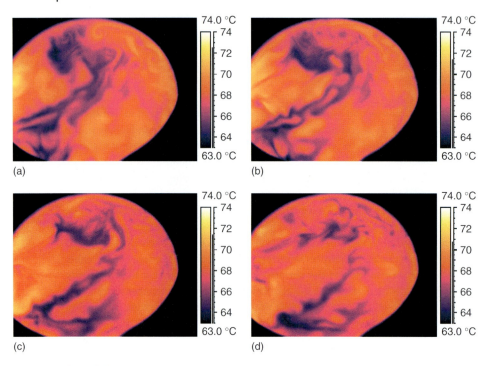

Figure 5.15 Convection features observed from a hot water surface, the water being heated from below.

demonstrates the heat conduction in the boundary layer. The line plot data in a "quiet" region show the expected gradual increase from the ice cube ($T \approx 0\ °C$) to the average water temperature of about 20 °C. In this case, the steep increase takes place over a distance of about 2 mm. For a nonmoving cube, the distance can easily be a factor of 2 larger. If an ice cube rests at the edge of the beaker in contact with the glass, it is also possible to directly observe cold convection currents from the outside of the beaker. This is due to cold water which starts to sink from the ice cube at the inside of the beaker thereby cooling the adjacent glass surface.

Convections with transport of larger volume elements of fluids are usually driven by larger temperature differences. A well-known example in nature is the convection cell structure, which can be observed at the surface of the sun. Figure 5.15 depicts a series of IR images, showing convection features of water within a large glass beaker being heated from below. Convection is a transient phenomenon and can be better observed in real time; however, the still images already show how structures are formed and transported across the surface of the water.

Similar convection cells are observable in everyday life when heating oil in a pan to very high temperatures. For appropriate temperature differences and oil thickness, this leads to so-called Bénard–Marangoni convections. Figure 5.16 illustrates how these convection structures form. Oil in the vicinity of the lower hot surface is heated and therefore starts to rise due to its lower density. Similarly, the

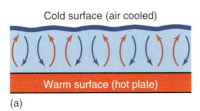 

Figure 5.16 Formation of Bénard–Marangoni convections of a liquid heated from below and cooled from above. Warm oil (red) starts to rise, whereas colder oil (blue) sinks (a). This can result in (ideally) hexagonal two-dimensional structures on the surface (b), where the rising oil is in the middle of the cells and the sinking one defines the cell boundaries.

colder and denser oil from the surface starts to sink down. Of course, this process cannot take place simultaneously everywhere in the pan. For given temperature difference, oil thickness, and diameter of the pan, regular cell structures start to form, which allow large quantities of hot oil to rise while simultaneously the same amount of colder oil sinks to the bottom of the pan. In the two-dimensional sketch of Figure 5.16, closed loops of flowing oil organize in such a way that neighboring loops rotate in opposite directions such that they do not disturb the flow of their neighbors. This process is self-organizing. The form and number of formed cells depend on the conditions, in particular the temperature difference. The complete theoretical modeling needs to take into account the buoyancy forces, temperature-dependent surface tension, and dynamic viscosity of the oil [11]. Since there are regions where colder oil sinks and others where warmer oil rises, a line plot of temperature across the surface will show regular structure.

Figure 5.17 shows an example of an experimental result, investigated with IR imaging (care must be taken to avoid oil vapor on the camera optics; either a mirror can be used or a thin transparent plastic foil acting as protective window). The

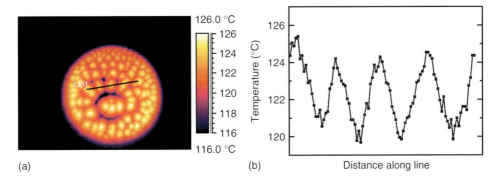

Figure 5.17 IR image of Bénard–Marangoni convection cells of oil in a pan, heated to about 120 °C (a). Convection leads to well-defined temperature variations across the surface (b).

oil thickness was 3 mm and the total diameter about 9.5 cm. The structures start to form well above 100 °C. For constant temperature the cell structure is stable; however, its geometry and number density changes with temperature. Temperature differences between rising and sinking oil amount to about 4.5 K in Figure 5.17. In a pan heated to 150 °C, we observed T variations of up to 9 K.

Some kitchen experts who want to prepare steaks in a pan use these convection cells as an indicator of the oil temperature. It is quiet easy to observe the convection cells with the naked eye when looking at grazing incidence at the oil surface. The oil is not hot enough unless the cells start to form.

5.3.4
Evaporative Cooling

The idea behind evaporative cooling can be guessed from the following description of an ancient cooling system: "In the Arizona desert in the 1920s, people would often sleep outside on screened-in sleeping porches during the summer. On hot nights, bedsheets or blankets soaked in water would be hung inside the screens. Whirling electric fans would pull the night air through the moist cloth to cool the room" [12]. The same article [12] by the California energy commission also emphasizes that many new technologies have been inspired by this principle of evaporative cooling.

The physics behind evaporative cooling is quite simple. One needs air with relative humidity below 100% (Section 4.3.6) which is directed over water, wet surfaces, or through wet blankets. While passing over the respective water surfaces, water molecules change their phase state from liquid to gas. Thereby the water molecules become part of the airflow, which will then have a higher humidity.

This phase change from liquid to gas does, however, require energy, to be specific, the heat of vaporization, which amounts to about 2400 kJ kg^{-1} at around 30 °C (sometimes, this number is also given as 43 kJ mol^{-1}, that is, as energy needed to vaporize 1 mol, here 18 g, of water or as 0.45 eV/molecule where one uses the fact that 1 eV $= 1.6 \times 10^{-19}$ J and 1 mol of water contains 6.022×10^{23} molecules). This is an enormous amount of energy that must come from either the water or the air or both. Therefore, there should be two observable effects: the water should cool down and the air should cool down. The latter effect was described above: the hot air being pulled through the wet blankets loses part of its thermal energy, which is transferred to the blankets, providing the energy for evaporation. In dynamic equilibrium, the blanket temperature would not change any more and the energy needed for the evaporation of water per time would be transferred to the blanket from the air.

Obviously, IR imaging should not try to detect air temperatures, rather, the temperature of wet surfaces that are exposed to airflow should be studied. For experiments, various liquids such as water and also aftershave lotion (which contains alcohol) were used. It is well known that a lotion containing alcohol will lead to a much more dramatic cooling effect compared to the same amount of

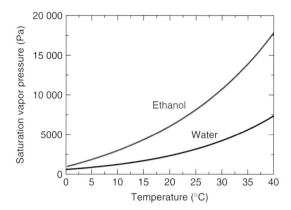

Figure 5.18 Saturation vapor pressure of ethanol and water. The differences are due to the lower boiling temperature of 78 °C, that is, the ethanol vapor pressure reaches the atmospheric pressure of about 1000 hPa already at 78 °C.

pure water. Although water and ethanol have about the same heats of vaporization (40–45 kJ mol^{-1}), they do, however, behave quite differently. This means that there is at least a second ingredient to evaporative cooling: the vapor pressure of the liquid at given ambient conditions. As shown in Figure 5.18, the saturation vapor pressure over a liquid increases steeply with temperature. The *saturation pressure* is defined as the equilibrium vapor pressure above a liquid. This means the following: some molecules evaporate per time interval from a liquid (from liquid to gas phase) and some gas molecules condense at the liquid again (from gas to liquid phase). For any given temperature, there is an equilibrium when equal amounts of molecules evaporate and condense. In this case, the gas pressure (being related to the number density of molecules in the gas phase), is the saturation vapor pressure. As shown in Figure 5.18, at any given ambient temperature between 10 and 30 °C, the ethanol vapor pressure is at least twice as large as the one of water. As a consequence, compared to evaporation of water molecules, twice as many molecules of ethanol vapor can evaporate that is, the evaporative cooling effect can be much larger.

Two more factors that have an influence on evaporative cooling are the relative humidity of the air and the speed of the airflow. If the air is already saturated with water vapor, it cannot accommodate more water. In this case, evaporative cooling cannot take place. The speed of the airflow can enhance evaporation quite appreciably. This is plausible since air in close contact with water will get higher values of relative humidity and hence it can accommodate less water vapor. Blowing fresh air of lower relative humidity toward the water surface will therefore enhance the evaporation. In addition, as was pointed out in Chapter 4 (Figure 4.4), the convective heat transfer coefficient increases with increasing airflow velocity. Therefore, larger amounts of energy for the vaporization of water vapor are available.

Figure 5.19a,b depicts an example with a water film on the wall surface of a model house. A warm air fan was directed onto the wall and the IR images were

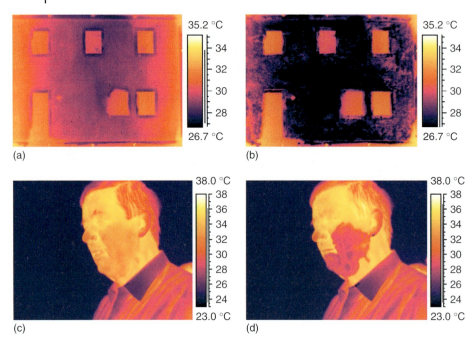

Figure 5.19 (a,b) Evaporative cooling due to water. A wet surface of a model house cools rapidly due to evaporative cooling, enhanced by a warm air fan (for details, see Section 6.4). (c,d) Evaporative cooling due to ethanol in aftershave, enhanced by a warm air fan.

recorded. Owing to evaporative cooling, the wet wall segment was strongly cooled initially, although warm air was used. Later on, after equilibrium was established, the wall temperature remained constant. Consequences for building thermography are discussed in Chapter 6.

Similarly, Figure 5.19c,d depicts evaporative cooling after using aftershave. The ethanol rapidly evaporates in particular if an airstream from a fan is used and gives rise to enhanced cooling.

5.3.5
Adiabatic Heating and Cooling

The state of any gas is usually characterized by three quantities. The most commonly used quantities are pressure, temperature, and volume; others include, for example, entropy. There are many ways of changing the state of a gas, for example, one may keep one of these quantities constant and change the other two. Two processes that are very important for technical applications are the so-called adiabatic expansions or compressions, the characteristic features of which are due to energy conservation.

Whenever a gas is compressed, work is done on the gas, which will lead to a change of its internal energy (which microscopically can be regarded as energies of the gas molecules). The first law of thermodynamics is a statement of energy conservation. It states that the internal energy of a gas can only change due to either heat transfer to/from the gas or work done on/by the gas. In most state changes of a gas, heat as well as work is exchanged. However, adiabatic processes are different. They take place too fast for thermal equilibrium to be established, and therefore adiabatic processes take place without exchange of heat. As an example, we consider a very fast compression of a gas. Assume that a gas is in a container with a movable piston. A typical everyday life example would be a pump for the bicycle tire with closed outlet valve. If the piston is moved inward very rapidly (in order to compress the gas), there is no time for exchange of heat, that is, the compression takes place as an adiabatic process. In such a case, the work done during compression is entirely transferred into internal energy of the gas. As a consequence, it will heat up rapidly. Those who use a bicycle hand pump know that the pump close to the valve gets warm quickly since the hot gas inside will finally also lead to a warming of the containing metal or plastic tube.

We illustrate the reverse process, the adiabatic expansion of a gas, with IR imaging. Expansion of a gas requires that work must be done (imagine the gas moves a piston outward upon expansion). Since heat exchange is not possible, the energy needed for this work must come from the gas itself, that is, the gas must reduce its internal energy. This goes along with a temperature decrease. This means that any adiabatic expansion will result in a decrease of gas temperature. Such adiabatic processes may be realized simply by using the tire of a bicycle, which is pressurized to 3 bar. Opening the valve leads to a rapid expansion through the valve. Therefore, the gas needs to cool down. The cold airstream touches the valve and therefore leads to a cooling of the valve. This is illustrated in Figure 5.20.

Holding a piece of paper originally at 23 °C (ambient temperature) in front of the expanding airstream leads to a rapid cooling of the paper to about 7 °C, which

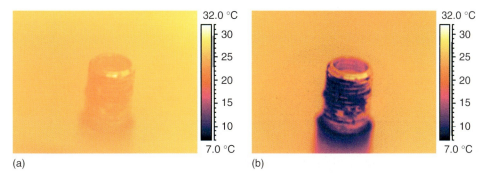

Figure 5.20 Valve of an automobile tire before (a) and after (b) opening the valve. Adiabatic cooling of expanding air from a bicycle tire leads also to a cooling of the valve with a large temperature drop.

is below its dew point at 50% relative humidity, that is, the paper gets wet during the cooling process.

5.3.6
Heating of Cheese Cubes

Several examples for heating and cooling of objects like metal cubes and light bulbs have been presented in Sections 4.4 and 4.5. In this Section and Section 5.3.7 we give additional examples related to heating and cooling of selected objects. Some of these examples illustrate the general physical principles of convection and radiation very effectively. In all cases, we assume small Biot numbers, that is, that the surface temperatures resemble something close to the average temperatures of the objects.

The first example [13] deals with cheese cubes. Imagine a piece of solid cheese, such as Gouda or Cheddar, that has no air holes in it. Cut the cheese into several, say 6–8, small cubes of sizes from 2 to 15 mm. Place the cubes in a circle on a small plate (Figure 5.21) and put the plate inside a conventional electric oven that has been preheated to 200 °C.

Question: What will happen to the cheese cubes? Will the small ones melt first or the large ones, or will all cubes melt at the same time or will some not melt at all? (The answer to this and the following question is given on the next page, to offer the possibility of thinking before reading the solution.)

After having dealt with this introductory problem, repeat the experiment, that is, prepare an identical set of cheese cubes on an identical plate, which, however, should be microwave proof. Then place the plate inside a microwave oven, which has been set at full power (e.g., 800 W). The heating should take place for an integer number of revolutions of the turntable; this ensures that all cubes experience the same microwave fields within the oven. The question will be the same as before: What will happen to the cheese cubes? Will the small ones melt first or the large ones or will all cubes melt at the same time or will there be some cubes not melting at all?

Figure 5.21 Cheese cubes of different sizes on a plate, which may be put into a conventional oven or a microwave oven.

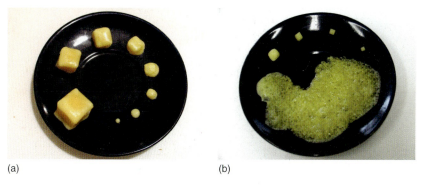

(a) (b)

Figure 5.22 In the conventional preheated electric oven, (a) the cubes were heated at 200 °C for about 70 s. The microwave oven (b) was operated at 800 W for about 30 s (for details see text).

Figure 5.22 shows the results of the cheese cube experiments heated in a conventional oven (a) and microwave oven (b). In the conventional oven the small cubes will start to melt first, which can be seen nicely by the rounding off of the corners. In contrast, the cubes behave totally differently in a microwave oven. The largest cubes melt first and it can even be observed that cubes with sizes below a critical size will not melt at all. Obviously, the different behavior must be due to the different heating and cooling processes involved.

The temperature of the air within the conventional oven is much larger than the cheese temperature. Therefore energy flows from the oven through the surface of the cubes into the cheese. The smaller cubes get heated throughout their interior much faster than the large ones and, hence, melt first. In contrast, heating in the microwave oven is realized via absorption of microwave radiation within the interior of the cheese cubes [14]. However, since the air temperature within the oven is about ambient temperature, the heated cheese cubes also start to cool via convection and radiation. The cooling power will be proportional to the surface area of the cubes, whereas the heating will be proportional to their volume. As a consequence, the surface-to-volume ratio determines a final maximum temperature of a cube. The smallest cubes will suffer the most effective cooling. Eventually, this can even prohibit melting. The final temperature will increase with cube size. This is shown in Figure 5.23, which depicts the situation after 10 s of heating. The quantitative analysis revealed a strong dependence of temperature on cube size and the temperatures of the largest cubes were above melting temperatures of the cheese.

The heating and cooling of cheese cubes in the microwave oven can also be easily treated theoretically [13]. The power absorbed by each cheese cube of size a is proportional to its volume:

$$\frac{dW_{abs}}{dt} = P_{absorb} \propto V = k_1 a^3 \qquad (5.1)$$

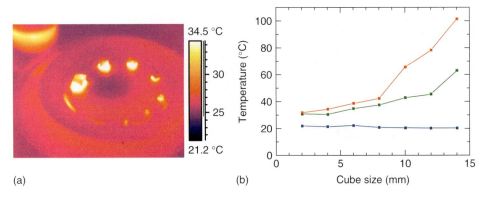

Figure 5.23 Cheese cubes after 10 s of heating (800 W) within a microwave oven (a) and measured maximum temperatures of the cubes before (blue, bottom), after 10 s (green, middle), and after 30 s (red, top) of heating (b).

where k_1 is a constant which depends on the absorption coefficient of microwaves in cheese. The cooling, which is due to convection and radiation (Section 4.2) can be approximated by

$$P_{\text{cool}} = k_2 \cdot a^2 \cdot (T_{\text{cheese}} - T_{\text{oven}}) \quad (5.2)$$

It depends linearly on the temperature difference between the cheese and its surrounding. The effective absorbed power leading to a temperature rise of the cheese is due to

$$P_{\text{eff.heating}} = P_{\text{absorb}} - P_{\text{cool}} \quad (5.3)$$

This leads to the differential equation

$$P_{\text{eff.heating}} = c \cdot m \cdot \frac{dT}{dt} = k_1 \cdot a^3 - k_2 \cdot a^2 \cdot (T(t) - T_0) \quad (5.4)$$

with the solution

$$T(t) = T_0 + \frac{k_1}{k_2} a \left[1 - e^{-\frac{(t-t_0)}{\tau}} \right] \quad (5.5)$$

with the time constant $\tau = 1/A = k_3 a$. Despite not knowing exact values for the constants, it is possible to plot the general form of $T(t)$ as shown in Figure 5.24 for different values of cube size a. It follows from Eq. (5.5) that the temperature of each cube eventually reaches the asymptotic value $T_0 + \frac{k_1}{k_2} a$. If this temperature is below the melting temperature, the cheese will never melt. An interesting feature of Eq. (5.5) is that since the time constant τ is proportional to cube size, the time until the maximum temperature is reached at equilibrium conditions is shortest for small cubes. As shown in Figure 5.24, only the smallest cubes have reached the maximum possible temperature, whereas the largest cubes are still far away from equilibrium.

More details on this experiment can be found in [13].

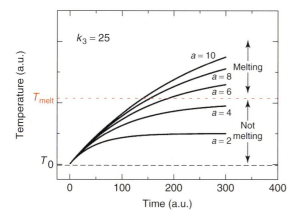

Figure 5.24 Temperatures of cheese cubes of different sizes a as a function of heating time in a microwave oven for a given value of $k_3 = 25$ (i.e., e.g., $\tau = 50$ for $a = 2$).

5.3.7
Cooling of Bottles and Cans

The second example concerns an everyday experiment using Newton's law of cooling (Section 4.5). Whenever an object of temperature T_{obj} is in a surrounding of lower temperature T_{surr}, it will cool down due to convection and radiation losses. Energy conservation requires that the heat losers lead to a decrease in thermal energy and hence in the temperature of the object. In this case, the cooling process is described by

$$mc\frac{dT}{dt} = -(\alpha_C + \alpha_R) \cdot S \cdot (T_{obj} - T_{surr}) \quad \text{where} \quad \alpha_R = \varepsilon \cdot \sigma \cdot k_{appr} \quad (5.6)$$

This differential equation has the solution

$$T_{obj}(t) = T_{surr} + (T_0 - T_{surr}) \cdot e^{-t/\tau} \quad \text{with} \quad \tau = \frac{\rho c V}{(\alpha_C + \varepsilon \cdot \sigma \cdot k_{appr})S} \quad (5.7)$$

Equation (5.7) means that if Newton's law of cooling is fulfilled, we expect an exponential decrease of the temperature difference with time, that is, a straight line in a semilogarithmic plot (Section 4.5). This expectation was checked by studying the cooling of soft drink cans and bottles. Particularly in summer time, the cooling of liquids in refrigerators is of importance. As two examples we measured the cooling of cans and bottles filled with water (or other liquids) as a function of time for different cooling methods. Figure 5.25 depicts the experimental setup.

The cooling power of the systems is expected to be quite different. The conventional fridge and the freezer both have objects surrounded by still air, since the temperatures are usually too low to generate natural convections. Hence, the heat transfer coefficients and the cooling time constants of both should be the same. However, the refrigerator has a smaller temperature difference than the freezer; therefore, the cooling power of the freezer is larger and the effective cooling times

(a) (b)

Figure 5.25 Cooling cans and bottles of liquids in a conventional refrigerator ($T_{final} = 6$ °C, not shown), a freezer ($T_{final} = -22$ °C), and an air convection cooler ($T_{final} = -5.5$ °C). A tape of known emissivity ($e = 0.95$) was attached to bottles and cans.

(times to reach a certain low temperature upon cooling) are smaller than in the refrigerator. The air convection cooler should have the fastest cooling since the convective heat transfer coefficient increases strongly with airflow velocity. Therefore, the time constant should also decrease.

As samples, we used glass bottles and aluminum cans. A (blue) tape was attached in order to ensure equal emissivity values for all samples. The containers were filled with water slightly above room temperature and placed inside the refrigerator, freezer, and air convection cooler. During temperature recordings with the IR camera, taken every few minutes, the cooling unit doors were opened for utmost 25 s each. Figure 5.26 shows the resulting plots of temperature difference between measured temperature and ambient temperature within the cooling system on a logarithmic scale. From Newton's law a straight line is expected. Obviously, this holds quite well for any cooling mechanism down to around 0 °C, where the phase transition water to ice imposes a natural limit. At the end of the experiments in the freezer and air convection cooler, we could indeed observe small pieces of ice, floating on the surface of the water in the bottle and the can.

The time constants τ from quantitative fits to the data nicely agree with theoretical expectations from Eq. (5.7). For the freezer and the refrigerator, $\tau \approx 8300-8400$ s, whereas for the air convection cooler, the value of τ is halved. This is due to the increased convective heat transfer coefficient. Theory also accounts for the differences (a factor in τ of about 1.2) between cans and bottles. It is, on one hand, due to the different amounts of water and, on the other hand, due to the differences in the stored thermal energy in the glass of the bottle as compared to that in the aluminum can.

The user of cold drinks is usually not interested in time constants, rather he/she would like to know at what time, a drink will reach a certain temperature. Figure 5.27 depicts the experimental cooling curves for the 0.5-l bottles

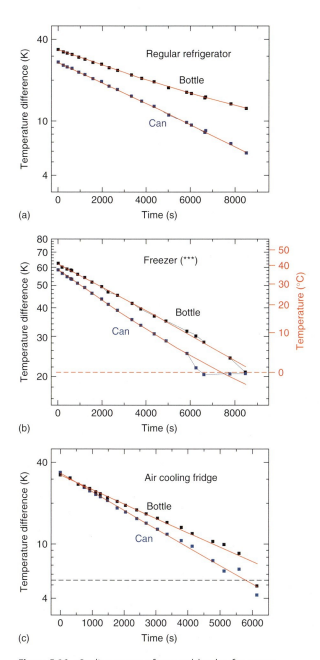

Figure 5.26 Cooling curves of can and bottles for a regular fridge, a freezer, and an air convection cooler. All plots can be fitted with a simple exponential, that is, they follow Newton's law of cooling.

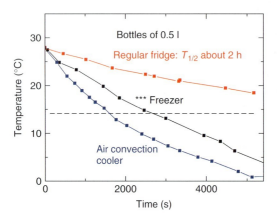

Figure 5.27 Cooling curves of 0.5-l bottles for a regular fridge, a freezer, and air convection cooler. The typical timescale for cooling from 28 °C to below 14 °C (which may be a suitable drinking temperature) is above 2 h for the fridge, about three-fourth of an hour for the freezer, but less than half an hour for the air convection cooler.

(linear scale). The initial temperature was about 28 °C. The fridge has the longest cooling time, whereas the air convection system cools fastest, for example, in about 30 min from 28 °C to below 13 °C.

Obviously, from daily experience, an even faster way of cooling would use forced convective cooling with liquids rather than gases, due to the large density difference between gas and liquid.

5.4
Electromagnetism

5.4.1
Energy and Power in Simple Electric Circuits

In any simple electric circuit that follows Ohm's law [5, 6], electrical energy is transferred into internal thermal energy within the resistor, revealing itself as a temperature rise of the resistor. Obviously, IR imaging can easily visualize this direct thermal consequence of electric currents through resistors. Figure 5.28 depicts the simplest electrical circuit: a wire is connected to a power supply. The wire itself is the resistor, which warms up while current is flowing through the wire.

Figures like 5.28 can also visualize that the heating of the wire before and after a coil is the same. This may help to get rid of misconceptions of students, that the current may have lost part of its "power" while traveling through the circuit, leading to less energy dissipation behind a coil.

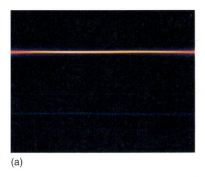

(a) (b)

Figure 5.28 The electrical energy associated with a current flowing through a wire is dissipated, leading to a temperature rise of the resistor, that is, the wire. The wire need not be straight. It can also be a coil or a wire spiral as in light bulbs. In the image, two Cu wires of different diameters (0.25 and 0.55 mm) were used with the same current of 0.5 A. As expected, the thin wire became very hot, whereas the thick wire only warmed up a little bit.

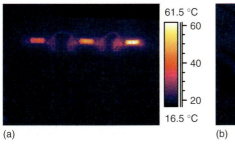
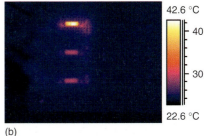

(a) (b)

Figure 5.29 Three different resistances (a) in series and (b) in parallel.

Usually, in electrical circuits, the electrical energy is not just dissipated in heating the metal wires, which are the connecting elements in the circuit. It is straightforward to design simple circuits with wires and resistors of varying size. Figure 5.29 depicts series and parallel circuits of several resistors. Obviously, the resistors heat up according to the power $P = R \cdot I^2$ dissipated within them. Measurement of the surface temperatures of these resistors then allows to sort them according to their size. Quantitative analysis, that is, to find the exact value of R from the surface temperature is a more complex problem. In this case, all heat transfer modes, conduction, convection, and radiation must be treated. In any case, studying various combinations of resistors with IR imaging may be a nice visualization of Kirchhoff's rules in simple electric circuits.

5.4.2
Eddy Currents

Faradays law of induction [5, 6] leads to a phenomenon called *eddy current*. Whenever a conducting material in the form of a loop is exposed to a changing magnetic field, Faradays law states that an electromotive force is induced. In a closed loop, this leads

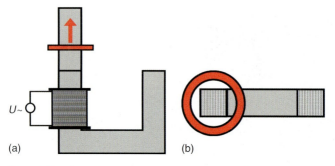

Figure 5.30 Setup for jumping ring experiment (a) from side and (b) from top. A conducting nonmagnetic metal ring (red) is placed over the extended vertical core of a demountable transformer.

to a ring current. Any current in a conductor dissipates energy according to $P = I^2 R$, that is, the current should heat up the object. The same happens if conductors of arbitrary geometrical form are exposed to changing magnetic fields. In any case, electromotive forces are induced, which lead to closed loop currents within the conductor, raising the temperature of the conductor. These currents are called *eddy currents*. These circulating currents by themselves create induced magnetic fields that oppose the change of the original magnetic field due to Lenz's law.

The effect of eddy currents can be made visible using IR imaging. Figure 5.30 depicts the setup for a popular physics demonstration, the jumping ring experiment [15, 16]. A nonmagnetic metal ring is placed on top of a solenoid over the core of a U-shaped demountable transformer unit. When AC power is applied to the solenoid, the ring is thrown off since the induced eddy currents induce secondary magnetic fields that are opposed to the primary magnetic field.

The heat generated by the eddy currents can be made visible by preventing the ring from being thrown off. Holding it by hand is not very wise; we arranged for a metal bar several centimeters above the solenoid to serve as a mechanical stop. Applying an AC power to the solenoid throws the ring to the stop, where it levitates for the rest of the experiment. Owing to the AC magnetic fields, eddy currents are permanently induced, that is, there is a continuous generation of heat according to $P = I^2 \cdot R$, which leads to a rapid heating up of the ring. Figure 5.31 depicts an example as observed after several seconds. One may study the heating as a function of time as well as differences due to different ring materials (e.g., copper vs aluminum).

5.4.3
Thermoelectric Effects

There are a number of thermoelectric effects [17] that are exploited in physics and technology. Temperature measurements with thermocouples use the Seebeck effect (Figure 5.32a).

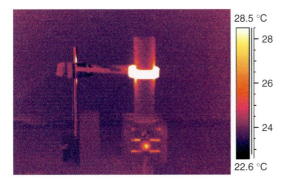

Figure 5.31 Eddy currents induced in a metal ring by the AC magnetic field in the solenoid of the open transformer unit lead to a temperature rise of the ring.

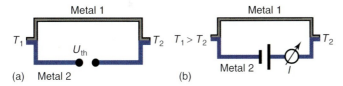

Figure 5.32 The thermoelectric Seebeck (a) and Peltier effects (b), (for details see the text).

Two different metals are joined at two points. If a temperature difference occurs between these two contact points, a small electric voltage U_{th} (typically, in the microvolts per Kelvin range) is produced, which drives a thermoelectric current. The physics behind the effect is as follows: for each metal, there exists a well-defined work function, which describes the minimum energy needed to remove an electron from the metal surface. If two different metals with differing values of their work function touch each other, there will be an electron transport from the metal with the lower work function to the one with the higher work function. This leads to a contact potential. If two metals are bent such that they touch each other at two ends, the same contact potentials will result, that is, they will cancel each other. However, the number of electrons transferred from one metal to the other depends on the temperature of the contact point. Therefore, a temperature difference between the two contact spots of two metals in Figure 5.32 will lead to a net potential difference U_{th} which depends on temperature. After calibration, this voltage is used for a quantitative measurement of temperature. In conclusion, the Seebeck effect creates a potential difference (i.e., a voltage) from a temperature difference.

The opposite effect, called *Peltier effect* (Figure 5.32b) uses an electric current to generate a temperature difference. In this case, an electric current is driven through a bimetallic circuit that is maintained at uniform temperature. Heat is generated at one junction, leading to an increase in temperature, and heat is extracted at the other junction, leading to a cooling of the junction. The direction of the current and the contact potentials determine which contact point is heated and which is

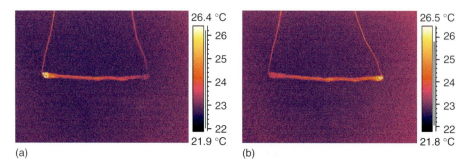

Figure 5.33 Demonstration of the Peltier effect with wires of two different materials (vertical wires, copper; horizontal wire, constantan), which produce a thermoelectric force of 42.5 µV K^{-1} in the temperature range from −200 to 500 °C.

cooled. Figure 5.33 depicts an example using two copper wires (vertical) and one constantan wire. A direct current of 3 A leads to heating of one connection and cooling of the other. The effect is reversed if the direction of the current is reversed.

In this (macroscopic) experiment, the constantan wire was made much thicker (several parallel wires) since it has a higher resistance compared to the copper wire. If a single wire were used, the dissipation of energy by its resistance alone (the I^2R joule heat) would lead to a homogeneous heating along the wire, which would cover up the small effect due to the Peltier effect. Nowadays, the Peltier effect is widely used in microscopic setups of cooling systems for microelectronics and detectors (Section 8.4.2).

5.4.4
Experiments with Microwave Ovens

Microwave ovens, which are also part of everyday life, combine electromagnetism, the general behavior of electromagnetic waves, and thermal physics in a unique way. The most common application is just heating of food, but industrial ovens are also used to dry a variety of goods [18, 19]. Here, some experiments with household microwave ovens are presented (for more information, see [14, 20–22]).

5.4.4.1 Setup

Figure 5.34 depicts the main features of a microwave oven. The microwaves are generated within a magnetron and guided into the cooking chamber, which has metal walls. There, the microwave energy is absorbed [14, 18] by the food or the object placed into this chamber.

To first order, a microwave oven with metallic walls resembles a three-dimensional resonator for electromagnetic waves. The microwaves of typical ovens have frequencies of about 2.45 GHz, giving wavelengths of about 12.2 cm. The problem is solved from the equations of electrodynamics for a chamber with lengths L_x, L_y, and L_z (typical lengths range between 20 and 30 cm, i.e., 8–12 in.).

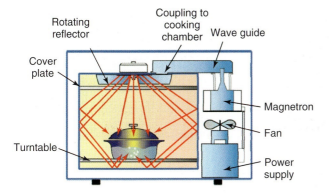

Figure 5.34 Schematic diagram of a microwave oven.

Similar to the one-dimensional case of standing waves on the string of a guitar, one finds three-dimensional standing waves, that is, there will be positions within the oven where there is a high energy density of the microwave field and there will be nodes of the standing waves, where there will be no energy density. In this respect, one speaks of horizontal and vertical modes of the microwave field.

An obvious consequence of nonhomogeneities of the microwave energy within the oven is that the absorption of the microwave energy by food or other products will strongly depend on the position. In order to reduce uneven heating of food, the effect of the horizontal modes is usually smeared out by using a rotating turntable and sometimes a top rotating reflector.

5.4.4.2 Visualization of Horizontal Modes

In order to visualize the undisturbed mode structure within a microwave oven (i.e., without turntable) using thermography, we place a thin glass plate of appropriate dimensions within the oven. Its height can be adjusted by placing Styrofoam below it. The glass does not absorb microwaves strongly. In order to measure the mode structure, we either put a wet paper on top of the plate or wet the glass plate by covering it with a thin film of water. The plate is then heated in the oven for a certain period (depending on the applied power). Directly after the heating, the door is opened and the plate is analyzed with the IR camera. Figure 5.35 depicts three examples of the observed mode structure of the otherwise empty microwave oven for the plate at the floor, in a height of 3.5 and 8 cm. In all cases, the oven was operating for 15 s at a power reading of 800 W. One clearly observes pronounced differences, that is, the horizontal mode structure also strongly depends on height.

Unfortunately, the situation is more complex for practical applications. Most importantly, the mode structure changes upon filling the oven. For example, when an object of given geometry, which can absorb microwave energy, is placed in the oven, the electrodynamic calculation of the loaded oven gives different mode structures compared to the empty oven, since the boundary conditions have changed.

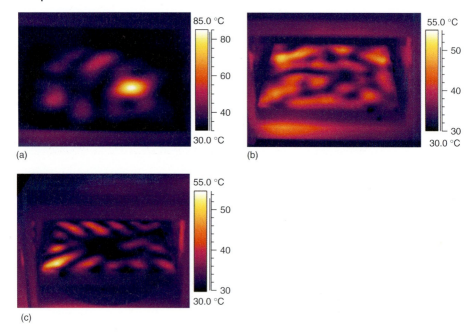

Figure 5.35 Visualization of the horizontal mode structure in a microwave oven. A glass plate with a thin water film was placed at the floor and heated for 15 s with a microwave power of 800 W without using the turntable. The plate was at the floor (a), at a height of 3.5 cm (b), and at a height of 8 cm (c).

5.4.4.3 Visualization of Vertical Modes

Although a turntable in a microwave oven may be useful in smearing out non-homogeneities of the horizontal mode structure, it does not have the same effect for the vertical modes. Figure 5.36 shows IR images of a tall glass cylinder of about 2-cm diameter filled with water before and after heating in the microwave oven. The cylinder was placed in the center of the turntable, which is what most people do with objects when placing them into a microwave oven. Obviously, the heating is quite uneven. There are large temperature differences of more than 20 K between the bottom, middle, and top of the glass. In this case, we found the temperatures to be 76 °C at the top, 43 °C in the middle, and 62 °C at the bottom of the glass. If baby food is heated in this way, and the cold part is on top, one may erroneously assume that the whole food is cold enough to eat. We conclude that all food in tall containers should be stirred before serving. Of course, the turntable may help partially, but only if the container is not placed at the center, since – at fixed height – the object may eventually also move through maxima as well as minima of the mode structure, which may lead to some averaging.

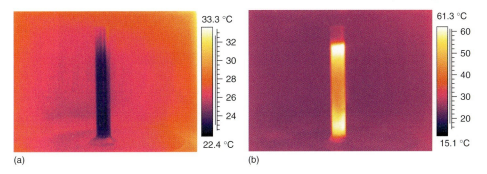

Figure 5.36 Visualization of the vertical mode structure in a microwave oven. A glass cylinder (diameter 2 cm), filled with about 30 ml of water was placed on the turntable and observed before (a) and after (b) heating for 15 s with a microwave power of 800 W.

5.4.4.4 Aluminum Foil in Microwave Oven

One often hears the statement that metals or objects with metallic parts should never be put into the microwave. Physicists know about the origin of this "wisdom;" however, also they are also aware of the limited range of validity. When microwaves interact with metals, they are not only effectively absorbed but also reradiate most of the energy. Since metals have a good thermal conductivity the fraction of the energy that is absorbed is rapidly distributed over the whole metallic body. If this body is very massive – as, for example, the walls of the microwave oven – the new equilibrium state, which depends on absorbed power, heat capacity, and heat losses corresponds to a very small warming. The behavior of smaller metal parts depends, however, strongly on their geometry and mass. Very thin metal sheets or similar bodies have only a very small heat capacity and can warm up quickly. This can even lead to glowing and evaporation, for example, from plates with golden edges. One should never put such plates in a microwave, unless the golden edge should be removed.

This leads to the typical question of what happens to thin metal foils like aluminum foil in a microwave oven. Thin strips of foil can heat up quickly, but what happens, if they have good thermal contact with another body that may absorb energy? Figure 5.37 shows two identical beakers filled with water before (a) and after (b) heating in the microwave oven. The right beaker in each image is surrounded by aluminum foil of about 30-µm thickness. This foil is thick enough such that no microwave radiation may penetrate through it, that is, in this beaker, only radiation from the top may reach the water. The foil does absorb a little bit of energy which is transferred to the water inside the beaker due to the good thermal contact. However, this energy transfer is much smaller than the energy that is absorbed in the other beaker by the water itself. Therefore, the beaker filled with water heats up much more quickly. Consequently, food should never be put in the microwave oven if it is surrounded by thick aluminum foil.

There are many more experiments that can be done with IR imaging and microwave ovens [14, 20–22].

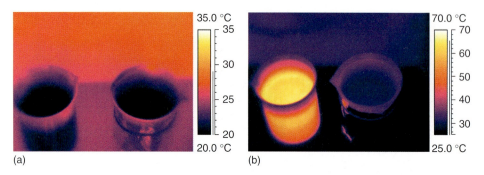

Figure 5.37 Two beakers are filled with water. The one wrapped in aluminum foil heats more slowly than the one without foil.

5.5
Optics and Radiation Physics

IR imaging can provide some fascinating insights into the optical properties of various materials and objects. Some objects are transparent in the Vis range, but opaque in the thermal IR and vice versa, and others are transparent in both spectral ranges. The general theoretical background has been discussed in Section 1.5; some simple experimental results are presented here. Infrared thermal imaging is based upon the laws of radiation by Kirchhoff and on Planck's law, describing the spectrum of thermal radiation. It also depends on emissivity and the fact whether gray or selective emitters are studied. Besides using these laws, one may, however, also use IR imaging to visualize these concepts.

5.5.1
Transmission of Window Glass, NaCl, and Silicon Wafer

Regular window glass or laboratory glass like BK7 show no transmission above $\lambda \approx 3\ \mu m$ (Figure 1.54). Therefore, any IR camera operating at longer wavelengths (LW cameras) will not be able to look through thick layers of glass, MW cameras may still see a tiny bit of radiation (See also Figure 3.2). This is known to thermographers doing outdoor building inspections; however, sometimes special care has to be taken since inhabitants of the houses can get the feeling that someone is observing them and even taking pictures through the window.

Figure 5.38 depicts someone holding a plate of glass (thickness of several millimeters) partially in front of his face. Obviously, it is not possible to look through glass, which is opaque in the IR spectral range. One may, of course measure the surface temperature of the glass plate, something which is regularly done in building inspections. In addition, Figure 5.38 visualizes one of the major problems encountered in thermography of flat surfaces; they may lead to thermal reflections, which can give rise to problems in quantitative analysis (Section 9.2).

5.5 Optics and Radiation Physics

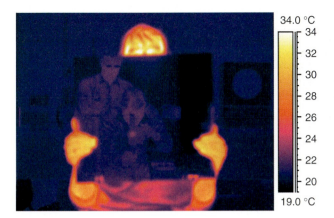

Figure 5.38 A room temperature glass plate is opaque to IR radiation. In addition, due to the flat surface, it serves as a source of thermal reflections, here two people standing behind the IR camera.

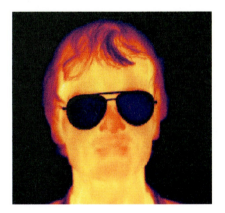

Figure 5.39 IR images of people, wearing glasses often suggest that dark sunglasses were used.

Objects made of glass are very often encountered in IR imaging, not only in building inspections but also when taking images of people. Everyone wearing regular glasses will appear in IR images as wearing very dark sunglasses (Figure 5.39) since the glass is opaque. But why is the glass temperature so much lower than the skin temperature? Glasses usually only have poor thermal contact with the face, and little thermal energy is conducted from the skin to the glasses at the three contact points at the nose and near the ears. Therefore, the heat transfer via convection from the ambient temperature air at the glass surfaces dominates and determines the surface temperature.

Figure 5.40 depicts Vis and IR images of a person using another pair of glasses. One lens is made of regular glass, the other of NaCl. From Figure 1.48, it is obvious that NaCl will transmit Vis and thermal IR radiation; therefore, one can readily look

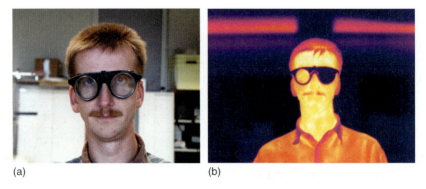

(a) (b)

Figure 5.40 A special pair of glasses, made of two different materials; visible radiation is transmitted by both materials, whereas thermal IR radiation is only transmitted by one of them.

through the lens and observe the higher temperature of the skin near the eye. It is easily possible to use the known theoretical transmission of NaCl (about 91%) and therefrom calculate correction factors for quantitatively measuring temperatures behind the lens [23].

In contrast to glass and NaCl which are both transmitting visible radiation, silicon wafers are opaque in the Vis spectral range (spectrum Figure 1.51). Therefore, it is not surprising that one cannot look through with the eye. IR imaging does, however, allow looking through matter (Figure 5.41).

In the experiment, a wafer of 0.362 mm thickness was placed directly in front of the IR camera lens. The real part of its index of refraction (3.42) leads to a transmission of about 53%.

5.5.2
From Specular to Diffuse Reflection

Usually, only the law of mirror reflection (here denoted as specular reflection) is treated (Eq. (1.2), Figure 1.9), when introducing reflection in optics. In contrast, diffuse reflection is encountered much more often in everyday life and technology, or at least a combination of diffuse and regular reflection, as illustrated in Figure 5.42.

The transition from pure specular reflection (e.g., from a mirror) to pure diffuse scattering (e.g., from a wall or blackboard) can be nicely studied using IR imaging with LW cameras. Diffuse scattering takes place if the wavelength of the electromagnetic radiation is comparable to the dimensions of the surface roughness. If the latter dimensions are small compared to the wavelength, regular reflection takes place. Analogously, a soccer ball will bounce back from a mesh wire according to the law of reflection, whereas a table tennis ball with similar dimension than the mesh will behave like a diffuse scatterer.

5.5 Optics and Radiation Physics | 319

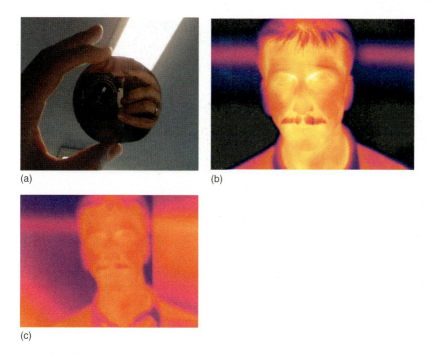

(a) (b)

(c)

Figure 5.41 A Si wafer – polished on both sides – is opaque for visible light (a), but transmits thermal IR radiation (c). The IR signal is just attenuated according to the Si transmission compared to the image without wafer (b)

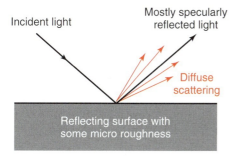

Figure 5.42 Real surfaces have surface roughness. Therefore, reflection consists of a superposition of specularly reflected and diffusely scattered light.

Using visible and IR electromagnetic radiation the transition from diffuse to specular reflection can be demonstrated directly. Consider, for example, a person in front of a brass plate which is oxidized and a diffuse scatterer in the visible ($\lambda = 0.4–0.8\,\mu m$): no mirror image can be seen (Figure 5.43). However, the wavelength of the IR radiation, detected in $\lambda = 8–14$-μm IR cameras is about a

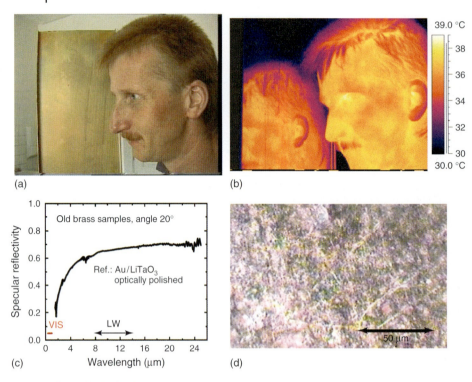

Figure 5.43 (a) Transition from specular to diffuse reflection, an oxidized brass plate scatters visible light diffusely, whereas LW IR radiation leads to a clearly observable specular reflection image. (b) This behavior is due to microscopic roughness (see electron microscope image (d)), which also shows up in the specular reflectance spectrum (c).

factor of 10 larger. Therefore, the IR image can demonstrate regular reflection (for more details, see Section 9.2 and [24]).

5.5.3 Blackbody Cavities

Blackbody cavities (Section 1.4.6) are considered to give the best possible approximations to blackbody radiation on earth. Therefore, many theoretical analyses were done on theoretical emissivities depending on properties of the used cavities. According to an old theory by Gouffé, the total emissivity of a cavity resembling a blackbody is given by [25]

$$\varepsilon = \varepsilon_0'(1 + \gamma) \tag{5.8}$$

where

$$\varepsilon'_0 = \frac{\varepsilon^*}{\varepsilon^*\left(1 - \frac{s}{S}\right) + \frac{s}{S}} \tag{5.9a}$$

and

$$y = (1 - \varepsilon^*)\left[\left(\frac{s}{S} - \frac{s}{S_0}\right)\right] \tag{5.9b}$$

In these equations, ε^* denotes the emissivity of the wall material of the cavity, s and S are the areas of the aperture and of the interior surface, and S_0 denotes the surface area of an equivalent sphere, which would have the same depth as the cavity in the direction normal to the aperture. Usually y is a small number; however, depending on the cavity shape it can be positive or negative.

From Eqs. (5.8) and (5.9) it becomes clear that even quite small numbers of material emissivity can give quite large values for the total emissivity.

Figure 5.44 shows the results of an experiment. A set of three cylindrical holes in a metal block could be covered by apertures to form cavities of different emissivities. IR images of the heated cavities revealed that the apparent temperature between the largest and the smallest aperture cavity evaluated for constant emissivity would

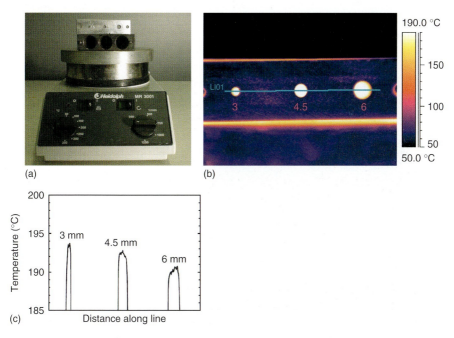

Figure 5.44 (a) A set of three cylindrical holes in a metal block, which may be covered by apertures of different sizes to form cavities of different emissivities. (b) IR radiation was detected when heating the cavities to a temperature around 200 °C. (c) The temperature along a line through the centers of the holes was evaluated for constant emissivity (for more details see the text).

differ by more than 2 K. Assuming, however, that the differences in detected IR radiation are due to changes in emissivity, the experiment nicely demonstrates the validity of Eqs. (5.8) and (5.9).

The cylindrical holes had an inner diameter of 18 mm and a depth of 36 mm each. The apertures used had diameters of 3, 4.5, and 6 mm, leading to values $(s/S) = 0.28$, 0.62, and 1.1% respectively. The values of (s/S_0) are even smaller; hence the correction term y is always below 0.01. The metal walls of the cavity were already slightly corroded and had $\varepsilon^* \approx 0.21$. This gives total emissivities of the three cavities of about 0.96 for the 6-mm aperture, 0.98 for the 4.5-mm aperture, and 0.99 for the smallest 3-mm aperture. These small differences in emissivity directly and quantitatively explain the observed results from the IR analysis (Figure 5.44).

We note that the front plate looks much colder than the holes in the IR image. This is due to the lower emissivity of the cover plate (see visible image). Repeating this experiment for a long time leads to oxidation of the front surfaces, which goes along with an increase in emissivity. Therefore, the actual IR image may change (i.e., the ratio of signal from the holes to the cover plate changes) when repeating the experiment. However, the amount of cavity radiation, which is studied by this experiment, is not changed.

5.5.4
Emissivities and Leslie Cube

The angular dependence of emissivity can be seen in Figures 5.45 and 5.46. In Figure 5.45 the aluminum cubes with high emissivity black paint (Figure 4.17) are depicted. The top face of the cubes is observed for a much larger angle than the side faces. Therefore, according to Figures 1.32 and 1.33, the emissivity is lower

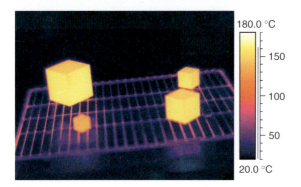

Figure 5.45 The normal emissivities of the faces of paint-covered metal cubes are the same. The two observable side faces are viewed from the same angle of about 45°, whereas the top face is seen from a larger angle. Owing to the angular dependence of emissivity, this leads to an apparently colder top face.

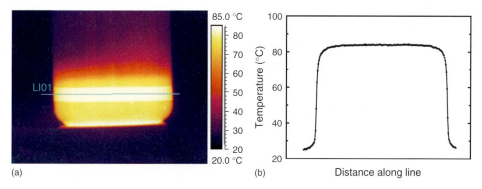

Figure 5.46 Glass cylinders filled with hot water allow to simultaneously observe emissivity effects due to a large variety of viewing angles. (a) IR image with line across a high emissivity tape. (b) Temperature profile along the line.

than for the side faces. As a result, the face appears to be cooler although it has the same temperature.

Figure 5.46 depicts a cylindrical container (large glass beaker) with an attached tape of high emissivity. One can clearly see that close to the edge, where the viewing angle is much larger, the apparent temperature drops with respect to the near normal observed areas of the object. A detailed analysis of the shape of the temperature profile is in agreement with the predictions of the drop of emissivity with observation angle (Figure 1.32).

Figure 5.47 shows an empty Leslie cube observed such that two side faces and the bottom face are all viewed from about the same viewing angle. If hot objects (finger of person) are close by, thermal reflections are clearly observable, which are dominant for the polished Cu metal surface and still detectable for the white and black paint covered surfaces.

The same Leslie cube is shown in Figure 5.48 while and after being filled with hot water. Now the surfaces are much hotter than the surroundings and no additional warm objects are around. Therefore, no thermal reflections are seen and the differences directly reflect the different surface emissivities at this fixed angle. The white and black paint surfaces show nearly the same emissivity for the LW camera, whereas the polished copper has still a lower emissivity compared to the diffusely scattering rough copper surface.

5.5.5
From Absorption to Emission of Cavity Radiation

In most experimental conditions in thermal physics, one has to deal with nonequilibrium conditions. An instructive experiment uses a small cavity, which resembles some kind of blackbody radiator. Such cavities are, for example, small graphite cylinders with an additional hole in the center of the side. Let us assume that the graphite surface may have an emissivity of say $\varepsilon = 0.9$, whereas the hole has a slightly larger

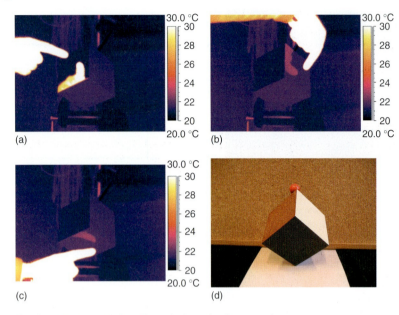

Figure 5.47 Empty Leslie cube with thermal reflections observed from an angle such that the sides (polished Cu, (a) white paint, (b) and black paint (c)) are viewed at the same angle.

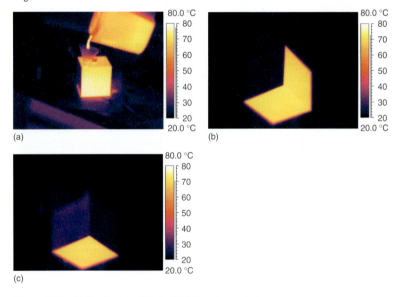

Figure 5.48 (a) Leslie cube filled with hot water and observed from an angle such that the sides are viewed at the same angle. (b) Polished Cu, white paint, and black paint (lowest segment). (c) Rough copper, polished copper, and black paint surface.

emissivity of, for example, around 0.98 (the exact values do not really matter, just the difference between surface and cavity values). Such graphite tubes are standard sample holders in atomic absorption spectroscopy (AAS). It is easily possible to heat the cavity by holding the end between fingers ($T > 30\,^\circ\text{C}$). After a dynamic thermal equilibrium between fingers and cavity is established, it is warmer than the surrounding, that is, the cavity is now not in thermal equilibrium with the colder surroundings. According to the laws of radiation, the temperature difference between cavity and surroundings will lead to a net emission of thermal radiation from the cavity, the amount being characterized by the emissivity. Since the cavity has a higher value of emissivity, it emits more radiation as clearly shown in Figure 5.49a.

The situation may, however, also be reversed by cooling the cavity. This was done by placing it between two ice cubes. After a (dynamic) thermal equilibrium between ice cube and cavity is established, the cavity is now much colder than the surrounding at room temperature. The cavity itself has high emissivity, that is, also high absorptivity (according to Kirchhoff's law). Therefore, it will absorb more radiation from the surrounding than the surface of the graphite tube. This energy quickly flows away to the ice cubes due to conduction, that is, we assume

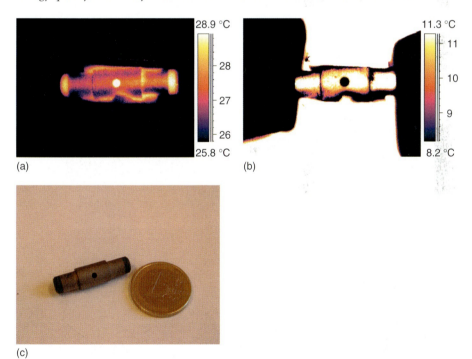

Figure 5.49 (c) Graphite tube (length = 3 cm, inner diameter 4 mm) with a small hole (diameter = 2 mm), typically used as sample holder in atom absorption spectroscopy (AAS) may serve as cavity. It can be used to record IR images while studying the transition from emission (a) to absorption (b). The tube can be heated by holding its ends between fingers (a) or cooled (b) by attaching the ends to ice cubes.

that the cavity temperature will stay low (this is why we speak of a dynamic equilibrium). The cavity will therefore emit radiation according to the cavity temperature, which is lower than the one of the surrounding. This has to be compared with the radiation from the graphite tube surface. Assuming the same temperature (thermal equilibrium within the tube) it should be lower than the one of the cavity due to the lower emissivity. However, since its emissivity is lower, its reflection coefficient is automatically much higher (Eq. (1.31)). Therefore the amount of thermal radiation from the much warmer surroundings, which will be reflected from the tube surfaces, add up to the pure thermal emission. This leads to a much larger total emission from the surface compared to the cavity. As a consequence, the cavity emits much less radiation than the tube surface as shown in Figure 5.49b.

5.5.6
Selective Absorption and Emission of Gases

The transition from absorption to emission of radiation by a selective emitter can also be demonstrated very nicely using selectively absorbing and emitting objects like molecular gases (for details see Chapter 7), Figure 5.50 depicts experimental results recorded with a LW camera using SF_6 [26]. SF_6 was filled in a plastic bag and cooled down to about $-20\,°C$ in an air convection cooler. The cold gas was taken out of the cooler, the valve of the bag was opened and cold gas pressed out of the valve. The process was observed with the IR camera using the wall at room temperature as background. As shown in Figure 5.50a, the IR radiation from the wall toward the camera is significantly attenuated due to absorption within the gas. This is due to the strong absorption bands of SF_6 in the wavelength range around 10–11 µm.

In order to observe emission at these wavelengths, we placed the gas-filled bag in our air convection heating system. The gas was heated to about $80\,°C$. It was taken out of the heater, the valve was opened, the gas was pressed out of the valve,

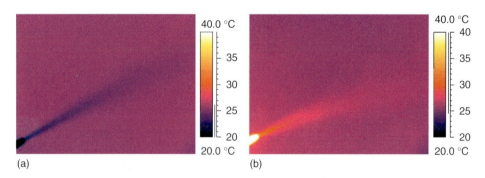

Figure 5.50 Absorption of cold SF_6 ($T \approx -20\,°C$ (a)) and emission from warm SF_6 ($T \approx 80\,°C$ (b)) in front of a wall at room temperature (detected with an LW camera).

and the process was again observed with the same wall at room temperature as background. The result of Figure 5.49b clearly demonstrates the emission of the hot gas, which leads to an increase of IR radiation from the streaming gas. Details of gas absorption and emission also with corresponding technological applications are discussed in Chapter 7.

References

1. Karstädt, D., Pinno, F., Möllmann, K.P., and Vollmer, M. (1999) Anschauliche Wärmelehre im Unterricht: ein Beitrag zur Visualisierung thermischer Vorgänge. *Prax. Naturwiss. Phys.*, **48** (5), 24–31.
2. Karstädt, D., Möllmann, K.P., Pinno, F., and Vollmer, M. (2001) There is more to see than eyes can detect: visualization of energy transfer processes and the laws of radiation for physics education. *Phys. Teach.*, **39**, 371–376.
3. Möllmann, K.-P. and Vollmer, M. (2000) Eine etwas andere, physikalische Sehweise – Visualisierung von Energieumwandlungen und Strahlungsphysik für die. *Phys. Bl.*, **56**, 65–69.
4. Möllmann, K.-P. and Vollmer, M. (2007) Infrared thermal imaging as a tool in university physics education. *Eur. J. Phys.*, **28**, S37–S50.
5. Halliday, D., Resnick, R., and Walker, J. (2001) *Fundamentals of Physics, Extended*, 6th edn, John & Wiley Sons, Inc.
6. Tipler, P.A. and Mosca, G. (2003) *Physics for Scientists and Engineers*, 5th edn, Freeman.
7. Bloomfield, L. (2007) *How Everything Works*, John Wiley & Sons, Inc.
8. Bellemans, A. (1981) Power demand in walking and pace optimization. *Am. J. Phys.*, **49**, 25–27.
9. Keller, J.B. (1973) A theory of competitive running. *Phys. Today*, **26**, 42–47.
10. Alexandrov, I. and Lucht, P. (1981) Physics of sprinting. *Am. J. Phys.*, **49**, 254–257.
11. Maroto, J.A., Pérez-Muñuzuri, V., and Romero-Cano, M.S. (2007) Introductory analysis of the Bénard Marangoni convection. *Eur. J. Phys.*, **28**, 311–320.
12. Consumer Energy Center of the California Energy Commission http://www.consumerenergycenter.org/home/heating_cooling/evaporative.html. (2010).
13. Planinsic, G. and Vollmer, M. (2008) The surface-to-volume-ratio in thermal physics: from cheese cubes to animal metabolism. *Eur. J. Phys.*, **29**, 369–384 and 661.
14. Vollmer, M. (2004) Physics of the microwave oven. *Phys. Educ.*, **39**, 74–81.
15. Baylie, M., Ford, P.J., Mathlin, G.P., and Palmer, C. (2009) The jumping ring experiment. *Phys. Educ.*, **44** (1), 27–32.
16. Bostock-Smith, J.M. (2008) The jumping ring and Lenz's law – an analysis. *Phys. Educ.*, **43** (3), 265–269.
17. Michalski, L., Eckersdorf, K., Kucharski, J., and McGhee, J. (2001) *Temperature Measurement*, 2nd edn, John Wiley & Sons, Ltd, Chichester.
18. Thuery, J. (1992) *Microwaves, Industrial, Scientific and Medical Applications*, Artech House, Boston.
19. Smith, B.L. and Carpentier, M.-H. (1993) *The Microwave Engineering Handbook*, vols. 1-3, Chapman & Hall, London.
20. Parker, K. and Vollmer, M. (2004) Bad food and good physics: the development of domestic microwave cookery. *Phys. Educ.*, **39**, 82–90.
21. Vollmer, M., Möllmann, K.-P., and Karstädt, D. (2004) More experiments with microwave ovens. *Phys. Educ.*, **39**, 346–351.
22. Vollmer, M., Möllmann, K.-P., and Karstädt, D. (2004) Microwave oven experiments with metals and light sources. *Phys. Educ.*, **39**, 500–508.
23. Vollmer, M., Möllmann, K.-P., and Pinno, F. (2007) Looking through matter: quantitative IR imaging when

observing through IR windows. Inframation 2007, Proceedings vol. 8, pp. 109–127.
24. Henke, S., Karstädt, D., Möllmann, K.P., Pinno, F., and Vollmer, M. (2004) in *Inframation Proceedings*, vol. 5 (eds R. Madding and G. Orlove), ITC, North Billerica, pp. 287–298.
25. Wolfe, W.L. and Zissis, G.J. (eds) (1993) *The Infrared Handbook*, revised edition, 4th printing, The Infrared Information Analysis Center, Environmental Research Institute of Michigan, Michigan.
26. Vollmer, M., Karstädt, D., Möllmann, K.-P., and Pinno, F. (2006) Influence of gaseous species on thermal infrared imaging. Inframation 2006, Proceedings vol. 7, pp. 65–78.

6
IR Imaging of Buildings and Infrastructure

6.1
Introduction

Infrared thermal imaging is considered to be an excellent noninvasive inspection tool for monitoring and diagnosing the condition of buildings by measuring the surface temperatures of the building envelope from either inside or outside (or both). The surface temperatures of buildings are due to three basic mechanisms: the flow of heat, air, and moisture through the building envelope. These three factors not only determine the building durability and its energy efficiency but also, most important, the feeling of comfort, health, and safety for the inhabitants of a building.

Heat flow is the primary mechanism for heating up the surfaces of a building. Heat flows from the warm interior to the cold exterior, giving rise to an increase in outer wall surface temperatures. The respective physical quantities like U-value have already been discussed in Section 4.3. In general, heat flow can lead to either warming up or cooling of building parts due to conductive differences, thermal bridges, and/or air infiltration or exfiltration.

Water intrusion into the walls/the insulation and surface moisture often reduce the measured surface temperatures due to evaporative cooling (Section 5.3.4). However, the huge heat capacity of water, that is, the possibility to store thermal energy, may (besides cooling) also lead to warming up of building parts, depending on the respective conditions. Also, water intrusions into the thermal insulation material may lead to an increased thermal conductivity from the inside to the outside with characteristic thermal signatures. Moisture is often present on flat roofs and may penetrate into buildings if the moisture barrier is not correctly installed.

Finally, uncontrolled airflows through air leaks may transport heat via additional convection losses. In addition, the airflows affect the temperatures in the vicinity of the air leaks, for example, in window frames. The reduction of the frame temperatures (mostly in corners) in wintertime to below the dew point can give rise to mold problems. Air leaks may also often occur close to wall openings for electrical wiring or plumbing and ducts.

Thermography can be used to find problems associated with heat, water, and airflows through the building envelope. In this manner, it may automatically test,

Infrared Thermal Imaging. Michael Vollmer and Klaus-Peter Möllmann
Copyright © 2010 WILEY-VCH Verlag GmbH & Co. KGaA, Weinheim
ISBN: 978-3-527-40717-0

for example, the energy efficiency of new homes or plan the restoration of old ones. In particular, it may help to

- locate heating and cooling losses;
- locate building envelope water leakages, that is, moisture sources;
- locate structural problems (missing insulation, degradation of old insulation, etc.);
- locate problems in floor heating systems;
- compare pre- and postrestoration conditions.

Since – besides irradiation with UV light – the flows of heat and water are also the most important damage functions for buildings, IR inspections may help to save a lot of money for restoration. With regard to energy efficiency, it can help save energy needed for the heating of a building and thus reduce the emission of greenhouse gases. In some countries like the United States or Australia, thermography is also a useful tool in pest control, for example, by localizing termites (Section 10.2.4).

6.1.1
Publicity of IR Images of Buildings

Infrared imaging of buildings is probably the best known application of thermography in the public. This is due to several reasons. First, buildings offer a standard textbook example of thermography (see also [1, 2]). In particular in wintertime, there is usually a large temperature difference between inside and outside of a house such that defects in the thermal insulation can be easily detected as thermal losses from the building envelope. Hence, thermography is a valuable tool in diagnosing thermal insulation problems of buildings. Second, thermography usually provides false color images of objects. In the modern communication age, false color representations of nearly anything are often welcomed by journalists as eye catchers. In particular, thermography false color images are often used in newspapers, on many web sites, and IR movie sequences are often shown in TV programs.

We believe that one reason may be that journalists, and also those looking at the images tend to think that they can immediately grasp the content and draw correct conclusions. And if those conclusions point out a problem previously unknown, the story will become widely known, which is good for the journalist. As a matter of fact, a red color in an IR image is usually considered to be a very bad sign (see Figure 6.1).

Unfortunately, it is very easy to produce false color images with IR cameras – and this will become even easier with the new class of low price thermal imagers. The lower the price of the cameras, the less experienced and educated the user will be, and the less they will think about training courses whose cost may be a large part of the price of the camera itself. This will lead to problems.

In this section, we demonstrate that thermal imaging is indeed a very valuable tool in building inspections and, in this sense, it is logical that buildings are the most common objects of published thermal images in newspapers, and so on.

Ein sogenanntes Thermogramm lässt durch Rotfärbung erkennen, dass die Hauswände Wärme abgeben   Foto: dpa

Figure 6.1 This infrared thermal image of a house was published a while ago in a daily newspaper (of 23 May 2009 in Berliner Morgenpost, www.morgenpost.de/berlin/article1097945). The image was shown in many different newspapers, sometimes even without temperature scale, the caption usually mentioning that the red color in the figure would indicate where the walls would lose heat. As a matter of fact, if one were to believe this statement, the whole house would need a lot of renovation. But is this really the case?

However, we also point out the large number of different external influences that may cause problems while one is trying to correctly interpret such thermal images. While doing this, we come back to Figure 6.1 and point out potential problems in a much-too-easy-interpretation of this nicely colored IR image.

6.1.2
Just Colorful Images?

In building thermography, the IR radiation signals contain information about emissivity and surface temperatures (this is the very first aspect, which is often neglected by amateur thermographers: they interpret any image as showing only temperatures and often use color scale and temperature span that exaggerate any slight differences seen). As a matter of fact, the choice of the color palette can strongly affect the interpretation as Figure 6.2 demonstrates.

6.1.2.1 Level and Span
The lower the span, the more dramatic the appearance of the IR images indicated by the images in Figure 6.2a–c for a 6 K temperature span (a) compared to a 10 K (b) or 30 K (c) span at same level. By watching only the largest span image (image (c) $\Delta T = 30$ K) one may not think of a problem at all due to the small color variations across the image. Keeping constant span, the change of level (d,e) also leads to appreciable color changes and if just being used with the notion that red or yellow

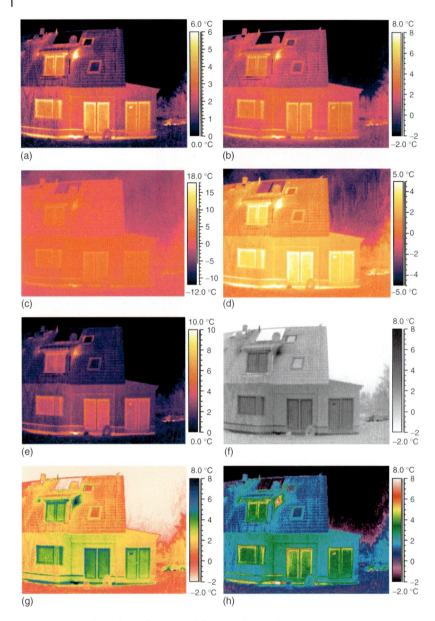

Figure 6.2 (a–f) IR thermal images of the same house displayed for a variety of temperature spans and color palettes. Emissivity was set to 0.96, distance 10 m, ambient air temperature outside was 0 °C. The whole house was heated inside to about 20 °C.

color is bad, one may interpret the house in (d) as a problem case, whereas the one in (e) as fine.

6.1.2.2 Sequences of Color and Color Palette

The color palette for the images Figure 6.2a–e was the same. Very often, color scales are used where red and yellow indicate higher temperatures. As a consequence, most people interpret all IR images accordingly, in particular, if images are shown without the color bar at the side (this may happen quite often in newspapers, etc.). Figure 6.2g shows the same image for a different color palette. In Figure 6.2h, the sequence of colors is inverted. Obviously these two images look very different. For example, Figure 6.2g would indicate that the two roof windows as well as the rectangular area close to the roof top (a solar thermal power heating system) may offer problems although they are cold.

One must keep in mind that owing to the variety of manufacturers and also the software within a camera type alone, there are usually a large number of possible color palettes to select from. The palettes from Figures 6.1 and 6.2a–e are different – however, in both cases, red indicates a higher temperature. In contrast, however, yellow areas in Figure 6.1 have lower temperatures than the red ones, whereas in Figure 6.2a–e, yellow refers to higher temperatures. Comparing the colored images of Figure 6.2 with an inverted gray scale image (Figure 6.2f) explains why color palettes are preferred. There are many more possibilities in the amplification of small signal differences. Nowadays, gray scales are – to our knowledge – used only in life image GasFind cameras as the standard palette. Observing life sequences of one image alone is not important but rather the change from one image to the next is important. Such changes are already seen in gray scale and considerably less memory space is needed for saving these files.

6.1.3
General Problems Associated with Interpretation of IR Images

Using the above discussion of color representation of data in IR images, we can have another look at the "bad (but published) example" of Figure 6.1. First, the temperature span is unusually large; more than 47 °C which means that any fine structure of the wall is completely lost. For any careful analysis, such a large color span is just impractical. As can be seen from the image, all relevant parts of the house have temperatures in a range above −5 °C; therefore the span should have been reduced accordingly. Second, the heat transfer through windows is usually much higher than through walls (Table 4.6), therefore one would expect higher temperatures of windows compared to the walls. In Figure 6.2, this is indeed observed, whereas surprisingly, the windows in Figure 6.1 seem to have lower temperatures than the walls. Since it is very unlikely that the walls should have a very bad thermal insulation (high U value), while at the same time, the windows in this house would have been of the absolutely best available type (low U-value), we may think of another cause for this observed anomaly (see below).

We will come back to the interpretation of Figure 6.2 later. So far, we note that in any image of Figure 6.2, the area at the right hand side of the dormer window as well as a spot in the lower left corner show up. A thorough investigation is needed to find out whether they indicate a real problem or whether they are due to structural or geometrical effects and hence, expected.

As a matter of fact, surface temperatures of objects may be deduced from a thermal analysis, provided the respective surfaces, walls, windows, roofs, and so on, are gray emitters and the emissivities are known. However, even measured surface temperatures alone do not mean anything. Too many environmental factors can have an additional influence. The most crucial question is whether measurements are done from inside or outside of a building. Outside thermography may provide an overview of a building envelope thus giving indication or orientation of where to look in more detail. For home inspections, this can always be the first step toward a thorough investigation. Indoor thermography is much less affected by external factors and should be performed whenever possible. Sometimes, outdoor thermography is even useless, for example, for rear ventilated wall constructions. The complications associated with outside thermography are manifold, that is, there are a large number of external factors that complicate any analysis considerably (see Table 6.1) and often lead to wrong conclusions. The most important environmental factors are direct or indirect radiation from the sun, wind of varying speed, and moisture, for example, rain. In addition, the sky conditions (cloudy or clear sky) and the view factor are particularly important when considering radiant cooling. Night sky radiant cooling and daytime exposure to sun radiation can cause strong variation of building envelope temperatures with long time constants. Therefore absolute

Table 6.1 External influences on outdoor thermography.

Influence of	Problem caused
Wind	$\alpha_{outside} = \alpha_{outside}(v_{wind}) \Rightarrow$ transient wall temperatures, see also Section 4.3.3
Sun radiation	Transient effects of wall heating, reflections of IR radiation (MW cameras)
Shadows	Transient effects of wall heating and cooling
Cloudy versus clear sky	Transient effects of radiant cooling of walls
View factor	Orientation dependent radiative contributions from surrounding
Rain/moisture	Evaporation cooling, change of thermal time constant, usually reduces thermal contrast

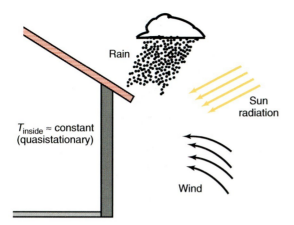

Figure 6.3 The inside building envelope can often be considered to be at quasi-stationary conditions. In contrast, the outside walls of the building are subject to sun radiation (and shadows), wind of varying speed as well as moisture due to rain.

temperatures as measured in outdoor thermography are often unimportant since a detailed quantitative analysis would be too difficult.

In the following sections, the problems introduced by these external factors are illustrated by thermal images. In contrast, indoor thermography has the advantage of nearly quasi-stationary conditions (Figure 6.3) provided a few general rules are followed while preparing an IR thermal imaging study of a building:

- need $\Delta T = T_{\text{inside}} - T_{\text{outside}} > 15$ K for more than a few days (this implies that most investigations take place in wintertime);
- try to enable quasi-stationary conditions inside by opening all doors, closing all windows;
- move furniture away from walls;
- let camera housing establish thermal equilibrium (at least 30 minutes).

In addition, recording outdoor thermography images require

- recording before sunrise after a cloudy night (small night-to-day temperature changes);
- avoiding fog, rain, or snow;
- avoiding strong winds ($v_{\text{wind}} < 1$ m/s)
- letting camera housing establish thermal equilibrium (at least 30 minutes).

The following conditions should definitely be recorded:

- inside and outside air temperatures;
- inside and outside reflected temperatures;
- inside and outside humidity;
- outside wind speed;

- distances from camera to walls: check geometrical resolution, maybe change lenses;
- emissivities of wall materials (tapes, etc.);
- orientation of surroundings with respective temperatures (view factor contributions);
- possibly record visible photos of areas that are investigated by IR.

Despite its problems, outside thermography may sometimes be the only useful method of investigating a building, for example, when studying half-timbered structures behind plaster, when trying to locate water intrusions in building fronts, or when it is not possible, for various reasons, to get inside buildings.

Whenever indoor or outdoor IR images – are interpreted, one first tries to exclude all possible error sources as mentioned above (sun, shadows, thermal reflections, etc.). The resulting temperature profiles may then point out problems. Sometimes, however, there are no problems at all, but the observed fluctuations are just due to structurally expected thermal bridges and so on, and within allowed limits.

To summarize, infrared thermal images from buildings may yield apparently significant signal differences between various locations. After corrections for emissivity, there may still be some temperature differences. Even if all other disturbing effects are eliminated, they may still be within the expected range, for example, owing to geometrical details. As a consequence, not all differences in thermal images of buildings imply structural problems or leaks in thermal insulation; rather a very careful analysis is needed.

6.1.4
Energy Standard Regulations for Buildings

In times of environmental awareness and sustainable development, the need for a reduction of greenhouse gases by reduction in the consumption of fossil fuels and the introduction of new renewable energies and technologies is obvious. In addition, over the years, the energy savings requirements for new homes regarding thermal insulation and heat losses through the building envelope have changed and the specific heating power per area of a home has been constantly decreasing (see Figure 6.4). Starting after the first severe oil crisis, regulations restricted the allowed annual energy need for heating of new buildings from around 200 to 300 kW h $(m^2\ a)^{-1}$ in the 1960s and 1970s to less than 70 kW h $(m^2\ a)^{-1}$ according to the 2002 regulation. The new regulations, starting from 2009, will further reduce these numbers by at least 30% and an end of this trend is not yet in sight.

Already in 1995, new standards for special buildings were defined, the so called low-energy houses or passive houses (e.g., [3]). The latter were only allowed to consume 15 kW h$(m^2\ a)^{-1}$, corresponding to a minimum of about 1.5 l of oil per m² annually. Nowadays, even zero-energy houses or energy-producing houses are available.

Such laws, if enforced, can lead to a drastic reduction of primary energy needed and also of greenhouse gas emissions. However, since the lifetime of old buildings is in the range of many decades and since less strict regulations apply to renovation

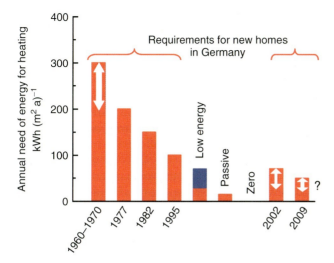

Figure 6.4 In many countries, the energy needed for the heating of a building per year is restricted by laws. In the given example of Germany, regulations started after the first oil crisis. They are valid for new buildings.

of old buildings, the average value of energy needed for heating for all buildings in the country decreases slowly. To accelerate this process, new energy standards for buildings have been defined in many countries; besides, so-called energy passports (see Figure 6.5) for buildings have been introduced and certificates issued for buildings (e.g., in Germany). In the United States, for example, the concept of home energy rating was introduced where the scale of 100 corresponds to a typical home built in 2006. Buildings needing less energy have a lower index.

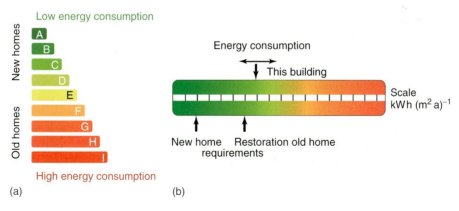

Figure 6.5 Possible energy passes for buildings. They may be qualitative only (a) with a highlighted symbol of the present home or include a quantitative scale, for example, for energy consumption per area and year (b).

Thermal imaging can play a major role in this field of energy saving, thus saving energy costs of private homes. In particular, thermography can test whether actions to save energy via better thermal insulation really work. The motivation of home owners is mostly of a financial nature: very often the cost for installing an improved thermal insulation is much less than what can be saved on energy costs within a few years, in particular, if energy costs of fossil energies are rising. Let us do a very naïve order-of-magnitude estimate of the energy cost savings when restoring an old house. Before restoration, the annual need of energy of a 100 m^2 house should be 270 kW h (m^2 a)$^{-1}$ and after restoration this should be reduced to only 70 kW h (m^2 a)$^{-1}$. Obviously, this leads to energy savings of 100 m^2 × 200 kW h (m^2 a)$^{-1}$ = 20 000 kW h a^{-1}. If heating is accomplished with gas having a price of €0.06/kW h, the annual cost savings after restoration would amount to €1200.

6.2
Some Standard Examples for Building Thermography

In general, building thermography is a tool to locate hidden structures due to thermal anomalies, which may be due to a variety of causes. In any case, the heat transfer through a wall or a building part causes differences in measured surface temperatures due to differences in heat conductivity and/or heat capacity of the building materials. Each thermal anomaly found via thermography must be discussed in detail with the background knowledge of the construction, the used materials, and so on, in order to find out whether the temperature anomaly is expected and within "normal" limits or whether it really represents an insulation problem concerning energy losses, structural damages, or even both. In this subsection, we first present a few qualitative examples of building thermography where the structures that give rise to specific thermal signatures are hidden beneath the surface, but are well known from the construction of the building.

6.2.1
Half-Timbered Houses behind Plaster

In Europe, there are still many old houses built as half-timbered structures. In Germany, timber framing was the most popular building technique from the twelfth to the nineteenth century. The idea is to create framed structures of heavy timber jointed together with special joints. Some diagonal frames are used to stabilize the structure. The spaces between the timbers are filled with brick or more often with a woven lattice of wooden strips daubed with a sticky material usually made of some combination of wet soil, clay, sand, animal dung, and straw. In the twentieth century many of these old half-timbered structures were covered with an additional layer of plaster, and sometimes with thermal insulation, often because the restoration of the original structure would have been too expensive. Nowadays, houses with half-timbered structures are often considered to represent valuable old

6.2 Some Standard Examples for Building Thermography

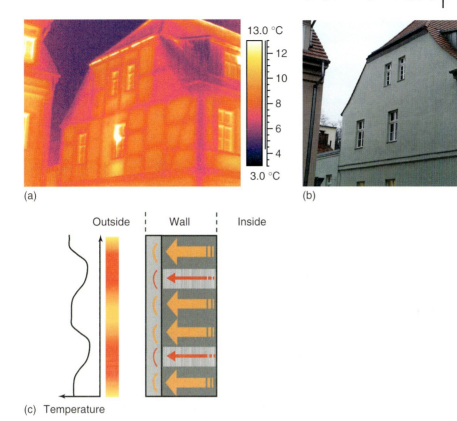

Figure 6.6 (a,b) Half-timbered structures of a house, which are hidden behind plaster, recorded during wintertime (no sunshine). The visible image (b) shows no structures at all. The white spot in the center of the IR image (a) corresponds to a small top window, which was opened after recording the visible image. (c) Scheme of heat flow through a structured wall. Shaded areas resemble wood, gray areas the filled compartments.

monuments that need to be preserved. In this context, infrared thermal imaging can be a valuable tool to identify such houses even if covered with a layer of plaster. Figure 6.6 depicts an example of a house in Brandenburg in Germany, recorded in wintertime with temperature differences from inside to outside of about 15 K. The visible image (Figure 6.6b) gives no indication of what is beneath the outer layer of plaster. The IR image (Figure 6.6a), however, clearly shows the structure of the wooden frames. Several features need to be discussed.

1) The clear view of the half-timbered structure in the thermal image results from the differing thermal conductivity of the wood and the filled compartments of the rest of the wall (see Table 6.2). As shown in the scheme of the heat flow (Figure 6.6c), the wood (brighter shaded areas) has lower thermal conductivity, therefore less heat flow results through the frames compared to the planar section of the filled compartments. The lateral heat flow is not

Table 6.2 Material properties of wood and typical sandstone relevant for interpreting IR images.

	Thermal conductivity W (m · K)$^{-1}$	Specific heat capacity kJ (kg · K)$^{-1}$	Density ($\times 10^3$ kg/m^3)
Dry wood	≈0.15	≈1.5	≈0.6
Typical sand stones	≈1.8	≈0.7	≈2

large enough to establish thermal equilibrium at the outer surface, as the temperature plot and the respective color scale (arbitrarily chosen to resemble the color scale of IR image) indicate. The most prominent feature in the thermal image is due to the open window in the center. The warm air from the inside heated up the parts adjacent to the window. Overall, the image shows the structures of the frames, however, the observed temperature differences across the wall surface do not indicate any energetic or thermal insulation problems.

2) The right-hand edge of the building seems to be colder than the plane wall. This is expected due to the geometrical thermal bridge (see below, Section 6.3) and does not pose a problem.
3) Only the middle section of the upper floor below the roof is heated and the wall section is tolerable. However, the rooftop suffers from missing insulation. This needs more attention. We also note that the frame of the right-hand side window is made of corroded metal with unknown emissivity, however, probably still a bit lower than that of the wall.
4) Furthermore, one can clearly observe a warmer wall section below the open window in the center either due to a heater or missing insulation. To a lesser extent, this can also be observed for the right window and the top floor windows. In order to decide whether this is energetically critical, more studies would be desirable. The present study focused only on detecting the half-timbered structures themselves.

In order to simulate half-timbered structures or, more generally speaking, building structures with differing thermal conductivities and heat capacities, students were given the task to build a model of such a hidden structure. The differing thermal properties were realized by using Styrofoam, air, metal, and wood. Figure 6.7a shows the front and Figure 6.7b the back of the final model with dimensions of 62 cm × 42 cm and a thickness of 2 cm. The IR imaging test was done by placing it in front of a planar, electrically heated plate of 0.6 × 0.9 m^2 (which, within minutes, reached surface temperatures of about 130 °C). Figure 6.7c depicts the result. The hidden structures became visible within a minute after starting to heat the plate. Other experiments with this model are discussed below in the context of wind and moisture effects.

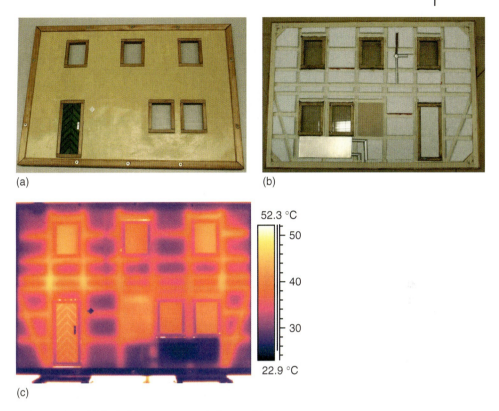

Figure 6.7 Model for hidden structure with front and back (a,b) as well as IR image (c) when heated from the back.

Half-timbered houses are examples of buildings that may not be investigated in wintertime alone (as shown in Figure 6.6). Since the heat capacity c and the density ρ of the used wall materials are quite different, thermal contrast can also be achieved after the wall is exposed to intense solar radiation even if inside and outside air temperatures do not differ (appreciably). Dry wood has a larger specific heat capacity, but much smaller density than those of stones or the filling material. A simple analysis relates the temperature rise ΔT to the incident energy ΔQ via $\Delta Q = c \cdot m \cdot \Delta T$, where the mass m is given by density times volume. Therefore, upon receiving the same amount of incident energy ΔQ, the temperature rise of a material with given volume depends inversely on the product of its specific heat and density. Dry wood has a smaller value of $c \cdot \rho$, therefore it will show up at a higher temperature. Of course, the structures are still behind the plaster, but the effect is clearly visible.

Depending on the recording conditions, the location of the wooden frames of half-timbered structures may thus show up in IR images either as colder or as warmer surface temperatures!

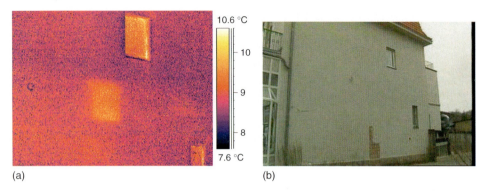

Figure 6.8 Old window, closed with masonry and covered behind plaster.

6.2.2
Other Examples with Outside Walls

Sometimes, buildings are modified, for example, building envelope openings like windows may be added or removed. Figure 6.8 depicts an example of the wall of a hotel building. As can be seen from the IR image, originally, there had been another window at the center of the wall, which was subsequently closed with masonry, before the wall was covered with plaster. If no more details about the work are available, one can only speculate about the reason. First, it may be possible that the thermal insulation was forgotten during the process; second, the thermal conductivity of the chosen brick could be different from the original brick used for the rest of the wall; or; third; heat was transferred to the wall (e.g., via radiation from the sun) prior to recording the image and the value of $c \cdot \rho$ of the chosen brick is smaller than for the rest of the wall, leading to a larger temperature. In this case, the owner immediately realized that the window was just filled with brick and covered with plaster without attaching any thermal insulation.

Many European houses are equipped with stoves (oil, coal, or wood) and therefore need chimneys. Sometimes, these are at the outside walls of a building. Figure 6.9 depicts an example of a three-story building with the chimney at the center of the wall. Obviously, the thermal insulation of the chimney is very bad or, at least, partially missing.

The observed temperature differences amounts to ≈2 K and the defect is energetically relevant. In addition, one can see that the connection of the ceiling of the second floor was also not properly insulated.

6.2.3
How to Find Out Whether a Defect is Energetically Relevant?

Interpreting outdoor IR images qualitatively is relatively easy (if all possible error sources are known and eliminated). However, the critical question of any home owner is whether observed defects that do not point to structural damages and that

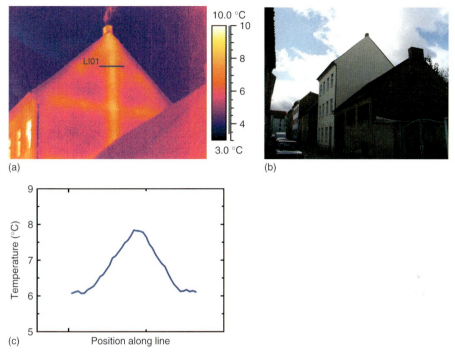

Figure 6.9 Infrared (a) and visible (b) image of a house. The location of the chimney can be easily seen through the outside wall. Temperature differences along the line in the IR image amount to about 2 K (c).

need to be dealt with anyhow, are energetically relevant. Solving this problem is not always an easy task.

First, any thermal bridge that points to increased heat transfer through the building envelope means that energy needed for heating is lost to the environment. If this energy comes from fossil energy, any such energy loss automatically implies that there is avoidable CO_2 emission: better thermal insulation means less primary energy needed for heating, that is, less CO_2 emissions. Hence any energy loss is CO_2 relevant.

Second – and usually more important for home owners – is the question of how much money could be saved per year, if the observed energy loss could be prevented. This savings in terms of money is then compared to the cost for restoration, and if the amortization time is, say, less than five years, the work may be done.

Restoration costs may depend on the wages of the workers, on the costs of the needed materials, on the type of house and so on, which is not discussed here. Rather, a very crude estimate is given of how much energy may be lost during a year due to an observed defect. Of course, much more sophisticated finite element models (computational fluid dynamics) are available [4], for a first-order estimate, the simpler approach is sufficient.

The idea behind the estimate is the concept of the U-value or R-value introduced in Section 4.3.3, equation (Eq. (4.11)).

$$\dot{Q} = U \cdot A \cdot \Delta T \tag{6.1a}$$

where

$$U = \frac{1}{R_{total} \cdot A} = \frac{1}{\frac{1}{\alpha_{conv,ins}} + \sum \frac{s_i}{\lambda_i} + \frac{1}{\alpha_{conv,out}}} \tag{6.1b}$$

Here $\alpha_{conv,ins}$ and $\alpha_{conv,out}$ represent the inside and outside wall heat transfer coefficients for convection and the sum represents the heat transfer through wall segments of thickness s_i having a thermal conductivity λ_i. In Chapter 4, model calculations are discussed, which allow to calculate all relevant temperatures, the respective U-values as well as total heat flux through a number of different walls.

If owing to a defect, one of these contributions changes, the heat flow through the respective part of the building envelope changes. Once the respective areas are known, one may estimate the total heat loss and therefrom calculate respective energy costs. The best way would be to directly measure U-values with thermal imaging. This is, in principle, possible (Section 6.7). Alternatively, one could start with the known insulated wall structure U-value and relate it to the respective outside wall temperature. Since adjacent thermal bridges have a higher wall temperature, one could vary the thermal resistance of the wall materials in Eq. (6.1b) such that the observed wall temperature is reproduced. The respective thermal resistance leads to a new, now larger U-value, which may be used to estimate the heat loss according to Eq. (6.1a).

Here, we would like to give a simpler rough estimate example based on typical U-values for walls and windows. From Table 4.6, concrete walls with no thermal insulation have U-values of around 3 W(m² K)⁻¹, brick walls of 1.5 W (m² K)⁻¹, and brick walls with insulation of about 0.5 W (m² K)⁻¹. Let us assume a family home with about 50-m² floor space (100-m² living space), two floors and overall (including roof) envelope area $A = 300$ m². For simplicity, we assume that all homes (not insulated and insulated) would have similar U-values all around the envelope (including windows and roof). Table 6.3 gives results using the following assumptions: a year has about $t = 3.15 \times 10^7$ seconds, 1 kW h = 3.6×10^6 J, 1 kW h costs €0.06 cts, a temperature difference (inside to outside) of only $\Delta T = 10$ K was assumed since the heating period amounts at most to half a year. The numbers could be used to estimate the improvement due to insulation, for example, from type 3 to type 2. If defects only occur at smaller areas, the respective numbers will be lower accordingly.

6.2.4
The Role of Inside Thermal Insulation

The importance of thermal insulation of walls was discussed in detail in Section 4.3.3. In short, the thermal conductivities of thermal insulation

Table 6.3 Rough estimates of annual heating energy and respective energy costs for three different homes assuming $A = 300 \text{ m}^2$, $\Delta T = 10 \text{ K}$ and 0.06 cts/KW h.

Type of house	U value in W $(m^2 K)^{-1}$	$\dot{Q} = U \cdot A \cdot \Delta T$ in W	Q/A in kW h $(m^2 a)^{-1}$	Q in kWh/a	Cost in €/a
1. Not insulated concrete	3	9000	263	7.9×10^4	4700
2. Not insulated brick	1.5	4500	130	3.9×10^4	2350
3. Insulated brick or wood	0.5	1500	44	1.3×10^3	790

of composite walls determine the temperature distribution within the wall (Figure 4.10). The largest temperature drop occurs in the thermal insulation layer, for example, made of Styrofoam. It is desirable that the freezing point should lie within this layer; therefore thermal insulation should be attached to the outside of a building. However, one needs to consider what happens if there is an additional thermal insulation attached to an inside wall. Figure 6.10 depicts an exterior wall of a classroom in a public school. The inner wall consists of studwork plates attached to metal poles. Just before recording the thermal image, a heavy and large whiteboard (dimensions about 1.6 m × 0.8 m) was removed from the wall. The vertical line structures are due to the thermal conductance of the metal poles to which the plates are attached. On both sides of the poles, air serves as thermally insulating material. The most prominent feature is that the whole wall area, which was covered by the whiteboard, has reduced surface temperatures by 2–2.5 K.

The explanation for this unusual behavior at first glance is simple. Figure 6.11 shows a schematic representation of a typical stone wall with inside plaster, masonry, and outside thermal insulation together with the respective temperature along a line through the wall (for details of calculation, see Chapter 4). The inside and outside air temperatures are fixed. As expected, the thermal insulation gives rise to the most dramatic temperature drop. The black lines show the situation for the uncovered wall in stationary equilibrium and the blue dotted line shows the same for an additional inside thermal insulation (Figure 6.10, abbreviated as whiteboard in the following).

In both cases, the inner surface temperature, that is, the plaster surface or the whiteboard surface (pink horizontal line) is the same. However, due to the thermal insulation properties of the whiteboard, the temperature drops by several Kelvin at the interface of whiteboard and plaster. When removing the whiteboard, this interface becomes the new wall surface showing lower temperatures. In conclusion, one should try to avoid covering the inside of exterior walls with thermally insulating objects; the better the insulation, the lower the real wall surface temperature. This may, in the worst situation, even lead to mold below the object.

346 | *6 IR Imaging of Buildings and Infrastructure*

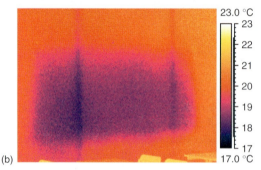

Figure 6.10 Visible (a) and IR image (b) of an exterior wall segment in a classroom. Prior to recording the IR image, a large whiteboard, which was attached to the wall (position indicated in Vis image) was removed.

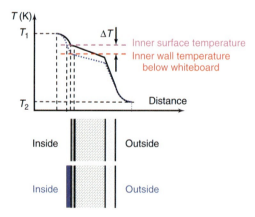

Figure 6.11 Scheme for a wall with additional thermal insulation (blue area) from the inside. The inner surface temperatures of the insulation surface and the neighboring wall without insulation are the same, however, the wall temperature below the insulation is reduced.

6.2.5
Floor Heating Systems

A typical indoor thermography application of homes concerns floor heating systems. Hot water radiant floor heating systems, which are quite common in many European countries, have also become more popular in North America lately [5–7]. While applications in North America also include exterior systems, for example, for melting rather than removal of snow in cold climates, the most popular use in northern and middle Europe and, particularly in Germany, are floor heating systems in private homes. Despite the higher initial costs during construction of a house, this technique is very competitive compared to conventional heating systems since it uses lower temperatures during operation, which means lower fixed running costs. In addition, many home owners prefer warm floors because they relate this to "just feeling good." Also the interior of a home, particularly near walls and windows, can be better used since heaters are absent.

Two typical laying patterns of the tubing of floor heating systems, spiral and meander, are common (Figure 6.12). They differ in the homogeneity of the surface temperatures: the meander-like structure usually shows a gradient from the left to the right as the flowing hot water cools down.

Figure 6.13 depicts an example of the spiral zone of a system in a living room (dimensions 3 m × 3.2 m).

Figure 6.13b shows the layout of the tubes before a floating floor screed of about 6 cm thickness was put on top Figure 6.13c. The IR image Figure 6.13a was taken several weeks later after the tubing was buried below the floating screed.

Usually, such images are obtained when the heating is turned on after the floor was initially cold [5]. The image presented here with just the screed on top of the tubing was, however, obtained after many hours of heating at high power [6]. Obviously, under normal conditions with less heating, the temperature variations would not be pronounced as shown here. Also, if additional floor coverage is used like wooden parquet floor, the temperature distribution at the surface of the floor

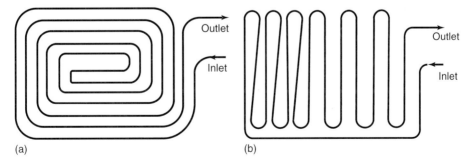

Figure 6.12 Spiral (a) and meander (b) laying of tubing of floor heating system. The tubes along the meander are mostly parallel, but need not be so.

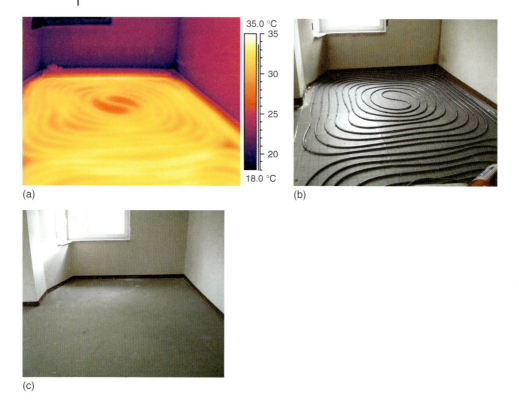

Figure 6.13 One zone of a spiral laying of a floor heating system (b). The tubing is on top of the concrete foundation slab. The IR picture (a) was taken after the tubing was covered by the floating floor screed (c).

is smeared out further. In this case, the tubes can indeed only be seen with good contrast in the heating-up period. Figure 6.13 shows that the heating system is working very well, the only problem may arise in the center of the spiral, where the temperature variation on top of the screed reaches 5 K over a distance of about 0.3 m.

Thermography can, of course, also be used to analyze the usual accessories of floor heating systems, like the boiler and its exhaust pipes, as well as the insulation of the connecting tubes between pump and floor heating laying [6].

6.3
Geometrical Thermal Bridges versus Structural Problems

Often, the structural differences beneath the building walls, which may give rise to thermal signature in the IR image, are not known in advance. In this case, a detailed knowledge of typical and unavoidable thermal bridges is the prerequisite in order to avoid wrong interpretations. In this subsection, we therefore present

6.3 Geometrical Thermal Bridges versus Structural Problems

first some qualitative examples of building thermography with geometrical thermal bridges and how they give rise to typically observed thermal features of the building envelope. Then we proceed to examples where the observed thermal bridges are due to structural defects like missing insulation.

6.3.1
Geometrical Thermal Bridges

As mentioned in Chapter 4, geometrical thermal bridges are present in corner sections of a building due to the details of the heat flow from a small warm inside area to a much larger, colder outside wall area (see Section 4.3.5 and Figure 6.14).

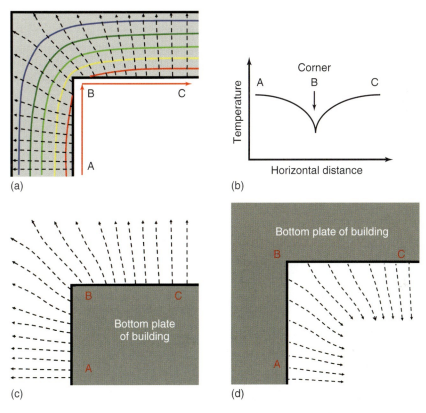

Figure 6.14 (a) Illustration of a geometrical thermal bridge in a corner of a house. The (colored) contours of constant temperature are curved and the heat flux is perpendicular to them (broken lines with arrows). (b) Plot of inside temperature of the wall along the line ABC. The temperature shows a pronounced minimum at the corner point B. (c) Geometrical thermal bridges observable with outdoor thermography. Increased heat flux from the outside corner (90° angle) of a concrete bottom plate of a building should lead to a lower temperature at the corner point B. (d) Reduced heat flux from the inside corner (270° angle) of a concrete bottom plate of a building should lead to a higher temperature at the corner point B.

As a consequence, the inside corner sections always show a minimum temperature compared to the adjacent flat area of an outside wall.

The conditions for corners of a building observed from outside are often different because of weather conditions, sunshine, shadows, and so on, (see below), which may have already lead to different wall temperatures at either side of a corner. However, under ideal outside conditions, late at night, with no wind and moisture in wintertime, it may also be possible to observe such geometrical thermal bridges due to corners with outdoor thermography. Figure 6.14c,d depicts schematically the situation for the bottom plate of a building (gray shaded area) with a heat transfer via convection to the outside air. Increased heat flux from outside corners will result in a lower temperature at the corner point, whereas reduced heat flux from the inside corner of a concrete bottom plate of a building will lead to a higher temperature at the corner point B.

The situation of Figure 6.14c,d was observed for a building under construction (see Figure 6.15). The concrete bottom plate ended at the boundary of the house. The plate was heated due to the floor heating system of the adjacent room. Since the patio was not yet built, the thermal insulation at the end faces of the plate had not yet been installed, and one could detect the plate in the overview thermal image. The geometry of the relevant portions (indicated by white circles) includes an outside corner with plate angle 120° and an inside corner with plate angle 240°. Although these angles differ from the schematic ones with 90° and 270° discussed above, the general argument still holds. Figure 6.16 shows IR images with expanded views of the respective corner sections.

The geometrical thermal bridge effect can be clearly seen. The plots of the measured temperature along the lines in the IR images reveal that temperature differences of as much as 3 K can result.

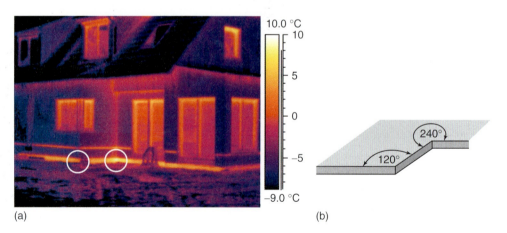

Figure 6.15 The geometrical thermal bridges of a concrete bottom plate of a building in the IR image (a) are due to inside and outside corners as shown in the sketch of the geometry (b).

6.3 Geometrical Thermal Bridges versus Structural Problems | 351

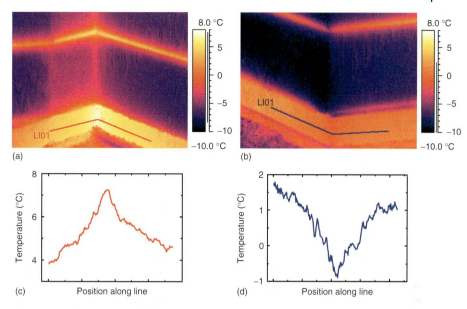

Figure 6.16 Thermal images (a,b) and temperature plots along the shown lines (c,d) for the geometrical thermal bridges of Figure 6.14c,d. Left: inside angle 240° and right: outside angle 120°.

Although such geometrical thermal bridges are sometimes observable from outside, the effect is better known from indoor thermography. Figure 6.17a depicts an example of the inside corner of a room. The thermal bridges lead to a temperature drop along the edges and an even larger one in the lower and upper corners. An expanded view of the lower corner was investigated quantitatively. The temperature plots along the lines, shown in the IR image, indicate temperature drops of about 2–2.5 K just around the edge and more than 5 K in the lower corner, reaching minimum temperatures below 11 °C. These low temperatures were critical with regard to mold formation. First, the investigation was done for an outside temperature of 1 °C. In midwinter, temperatures often drop to below zero, and sometimes stay for several days to weeks below −5 or −10 °C. This will obviously lower the respective inside wall temperatures and those of the corner sections. Second, the dew point temperature (see Section 4.3.6) for 50% humidity and an air temperature of 20 °C corresponds to 9.7 °C, that is, water will start to condense at wall areas with $T \leq T_{\text{dewpoint}}$. However, it was shown that in order for mold to grow, wall temperatures need not reach dew point temperatures; typically regions with 80–90% relative humidity are sufficient [8]. Owing to the low corner temperatures, the adjacent air will reach 80 or 90% relative humidity, which is already well above dew point temperature, in this case, between 14 and 12.5 °C. This means that the edge, in particular the corner regions may give rise to problems if air movement is restricted. As a consequence, the inside corner was kept free of shelves or other furniture to allow for sufficient air ventilation.

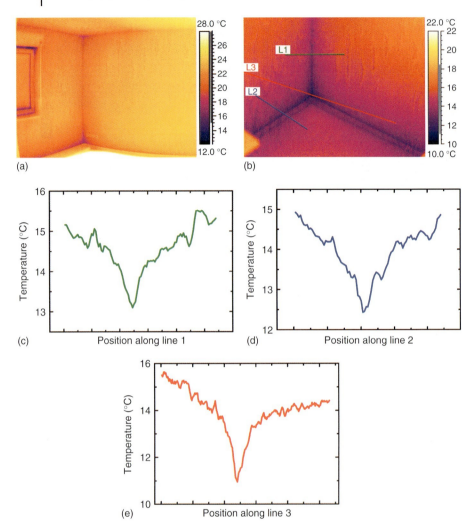

Figure 6.17 Geometrical thermal bridge observed for an inside corner of a basement room (no cellar) with two outside walls. The dark blue areas (a) indicate the lowest temperatures. An expanded view of the lower corner section was analyzed in more detail as indicated by the three lines (b). The respective temperature plots along the lines are shown in c–e. Outside air temperature is 1 °C.

Figure 6.18 shows another uncritical example of such geometrical thermal bridges.

6.3.2
Structural Defects

In addition to geometrical thermal bridges observed at corners or edges, there is another class of common thermal bridges due to structural defects. Figure 6.19

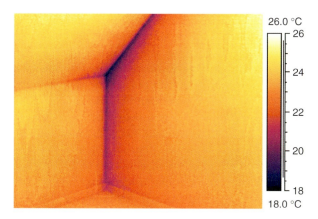

Figure 6.18 Another typical geometrical thermal bridge observed for inside corners of a bedroom below an inclined roof. The room was heated to around 26 °C for this study.

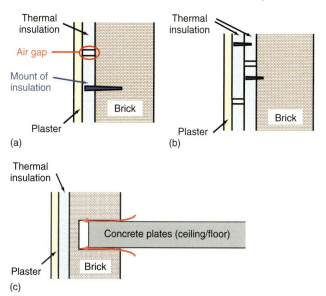

Figure 6.19 Structural defects can, for example, be air gaps between adjacent parts of thermal insulation or plastic anchor bolts with steel core for attaching the insulation to the masonry (a). Problems may be avoided by using two displaced layers of insulation and shorter anchor bolts (b). Other potential problems concern badly insulated concrete plates, serving as ceiling or floor in large apartment buildings (c).

depicts two examples that may show up in IR images. First (a), air gaps between adjacent parts of thermal insulation can give rise to linelike features or plastic anchor bolts with steel core for attaching the insulation to the masonry can show up as a series of spotlike features in IR images. Such problems may be avoided by using two laterally displaced layers of insulation and shorter anchor bolts (b). Other potential problems concern badly insulated concrete plates, serving as ceiling or floor (c) in large apartment buildings giving rise to air-filled cavities. They also lead to linelike features seen through the outside wall along the whole length of the plate. Such defects are energetically relevant. However, in addition to heat flow, moisture can leak into these air cavities, which also drastically reduces the lifetime of the whole structure, that is, they can easily lead to very expensive structural damages.

Figures 6.20 and 6.21 depicts examples of a nine-floor apartment building made of precast concrete slabs. This building technique was quite common in the 1960s, in particular also in the German Democratic Republic. Figure 6.20 shows the visible (b) and the IR image (a) of part of the window front of the building. The IR images were recorded in the early morning hours before sunrise. The central window had remained closed for a long time and also during the recording (only the window partially visible in the lower right corner was open). One can clearly identify the structural problem of missing insulation of the concrete plates (floor/ceiling) as described in Figure 6.19. In addition, the window frames are particularly badly insulated since not only the window seals but also parts of the adjacent concrete plates are heated appreciably. The worst problem is the missing thermal insulation of the concrete wall plate below the window. It just does not exist; one can even see the position of the heater through the wall. In these old buildings, the heater could not be turned off or on at all, regulation was usually done by opening the window. This building suffered enormous heat losses and heating costs were extraordinary. The building was however renovated in the late 1990s and a thermal insulation was added on the exterior wall.

Figure 6.21 shows two IR images of the windowless east wall of the same building recorded before and after restoration of the building envelope. The thermal imaging analysis demonstrated that the restoration work was indeed successful. The temperature span of the images after repair was reduced to only 3 K, and still no structural thermal bridge could be detected.

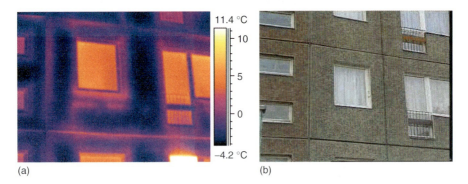

Figure 6.20 (a) A section of the window front of an apartment building made of precast concrete slabs before restoration work. (b) One can see the position of the concrete plates (ceiling/floor), the bad insulation of the window as well as the heater below the window through the wall.

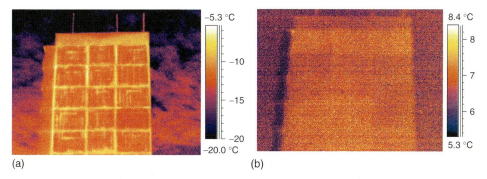

Figure 6.21 East side wall of building from Figure 6.20 before (a) and after (b) adding thermal insulation to the wall. The thermal bridges due to structural defects could not be detected after the repairs.

Finally, we want to mention another common thermal bridge effect. In the example of Figures 6.20 and 6.21, no thermal insulation was present anywhere, it was just not planned in the original design. In contrast, a large number of structural thermal bridges occur in buildings where thermal insulation is of course planned, but not realized properly.

Figure 6.22a depicts again the overview image of a family home (Figure 6.2). Besides the geometrical thermal bridges in the lower part, the feature near the dormer window needs attention. The window section is shown in more detail in Figure 6.22b, clearly indicating an obvious problem. It is energetically relevant. The problem consists in partially missing insulation at the connection of the wooden ceiling to the attic. The possible reasons are manifold: first it may (and does) happen that construction workers get rid of their waste (soft drink cans, etc.) by just stuffing them into the insulation material. Second, the density of the insulation material, which is stuffed into openings, is still low when the morning break starts, but this is forgotten after the break, or third, some additional installations (electrical, plumbing, etc.) are needed, that is, insulation is removed but not added again after the work is finished and so forth.

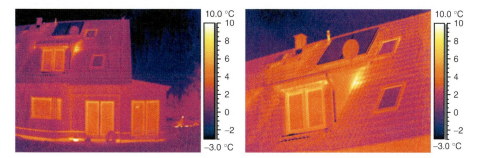

Figure 6.22 (a,b) Thermal bridge at right side of dormer window due to missing thermal insulation material.

6.4
External Influences

6.4.1
Wind

The surface temperature of objects is strongly influenced by an airflow around them. This is due to the fact that the heat transfer coefficient describing convection between the object surface and the surrounding air depends on the flow conditions, in particular, the flow velocity [9–11]. Large flow velocities can result in a strong cooling effect. This has an effect, for example, on the results of outdoor infrared imaging of buildings or electrical equipment. Moreover, the wind not only reduces the surface temperature but also changes the thermal signature of the object surface. For most practitioners, this behavior is known and they use wind speed limits for their qualitative or quantitative analysis. Unfortunately, these limits are not standardized and the theoretical connection between the wind speed and the heat transition due to forced convection at the surface is very complicated.

In any case, the heat transfer coefficient α_{conv} increases with flow velocity (Figure 6.23a). This leads to changes of observed surface wall temperatures as shown in Figure 6.23b. It depicts the calculated temperatures at the surface of an outside wall (calculation similar to the one in Section 4.3.3) for given conditions $T_{inside} = 20\,°C$, $T_{outside} = 0\,°C$, 24-cm brick wall with $\lambda = 1.4$ W (m K)$^{-1}$ and $\alpha_{conv.inside} = 7.69$ W (m^2 K)$^{-1}$.

Obviously, both wall surface temperatures change upon wind speed. For the outside wall, this change also leads to a strong decrease in thermal signature (i.e., the detected temperature difference between two adjacent spots, which are due to differences in thermal conductivity of the wall, will decrease for increasing wind speed) [11]. This behavior was demonstrated for our model half-timbered wall structure of Figure 6.7. The experimental arrangement is shown in Figure 6.24. The

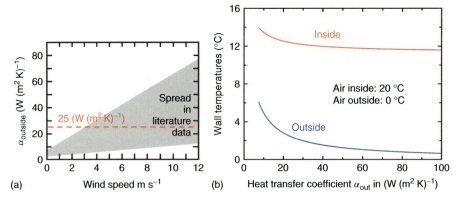

Figure 6.23 (a,b) Theoretical wall surface temperatures for a single layer wall with $\lambda = 1.4$ W (m K)$^{-1}$ as a function of the outside convective heat transfer coefficient.

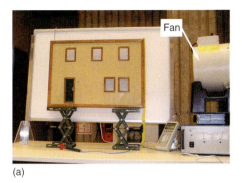
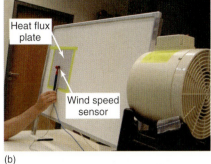

Figure 6.24 Experimental setup for analyzing the effect of wind speed on surface temperatures and heat transfer rates.

wall model was placed in front of a heating plate and a room temperature airflow was generated by a large fan. The wind speed was measured using a calibrated sensor and the heat flux (in W m^{-2}) was determined using a calibrated heat flux plate. Determination of its surface temperature revealed the expected temperature drop with increasing wind speed as in Figure 6.23.

A demonstration of the wind speed effect for the house wall model is shown in Figure 6.25. The Figure 6.25a depicts the rear heated wall at zero wind speed and Figure 6.25b refers to a maximum wind speed of 7.3 m s^{-1}. A quantitative analysis of two spots (good insulation of the wall, SP01; bad insulation of the wall, SP02) clearly demonstrates the decrease in thermal contrast (Figure 6.26). With increasing wind speed, both spot temperatures are decreasing as expected due to forced convection. However, the temperature difference between the two spots – describing thermal contrast – also decreases (Figure 6.26b).

This means that outside thermography becomes less sensitive to detect thermal bridges/temperature differences for high wind speeds.

The increase of the convective heat transfer coefficient with wind speed is also one of the physical principles underlying the so called wind chill factor, well known from weather reports. The *wind chill temperature* is defined as the apparent

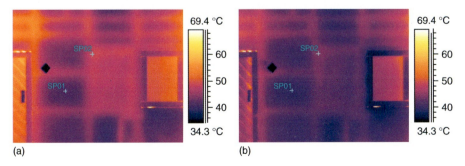

Figure 6.25 Change of thermal images with increasing wind speed. (a) $v_{wind} = 0$ m/s and (b) $v_{wind} = 7.3$ m/s.

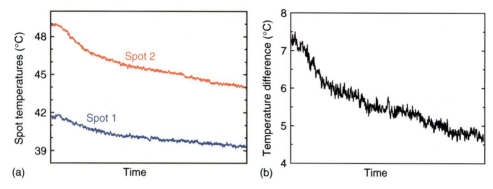

Figure 6.26 (a) Wall surface temperatures at spots 1, 2 of Figure 6.25 as a function of time while increasing the wind speed from 0 to 7.3 m/s (nonlinear scale). (b) According difference of spot temperatures which is a measure for thermal signature.

temperature felt on exposed skin due to the combination of air temperature and wind speed. Humans do not sense the temperature of the air but sense heat flow due to the temperature difference between the skin temperature and the temperature of the surrounding air. When there is wind, the increasing α_{conv} between the skin and the air results in larger heat loss, that is, the temperature of the skin starts coming closer to the air temperature. This is interpreted as "it feels colder." However, the exact definition of the wind chill index [12] also includes the human temperature perception and is therefore more complex.

6.4.2
The Effect of Moisture in Thermal Images

Water damages can result in very large repair costs. Therefore, it is very important to know how to detect water in buildings and to know how water changes thermal signatures. As noted before, surface moisture on walls often reduces the measured surface temperatures due to evaporative cooling. This effect must be considered separately from the water condensation on wall areas in corners or at edges, where the wall temperature is, in any case, very low due to geometrical thermal bridges. Besides, the huge heat capacity of water may also lead to warming up of building parts, depending on the respective conditions.

Evaporative cooling occurs whenever a gas flows over a liquid surface [13]. This effect is well known and has been used for cooling since a long time. "In the Arizona desert in the 1920s, people would often sleep outside on screened-in sleeping porches during the summer. On hot nights, bedsheets or blankets soaked in water would be hung inside of the screens. Whirling electric fans would pull the night air through the moist cloth to cool the room" [14]. Evaporative cooling is a very common form of cooling buildings for thermal comfort since it is relatively cheap and requires less energy than many other forms of cooling.

In brief, all solids and liquids have a tendency to evaporate to a gaseous form, and all gases have a tendency to condense back. At any given temperature, for a particular substance such as water, there is a partial pressure at which the water vapor is in dynamic equilibrium with liquid water. With increasing wind speed the number of water molecules of the liquid water that experience collisions with the gas molecules increases. These collisions increase their energy and they are able to overcome their surface-binding energy of the liquid. This results in an increasing evaporation effect. The energy necessary for the evaporation of the liquid comes from the internal energy of the liquid. Therefore the liquid must cool down. This effect must be taken into account for all temperature measurements at objects with moist surfaces where the measured temperature depends on wind speed.

To demonstrate the consequences of evaporative cooling for buildings, laboratory experiments with the house wall model were carried out (Figure 6.27).

First, the dry surface of the house wall was analyzed (Figure 6.27a). Temperature differences due to different heat insulation quality are clearly seen as a thermal signature. Subsequent to moistening the surface, the thermal signature of the wall is changed. The thermogram shows a more homogeneous temperature distribution (Figure 6.27b). This effect is due to the evaporation cooling. Water evaporation increases at wall locations with higher temperatures, causing a higher cooling effect at these areas. As a consequence, a more homogeneous temperature distribution across the model wall can be observed. Additional air flows over the wall increase

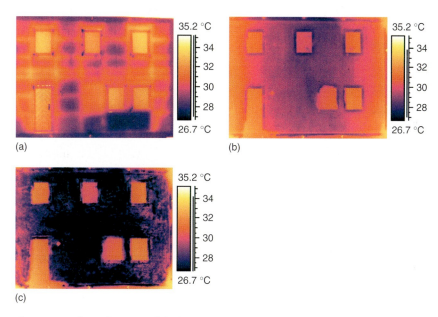

Figure 6.27 Thermal images of the house wall model showing the effect of evaporative cooling. (a) Dry surface; (b) moistened surface, no wind; and (c) moistened surface, wind speed 7.3 m/s.

the evaporation effect (Figure 6.27c). This results in a further cooling of the wall surface by 5–6 K at a wind speed of 7.3 m s^{-1}.

It can be concluded that first, an overall reduction of the absolute wall temperature happens and second, evaporative cooling decreases the thermal contrast considerably, much more than wind over dry surfaces alone. On the one hand, this means that outdoor thermography during rain or with moist wall surfaces is very insensitive regarding the detection of thermal bridges. On the other hand, one may just utilize the reduction of the wall temperature for increased airflow as a means to detect moisture.

We successfully tested an idea to use evaporative cooling consciously to check if a cold spot at a wall is dry or moist. Figure 6.28 depicts the result of this experiment. At the wall in the lab some areas were moistened with cold water (Figure 6.28b within the black circle in the IR image) and other areas were just cooled down using some pieces of ice within a plastic bag (within the white circle in Figure 6.28b) in order to avoid the moistening of these areas. The water temperature was chosen such that approximately the same initial wall temperatures (without airflow) were measured. They were slightly below room temperature. It was not possible from the initial thermal image alone (Figure 6.28b) to decide which area was moist and which area was dry.

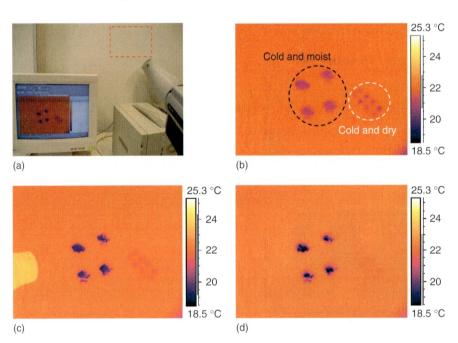

Figure 6.28 Analysis of dry and moist cold spots on a wall while applying airflow. (a) Experimental set up, the red rectangle indicates the observed area; (b) IR image of prepared wall without airflow; and (c,d) Thermal images using an airflow of 7.3 m/s after 2 seconds (left) and 11 seconds (right).

If an additional airflow is directed onto the surface by using a fan, the rate of evaporation of the water at the moist areas is increased resulting in a cooling down of these areas by 3–4 K within 10 s (Figure 6.28c,d). In contrast, the dry areas are heated by the room temperature airflow. With such experiments, a qualitative analysis to find moist areas on walls becomes possible with thermography by directing a fan toward the wall.

Quite a few investigations have been reported for moisture detection in building envelopes and on roofs using thermography [15–19]. The idea behind this is usually to utilize evaporative cooling to detect moisture within a wall or ceiling. Thermography was used to observe the walls and the ceiling prior, during, and following water testing. In particular, the differences of moisture for roof assemblies, above grade assemblies, and foundation wall assemblies were discussed in detail together with a methodology for exterior and interior inspections [17]. Another study used solar loading to detect suspect areas [18]. In this case, the heating and cooling of the walls were observed. If water, which has a much higher heat capacity than other building materials, is trapped in the walls, the respective regions will warm up more slowly.

6.4.3
Solar Load and Shadows

The outer walls of buildings are often subject to the radiation of the sun. This solar load can lead to appreciable temperature rises within a wall [18]. However, even for clear sky conditions, the radiation flux incident on a wall changes during the day due to the changing sun elevation and direction. Additional short-term changes may happen due to cloud shadows or shadows from trees or neighboring buildings, which can move across the wall of a house. As a consequence, these external effects of solar load and shadows inevitably lead to transient effects: the surface temperatures of sun-facing walls and roofs will change continuously. In addition, all walls and roofs can experience night sky radiant cooling (see Section 6.4.5), which also leads to transient effects. Following the description of simple model results of these effects, respective experiments and observations are presented.

6.4.3.1 Modeling Transient Effects Due to Solar Load

Measurements were done for two buildings with different but typical wall systems in Germany. Simple model calculations for these walls were studied. The walls (Figure 6.29) consist of a layer system of gas concrete and exterior plaster (wall 1, Figure 6.29a) and the same system with an additional layer of 6 cm Styrofoam, also called *expanded polystyrene* (wall 2, Figure 6.29b). Styrofoam is a very good insulating material with low thermal conductivity. The table in Figure 6.29c gives a summary of the relevant properties of the materials used. The time dependence of wall heating and wall cooling depends on the construction and the properties of the wall materials. Owing to its insulation properties, it is expected that the Styrofoam coating will considerably affect the temperature distribution in the

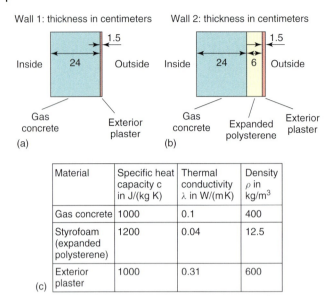

Figure 6.29 Wall models. (a) Wall 1, (b) wall 2 = wall 1 with additional thermal insulation, and (c) table of relevant material properties used in the model.

walls. Therefore, a difference in the time dependence of heating and cooling processes between the two wall models is expected.

The surface temperatures of the wall and the temperature distribution within the wall were simulated using a freeware Excel-program DynaTherm2000 [20]. The program uses the finite element method to calculate the temperature distribution dependent on the boundary conditions. It allows the calculation of the transient heat conduction in any composite wall while taking into account the main meteorological influences, that is, geographical position, solar conditions (sun elevation as function of time during day, day of year, geographical latitude, and longitude), and wind speed. In particular, it automatically includes the solar-load-induced wall heating. Figure 6.30 depicts the temperature distributions in the cross section of the wall models for several time periods after sunset and after the wall had been exposed to sun radiation for 5 h (this resembles a house, which only receives sunshine in the afternoon). The steady-state conditions for the air temperatures were $T_{inside} = 20\,°C$ and $T_{outside} = 6\,°C$. Owing to the solar load, inner wall regions of gas concrete are heated up in Figure 6.29a. The Styrofoam insulation in wall 2, however, act as a barrier for the solar-load-induced thermal flux. Therefore, the maximum gas concrete temperatures are below 21 °C in Figure 6.29b, which is much lower than in Figure 6.29a.

The larger amount of stored thermal energy within wall 1 will obviously lead to longer cooling times compared to wall 2. This cooling process will be governed by heat diffusion from the inside of the wall to the surface. In order to compare the results of simulations with the experimental results, it is necessary to calculate

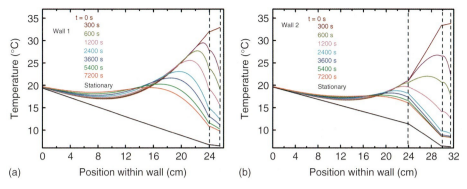

Figure 6.30 Simulation of temperature distributions within the model walls for different time periods after sunset. Stationary conditions (lowest curves) are $T_{in} = 20\,°C$ and $T_{out} = 6\,°C$ without any solar irradiance. The curve ($t = 0$ second) reflects the temperature distribution after 5 hours sunshine duration immediately after sunset.

the outside wall surface temperatures due to the solar load. With the appropriate conditions of solar irradiance, the modeled surface temperatures during the heating and the cooling process are shown in Figure 6.31.

At first glance, the temperature rise as well as the cooling of the wall surface seems to follow simple exponential behavior. This is motivated by the fact that it is possible to characterize a one-dimensional time-dependent cooling process for an

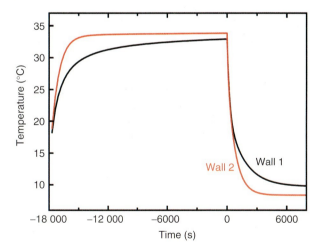

Figure 6.31 Simulation of the surface temperatures of the model walls 1 and 2 for $T_{in} = 20\,°C$, $T_{out} = 6\,°C$. Solar irradiance started at $t \approx -18\,000$ seconds, yielding 5 hours solar load. Time $t = 0$ corresponds to the end of solar irradiance, that is, the start of the subsequent cooling process. Time corresponds to those from Figure 6.30.

infinite one-dimensional wall with one material by

$$T = A_1 \cdot \exp(-t/\tau) + T_0 \qquad (6.2)$$

where T is the surface temperature of the wall, T_0 is the wall temperature in stationary thermal equilibrium, and τ is the corresponding time constant.

The cooling process in walls with multilayer systems, like for walls 1 and 2, is more complicated. From the simulations, it is however found that the model results may be fitted with the following equation:

$$T = A_1 \cdot \exp(-t/\tau_1) + A_2 \cdot \exp(-t/\tau_2) + T_0 \qquad (6.3)$$

From these modeling results, we conclude that due to the complex interplay between cooling due to convection and radiation from the outside and heat flow from stored energy within the wall to the outside surface, no simple Newtonian exponential cooling is expected [21]. Rather, we will try to fit the experimental results by the more complex double-exponential function of Eq. (6.3).

6.4.3.2 Experimental Time Constants

To analyze the cooling process of a wall experimentally [22], it was necessary to provide either solar load or an alternative wall-heating mechanism. In a first test, wall heating was realized with an electrical heating plate ($A = 1.2 \text{ m}^2$) which could be attached to the outside surface of the walls. It was thermally insulated to the outside (Figure 6.32).

The cooling of the two walls after electrical heating was measured with IR cameras in the MW and LW region. As suggested in Section 4.5 we do not expect

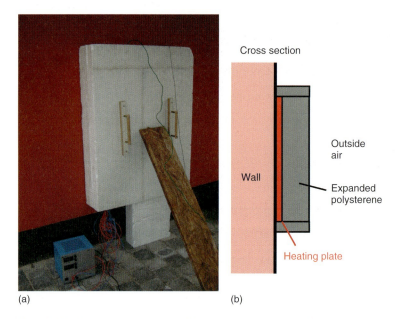

Figure 6.32 Experimental set up for electrical heater induced wall heating.

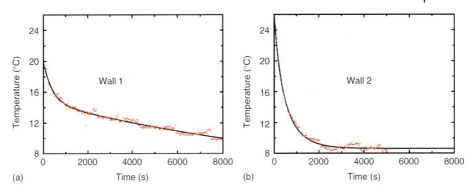

Figure 6.33 Measured surface temperatures of wall 1 and 2 immediately after removing the heater. The reason for the occasional drops in the data points are automatically performed camera calibrations.

simple exponential cooling, but rather fit the data with Eq. (6.3). The results for the LW measurements and both walls are shown in Figure 6.33. When comparing these results with the rather crude infinite one-dimensional wall model of the simulation, one finds the same trends between wall 1 and wall 2, that is, the cooling times of wall 1 are much larger than for wall 2.

Measurements using solar heating were done after 4 h of sunshine onto wall 2 in the late afternoon. Figure 6.34 shows the measurement with analysis regions AR01 and AR02. The average temperature of AR02 is higher due to the reflected irradiation by the lateral window. Figure 6.35 depicts the measured temperature differences $\Delta T = T_{Area} - T_{amb}$ as a function of time, after the sun has stopped to irradiate the wall (hidden behind forest trees). At $t = 1600$ s, part of the wall around AR01 was briefly irradiated again (suitable geometry of sun, clearing in forest, and house). Obviously, solar load can easily lead to temperature differences of more

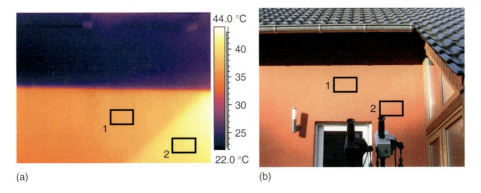

Figure 6.34 IR image of solar radiation heated wall 2 (a) and visible image (b) after 4 hours of solar load in the afternoon. LW camera, T(AR01) = 36 °C, T(AR02) = 42 °C, $T_{amb.air} = 20$ °C.

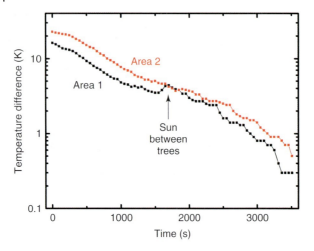

Figure 6.35 Temperature difference ΔT between $T(AR01)$, $T(AR02)$ of Figure 6.34 and $T_{amb.air}$ as a function of time.

than 20 K on walls. The respective time constants can easily amount to 1000 s. Since a signal has decayed to less than 1% of its original value only after at least five time constants, this means that whenever solar load was present on a wall, one should wait at least 1–2 h before the solar load effect can no longer show up in the IR image.

6.4.3.3 Shadows

Figure 6.36 depicts a typical example for outdoor building thermography. A house wall is illuminated by the sun but part of it is already in the roof shadow. In addition, on top of the right window, one can clearly observe thermal reflections. The shadows lead to transient effects concerning solar load heating and cooling of the wall.

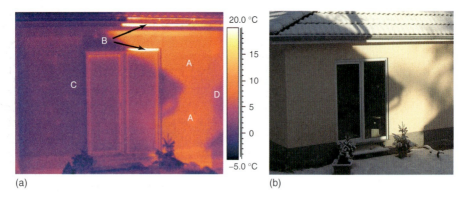

Figure 6.36 MW camera IR image (a) and visible image (b) of a wall with solar induced wall heating (A), solar reflections (B), shadowing effects of neighboring building (C), and tree branches (D).

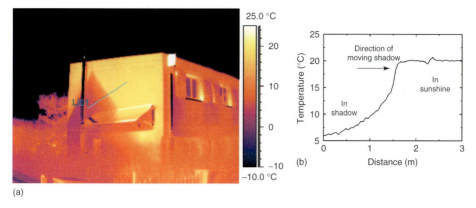

Figure 6.37 Moving shadow on house wall leads to temperature gradient due to transient cooling and heating effects.

A similar example, recorded in wintertime is shown in Figure 6.37. The shadow of the rooftop of the neighboring house is clearly visible on the house wall. Owing to the change of the position of the sun as a function of time (1° in 4 min), the shadow moves across the wall. From the geometry, one may estimate a typical change of, say, 20 cm in 5 min. In the center of the shadow, the house wall is about 15 K colder compared to locations that were fully exposed to solar radiation for hours.

Such transient effects may lead to misinterpretation. Imagine, for example, that the shadow of a tall chimney had been on a wall, and then the sun had disappeared behind clouds. Still, the wall would present a thermal image of this shadow for more than an hour. If one were not aware that the thermal feature would be due to solar load and shadows, one may think about structural defects, which would be a totally wrong interpretation.

6.4.3.4 Solar Load of Structures within Walls

Solar load effects do not only affect plane walls, but they may also change thermal features of structures within walls as a function of time. This has been studied for the house shown in Figure 6.38. As an example, Figure 6.39 illustrates the difference of time-dependent cooling processes between a thermally insulated plane wall section (B) and an adjacent closed blind (A).

Figure 6.39 shows measurement results before sunrise (a), 30 min (b), and a longer time (c) after solar load induced heating (4 h with interruptions). Average surface temperatures were analyzed. Initially the temperature of the two areas was the same. Solar load led to an increased wall temperature compared to the blind. This temperature difference ΔT was slowly decreasing with cooling time (d). Owing to the very low thermal heat capacity and the good insulation from the window by the air gap, the closed blind could cool down after sunset more efficiently than the solid wall. Absolute temperatures depend on material properties. From this, we can conclude that solar load can also affect thermal contrast between a wall and its composite structures.

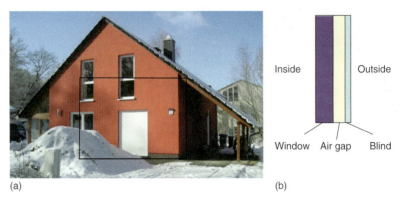

Figure 6.38 House wall with windows and a closed blind (a) in front of a window with an air gap. This resembles a composite wall structure (b) similar to Figure 6.29, however, with different material properties.

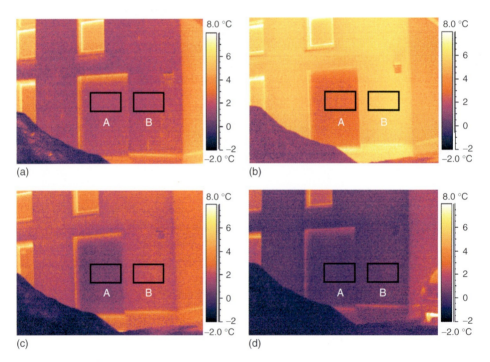

Figure 6.39 MW camera thermal images of closed blind (A) and thermally insulated wall (B). (a) At 7:30 a.m. before sunrise, $\Delta T = 0$ K; (b) at 3:50 p.m., 30 minutes after 4 hours of solar load, $\Delta T = 2.6$ K; (c) at 4:50 p.m., $\Delta T = 1.3$ K; and (d) at 5:50 p.m., $\Delta T \approx 0$ K again.

6.4.3.5 Direct Solar Reflections

Solar irradiation is not absorbed 100% by a wall, therefore, it does not only lead to a heating up but it also results in a diffuse reflection from rough surfaces and specular reflections from polished surfaces. Owing to the spectrum of the radiation from the sun, this effect should be more pronounced for MW cameras than for LW systems. At first, we neglect specular reflections from polished surfaces (as are indicated by B in Figure 6.36) but focus on rough wall reflections.

There is a simple way to test the effect of wall reflection. A wall is exposed to solar irradiation until nearly steady-state conditions are reached after several hours. If, at time t_0, some object is placed between the sun and wall such that part of the wall surface is in shadow, the apparent surface temperature measured with IR cameras will exhibit a behavior as shown in Figure 6.40. At first, there is a very rapid, instantaneous drop, owing to the now missing direct reflection of the solar radiation. Later, the wall area in the shadow starts to cool down with time constants as discussed above.

The experimental setup and measurements are shown in Figure 6.41. A wall was exposed to solar radiation for about 3 h. The shadow was produced on the wall by using a large Styrofoam block. Detecting surface temperatures (around 35 °C for $\varepsilon = 0.94$) with the MW camera, we found a temperature drop $\Delta T = 1.8$ K immediately after shadowing. Using a LW camera the instantaneous drop is nearly nonexistent ($\Delta T \leq 0.4$ K). However, the subsequent cooling due to the persistent shadow became visible later (1.5 K after 35 s).

From Figure 6.41, we conclude that solar reflections can become very important for MW camera systems with apparent temperature effects of up to 2 K. The size of this effect is surprising since walls have very rough surfaces, that is, high emissivity and very low diffuse reflectivity. Obviously, roof tiles with usually more flat surfaces may exhibit much stronger directional reflectivity, thereby amplifying the solar reflection contributions.

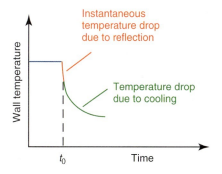

Figure 6.40 Detecting wall reflections. The temperature of a wall exposed to solar irradiation for a long time is nearly described by its steady state temperature (blue). A shadow on the wall leads to an instantaneous drop of apparent temperature (red) due to solar reflections before cooling processes start (green curve).

6 IR Imaging of Buildings and Infrastructure

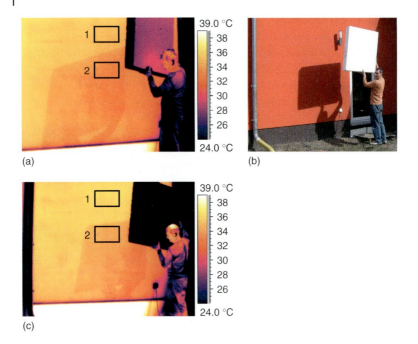

Figure 6.41 VIS image (b) and MW camera IR image of wall 2 immediately after shadowing recorded (a), indicating detection of direct solar reflections. Detection with a LW camera (c) was less sensitive to reflections, however, the shadow could be detected via the wall cooling (here after 35 seconds).

Figure 6.42 depicts an example of a roof observed with an MW and an LW camera. The geometry is shown in Figure 6.43. From the known angles (solar elevation, roof angle, camera observation angle), it was found that the observation was in a direction rather close ($\Delta\varphi \leq 15°$) to the specular reflection of sun radiation.

Therefore, one may expect a strong diffuse reflection contribution (see Figure 6.43b). Analyzing the geometry in detail, we found the observation direction to be at $\approx 75°$ with respect to the surface normal of the roof tile. For such large angles, the emissivity has decreased to a value of about $\varepsilon(75°) \approx 0.8$ as has been tested in the laboratory giving a reflectivity of $\rho = 0.2$ for the subsequent roof analysis.

In order to quantitatively estimate the influence of the reflected solar radiation on the wall and roof signals in the MW and LW region, we need to introduce the blackbody radiation functions from band emission (Section 1.3.2.5). $F_{(0\to\lambda)}$ gives the fraction of total blackbody radiation in the wavelength interval from 0 to λ (see Eq. (1.20)), compared to the Stefan–Boltzmann law, that is, the total emission from 0 to ∞ (M_λ denotes, spectral emissive power). With the assumptions of constant detector sensitivity in a wavelength interval (λ_1, λ_2) and a constant emissivity ε, $F_{(0\to\lambda)}$ can be used to calculate the fraction $\Delta F(\lambda 1, \lambda 2)$ of radiation in the MW and LW range (see Eq. (1.21)). For a sun temperature $T_{\text{sun}} = 6000$ K, reflectivity

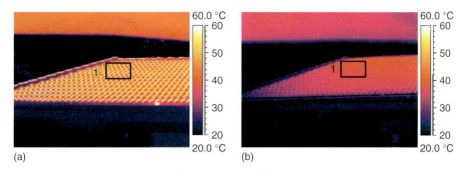

Figure 6.42 IR images of a roof, recorded with a MW camera (a) and a LW camera (b) at $T_{amb} = 20\ °C$. The MW camera detects a higher IR signal due to solar reflections. $T_{ave}(AR01) = 48.6\ °C$ for MW (a) and $T_{ave}(AR01) = 38.1\ °C$ for LW (b) using $\varepsilon = 0.8$.

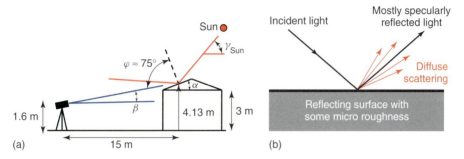

Figure 6.43 (a) Geometry of measurement, $\gamma_{sun} = 53°$ at noon, $\alpha_{roof} = 24°$, $\beta_{camera} = 9°$. (b) Scattering of radiation from surface with micro roughness (after [23]).

of the wall $\rho = 0.06$, solar irradiance in winter $I = 200\ \text{W m}^{-2}$ and in summer $I = 800\ \text{W m}^{-2}$, and defining the fraction of reflected solar radiation

$$S_{\text{refl}} = \rho \cdot \Delta F \cdot I \quad \text{in W m}^{-2} \tag{6.4}$$

one finds the results shown in Table 6.4.

Table 6.4 Results of fraction of reflected solar radiation in MW and LW range.

IR-region in μm	ΔF (6000 K)	S_{refl}: fraction of reflected solar radiation in W/m²	
		Winter $I = 200\ \text{W/m}^2$	Summer $I = 800\ \text{W/m}^2$
MW 3–5	1.45×10^{-2}	0.174	0.696
LW 8–14	9.88×10^{-4}	1.18×10^{-2}	4.74×10^{-2}

These numbers depend strongly on the actual value of ρ. Larger reflectivities (i.e., smaller emissivities) can drastically increase the reflected solar radiation particularly in the MW band, since $\Delta F_{MW} \approx 15 \times \Delta F_{LW}$.

In order to judge the effect of these reflected solar radiations S_{refl} on the IR camera readings, we estimate the necessary thermally induced radiant signal changes $\Delta S_{thermal}$, which result in an apparent $\Delta T = 1$ K increase in temperature of a wall from, say $T_1 = 307$ K to $T_2 = 308$ K. It is calculated from

$$\Delta S_{thermal,MW} = \varepsilon \cdot (\Delta F_{MW}(T_1)\sigma T_1^4 - \Delta F_{MW}(T_2)\sigma T_2^4) \tag{6.5}$$

$$\Delta S_{thermal,LW} = \varepsilon \cdot (\Delta F_{LW}(T_1)\sigma T_1^4 - \Delta F_{LW}(T_2)\sigma T_1^4) \tag{6.6}$$

With an emissivity of the wall of $\varepsilon = 0.94$, we find, $\Delta S_{thermal}(MW) = 0.25$ W m^{-2}. The spectrally resolved irradiances from a wall surface ($\varepsilon = 0.94$, $\rho = 0.06$) for reflected solar radiation S_{refl} and the thermally induced radiant signal change $\Delta S_{thermal}$ from 307 to 308 K is plotted in Figure 6.44. The blue and red regions denote the MW and LW camera ranges. The hatched areas below the plots of $S_{refl}(\lambda)$ and $\Delta S_{thermal}(\lambda)$ correspond to the computed values. For example, for 200 W m^{-2}

$$S_{refl} = \int_{3\mu m}^{5\mu m} S_{refl}(\lambda)d\lambda \approx 0.174 \text{W m}^{-2} \text{ and}$$

$$\Delta S_{thermal} = \int_{3\mu m}^{5\mu m} S_{thermal}(\lambda)d\lambda \approx 0.25 \text{W m}^{-2} \tag{6.7}$$

As is obvious from Figure 6.44, solar reflections are negligible for LW camera systems. Of course, the graphs for $\Delta S_{thermal}$ in Figure 6.44 will change if the object temperature T is changed. The interpretation will, however, be the same.

The above analysis can help understand the differences in solar reflection from the wall and the roof. In the wall measurement (Figure 6.41), the reflected portion of the sun's radiation amounted to an apparent temperature drop $\Delta T = 1.8$ K, which (at $T_1 = 307$ K) corresponds to 1.8×0.25 W m$^{-2} = 0.45$ W m^{-2}. If this additional

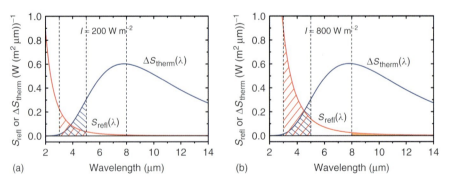

Figure 6.44 Reflected spectral radiance of the sun S_{refl} for $\varepsilon = 0.94$, that is, $\rho = 0.06$ for a given irradiance of 200 (W m^2)$^{-1}$ (a) and 800 (W m^2)$^{-1}$ (b), compared to a signal change of the thermal radiation of the object at 307 K. ΔS_{th} induced by $\Delta T = 1$ K change.

irradiance were due to reflected solar radiation, one would need a solar irradiation of $(0.45~\text{W m}^{-2}/0.174~\text{W m}^{-2}) \times 200~\text{W m}^{-2} \approx 520~\text{W m}^{-2}$, which is reasonable for a clear day in May at noontime.

For the roof, however, the angle of emission of around 75° with respect to the surface normal led to a reduced emissivity ($\varepsilon(75°) = 0.8$), that is, increased reflectivity ($\rho = 0.2$). Repeating the above calculation for the higher basis temperature of the roof (38 °C), we do find for the same solar irradiation of 520 W m^{-2} that the reflected portion of MW radiation will cause a temperature difference of about 7 K. Experimentally, the temperature difference between the LW and the MW camera amounted to about 9 K, which is reasonably close.

6.4.4
General View Factor Effects in Building Thermography

Besides transient effects due to solar loads, shadows as well as changes of wind and rain with time, there is another very important fundamental parameter, which has a strong impact on outside building temperatures: the geometry of the surrounding. In Chapter 1, the laws of blackbody radiation and radiative transfer between objects of different temperatures were introduced. Consider the simplest arrangement, which deals with a gray object of given temperature T_{obj} within an isotropic surrounding of constant temperature T_{surr}. For simplicity, the space between object and surrounding should be a vacuum such that conduction and convection can be neglected. In this case, the radiation exchange between object and surrounding is due to the radiant power $\varepsilon \cdot \sigma \cdot T_{\text{obj}}^4 \cdot A_{\text{obj}}$ emitted by the object and the respective radiant power $\varepsilon \cdot \sigma \cdot T_{\text{surr}}^4 \cdot A_{\text{obj}}$ received by the object from the surrounding. Hence the net radiant power emitted ($T_{\text{obj}} > T_{\text{surr}}$) or received ($T_{\text{obj}} < T_{\text{surr}}$) by an object is given by

$$\Phi = \varepsilon \cdot A_{\text{obj}} \cdot \sigma \left(T_{\text{obj}}^4 - T_{\text{surr}}^4\right) \tag{6.8}$$

However, when an object is not surrounded by an isothermal surface, the situation becomes more difficult. Consider, for example, the situation depicted in Figure 6.45a. An object is surrounded by two hemispheres of different temperatures. In this case, the left side of the object (in the center) emits more radiation than it receives, whereas the right side of the object receives more than is emitted. As a consequence, the left and right object surfaces may have different temperatures unless the thermal conductivity of the object is extremely high. The same may happen to buildings. Consider the situation depicted in Figure 6.45b. House number 1 under study has two neighbor buildings of different temperatures. House 2 has colder and house 3 has warmer outside surfaces than house 1. Assuming again only a finite thermal conductivity of the wall material (which is reasonable), the wall facing house 3 should become warmer than the wall facing house 2. Figure 6.45 is, of course, only a simplified description. Usually any object (e.g., a building) is surrounded by many different objects (houses, trees and plants, ground, sky, clouds, etc.) and one must consider the radiation balance to all parts. Quantitatively

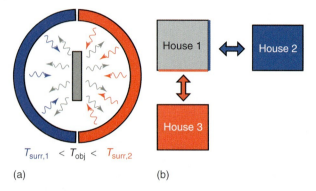

Figure 6.45 An object which is surrounded by objects of different temperature will have different radiative transfer in the various directions (a). This is similar to a house, facing two neighbor buildings of different temperatures (b). This will lead to differences in the temperatures of the respective object sides.

the radiative contribution from each part of the surrounding is described by the respective view factor, introduced in Section 1.3.1.5.

Rather than presenting such complex calculations, its effect on building thermography is illustrated schematically in Figure 6.46 for three different typical situations in the field of outdoor building thermography.

For buildings, there are usually three elements of different temperature around, first the ground including vegetation with T_{ground}, second, neighboring objects (buildings) which – due to internal heating – may have different surface temperatures T_{object}, and third, the sky which may be clear or covered with clouds having a temperature T_{sky}. The question of how much each part of the surrounding contributes is given by the view factor. Qualitatively, it is larger for two objects facing each other compared to the case when there is an angle involved. Figure 6.46a depicts a flat roof building that is facing the sky. The radiative exchange of the roof with the surrounding is described by the view factor. It is calculated by integrating over the half sphere above the roof. Therefore, only the sky contributes. For a building with a tilted roof (Figure 6.46b) the sky can only contribute with a

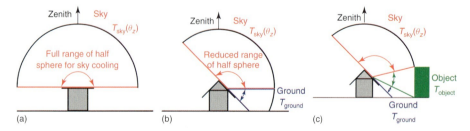

Figure 6.46 (a–c) Illustrating the view factor for different situations for the radiation exchange (details see text).

reduced angular range to the radiative exchange of the roof. In addition, part of the roof also faces the ground of different temperature. In this case, two view factors must be calculated and the contributions need to be summed up accordingly. Qualitatively, one could at once argue that the overall effect would be a larger or smaller radiative exchange depending on whether the additional ground temperature would be larger or smaller than the sky temperature. Finally, Figure 6.46c depicts again a tilted roof building which in addition faces another neighboring building. Now, there are three contributions to the radiative exchange of the roof surface, a further reduced sky contribution, a reduced ground contribution, and an additional object contribution. Obviously, the situation becomes more complex as more additional objects of different temperatures are included. Different parts of the building, such as the walls, will have different contributions due to their differing orientation.

A test with a heat flux plate showed [24], that for a clear night sky (see below) the measured heat flux from the plate toward the surrounding could amount up to 60 W m^{-2} if oriented horizontally (i.e., the plate normal facing the zenith angle 0°). Changing the angle between plate normal and zenith from 0 to 90° (the latter corresponds to a vertically oriented plate, the surface normal being horizontal) led to a decrease of observed heat flux to only 10 W m^{-2}.

Owing to the fact that building walls or roofs usually have only low values of thermal conductivity, the different surrounding temperatures may easily lead to temperature differences on the building envelope, even if wall thicknesses are the same and no thermal insulation problems exist. Examples are given below after a more detailed discussion on night sky radiant cooling.

6.4.5
Night Sky Radiant Cooling and View Factor

The three typical temperatures T_{ground}, T_{obj}, and T_{sky} of the surroundings of buildings vary within different limits. Typical building temperatures (T_{obj}) can vary due to solar load and shadows during the daytime and due to cooling effects during the night, however, since there is internal heating in wintertime, the building envelope temperatures will usually not drop below the outside air temperature. Similarly, ground temperatures will not drop significantly below the air temperatures (see also below). In contrast, the sky can show the strongest variation in temperature. The largest temperature difference occurs between clear and cloud-covered skies. This is due to the fact that clouds are opaque objects in the IR and have temperatures according to their minimum height. In contrast, measured temperatures of clear skies are due to contributions from all atmospheric heights since the emissivity of thin layers of air within the atmosphere is quite small.

Nearly everyone having access to an IR camera has probably once directed it toward a clear sky, maybe just out of curiosity. Figure 6.47 depicts one such thermal image, which includes the horizon. The emissivity was chosen near unity in order to get a reasonable measure for the objects close to the horizon within a distance of about 100 m (air temperature 10 °C). The sky shows a colored band (in the chosen

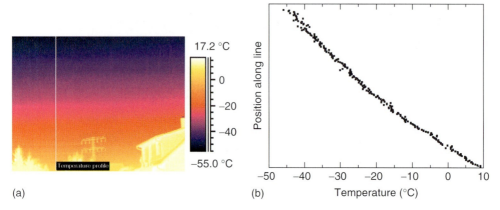

Figure 6.47 Typical thermal image of a clear night sky, angular range 24° (a) and apparent temperature profile for constant emissivity along the chosen line (b).

color palette) with apparent temperatures (along the line of IR image) decreasing to about −40 °C at the top of the image.

Of course, these apparent temperatures are absolutely useless; they refer to some kind of effective sky temperature averaged over the whole line of sight (more details, see Section 10.6).

A more thorough quantitative analysis of IR emissivities of sky radiation leads to emissivity values of the order of 0.8 [25]. Using these emissivities, effective sky temperatures were defined using the net radiant heat flux from an object of temperature T toward the hemispherical sky with effective temperature T_{sky}. In a simplified model, the effective sky temperatures depend on dew point temperature, ambient temperature, and cloud cover [25].

Figure 6.48 depict VIS and IR images of clouds and a clear sky in the background. Owing to their low height, the clouds have much higher temperatures than the clear sky in the background.

The possible very low effective clear sky temperatures can have a drastic influence on outdoor IR thermal imaging of buildings, in particular if the rule for outside thermography is obeyed that recordings should be made in the early morning hours before sunrise. Clear skies strongly increased radiative losses compared to completely cloud-covered skies. They can therefore lead to large time-dependent temperature drops of walls, roofs, and windows of buildings. In this case, weather conditions of the previous night define the thermal behavior of the outside building envelope. For hot climates, this effect has been proposed as a cooling mechanism [26].

From everyday experience (in climates that occasionally suffer freezing temperatures), it is well known that the air temperatures can get unusually low after nights with clear skies in contrast to the case of completely cloud-covered skies. Therefore, the physical process is indicative of night sky radiant cooling [24] (of course, the process also takes place during clear days, however, it may then be obscured by solar load effects, since clear daytime skies automatically correlate with daytime

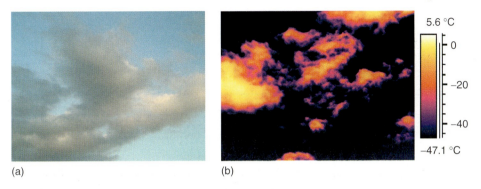

Figure 6.48 Influence of clouds on detected IR camera temperature (VIS – image taken during the day, IR – image taken during the night). Clouds can have much larger apparent temperatures than the clear sky.

solar loads). In the following sections, a number of examples are discussed, which illustrate the importance of the view factor in building thermography, in particular when night sky radiant cooling is involved.

6.4.5.1 Cars Parked Outside or Below a Carport

It is known that parking a car in a carport can prevent the cooling down of the car windshield to a temperature below the dew point or the freezing point of water. This behavior is shown in Figure 6.49. In Figure 6.49a,c two cars are parked: one inside and the other outside of a carport. The carport roof consists of 3.5-mm thick plastics and is opaque in the infrared. The effective clear night sky temperature is much lower than the ambient air temperature and cools down the carport roof and the outside car. The windshield of the car below the carport roof, however, does see only a small angular part of the clear night sky, and a much larger part of the inner roof with a temperature above the sky temperature geometry similar to (Figure 6.46b). Therefore, it has lower radiative losses to the surroundings than the car outside.

During a clear night sky the areas of the car parked outside (Figure 6.49c,d) directed to the sky are strongly cooled down to $-12\,°C$. In contrast, the temperature of the car within the carport equals approximately the ambient temperature of $-4\,°C$.

A well-known and often practiced possibility to avoid a frozen windscreen of a car parked outside is to attach a foil blanket on top. Figure 6.49d depicts a partially covered windshield of the outside car with an ordinary aluminum foil with a thickness of about $30\,\mu m$. The high reflectivity ($R \geq 95\%$) of the aluminum foil causes a low emissivity $\varepsilon \leq 0.05$ of the covered area. The radiant energy loss during the night is reduced to a minimum and the corresponding infrared image depicts the higher temperature of the previously foil-covered area. The thermal conductivity of glass is very low with $\lambda_{\text{glass}} \approx 0.7\,W\,(m\,K)^{-1}$. Therefore, the temperature difference of about 5 K between the two areas of the windshield can be observed. For comparison aluminum foil was also placed on an area of the

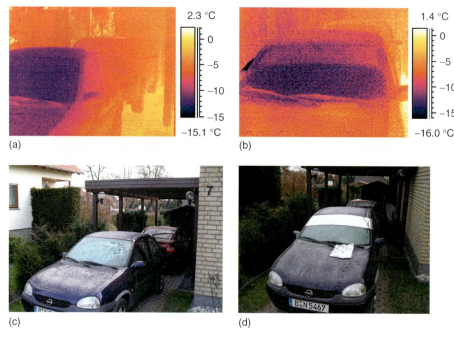

Figure 6.49 (a,c) Two parked cars (one in and one outside a carport with a plastics roof) after a clear sky night during winter. (b,d) During another night the windshield and the engine hood of the car parked outside (d) was partially covered with an aluminum foil which was removed directly before recording of the IR image.

engine hood. The infrared image does not show a temperature difference on the engine hood because the thermal conductivity of the metal is much larger with $\lambda_{steel} \approx 50$ W (m K)$^{-1}$ which leads to a fast temperature equalization.

Figure 6.49 also demonstrates the fact that objects can indeed cool down much below ambient air temperature during clear night sky conditions since the large loss in radiant heat cannot be accounted for fast enough by heat transfer due to convection.

6.4.5.2 Walls of Houses Facing Clear Sky

For vertical walls of buildings the view factor for the night sky cooling is much smaller than for the car windshield because there is a 90° angle to the zenith. To analyze the influence of clear sky radiant cooling on the temperature of a wall without any neighboring objects (situation as depicted in Figure 6.46b for vertical surfaces), two areas of the wall were covered with aluminum foil at sunset, see Figure 6.50. Four hours after sunset, the foil was removed and a thermal image was taken. Similar to the windshield of the car, the previously covered areas exhibit a higher temperature with a ΔT reading up to 6 K caused by the reduced night sky cooling.

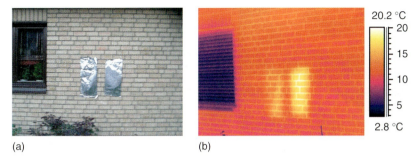

Figure 6.50 Areas of an outside building wall (facing no other warm objects/buildings) covered with aluminum foil (a), thermal image of the wall directly after removal of the aluminum foil 4 hours after sunset (b).

6.4.5.3 View Factor Effects: Partial Shielding of Walls by Carport

Carport roofs also present examples to study the differences in radiant night sky cooling of building walls. Figure 6.51 depicts again the carport of Figure 6.49. As mentioned above, the temperature of its roof has decreased during the night sky radiant cooling, but remains at a higher temperature level than the effective night sky temperature. It is interesting to compare the three house wall segments indicated by the areas AR01, AR03, and AR04. They all belong to the same wall with the same construction and no geometrical thermal bridges. The radiant energy loss of the wall below the carport roof (AR01) is much smaller than the energy loss of the wall above the roof (see temperatures of AR04) due to the differences in view factor to the clear sky and the long thermal time constants of the wall. The temperature differences 4 h after sunset amount to about 3 °C. The wall temperatures of AR03 are in between, since this wall segment is still partially shielded from the night sky (resembling a smaller viewing angle as in Figure 6.46) by the carport roof.

The small wooden house at the rear end of the carport is also completely covered by the carport roof. The temperature of this wooden house (AR02) is however

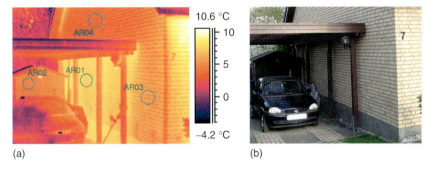

Figure 6.51 Influence of different view factors on the radiant cooling rates of building walls. The thermal image was taken four hours after sunset.

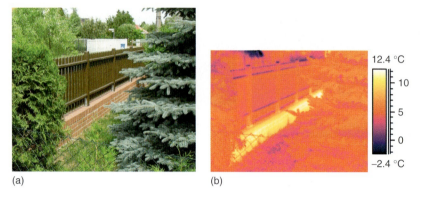

(a) (b)

Figure 6.52 Transient temperature distribution caused by different thermal time constants. The IR image was recorded shortly before sunrise.

smaller than the temperature of the adjacent building wall (AR01). This effect is caused by the much smaller thermal capacitance (smaller thermal time constant).

The difference in thermal time constants of brick versus wood is also shown in Figure 6.52. It depicts a wooden garden fence on a base of concrete and bricks. A typical wrong interpretation of such an image could be the assumption that heating pipes are below the fence. In reality, the garden fence cools down very quickly due to its low thermal mass. The base, however, can store much more thermal energy and cools down more slowly. The cooling time constants of objects can be estimated from their thermal properties [21, 27]. In the present case, for the base of the fence, one finds a cooling time constant of about 3 h.

It is to be noted that without consideration of the view factor, the thermal image of Figure 6.51 could be erroneously interpreted as a typical example of very bad thermal insulation of the wall segment below the carport roof.

6.4.5.4 View Factor Effects: The Influence of Neighboring Buildings and Roof Overhang

Figure 6.53 depicts another example of possible erroneous interpretation of IR images if the view factor is not correctly taken into account. The figure shows two neighboring houses. Toward the west of both the houses, there is no surrounding property, that is, there is a large view factor contribution to the clear sky. The southern side wall of the more distant white colored house faces the house in the foreground of the VIS image. Therefore, its view factor contribution to the cold clear night sky is reduced and instead replaced by a contribution due to the warmer neighboring house. Therefore, this wall (AR03) is warmer than the west-facing wall (AR01) (it was checked experimentally that there is no influence of an angle-dependent emissivity on the thermal imaging results).

The marked areas A02 and A04 in Figure 6.53 depict the effect of the roof overhang on the radiant cooling rate. Just below the overhang (AR02) the temperature is much higher due to the strongly decreased radiant cooling. The decrease of the view factor

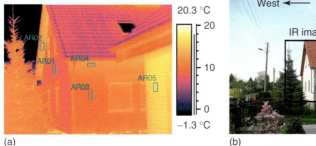

(a)   (b)

Figure 6.53 Influence of radiant night sky cooling and view factor on apparent wall temperatures (details see text).

for the cold sky in the upper region of the side wall by the overhang (AR04) also causes an area with a higher temperature level compared to the lower wall (AR03).

Such experimental results may lead to misinterpretations of thermal images. The higher temperature areas caused by reduced radiant cooling are often misinterpreted as thermal leakages.

Another example of possible misinterpretation of thermal images concerns the temperatures of the western walls of the two buildings. The measured wall temperature of the foreground house is significantly higher (AR05) than the background house (AR01). This is caused by different wall construction and not by a different thermal insulation. The house wall exhibiting the higher temperature (foreground house) consists of a brick wall at the outside, an air gap, and an insulated inner wall. Owing to the high thermal capacitance, the heat storage capacity is large. The wall of the background house consists of a thin (2 cm) plaster followed by thermal insulation and an inner wall. The thermal storage capacitance of the plaster is much smaller than that of the brick wall. Therefore the wall of this house exhibits a lower temperature at the surface if there is a strong heat loss at the surface due to the radiant cooling. The measured temperatures do not represent the energy loss of the houses but rather the transient effect of cooling down during the night after heating during the day (sun and air).

6.5
Windows

Windows belong to the most prominent features on any IR image of buildings. This is due to two reasons. First, glass surfaces are usually very flat, providing excellent conditions for mirror images, that is, thermal reflections. Second, the connection of window frames to the building wall is critical concerning thermal insulation. Figure 6.54 depicts a schematic very simplified view of a window frame attached to a building wall. Nowadays, double- and triple-pane windows are available with U-values that are quite low, however, the critical energy loss comes with the frame. There are associated possible problems, for example, bad insulation (hollow space

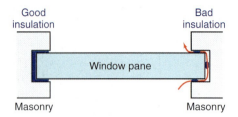

Figure 6.54 Schematic view of a window and its connection to the building wall.

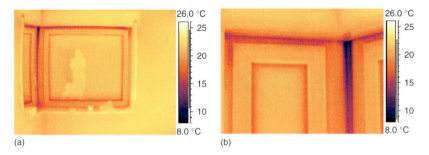

Figure 6.55 (a) Overview of living room window showing thermal reflections. (b) Expanded view of critical section. Conditions: outside air: 0 °C; inside air: 21 °C; and minimum at leak: 8 °C.

between frame and masonry incompletely filled with insulating foam) leading to airflow (red arrow) to the outside. In addition, the corner geometry of windows in their frames also represent geometrical thermal bridges. Therefore, window frames usually show up in any thermal image. The important question is whether the associated temperatures are above or below the typical dew point temperatures. Therefore, any careful IR analysis from the inside of a building will always check window frames with regard to dew point temperatures.

Figure 6.55a shows the inspection of a living room window. It is a split window with a 120° angle of the wall corner. The most prominent feature is apparently the thermal reflection of the thermographer himself (for details on thermal reflections, see Section 9.2). The real prominent feature is the cold spot at the upper corner between the two window parts. Figure 6.55b is an expanded view of this critical section. Owing to missing thermal insulation in the frame at the edge, the minimum temperature amounted to only 8 °C which is below the respective dew point temperature for 50% relative humidity. The leak is relevant in terms of mold formation.

Another example of a bad window thermal insulation in the frame is shown in a skylight window within a tilted roof (Figure 6.56). For the same inside–outside conditions, the lower right corner had even lower temperature of only 7.3 K, clearly below the dew point temperature. The window often showed a wet frame after cold winter nights and was replaced.

In contrast, Figure 6.57 shows an example of a large window front at the top floor of a house from outside as well as inside. Although the window frames

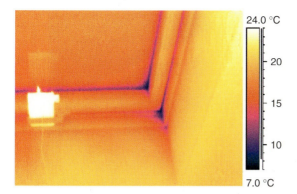

Figure 6.56 Sky light window with bad thermal insulation of frame. The lower corner edge had a minimum temperature of only 7.3 °C, well below the respective dew point temperature.

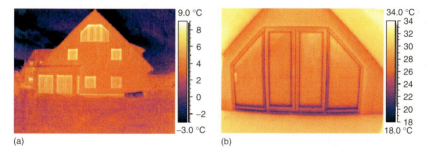

(a) (b)

Figure 6.57 Outside (a) and inside (b) view of a four-split window section at the top floor of a house for very large ΔT values.

clearly show up in the images, the inspection showed no problems. The features are due to the expected geometrical thermal bridges of the frame as well as the slightly lower U-values of the frames compared to the panes. The double-pane windows themselves are slightly warmer than the brick wall due to their slightly larger U-value (see Table 4.6). Even close-up imaging of the window frames from inside revealed no problem and frame temperatures were well above dew point temperatures also for lower inside temperatures.

Shadows of neighboring buildings can also have a drastic influence on the appearance of windows in thermal images [28]. Figure 6.58a shows an indoor thermal image of two adjacent double pane windows A and B. Without considering the outside conditions ((Figure 6.58b) with B being in shadow), one could think that the window (B) was defective having, maybe, a gas leak.

In contrast, Figure 6.59 shows a real gas leak of a double-pane window. Such insulating glasses are usually filled with dry air (to avoid condensation) or special gases (mostly argon). The window part A in Figure 6.59 had a gas leak. Theoretically

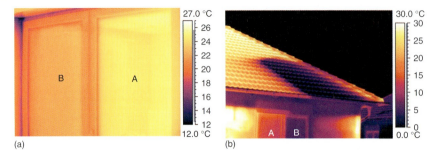

Figure 6.58 (a) Inside IR image of two window panes, showing quite different temperatures. (b) Outside analysis of the same window. Part (A) was within sunshine whereas part (B) was in the shadow of a neighbor building.

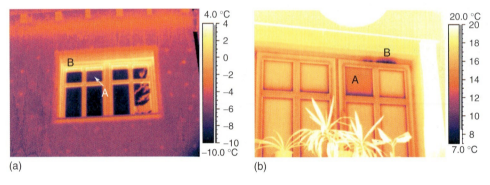

Figure 6.59 A gas leak (A) in a window detected using a combination of inside (b) and outside (a) thermography. The feature B is an additional defect in the rubber gasket.

the low U-value of the double-pane window must be replaced by larger U-values of two single-pane windows, which means that the respective glass surfaces will be colder from inside and warmer from outside. Besides the gas leak in the window, Figure 6.59 also shows an additional defect in the rubber gasket of the window frame itself (B). This could only be detected from the inside. In the outside image, thermal reflections from the window frame overhang were dominating and hiding this thermal feature.

A similar example with an apparently warmer strip below the roof is shown in Figure 6.60b. It results from reduced radiative losses (view factor effects) due to the roof overhang. In addition, one clearly sees the effect of less insulation directly above the windows due to reduced thermal insulation within and around the roller shutter housing. Figure 6.60 shows an additional effect: the right-hand side window reflects the night sky (IR recorded during night, VIS during daytime), whereas the left-hand side windows reflects radiation from a neighbor building. Therefore, the various view factor contributions of night sky and warmer objects

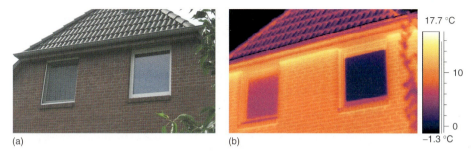

(a) (b)

Figure 6.60 Influence of differing view factor contributions of cold sky and warm nearby objects (measurement 4 hours after sunset).

Figure 6.61 Two sections A, B of normal windows and sky light windows (C). Differences in temperatures are due to view factor effects of the night sky and the roof overhang.

around differ, leading to a colder right window. Again, this has nothing to do with a badly insulated window.

Figure 6.61, left shows vertical windows (areas A, B) and a skylight window (C). Owing to the different tilt angle of the windows with respect to the zenith, window C will suffer more night sky radiant cooling and is much colder. In addition part B of the left window appears warmer due to a reflection of the roof overhang.

A final example of night sky radiant cooling on the appearance of windows is shown in Figure 6.62.

It illustrates the use of rolling shutters for the skylight windows in order to reduce the heat transfer due to night sky radiant cooling. A roof with two dual-pane skylight windows was analyzed. During a night with a clear sky, the left rolling shutter in Figure 6.62 was open and the right one was closed during the night. The closed shutter of the right window was then opened directly before the IR

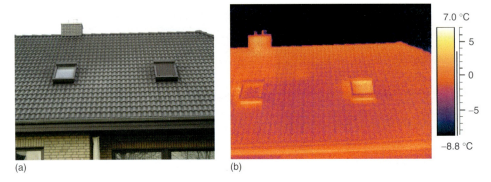

Figure 6.62 Influence of a rolling shutter on the outside temperature of skylight window panes. The shutter at the right skylight window was opened directly before recording of the IR image measurement at sunrise.

image was taken (for the thermal imaging, an MW camera was used. For detailed information about temperature measurements at window panes with MW cameras see [28]).

The temperature of the skylight window previously shielded by the closed roller shutter exhibits a higher temperature with a ΔT reading up to 5 K due to the reduced night sky cooling. Consequently, the temperature difference between the inside and the outside window pane temperature is lower than for the other window. This causes a lower heat loss through the skylight window. One may conclude that shutters may indeed help to conserve energy, in particular in cold clear night sky conditions.

6.6
Thermography and Blower-Door-Tests

No building is 100% airtight. The air exchange rate is a quantitative number, describing how much air is exchanged in a building per time with regard to the volume of the building. For example, an air exchange rate of 3 means that air with 3 times the inside volume of the building is exchanged per hour. If all natural openings in a building (windows and doors) are closed, air exchange will take place only through air leaks in the building envelope which are, for example, small channels of missing insulation or fissures and crevices in seals of windows or doors, and so on. Figure 6.63 depicts an example of an air leak within a wall socket in an outside wall. The problem was due to missing thermal insulation in the wire channel and socket hole that could be easily resolved.

In such cases, that is, when air is streaming through channels, it is well known from aerodynamics and fluid dynamics that the volume flow through pipes or tubes

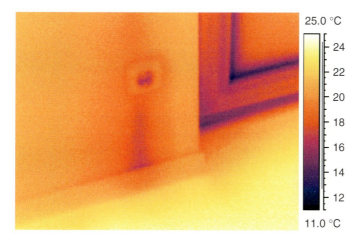

Figure 6.63 Leaks within a wall socket in an outer wall leads to a steady flow of cold air, which flows downward and thereby colds the adjacent wall.

depends linearly on the pressure difference of both sides (law of Hagen Poiseuille). This linear dependence of the air exchange rate is easily measured as a function of pressure difference. Typically, a standard pressure difference of 50 Pa between the inside and the outside of a building is chosen for localizing the leaks and giving standardized air exchange rates. The corresponding method is called the *blower door technique* [29] (see Figure 6.64).

A powerful fan is attached to a metal frame that can be adjusted airtight to any door opening. The fan can be operated at variable pressure differences

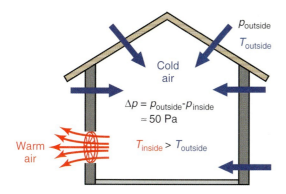

Figure 6.64 Blower door technique for measuring air exchange rates: a powerful fan creates a pressure difference of typically 50 Pa. As a consequence, cold air from outside is flowing in through holes in the building envelope. Besides measuring the volume flow at the fan, the technique can be combined with thermography to locate the leaks. The cold airflow will lead to a temperature decrease at the respective parts of window frame, wall, and so on.

to the outside atmospheric pressure, which is usually measured at a distance of at least 10 m from the building. Both higher and lower pressure is possible within the building though mostly a lower pressure of 50 Pa is used. Once every window and door is closed, the resulting low pressure leads to the inflow of air through holes in the building envelope. Large airflows can often be detected by the skin. Alternatively, smoke sources were used from the inside or outside to visualize the airflow directly. The best visualization is, however, possible with infrared thermal imaging, provided that measurements are done while large enough temperature differences exist between outside and inside air. This means that measurements are typically performed in winter (see Section 6.1.3 for precautions and requirements of thermal imaging of buildings). In this case, cold air is flowing into the building, thereby cooling parts adjacent to the leaks. The problem is that leaks often occur, for example, close to window frames or parts of the wall, or roof constructions which, in any case, show up in IR images as thermal bridges. Therefore, the proper way of analyzing thermal images while applying the blower door technique is to record images with and without pressure difference and to subtract the IR images [8]. The resulting image only shows the additional changes induced by the airflow. A few examples of blower door thermal imaging are reported in the literature [4, 30–32].

Figure 6.65d depicts a visible image of a room in a penthouse with a pitch of the roof. The most prominent features are two skylight windows and the rafter at the roof edge. The IR image (a) was recorded at normal pressure. The fan was then operated to reduce the inside pressure and the image (b) was recorded at a pressure difference of 50 Pa to atmospheric pressure. Obvious differences have occured. However, as mentioned above, all window frames already have typical edge structures due to thermal bridges. In order to find the effects solely due to the airflow, one needs to subtract the two images as is shown in the image (c).

The air leaks become visible instantaneously: they are most pronounced at the left edge of the left window and the right edge of the right window and partially also at the lower part of the rafter.

Another set of IR images of the roof area of the upper floor of a house is shown in Figure 6.66. The most prominent features in these images are the rafters, the ridge at the rooftop, as well as the upper part of a round arc window. Similar to Figure 6.65, Figure 6.66a was recorded at normal pressure and Figure 6.66b, at a reduced pressure of 50 Pa below atmospheric pressure. The subtracted image (Figure 6.66c) proves that the roof itself is fine, whereas the connection between the roof and wall has leaks all along the edge. Similarly, the ridge connection to the wall has problems of airflow. In addition, the window frame and window middle seal show air leaks.

These two examples clearly demonstrate the usefulness of the blower door technique if combined with thermal imaging: only the areas with air leaks are visualized.

6.6 Thermography and Blower-Door-Tests

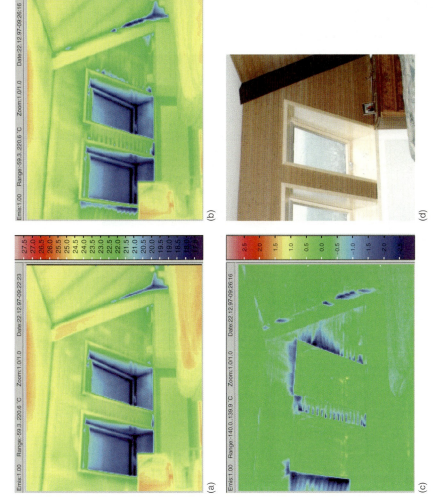

Figure 6.65 Visible (d) and thermal images (a,b) of a room in a penthouse with a pitch of the roof. Images were recorded at normal pressure (a) and at an inside pressure lowered by 50 Pa (b). The lower IR image (c) results from image subtraction of the two upper images. Thermal images were recorded with a LW Nec TH 3101 camera (256 × 207 pixels, HgCdTe detector). Images courtesy Christoph Tanner, QC-Expert AG.

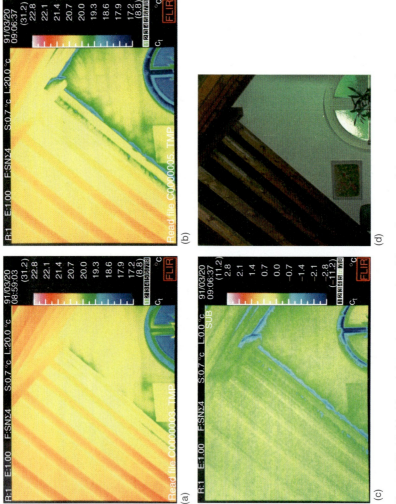

Figure 6.66 Visible (d) and thermal images (a,b) of the roof section of the upper floor of a house. Images were recorded at normal pressure (a) and at an inside pressure lowered by 50 Pa (b). The lower IR image (c) results from image subtraction of the two upper images. Thermal images were recorded with a FLIR 6200 camera. Images courtesy Christoph Tanner, QC-Expert AG.

6.7
Quantitative IR Imaging: Total Heat Transfer Through Building Envelope

So far, thermography has mostly been utilized in building diagnostics to qualitatively locate thermal bridges and interpret them in terms of air leakages, geometrical effects, missing insulation, design flaws, moisture problems, and so on. It is then more or less a matter of experience to directly judge the importance of any detected thermal signatures with respect to their being energy relevant or even relevant concerning building damage. In many cases quantitative analysis means at most that the measured surface temperatures are usually compared to respective dew point temperatures in order to judge whether problems of mold and/or condensation may take place. Of course, this difficulty in direct quantitative analysis is due to the possible large variety in building types, wall constructions, and so on. Obviously, it would be highly desirable to use thermography to directly and quantitatively judge the heat transfer rates through the building envelope.

As shown in Section 4.3.3, heat transfer through any composite wall may be described by a single number, the so called U-value in Europe or its reciprocal value, the R-value ($R = 1/U$) in the United States. The U-value is given in W $(m^2 K)^{-1}$ and describes the amount of energy per second, which is transmitted through a surface of $1 m^2$ through a wall for steady-state conditions, if the temperatures on both sides of the wall differ by 1 K. Hence, the overall heat transfer rate (in W) through a wall of area A and a temperature difference ΔT is given by

$$dQ/dt = U \cdot A \cdot \Delta T = (1/R) \cdot A \cdot \Delta T \tag{6.9}$$

We note that the R-value is often not given in SI units (m^2 K W^{-1}) but in the old units ($ft^2 \cdot F/(BTU/h)$).

In the following sections we will use $1/U$ when referring to the R-value because the thermal resistances are also usually abbreviated by the symbol R.

These values (U or $1/U$) are easily calculated from the respective heat conductivities of the building materials while assuming a standard convection coefficient for the outside wall (Section 4.3.3). Therefore, architects must promise that any new building will at least fulfill the requirements for U-values given by the respective national laws. However, till recently, it was very difficult to directly measure these values. The situation has changed recently, when Madding proposed a simple method based on IR thermal imaging to directly evaluate the R-values for exterior wall segments [33]. The good news is that this method has been so far tested for a few standard wall constructions and seems to be successful and reliable. The drawback is that a user cannot just hit a button on the camera and gets the result, rather, some background knowledge is needed for a correct measurement and analysis. In the following, the idea behind the method is briefly described.

According to the definition Eq. (6.9) (see also Figure 4.9), the R-value can be written as

$$\frac{1}{U} = \frac{A \cdot \Delta T_{in-out}}{\dot{Q}} \tag{6.10}$$

Therefore, we need to know the area A of a wall (typically $1\,m^2$), the temperature difference ΔT_{in-out} between inside air and outside air (easy to measure), and the total heat transfer rate dQ/dt through the wall. The latter is the same, however, anywhere within the wall, since the energy that flows from inside to outside must pass through all components of the wall, which is in steady-state conditions. Therefore (Eq. (4.9)),

$$\dot{Q} = \frac{\Delta T_i}{R_i} = \frac{T_{inside,air} - T_{outside,air}}{R_{total}} \tag{6.11}$$

where ΔT_i and R_i denote any selected temperature drop due to the respective thermal resistance (see Section 4.3.3) and R_{total} the total thermal resistance across the wall. Obviously, one just needs to know the thermal resistances to calculate the heat transfer from temperature measurements alone. Madding proposed the use of the thermal resistance between the inside room temperature and the inside wall surface temperature to calculate the heat flow. The argument is described as follows:

The heat transfer between the inside wall surface and the surroundings inside the room is due to radiation and convection. Both contributions are theoretically well known. The heat transfer rate due to radiation is given by the equation

$$\dot{Q}_{Rad} = \varepsilon \cdot \sigma \cdot A \cdot (T_{inside,wall}^4 - T_{inside,surr}^4) \tag{6.12}$$

where $T_{inside,surr}$ is the temperature of the inner surfaces of the room. It is assumed that these inner surfaces act as a large thermal reservoir (no change of temperature) and thus serve as the reference temperature. In thermography, this temperature is also known as *apparent reflected temperature*. As shown in Chapter 4, Eq. (6.12) can be linearized for small temperature differences to give

$$\dot{Q}_{Rad} = 4\varepsilon \cdot \sigma \cdot A \cdot T_{mean}^3 \cdot (T_{inside,wall} - T_{inside,surr}) \tag{6.13}$$

where T_{mean} is the average temperature $(1/2) \times (T_{inside,wall} + T_{inside,surr})$.

The heat transfer rate between the inside wall surface and the air inside the room due to convection is given by the equation

$$\dot{Q}_{Conv} = \alpha_{Conv} \cdot A \cdot (T_{inside,air} - T_{inside,wall}) \tag{6.14}$$

where α_{conv} is the convective heat transfer coefficient.

Combining Eqs. (6.10), (6.13), and (6.14), one finally arrives at the following equation:

$$\frac{1}{U} = \frac{T_{inside,air} - T_{outside,air}}{4\varepsilon \cdot \sigma \cdot T_{mean}^3 \cdot (T_{inside,wall} - T_{inside,surr}) + \alpha_{Conv} \cdot (T_{inside,air} - T_{inside,wall})} \tag{6.15}$$

Equation (6.15) is the basis to determine the R-value $R = 1/U$ of a wall (it would be even more straightforward to use the nonlinear relation Eq. (6.12) rather than (Eq. (6.13)). One needs to know combinations of the following quantities:

ε: emissivity of wall
σ: Stefan–Boltzmann constant 5.67×10^{-8} W (m² K⁴)$^{-1}$
α_{conv}: convective heat transfer coefficient
$T_{inside,air}$: inside bulk air temperature (distance from wall large enough that T = const)
$T_{outside,air}$: outside bulk air temperature
$T_{inside,wall}$: inside wall surface temperature
$T_{inside,surr}$: reflected apparent temperature, that is, temperature of the inner surfaces of the room
T_{mean}: average temperature $(1/2) \cdot (T_{inside,wall} + T_{inside,surr})$.

Three temperature differences and one mean temperature value need to be measured with thermography. Bulk air temperatures can be measured by placing a cardboard at a safe distance (e.g., 0.5 m) from the walls, wait until thermal equilibrium is established, and then measure the surface temperature of the cardboard. The inside wall temperature is directly measured with the camera. One may use a trick to measure the reflected apparent temperature by crumbling a piece of aluminum foil in an area large enough to get spatially resolved by the IR camera. This piece can be attached to the cardboard (at a safe distance from the wall) used for measuring the inside air temperature. The foil has very low emissivity and is strongly reflecting (however, diffuse due to the crumbling). Therefore, the measured temperature from the foil directly yields the temperatures of the inner walls. By measuring ΔTs for interior wall surface to interior bulk air and to the reflected apparent temperature in a single IR image, one reduces measurement uncertainty significantly. Using the same IR camera to measure the bulk outdoor air temperature has a similar advantage.

Besides the temperatures and the emissivity, one needs to know α_{conv}. This is probably the most critical input parameter, since such coefficients depend on the airflow conditions (laminar vs turbulent) and the actual temperature difference between wall surface and the air. Fortunately, one may safely assume laminar flow, nevertheless, different formulas are known. Typical values for α_{conv} of inside walls are of the order of 2–8 W (m² K)$^{-1}$.

Further, steady-state or near steady-state conditions are crucial, that is, the inside to outside temperature difference should be stable, the outside walls should not suffer solar and/or wind loading. In addition, the inside walls must be free of pictures, and so on, avoiding any additional thermal insulation. The temperature differences between inside air and inside wall, as well as inside wall to inside surroundings can be very small (e.g., 0.3 K for good insulation and up to 6 K for bad insulation). In order to test good insulation, the IR camera should therefore have a low NETD of less than 50 mK. Cameras with much larger NETDs are not suitable.

Madding successfully measured R-values of Stud frame constructed houses including laboratory models [33] and structural insulated panel as well as insulated concrete form wall constructions [34]. He could show good agreement with measurement deviations amounting to less than 5%, and less than 12%

compared to values calculated by summing the *R*-values of materials used in construction. The latter differences are also influenced by uncertainties in stated material *R*-values compared to actual as-built conditions. The biggest problem while experimentally estimating *U*-values is studying the measurement under only near steady-state conditions as external ambient conditions can vary significantly over time.

6.8
Conclusions

Building thermography is probably the most popular field in IR thermal imaging, however, it is not the easiest with regard to quantitative analysis, and it is often even difficult to extract useful qualitative information. As discussed above, several external factors are important such as, wind, solar load, shadows, moisture, view factors, and night sky radiant cooling. If outdoor thermography is needed, the following are the best rules: avoid solar load, that is, measure during the night; avoid transient effects due to night sky radiant cooling, that is, cloudy nights are better than clear sky nights; try to avoid transient effects due to large temperature fluctuations of the buildings, that is, it is best to measure after cloud-covered night that followed a cloud-covered day; and avoid rain and wind.

Knowing all the problems, we finally come back to the example of Figure 6.1, discussed in the beginning. The mentioned comment that the red color indicates where this building is loosing energy is probably absurd. The large signal from the roof section can only be understood if it is a tilted roof, probably directly exposed to the sun. The solar load leads to absorption of energy in the roof tiles, heating it up considerably. In case the image was taken with an MW camera, there may also be an appreciable amount of direct solar reflections involved. Owing to differences in view factor, the solar load induced warming of the walls is lower compared to the roof (vertical wall, tilted roof). This solar load wall heating also explains why the windows seem colder than the wall. Usually, due to *U*-values, it should be the other way around.

The fact that the basement and first floor windows show different temperatures is most probably due to the fact that the basement was heated whereas the first floor was not heated.

This outdoor infrared thermal image is absolutely useless unless one wants to indicate all possible errors in interpretation of IR images.

References

1. Holst, G.C. (2000) *Common Sense Approach to Thermal Imaging*, SPIE Optical Engineering Press, Washington, DC.
2. Kaplan, H. (1999) *Practical applications of infrared thermal sensing and imaging equipment*, Tutorial Texts in Optical Engineering, 2nd edn, vol. TT34, SPIE Press, Bellingham.
3. Feist, W. and Schnieders, J. (2009) Energy efficiency – a key to sustainable

housing. *Eur. Phys: J. Special Top.*, **176**, 141–153.
4. Fronapfel, E. and Kleinfeld, J. (2005) Analysis of HVAC system and building performance utilizing IR, physical measurements and CFD modeling. Inframation 2005, Proceedings Vol. 6, pp. 219–229.
5. Amhaus, E.G. (2004) Infrared applications for post construction radiant heating systems. Inframation 2004, Proceedings Vol. 5, pp. 1–8.
6. Karstädt, D., Möllmann, K.P., Pinno, F., and Vollmer, M. (2005) Using infrared thermography for optimization, quality control and minimization of damages of floor heating systems. Inframation 2005, Proceedings Vol. 6, pp. 313–321.
7. Consumer information of the US Department of Energy http://www.eere.energy.gov/consumerinfo/factsheets/bc2.html (2010).
8. Tanner, Ch. (2000) *Die Gebäudehülle - Konstruktive, Bauphysikalische und Umweltrelevante Aspekte*, EMPA Akademie, Fraunhofer IRB Verlag, Stuttgart, p. 2437.
9. Madding, R.P. and Lyon, B.R. (2000) Wind effects on electrical hot spots – some experimental data. Thermosense XXII, Proceedings of SPIE, Vol. 4020, pp. 80–84.
10. Madding, R.P., Leonard, K., and Orlove, G. (2002) Important measurements that support IR surveys in substations. Inframation 2002, Proceedings Vol. 3, pp. 19–25.
11. Möllmann, K.-P., Pinno, F., and Vollmer, M. (2007) Influence of wind effects on thermal imaging results – Is the wind chill effect relevant? Inframation 2007, Proceedings Vol. 8, pp. 21–31.
12. Oscevski, R. and Bluestein, M. (2005) The new wind chill equivalent temperature. *Bull. Am. Meteorol. Soc.*, **86 (10)** 1453–1458.
13. Incropera, F.P. and DeWitt, D.P. (1996) *Fundamentals of Heat and Mass Transfer*, 4th edn, John Wiley & Sons, Inc., New York.
14. http://www.consumerenergycenter.org/home/heating_cooling/evaporative.html (2010).
15. Grinzato, E. and Rosina, E. (2001) Infrared and thermal testing for conservation of historic buildings, in *Nondestructive Testing Handbook*, Infrared and Thermal Testing, Vol. 3, Chapter 18.5 (ed. P.O. Moore), 3rd edn, American Society for Nondestructive Testing, Inc., Columbus.
16. Wood, S. (2004) Non-invasive roof leak detection using infrared thermography. Inframation 2004, Proceedings Vol. 5, pp. 73–82.
17. Colantonio, A. and Wood, S. (2008) Detection of moisture within building enclosures by interior and exterior thermographic inspections. Inframation 2008, Proceedings Vol. 9, pp. 69–86.
18. Kleinfeld, J.M. (2004) IR for detection of exterior wall moisture and delamination: a case study and comparison to FEA predictions. Inframation 2004, Proceedings, Vol. 5, pp. 45–57.
19. Royo, R. (2007) Looking for moisture: inspection of a tourist village at the south of Spain. Inframation 2007, Proceedings Vol. 8, pp. 39–49.
20. Internet source for download: http://www.holznagels.de/DYNATHERM/download/index.html DynaTherm2000.zip (2009).
21. Vollmer, M. (2009) Newton's law of cooling revisited. *Eur. J. Phys.*, **30**, 1063–1084.
22. Pinno, F., Möllmann, K.-P., and Vollmer, M. (2009) Solar load and reflection effects and respective time constants in outdoor building inspections. Inframation 2009, Proceedings Vol. 10, pp. 319–330.
23. Henke, S., Karstädt, D., Möllmann, K.P., Pinno, F., and Vollmer, M. (2004) Identification and suppression of thermal reflections in infrared thermal imaging. Inframation Proceedings Vol. 5, pp. 287–298.
24. Möllmann, K.-P., Pinno, F., and Vollmer, M. (2008) Night sky radiant cooling – influence on outdoor thermal imaging analysis. Inframation 2008, Proceedings Vol. 9, pp. 279–295.
25. Martin, M. and Berdahl, P. (1984) Characteristics of infrared sky radiation in the United States. *Solar Energy*, **33**, 321–336.

26. Moyer, N. (2008) Using thermography in the evaluation of the NightCool nocturnal radiation cooling concept. Inframation 2008, Proceedings Vol. 9, pp. 309–319.
27. Vollmer, M., Möllmann, K.-P., and Pinno, F. (2008) Cheese cubes, light bulbs, soft drinks: An unusual approach to study convection, radiation and size dependent heating and cooling. Inframation 2008, Proceedings Vol. 9, pp. 477–492.
28. Pinno, F., Möllmann, K.-P., and Vollmer, M. (2008) Thermography of window panes – problems, possibilities and troubleshooting. Inframation 2008, Proceedings Vol. 9, pp. 355–362.
29. http://www.energyconservatory.com/ (2010).
30. Streinbronn, L. (2006) Building air barrier testing and verification using infrared thermography and blower doors as part of the building commissioning process. Inframation 2006, Proceedings Vol. 7, pp. 129–144.
31. Amhaus, E.G. and Fronapfel, E.L. (2005) Infrared imaging and log construction thermal performance. Inframation 2005, Proceedings Vol. 6, pp. 285–292.
32. Coloantonio, A. and Desroches, G. (2005) Thermal patterns due to moisture accumulation within exterior walls. Inframation 2005, Proceedings Vol. 6, pp. 249–260.
33. Madding, R. (2008) Finding R-values of stud frame constructed houses with IR thermography. Inframation 2008, Proceedings Vol. 9, pp. 261–277.
34. Madding, R. (2009) Finding R-values of SIP and ICF wall construction with IR thermography. Inframation 2009, Proceedings Vol. 10, pp. 37–47.

7
Industrial Application: Detection of Gases

7.1
Introduction

In most applications of IR thermal imaging, the IR radiation from objects to be studied passes through gases before reaching the camera detector. The most common gas is air and, therefore, IR absorption features of atmospheric air are important for any quantitative analysis using thermography (Section 1.5.2). The most prominent spectral absorption features in the range of commercial IR cameras are due to water vapor and carbon dioxide (Figure 1.46), and therefore relative humidity and object distance are needed as input parameters to correct for transmission losses between the object and camera.

In this section, we want to focus in more detail on the influence of gaseous species on IR imaging. After a short introduction of spectral features of gases in general, we focus on absorption, emission, and scattering of radiation by molecular gases due to rotational–vibrational excitations in the thermal infrared spectral range. These lead to changes in detected IR radiation from objects passing through gases. Although a quantitative description is difficult, qualitative and semiquantitative applications are possible, which has led to the development of commercial qualitative gas-detecting cameras. Nowadays, thermography is an established technique for leak detection of volatile organic compounds (VOCs), sulfur hexafluoride, and lately carbon monoxide. The influence of atmospheric absorption and scattering on the range of IR cameras is a related topic which is discussed in the context of surveillance systems (Section 10.5.2).

7.2
Spectra of Molecular Gases

One of the most prominent manifestations of atomic physics in optics is the observation of absorption or emission of radiation. In the visible spectral range, an illustrative example is shown in Figure 7.1a, depicting the Fraunhofer lines in the solar spectrum. They are a result of absorption of solar radiation in the outer atmosphere of the sun, due to the presence of colder elemental gases. In atomic theory, a ladder of possible energy levels of atoms can be computed. Absorption (excitation) or emission (de-excitation) of light corresponds to electronic transitions

Infrared Thermal Imaging. Michael Vollmer and Klaus-Peter Möllmann
Copyright © 2010 WILEY-VCH Verlag GmbH & Co. KGaA, Weinheim
ISBN: 978-3-527-40717-0

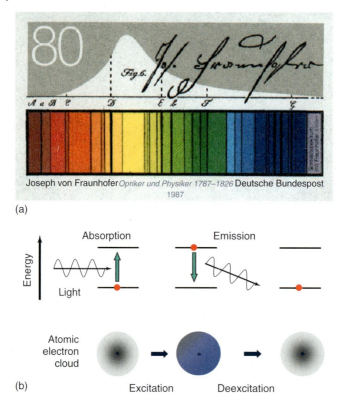

Figure 7.1 German stamp, illustrating the famous Fraunhofer absorption lines in the spectrum of the sun, due to electronic excitations in atoms (a) and scheme of electronic excitation and deexcitation in atoms (b).

between two of these levels. While changing the energy level, the electron charge distribution around the atomic nucleus changes too, as showed in Figure 7.1b, right.

Most electronic excitations of atoms and also of molecules are in the ultraviolet or visible spectral range. Therefore, electronic excitations are only of minor importance when considering the influence of gaseous species in the thermal infrared with wavelengths ranging from 1 to 14 μm. This spectral range is, however, the region of spectroscopic fingerprints of thousands of molecules, due to vibrational and rotational excitations.

The easiest molecules are diatomic ones. Molecules that are composed of two atoms of a kind (N_2, O_2, etc.) are called *homonuclear molecules*, those which are composed of different atoms are heteronuclear molecules (NO, CO, HCl, etc.). In addition to electronic excitations, molecules can exhibit vibrations and rotational excitations. Figure 7.2a depicts a schematic view of the vibration and the rotations of a diatomic molecule. In a semiclassical description, the two atoms of the

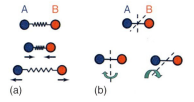

Figure 7.2 Vibration and rotations of a diatomic molecule.

molecule can be considered to be connected by a spring (a), whose spring constant determines the oscillation frequency (in introductory mechanics, the oscillation of a mass m attached to a spring with spring constant K is given by $\omega = (K/m)^{1/2}$). In addition, the molecule may rotate around the two axes, perpendicular to the chemical bond. From quantum mechanics, it is known that with each of these possible states – vibration as well as rotation – energy levels are associated, similar to the ones of the electronic excitations from Figure 7.1b, but with much lower energy separation. In particular, the vibrational frequencies of many molecules lie in the thermal infrared spectral range and are important for IR imaging.

Whether vibrations of molecules can lead to absorption of IR radiation depends on so-called selection rules. The most important rule is that the electric dipole moment of a molecule must change during a vibrational excitation if IR radiation is to be absorbed. An electric dipole consists of two opposite charges separated by a distance d. Each heteronuclear diatomic molecule resembles such an electric dipole, since different atoms have different abilities to attract the electron cloud around the nucleus. For example, in a HCl molecule, the electron charge from the hydrogen atom is more strongly attracted by the chlorine atom which means that overall, the chlorine end of the molecule has more negative charge and the hydrogen end more positive charge. In this case, the molecule has an electric dipole moment defined by the charge times the bond length. If during a vibration of this molecule, the bond length increases and decreases periodically, the dipole moment also changes periodically, that is, heteronuclear diatomic molecules may absorb IR radiation.

In contrast, homonuclear diatomic molecules like N_2 or O_2 do not have a dipole moment in the beginning, and hence there can be no change in dipole moment during a vibration of the molecule. Therefore, such gases are not able to directly absorb thermal IR radiation. This is, of course, very good for the Earth: if the main constituents oxygen and nitrogen would absorb IR radiation, the Earth's atmosphere would resemble a gigantic greenhouse, which, similar to the greenhouse planet Venus (atmosphere with 96.5% of the greenhouse gas CO_2), would have temperatures well above the boiling point of water, making human life impossible.

The next simple kind of molecules is triatomic molecules. They present a new possibility: the molecules can be either linear or nonlinear. As an example, consider the H_2O and CO_2 molecules, which act as greenhouse gases in the atmosphere. Figure 7.3 schematically depicts the fundamental vibrational modes, which are

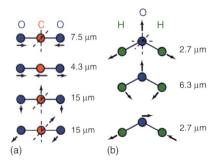

Figure 7.3 Examples for vibrational modes of linear (CO$_2$) and nonlinear (H$_2$O) three-atomic molecules. Oscillation frequencies are often given as so-called wavenumber in cm^{-1}. To find the wavelength in micrometers, divide 10 000 by the wavenumber in cm^{-1}. Example: 10 000/2349 cm^{-1} = 4.26 μm.

possible for these triatomic linear (CO$_2$) and nonlinear (H$_2$O) molecules (after [1]). Besides vibrations, each molecule can also have two (CO$_2$) or three (H$_2$O) modes of rotational motion, as indicated by the axes (broken lines) in Figure 7.3.

Applying the dipole moment rule to the four vibrational modes of CO$_2$, it is obvious that the mode for $\lambda = 7.5$ μm cannot be excited by absorption of IR radiation, since CO$_2$ is a symmetric molecule which does not have a dipole moment at rest (small negative charges at the ends, small positive charge at center ⇒ two dipoles cancel each other). Therefore, a breathing-type oscillation (top vibration in Figure 7.3a) cannot change the dipole moment. In contrast, the modes where the oxygen and the carbon atoms move out of the linear geometry do lead to a finite dipole moment, and hence these modes can absorb IR radiation. The water molecule, on the other hand, can absorb IR radiation with all three possible fundamental oscillation modes. Figure 7.4 depicts the corresponding IR absorption spectra of water and carbon dioxide molecules. The spectra reveal quite complex shapes, since, in reality, vibrations are coupled to rotational excitations, and therefore one usually speaks of rotational–vibrational bands.

Obviously, the situation for triatomic molecules is more complex than the one for diatomic molecules. Molecules with four atoms exhibit another possibility: geometrically they can be linear (C$_2$H$_2$), planar (SO$_3$), or even three-dimensional in shape (NH$_3$). If more atoms are added, the molecule has usually three-dimensional shape. For example, methane (CH$_4$) has tetrahedral shape with the C atom being in the center and the hydrogen atoms at the four corners.

In general, a molecule composed of N atoms has $3N$ different ways to store energy (also called *degrees of freedom*). Three ways are the kinetic energies in the three directions of space. Then linear molecules can rotate around two axes for linear molecules (Figure 7.3) or around three axes for nonlinear molecules, giving two or three additional ways to store energy (the rotation around the bond axis of linear molecules is forbidden according to the quantum mechanical treatment; in the semiclassical description, the necessary excitation energies would be too large

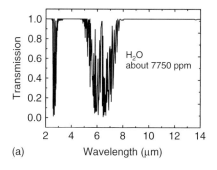
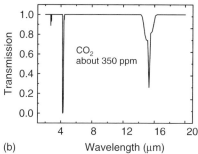

Figure 7.4 Survey spectra of H_2O vapor (a) and (b) CO_2 vapor over a path length of 10 cm as a function of wavelength. The features reflect the IR active vibrational modes of Figure 7.3. The arrows indicate the spectral ranges of MW and LW IR camera systems. If CO_2 spectra were recorded with higher resolution (Figure 7.10), each absorption feature would be further divided into a multitude of individual absorption lines.

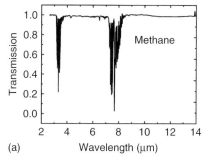
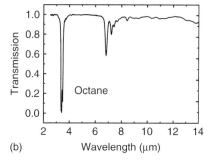

Figure 7.5 Survey spectra of methane (CH_4, high resolution, (a), own FTIR measurements) and octane (C_8H_{18}, low resolution, (b), after NIST (National Institute of Standards) data) as a function of wavelength between 2 and 10 μm (these and other spectra only indicate the resonance positions; for quantitative information about concentrations, see, for example, [2]).

to be excitable). This finally gives either $3N$-5 (linear) or $3N$-6 (nonlinear) different ways to store energy in vibrations, respectively. For example, triatomic molecules ($N = 3$) can have four vibrational modes like CO_2 or three vibrational modes like H_2O. A molecule like methane with $N = 5$ atoms will therefore, in principle, have nine different vibrational modes and larger molecules even more.

Fortunately, the respective IR spectra are, however, still very simple (e.g., methane and octane spectra in Figure 7.5). Although methane should have nine different vibrational modes and octane even 72 different modes, the spectra look astonishingly simple and – since both belong to the same group of simple hydrocarbons with single bonds – even similar.

The spectrum for methane may be understood in the following way: being a tetrahedron, methane is a very symmetric molecule, therefore some of the modes which resemble "breathing modes" cannot be excited (no change of dipole moment). In addition, the symmetry forces some of the allowed modes for IR excitation to lie very close to each other and some of them may be suppressed by other quantum mechanical selection rules.

Obviously, the larger the molecule, the more the vibrational modes that are possible. But why does the octane spectrum from a molecule with 26 atoms look too simple and even similar to the one from methane, in particular concerning the spectral feature around $\lambda = 3.3$ μm? As a matter of fact, many different kinds of organic hydrocarbon molecules (spectra in Appendix 7.A) show similar spectral features around the wavelength range of 3.1 to about 3.5 μm, which by the way is also the reason why GasFind cameras for these species were successfully developed.

Each absorption feature in a spectrum corresponds to a specific vibrational excitation of the molecule. The interactions of the various vibrations with each other and with rotations can lead to very complex spectra, however, with very characteristic spectral features. According to the above discussion on diatomic molecules, the vibrational frequency depends on the "spring constant," that is, the strength of the chemical bond between the two atoms and their masses. Therefore, if a molecule has atom pairs with different chemical bonds (e.g., single, double, or triple covalent bonds or ionic bonds of different strengths), vibrations between adjacent atoms will take place at quite different frequencies [1, 3]. This allows us to characterize a complex molecular spectrum in terms of the vibrations of some of its constituting atomic pairs. For example, the spectral regions due to the C–H bonds, C–C bonds, C–O bonds, or C–N bonds, and so on are quite well separated in a spectrum, since the mass of the carbon atom partner changes appreciably (the larger the mass, the lower the oscillation frequency, that is, the larger the wavelength [1]).

All simple hydrocarbons, for example, octane, have two different functional groups within the carbon chain, CH_2 groups in the middle of the chain and CH_3 groups at the ends of the chain. Some other hydrocarbons have CH-groups only. Owing to the similarity of the molecules, these functional groups dominate the spectral features by their various stretching and bending vibrational modes. In particular, the stretching modes of the CH, CH_2, and CH_3 groups contribute considerably around wavenumbers between 2800 and 3000 cm^{-1}. In addition, deformation modes (bending vibrations) give rise to spectral features around 1300–1500 cm^{-1}. This explains why quite different molecules (methane and octane) exhibit similar spectra. The same holds for numerous other hydrocarbon species (spectra in Appendix 7.A). These similarities are utilized in IR thermal imaging, which uses a broad spectral region (see below) for mostly qualitative analysis. However, it should be emphasized that the spectra of these different molecules – though being similar – also differ enough from each other such that high-resolution detection enables to clearly identify each molecule separately via its spectroscopic fingerprint. In this respect, high-resolution IR spectroscopy is an important quantitative tool in gas detection as well.

In Appendix 7.A, we present a selection of IR spectra of inorganic as well as organic gases (Table 7.A.1). Some of them were recorded with conventional prism or grating spectrometers, others with FTIR [4, 5]. These spectra indicate which IR camera may be used for the detection of a specific gas. The commercial cameras only operate in a narrow predefined spectral band, whereas research cameras with exchangeable filters allow detection of a larger variety of species. More information on spectra can, for example, be found in [2].

7.3
Influences of Gases on IR Imaging: Absorption, Scattering, and Emission of Radiation

7.3.1
Introduction

For many years, the importance of gases in IR imaging was restricted to being the medium through which the IR radiation of objects was transmitted before reaching the detector of the camera. Therefore gases should neither absorb nor emit any IR radiation in order not to change the camera signals. These restrictions for the gas encountered in most applications, that is, atmospheric air, define the typical spectral ranges of IR cameras (Figure 7.4, Figure 1.8). In the wavelength ranges of commercial IR camera systems, the extinction of atmospheric gases can be nicely compensated and quantitative measurements are possible even for large distances [6].

In contrast, the qualitative detection of gases themselves by thermal imaging is a rather new application – first reported to our knowledge in 1985 [7]. This has meanwhile led to the development of new IR cameras. They all utilize absorption or emission of IR radiation [8] in rovibrational bands of molecules contributing in the thermal IR region between 1 and 15 µm. Although strong absorption/emission features allow detection of gases with broadband IR cameras operating in the MW and LW range [8], the more sensitive commercial GasFind cameras [9] use narrowband cold filters in front of the detector. At present, GasFind cameras are available for the detection of VOCs [10], SF_6 [11], and most recently CO. Alternatively, warm filters may also be used for sensitive CO_2 detection [12, 13].

One of the key issues associated with any kind of IR imaging of gases is whether it gives qualitative results in terms of visualizing the gas or quantitative numbers for the respective gas concentrations. All commercially available GasFind cameras are only operated in the qualitative mode. The problems associated with quantitative imaging are discussed in Section 7.4.

7.3.2
Interaction of Gases with IR Radiation

As outlined in Section 7.2, infrared radiation may be absorbed by molecules due to excitation of rotational–vibrational bands. For simplicity we assume just two discrete energy levels, representing ground and excited states of the molecule within such bands. Thus we may distinguish three different possibilities for gas molecules to interact with IR radiation (Figure 7.6).

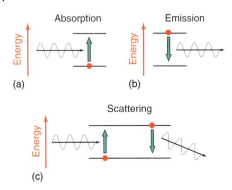

Figure 7.6 Each molecule has specific energy levels, the transitions between them in the IR range being characterized by spectra of rotational–vibrational excitations. The three basic modes of interaction between gas molecules and IR radiation are (a) absorption: radiation may be absorbed from an initially low-lying state. The molecule ends up in a different rotational–vibrational state (b) emission: radiation may be emitted from an initially excited state, or (c) scattering: radiation may be scattered (absorption–emission cycle).

- First, suitable IR radiation may be absorbed if the molecules are initially in the ground state (a). In this case, directed IR radiation from an object behind the gas is attenuated by the gas, that is, less radiation is detected than without gas.
- Second, suitable IR radiation may be emitted isotropically if the molecules are initially in the excited state (b). In this case, one would detect more IR radiation in the direction of an object than would be detected without the gas.
- Third, IR radiation may be scattered by molecules. This process happens if molecules are initially in the ground state (c). They may then be excited by suitable IR radiation, for example, due to hot objects that are close by. Immediately after excitation they re-emit this IR radiation more or less isotropically. In this case, one would also detect more IR radiation in the direction of an object than would be detected without the gas.

Both the emission and the scattering processes lead to additional IR radiation due to the gas. If the population of the excited state of the gas is due to its temperature (thermal excitation) and the gas is optically thick (see below), the spectrum or the emission process can be described by Planck's law of blackbody radiation. If it is optically thin, the spectrum is obtained by multiplying Planck's law with a suitable emissivity which may be wavelength dependent (Section 1.4).

Scattering – on the other hand – does not need an initial population of an excited state. Scattering of electromagnetic radiation from atoms is well known. One usually distinguishes resonant and nonresonant scattering. In resonant scattering, the energy of the incident photons matches the required energy for excitation of the molecule to an excited state. The respective process is called *resonance fluorescence*. The best-known example of nonresonant scattering, that is, when the photon energy does not match the energy for an excitation, is Rayleigh scattering, which is responsible for the blue skies (due to visible light scattering with energy lower than that for electronic excitations and higher than that for vibrational excitations of air molecules). If nonresonant radiation is incident on molecules, one may also observe the additional phenomenon that the scattered light has a slightly different energy compared to the incident light. This process is called *Raman scattering* and serves as an established technique of IR spectroscopy.

In the thermal IR range, the probability for nonresonant scattering from molecules is very small compared to resonant scattering, and therefore only scattering processes at the respective transition wavelengths must be considered. Such scattering processes from molecular gases may be thought of as resembling something like thermal reflections known while studying solid objects.

This explains an additional difficulty when trying to quantitatively measure the influence of gases: one needs to know the contribution of scattering. This may be solved similar to reducing thermal reflections in conventional thermography: one must reduce the amount of scattered IR radiation by shielding the observed gases from any hot objects which may provide the IR radiation, needed for the excitation of the scattering process.

7.3.3
Influence of Gases on IR Signals from Objects

The impact of gas absorption, emission, or scattering on IR thermal imaging is schematically illustrated in Figures 7.7–7.9. The top row of Figure 7.7 depicts the situation of an object which is observed by an IR camera through a cold gas of temperature T_{gas}. The object should have a temperature $T > T_{gas}$. The middle row graphs show (from left to right) the object radiation (e.g., radiance) whose spectrum for a black body would follow Planck's law. An IR camera only uses a predefined spectral range, which is indicated by the pink-colored area below the curve. This area reflects the object signal seen by a detector if no additional attenuation of the radiation would take place on its path from object to detector. If, however, the cold gas has absorption features within the IR camera spectral range (indicated by the transmission spectrum with a well-defined minimum), the detected object signal is lower (smaller pink area). This means that observation through a gas stream leads to a decrease of the object signal.

If a narrowband spectral filter is used in front of the camera detector (bottom row), the object signal (green-colored area) is on the one hand smaller than that obtained by broadband detection. On the other hand – if it is tuned to the absorption feature of the gas – the relative signal change due to absorption by the gas is much larger, that is, one may improve the signal contrast and detect more sensitively. In addition, a detailed analysis must also consider the respective change of signal-to-noise ratio (Section 3.2.1).

Figure 7.8 illustrates the measurement situation if, in addition to the object under study, other warm or hot objects that emit thermal IR radiation are close by. The hot surrounding objects emit thermal radiation which may lead to resonant scattering processes from the molecular gas. This closely resembles the effect of "thermal reflections" known from solid objects (Section 9.2).

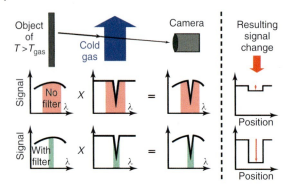

Figure 7.7 Detecting cold gases in front of a warm background with or without filter. The shaded regions reflect the detected spectral range of the camera system (for details, see text).

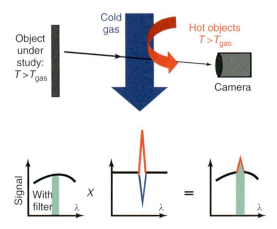

Figure 7.8 Detecting warm objects through cold gases if additional hot objects are close by. These may lead to a kind of "thermal reflection" thereby altering the signal strength (for details, see text).

The detected camera signal may change according to the relative importance of gas absorption (blue, signal decrease) and resonant scattering (red, signal increase). Scattering effects can often be observed using GasFind cameras, if the geometry/orientation of the camera is changed and, as a consequence, the amount of "thermal reflections" from the gas varies.

The top row of Figure 7.9 depicts the ideal situation of an object which is observed by a broadband IR camera through a warm or hot gas of temperature T_{gas}. The object should have a temperature $T < T_{gas}$. If the gas would be optically thick, the thermal emission from the gas would be described by Planck's law using the gas temperature. However, for realistic situations, the gases under study are optically thin, that is, they are spectral emitters with emissivities below unity. In any case,

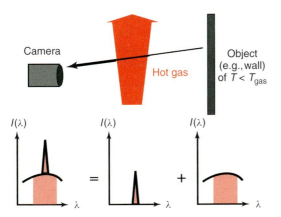

Figure 7.9 Detecting hot, thermally emitting gases in front of a colder background with a broadband IR camera. The shaded regions reflect the detected spectral range of the camera system (for details, see text).

the hot gas radiation may add up to the object signal. This means that observation through a hot gas stream leads to an increase of the object signal.

We note that Figure 7.9 can be extended to also illustrate the use of narrowband filters (as in Figure 7.7), and it is furthermore also possible for this setup that additional hot objects can lead to resonant scattering (like in Figure 7.8), thereby changing object signals.

As demonstrated in Figures 7.7–7.9, absorption, emission, and/or scattering of IR radiation by gases can have an influence on IR thermal imaging signals. The improvement in signal contrast and sensitivity by using narrowband filters is common to many cameras, optimized for gas detection. As an example, Figure 7.10

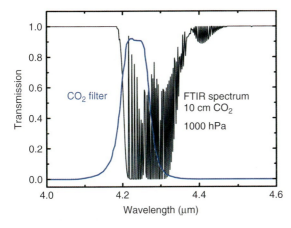

Figure 7.10 The 4.3-μm band spectrum of CO_2 with superimposed filter transmission curve.

shows an expanded high-resolution view of the 4.3-μm absorption band of CO_2 (Figure 7.4), together with the transmission spectrum of a typical commercially available filter in this wavelength region. Such a filter does easily serve the purpose of IR imaging. A multitude of narrowband and broadband filters are available in the thermal IR range [14].

Further improvement in sensitivity results when using cold rather than warm filters since the detected background signal will also be reduced in particular if the filter material is also absorbing IR radiation (Section 3.2.1). This gas detection method with cold filters is commonly used with GasFind cameras for VOCs. In addition, there are also some cameras which do not need any filter at all: cameras with quantum well narrowband detectors (QWIP systems), their wavelength being tuned to molecular absorption bands, for example, to the one of SF_6, do not need any additional filter.

7.4
Absorption by Cold Gases: Quantitative Aspects

In principle, the quantitative analysis of absorption of IR radiation by gases is possible as outlined below. However, we only give a schematic description, since any adaptation to a specific gas measurement requires us to take into account quantitative details of the respective gas absorption, specifics of the geometry of the surroundings, measurement conditions, and so on [15]. Other preliminary studies to quantitatively measure gas sensitivities of thermal imagers for methane have been reported recently [16].

The reasoning for an analysis is the following:

- First, quantitative gas spectra must be known;
- second, one may therefrom calculate the decrease of IR radiation (e.g., its radiance) induced by a certain gas as a function of gas parameters and the used spectral range of the IR detector; and
- third, this leads to quantitative signal estimates, that is, kind of a calibration curve for an IR camera.

7.4.1
Attenuation of Radiation by a Cold Gas

The attenuation losses of IR radiation passing through a gas are described by Bouguer's law (sometimes also called *Lambert–Beer's law*), which states that the change of radiance passing through matter as a function of length L is given by

$$I(\tau) = I_0(\lambda) \cdot e^{-\tau(\lambda, c, L)} \tag{7.1}$$

Here I_0 defines the radiance entering the matter. Figure 7.11 depicts a graphical representation of Eq. (7.1) in terms of transmission $T(\tau) = I(\tau)/I_0$. The quantity $\tau(\lambda, c, L)$ is the optical thickness of the gas. It depends on wavelength λ, gas concentration c, and length L of optical path through the gas. The gas concentration

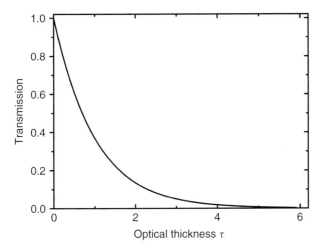

Figure 7.11 IR radiation, passing through a gas is attenuated. The relevant quantity is the optical thickness τ of the gas.

may be written in a number of ways. It may be given as $\rho = m/V$, that is, mass per volume (kg m^{-3}), as number density $n = N/V$, that is, number of molecules per unit volume (in m^{-3}) or as pressure p (in hPa or atm), since according to the ideal gas law, p is proportional to number density n.

For a gas of constant density n (molecules per unit volume) over the whole optical path, τ can be written in two different ways as

$$\tau(\lambda, c, L) = \sigma_{\text{ext}}(\lambda) \cdot n \cdot L = k(\lambda) \cdot p \cdot L \tag{7.2}$$

where the extinction cross section $\sigma_{\text{ext}}(\lambda)$ accounts for scattering as well as absorption losses. In the following, we only deal with absorption features. $k(\lambda)$ is the spectral gas absorption coefficient, usually given in units of 1/(atm m). Once $k(\lambda)$ is known, one may quantitatively calculate the transmission $T(\lambda, p, L) = I(\lambda, p, L)/I_0(\lambda)$ of radiation through any gas of given pressure p and dimension L from Eq. (7.1). Obviously, Eqs. (7.1) and (7.2) are the key for any quantitative analysis of gas absorption: one needs to know the absorption coefficient $k(\lambda)$ very precisely. Then, it is straightforward to compute the portion of transmitted radiation through a gas of given length at given pressure.

We note that the situation may be more complex, for example, if the gas is not distributed homogeneously. In this case the optical thickness is calculated from

$$\tau(\lambda) = \int_0^L \sigma(\lambda) \cdot n(x) dx = \int_0^L k(\lambda) \cdot p(x) \, dx \tag{7.3}$$

that is, contributions must be summed up over the total length of the optical path, which is considered. In all of the following, we use the description with the absorption constant k.

7.4.2
From Transmission Spectra to Absorption Constants

The main problem in quantitatively evaluating camera signal changes due to the gas absorption is the precise knowledge of the absorption coefficient k. Starting points are theoretical or experimental literature data for transmission spectra. In order to extract k values from such spectra according to Eqs. (7.1) and (7.2), the precise values of pressure p and path length L through the gas must be known. It is sometimes not very easy to get hold of reliable data, since literature data may deviate from each other appreciably [15].

As an example, Figure 7.12 depicts CO_2 spectra (the analysis scheme in this chapter is presented for CO_2). The various literature data [2, 8, 17, 18] may differ from each other, the main differences usually being due to the chosen spectral resolution. Low-resolution spectra only show averaged absorption and do not allow us to resolve individual rovibrational lines within the spectrum. The spectrum presented in Figure 7.12 was originally recorded with high resolution to resolve fine structures. Then, a low-resolution spectrum was generated by averaging with neighboring data points (smoothed spectrum). If recording conditions (gas pressure p, length of optical path through gas L) are chosen unfavorably, the spectrum may show saturation behavior (zero transmission) at the center of the resonance between $\lambda = 4.2$ and 4.35 μm, which complicates the analysis.

Using Eqs. (7.1) and (7.2), the transmission spectra of Figure 7.12 are used to extract the absorption constants if pressure p and length L of the recorded spectrum are known. Obviously, extraction of data from regions in the spectrum where transmission is close to zero is difficult, since errors in k get very large. Similar arguments hold for transmission close to unity. Regarding additional uncertainties of 0 and 100% lines (in particular for old literature spectra), we

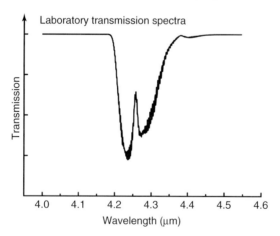

Figure 7.12 Starting point for quantitative analysis: a typical transmission spectrum of a gas, here CO_2, recorded, for example, with Fourier transform infrared spectroscopy.

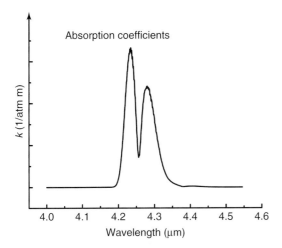

Figure 7.13 Absorption coefficients k extracted from the transmission spectra of Figure 7.12.

propose that reasonable transmission values for k-value extraction should lie in the interval {0.1, 0.9} for the transmission data. Figure 7.13 depicts the absorption coefficient k evaluated from the spectra of Figure 7.12 by using Eqs. (7.1) and (7.2).

7.4.3
Transmission Spectra for Arbitrary Gas Conditions and IR Camera Signal Changes

Once the absorption constant k of a gas is known, the transmission of the gas for any pressure and optical path length may easily be computed according to Eqs. (7.1) and (7.2). As an example, Figure 7.14 depicts a number of transmission spectra for selected $p \cdot L$ values.

Modeled transmission spectra like these, due to given concentrations and path lengths, can be used to estimate the changes, induced on the object radiance of hot objects, observed through cold gases. In addition, one needs to know (Figure 7.15)

1) the object radiation (here the $M_\lambda d\lambda$-spectrum) referring to object temperature;
2) the emissivity of the object;
3) the detector sensitivity (here of the mid-wave SC6000 camera, used in some experiments), determining the detected wavelength band of the used camera; and
4) the transmission of optics, in particular of narrowband filters (here denoted 4235 and 4480).

Figure 7.15a illustrates the idea behind the quantitative analysis of an object (350 K) observed through a gas (gas + background at 300 K). The region of interest is defined by the absorption bands of CO_2. The detected radiance as a function of wavelength is proportional to the product of object Planck function, object emissivity, detector sensitivity, filter transmission, as well as transmission through

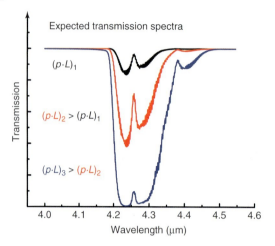

Figure 7.14 Transmission spectra for various conditions of $p \cdot L$, computed from the k-spectra of Figure 7.13.

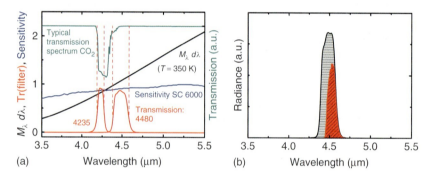

Figure 7.15 (a) Planck black body radiance $M_\lambda d\lambda$, detector sensitivity for SC 6000, two commercially available filters [14] as well as a typical transmission spectrum for CO_2, all given in arbitrary units. (b) Schematic camera signals with and without gas absorption when using the narrow band filter 4480.

the gas under study. The filter transmission changes most rapidly, and hence it dominates the shape of the resulting detected spectral radiance function. The total detected IR radiation (Figure 7.15b) is proportional to the area under the spectral radiance function.

In case there is no absorbing gas (CO_2 transmission 100%), the gray-shaded area determines the maximum possible signal. If, however, there is attenuation due to absorption by CO_2, the spectral radiance is reduced and the new signal is determined by the red-shaded area. This reduction in area corresponds to the signal reduction of Figure 7.7.

7.4.4
Calibration Curves for Gas Detection

The change in IR signal due to the gas is determined by the value of $p \cdot L$ (pressure times path length in gas). Figure 7.16 illustrates the fact that for very small absorption (small value of $p \cdot L$), both signals will be about equal (plateau region to the left), whereas for very large values of $p \cdot L$, nearly all of the IR radiation from the object is absorbed (signals drop close to zero). For any intermediate value of $p \cdot L$, an intermediate signal ratio will result. Hence, if such a calibration curve is known, one may easily find the value $p \cdot L$ from the measured signal attenuation (blue arrows).

How sensitive the method can be depends on the gas under study (value of k-constants) and the use of the filters. A camera is considered to be sensitive to a certain amount of gas if small changes of $p \cdot L$ induce large changes of the signal. The two filters in Figure 7.16 lead to two separated sensitive regions which lie between 10 and 90% of the respective saturation value. Filter 1 may detect smaller $p \cdot L$ values, that is, it allows detection of smaller amounts of gas.

Obviously, the lowest $p \cdot L$ value for a gas is defined by the given k-value of the gas and the used narrowband filter. If the filter is on resonance, that is, completely overlapping the rovibrational absorption bands, the signal decrease for fixed $p \cdot L$ has its maximum value and the lowest detectable $p \cdot L$ values are reached (e.g., graph of filter 1 in Figure 7.16). This is nearly the case for the 4235 filter in Figure 7.15. If, however, the filter is slightly off resonance (e.g., the filter 4480 in Figure 7.15), less radiation passing through the filter may be absorbed. Therefore the $p \cdot L$ value belonging to the same signal attenuation must increase. As a consequence, the calibration curve is shifted to larger $p \cdot L$ values (e.g., graph of filter 2 in Figure 7.16),

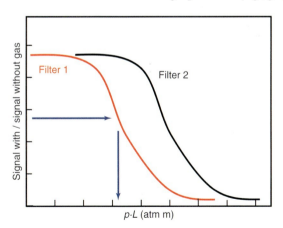

Figure 7.16 Schematic normalized plots of signal changes expected upon CO_2 gas absorption for given T_{object} as a function of gas pressure and optical path length through gas. The range of sensitive gas detection can be shifted by the use of appropriate filters.

resembling a less-sensitive detection. This may be necessary if a gas has very strong absorption bands and is present at high pressure and over a long path. In practice, it may be desirable to switch from a high sensitivity mode to a lower sensitivity mode in a camera, for example, when studying very large gas leaks (in this case, the camera would probably not receive any background signal since strong gas absorption would block any object radiation from being detected).

The only remaining question is "how can a 'calibration' curve as shown in Figure 7.16 be obtained for given gas conditions?" The easiest way is to use calibration procedures. In principle, it seems possible to include a database in the camera with signal attenuations as a function of $p \cdot L$ at certain background object temperatures and distances. This could be done by calibration upon production of the camera. One would need several gas containers with IR windows such as heatable cells of various lengths, say, between 20 cm (for later close up studies) up to, say, 10 m. Then the procedure for evaluating the gas absorption effects would be similar to the study by Richards [6], who studied the influence of long atmospheric paths (up to 1 km). The camera needs to be directed onto a blackbody source of given temperature, and the signal changes are recorded as a function of increasing pressure and/or length. There would be a multitude of such graphs as a function of temperature.

7.4.5
Problem: the Enormous Variety of Measurement Conditions

So far, no commercial quantitative gas-detecting camera is available. This is due to the fact that in practice, there is such an enormous variation in measurement parameters (geometrical distribution of gas densities, background object temperatures, ambient temperatures, etc.) that for each measurement situation, an own calibration curve would be necessary.

Even if calibration curves were available and measurements would give as a result a certain $p \cdot L$ value, there would still be the problem of interpreting it.

In order to estimate concentrations from the product $p \cdot L$ (p is proportional to volume concentration according to the ideal gas law), one needs to have a good estimate of the length. Three measurement conditions seem possible, first homogeneous gas distribution, second localized gas leak with rather well-defined gas flow, and third a complex inhomogeneous distribution of gas, which may be due to localized leaks, in combination with diffusion processes which can lead to turbulent mixing with the surrounding air. These three conditions are depicted schematically in Figure 7.17.

For a homogeneous distribution of the gas between camera and the observer (Figure 7.17b), the analysis is straightforward since L_{total} can be guessed easily, for example, with a laser range finder. From the measurement of the signal change, $p \cdot L$ is extracted which means that p follows directly and therefrom the volume concentration.

The most general and also the most difficult case is depicted schematically in Figure 7.17c. Such a situation may be the result of localized leaks; however, diffusion

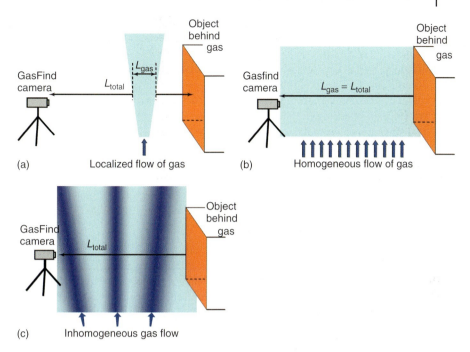

Figure 7.17 (a–c) Three typical measurement situations when studying gas absorption with IR cameras.

processes and turbulent mixing with the surrounding air can easily obscure the localized signature of laminar flow from a leak, in particular if observed from a distance (this may be advisable for security reasons). In this case, the change in signal will just give the optical thickness, which is due to an averaged value $p_{average} \cdot L_{total}$ and which results from integration over the optical path. In this case, only an average concentration may be deduced and high-concentration gas leak spots may be hidden.

For localized gas leaks with gas streaming out from a nozzle or a hole in some pipe and so on, one usually expects an inhomogeneous distribution which, close to the leak, would still resemble a laminar flow like the one shown in Figure 7.17a (see also examples in [12, 13]). If it is possible to estimate the actual gas flow diameter L_{gas} while measuring over a distance of L_{total}, the related localized pressure value p_{gas} follows from the measured average pressure p over L_{total} according to $p_{gas} \cdot L_{gas} = p \cdot L_{total}$. In any case, the net result would be at best a volume concentration of a gas.

Obviously many practical applications require the knowledge of gas flows. From the above discussion, it follows that these cannot be estimated *per se* by IR imaging. However, as has been demonstrated for the example of artificial CO_2 gas leaks [12, 13], sensitive detection with small gas flows is possible and for given leak sizes, calibration measurements may become feasible.

7.5
Thermal Emission from Hot Gases

Although in practice, one will probably encounter mostly cold gases which were discussed in the previous subsections, we also want to briefly focus on the case of colder objects observed through hot gases, in particular for the case when the gases are thermally emitting IR radiation. In this case (Figure 7.9), gas thermal emission adds up to the object radiation, that is, the original signal (without gas) is enlarged. Some experimental examples are described in Section 7.6; here, we focus on the problem of the quantitative analysis for which one needs to know the emissivity of the hot gas. This problem is raised quite often by practitioners who search for a method to estimate gas temperatures from hot flames with IR imaging.

The absorption constant $k(\lambda)$ is related to the directed emissivity of a gas. The total averaged volume emissivity ε_{gas} is then given by (e.g., [19])

$$\varepsilon_{gas}(T, p_{tot}, p_{gas}, L) = \frac{1}{\sigma T^4} \int \left(1 - e^{-k(\lambda, T) \cdot p \cdot L}\right) M_\lambda(T) d\lambda \qquad (7.4)$$

Results for gases depend on the optical thickness; usually they are depicted as a function of $p \cdot L$. They are well known for H_2O vapor and CO_2 at elevated temperatures since these gases are by-products in combustion processes and it is important to know the radiative heat load on the interior walls of furnaces. This was the original motivation for calculating emissivities of CO_2, and therefore, data are mostly available for elevated temperatures. Still, they can give some insight into the general behavior. Figure 7.18a (after [20] using data of Hottel) shows $\varepsilon_{gas}(CO_2)$ for a variety of $p \cdot L$ values [17, 19, 21–23]. The parameter is given in feet atmosphere with 1 ft = 0.3 m. Slightly above room temperature and at standard pressure, CO_2 with $p \cdot L = 1$ atm m (≈ 3 atm ft) gives $\varepsilon_{gas} \approx 0.18$. The peculiar form of the $\varepsilon(T)$ curves can be understood from basic physics of blackbody radiation and respective band absorption.

Figure 7.18b illustrates which absorption bands of CO_2 do contribute as a function of temperature for $p \cdot L = 1$ ft $\cdot$ atm. For low temperature, the $M_\lambda(T)$ spectra peak in the LW region, and hence the 15-μm absorption band, will dominate absorption for low temperatures. For larger temperatures, the contribution of the 15-μm band will decrease and the 4.2-μm band contribution will increase as $M_\lambda(T)$ shifts to the MW range. For still larger temperatures, also the 4.2-μm band contribution decreases, whereas, now, the 2.7-μm contribution can still increase. The changes of the various contributions from different absorption bands to the total emissivity as a function of temperature can be calculated from transition matrix elements, which give the respective oscillator strengths. Changes result more or less from the fact that the total oscillator strength is finite and transitions from the various bands compete with each other.

When discussing measurements with MW cameras, we would not deal with the total gas emissivity from Figure 7.18a, but rather with the smaller pink-shaded contribution of the 4.3-μm band as indicated in Figure 7.18b. This means that any MW camera would be dealing with much smaller emissivities, for example, $\varepsilon_{4.3\text{-μm band}} \approx 1/4 \, \varepsilon_{gas}$ at $T \approx 500$ K, and so on.

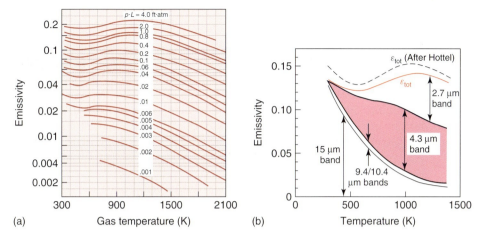

Figure 7.18 (a) Volume emissivities of CO_2 for various values of $p \times L$ as a function of temperature (a) and correction factors for variations of pressure (b), (from [20] after Hottel).(b) The total emissivity has contributions from absorption of the various rovibrational bands, shown here for $p \times L = 1$ ft $\times$ atm (after [17]).

Obviously, Figure 7.18 indicates that the emissivities of hot gases and dimensions of the order of 1 m are still quite low, that is, the respective thermal gas radiation is far from resembling a Planck black or gray body source (for a discussion of typical gas dimensions needed to reach thermal equilibrium and emissivity $\varepsilon = 1$ for atomic hydrogen gas, see [24]).

7.6
Practical Applications: Gas Detection with Commercial IR Cameras

In this section, we present a selection of IR images of different gases, detected via absorption or emission processes of the gas between camera and a background object under study. Some studies were performed using broadband cameras, others using cameras equipped with narrowband filters. The spectra of all investigated gases are given in Appendix 7.A. Most examples shown here are qualitative only with the exception of a few studies on CO_2 detection.

7.6.1
Organic Compounds

The first commercial IR camera which was developed only for the purpose of detecting gases was the GasFind camera [9] sensitive for VOCs.

Although some countries have definitions what a VOC is, there is no generally accepted one. Qualitatively, VOCs are volatile organic compounds with rather high vapor pressure under normal conditions such that there is significant vaporization. In the European Union, for example, a VOC is any organic compound having an initial boiling point less than or equal to 250 °C measured at a standard atmospheric pressure of 1013 hPa. Obviously, this vague definition explains why there are millions of different compounds – natural as well as synthetic – which may be classified as VOCs. Typical examples are fuels, solvents, drugs, pesticides, or refrigerants. Since many VOCs are toxic (e.g., benzene compounds) or dangerous for the earth climate (e.g., greenhouse gases, chlorofluorocarbons for ozone layer), they are often regulated and efficient means of detecting them are needed. The most important single application field of IR gas imaging of VOCs is probably the petrochemical industry.

Probably the most simple but also the most well-known VOC molecule is methane. According to the spectrum shown in Figure 7.5, there are two major absorption bands, one centered around 3.3 µm in the MW band and the other between 7 and 8 µm. This means that MW cameras are better suited for detection, and the commercial GasFind cameras are MW cameras with narrowband filters adjusted in the wavelength range between 3 and 3.5 µm. However, some LW cameras start detecting radiation even at 7.5 µm wavelength, that is, their detectors can sense a bit of the methane absorption band around 7.5 µm. Propane, on the other hand, shows similar spectral features; however, the relative strengths of the bands are changed. Compared to methane the absorption in the LW band is reduced with respect to the MW band.

Figure 7.19 compares two camera types (GasFind MW and typical LW) for the detection of methane and propane, flowing out at atmospheric pressure from a nozzle in front of a black body source at around 55 °C. The two images of each gas were recorded nearly simultaneously – the tube with the gas flow was sometimes moved for better view in life images. Three things are obvious:

1) It does not really matter which wavelength region is used for the detection of gases, the only prerequisite being that there must be an absorption band.
2) These images indicate the improvement which is due to the narrowband filter in the GasFind camera. Its images (a,b) have a much better signal contrast. In real-time observations this is even more obvious. We note, however, that here the two cameras have different spectral ranges. Below, we present another example with CO_2 where the comparison between broad- and narrowband filter detection is shown for the same spectral range.
3) The differences between the spectra show up in the images. For propane, the MW detection improved considerably due to the much higher absorption constants (the two spectra in the Appendix 7.A refer to the same gas pressure and sample length).

The commercial VOC GasFind cameras shall detect a variety of gaseous species. Since the respective spectra sometimes shift around the fixed filter region, some gases may be detected slightly better than others. This may give rise to additional changes when detecting different gases.

7.6 Practical Applications: Gas Detection with Commercial IR Cameras

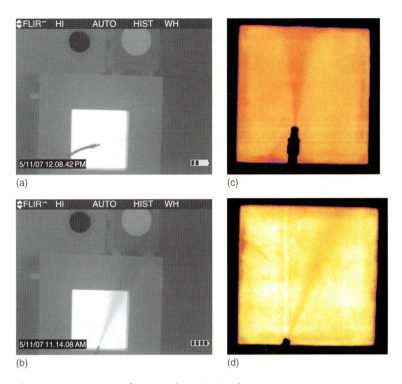

Figure 7.19 IR images of pure methane (a,c) and propane (b,d) flowing out of a nozzle in front of a black body source. (a,b) MW FLIR® GasFind camera images and (c,d) broadband LW FLIR® P640 camera images. Images recorded at laboratory in Brandenburg.

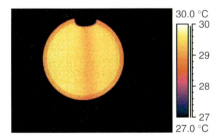

Figure 7.20 IR image of the gas of a camping burner, recorded with an LW FLIR® SC 2000 camera.

Together with propane, butane is another very important everyday life gas. In Europe, mixtures of these two gases are commonly used in lighters or camping burners. Butane shows absorption features very similar to those of propane. Figure 7.20 shows an image of the gas (no flame) flowing out of a camping burner. The burner was turned upside down since propane and butane are both heavier than air.

A variety of other organic compounds were investigated experimentally using broadband LW cameras [8], mostly in order to demonstrate the possibility for detection of these species with broadband systems. Of course, narrowband detection is more sensitive. Here, ethanol and gasoline are discussed as two more examples.

Ethanol is probably the most important alcohol species. In order to solely detect the effects of pure ethanol, we used laboratory grade ethanol (Figure 7.21). Upon opening the flask at room temperature (about 18 °C) in front of a blackbody source at about 35 °C, one immediately observes the absorption effects of the vapor. This becomes even more obvious when pouring the alcohol into a shallow dish and slightly moving the dish.

The IR camera detection of ethanol is sensitive enough to detect alcohol vapor in the breath directly after consumption of high-percentage alcohol (of course, there are more sensitive spectroscopic methods available). The IR images in Figure 7.22 depict the results of the following experiment: a volunteer drinks either pure water or alcohol and then exhales (it is sufficient to flush the mouth without swallowing and care must be taken to exhale gas and not liquid droplets). Exhaling after drinking water does not show any noticeable effect (a), whereas the alcohol within the breath is easily observable (b).

Figures 7.23 and 7.24 depict two typical examples for on-site analysis of VOCs recorded with GasFind cameras. A very common use of VOCs is as gasoline for cars. While pumping gas at gas stations, the VOC fumes, due to the vapor pressure of the liquids, may be emitted into the environment. This can be clearly seen in Figure 7.23, which depicts the situation before (a) and after starting the fill up. In order to reduce these emissions, the hose may be equipped with a surrounding tube, which pumps the fumes away. Figure 7.24 depicts an example of a refinery where one could detect VOC fumes leaking out from below the cover of a wastewater tank.

7.6.2
Some Inorganic Compounds

Ammonia is a very common substance used in the chemical industry. Opening a bottle with pure ammonia leads to the escape of the vapor pressure above the liquid. Figure 7.25 depicts this vapor in front of a temperature-stabilized blackbody source at around 41 °C, used as background object. The strong absorption of the vapor in the LW range (spectrum in Appendix 7.A) immediately leads to clearly observable features in the IR images.

Whereas ammonia can be sensed by humans due to its bad smell, sulfur hexafluoride (SF_6) cannot be detected with human senses. Whereas ammonia also shows small absorption features in the MW band, SF_6 – due to its high symmetry – only shows one pronounced and very strong absorption feature in the thermal IR range around 10 µm.

In the following experiments (after [8]), SF_6 was filled into a balloon and its effects on IR signals were studied by positioning it in front of a blackbody source.

7.6 Practical Applications: Gas Detection with Commercial IR Cameras | 421

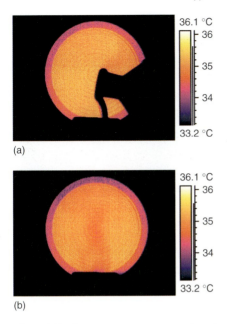

Figure 7.21 Room temperature pure ethanol poured into a dish in front of temperature-stabilized blackbody source (a) and ethanol vapor, above the dish (b)

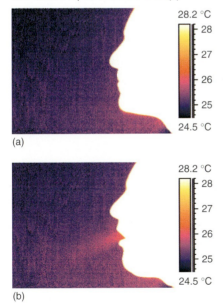

Figure 7.22 Normal breathing and exhaling after drinking water does not have any observable effect on LW IR thermal imaging (a). In contrast, drinking alcohol does lead to alcohol vapor in the breath which is detectable (b).

422 | 7 Industrial Application: Detection of Gases

(a)

(b)

(c)

Figure 7.23 Three snapshots of the fill up of a car tank with gasoline at a regular gas station. Before starting the fill up (a), no fumes can be seen. (b) Immediately after starting to pump gas, fumes are detected. (c) Although being heavier than air, they are driven by air currents and may also be blown toward the customers. (Courtesy Infrared Training Center, FLIR® Systems Inc.)

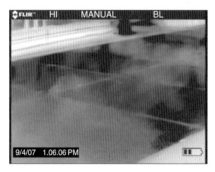

Figure 7.24 Fumes of VOCs leaking through the covers of waste water holding tanks in a refinery. Such tanks shall separate the oil from the water. (Courtesy Infrared Training Center, FLIR® Systems Inc.)

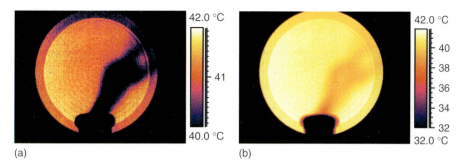

Figure 7.25 Room temperature ammonia vapor in front of warmer object is easily visible with an LW camera not only for a 2-K span (a) but also for a larger 10-K span (b).

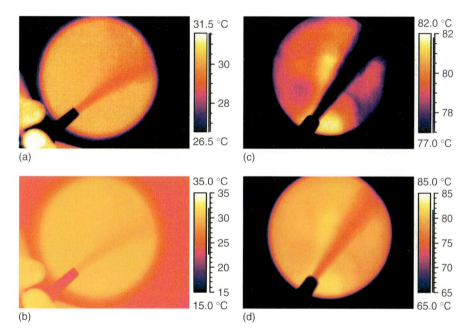

Figure 7.26 Room temperature SF$_6$ in front of a blackbody of 32 °C (a,b) and 82 °C (c,d). Both images were recorded with a broadband LW camera and a 5-K span (a,c) and 20-K span (b,d).

Then a valve was opened and the gas was pressed out of the balloon. Figure 7.26 depicts some examples with room temperature gas (about 24 °C) and varying object temperature for spans of 5 and 20 K. These images show the increase in contrast with increasing temperature difference between the gas and the object. Furthermore, the good signal contrast also for a large span clearly demonstrates that SF$_6$ can easily be detected without carefully adjusting the level and span.

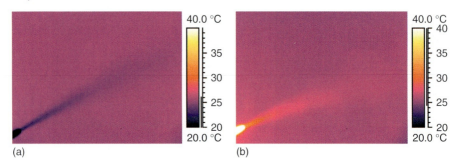

Figure 7.27 Cold (a) and hot (b) SF_6 in front of a room temperature wall.

As a consequence, a new type of LW GasFind camera for SF_6 and species with absorption around 10-μm wavelength was introduced [11].

The gas-filled bags with SF_6 were quite easy to handle and allowed us to study the changes in IR images when varying the gas temperature with regard to the surrounding. This was done by placing the gas balloons in a temperature-stabilized oven for time enough to establish thermal equilibrium. One balloon with SF_6 was cooled down to $-20\,°C$ and a second was heated to $+80\,°C$, that is, one gas bag was much colder than the ambient temperature while the other was much warmer. Figure 7.27 depicts the respective gas flows observed in front of a room temperature wall. These images nicely demonstrate the change from absorption of IR radiation by the cold gas to the thermal emission of IR radiation by the hot gas.

7.6.3
CO_2 – Gas of the Century

The final section of gas detection with IR cameras discusses applications for probably the most important gas of the twenty-first century. Carbon dioxide (CO_2) is a natural gas which has found numerous applications in industries. Besides, it is a by-product of the combustion of carbon-based energy sources. As a result, it has at present the most important contribution to the anthropogenic greenhouse effect [25, 26]. In order to mitigate climate change due to anthropogenic emission of greenhouse gases, many big corporations in the field of energy plan to introduce so-called carbon capture and storage (CCS) technologies. The concept of CCS is briefly summarized in Figure 7.28 (from [27]).

CO_2 from primary sources, for example, large power plants, is captured and transported to several storage sites (mostly in pipelines). Storage can, for example, occur in deep geological formations, oil wells, the deep ocean, or in the form of mineral carbonates.

The introduction of CCS technologies will raise the issue of

- verification that the power plants do indeed no longer emit CO_2;
- verification that pipelines do not leak;

7.6 Practical Applications: Gas Detection with Commercial IR Cameras | 425

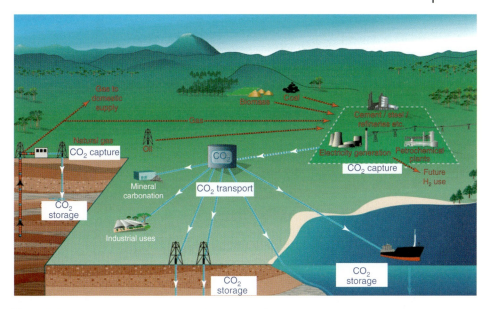

Figure 7.28 CCS systems showing the carbon sources for which CCS might be relevant and options for the transport and storage of CO_2. (Reproduced from [27], Chapter 4, Energy supply, Figure 4.22.)

- verification that underground storage is safe, that is, that there are no leakages.

For all of these verification issues, analytical tools are needed which easily allow monitoring of pipelines or sites by sensitive detection systems for gaseous CO_2.

As is demonstrated by the examples below, IR thermal imaging provides an excellent tool for this purpose [12, 13, 28]. As an imaging device it offers in particular the possibility of fast leak detection in large areas. Once leaks are localized, the method may be combined with other conventional and very sensitive spot measurement techniques for CO_2. We also note that recently, a passive technique was developed to detect large amounts of CO_2 degassing from the earth surface by secondary thermal effects [29].

In the following, a number of different thermal imaging experiments with CO_2 are presented [12, 13]. These include comparison of broadband detection versus narrowband detection in the MW range, detection of low concentrations of the order of 500 ppm over short distances, detection of CO_2 in exhaled air and in combustion processes, as well as visualization of absorption versus scattering and emission of radiation by gases. Most importantly, detection of well-defined gas flows, simulating gas leaks, is presented, and it is demonstrated that CO_2 gas flows as small as $1\,\mathrm{ml\,min^{-1}}$ corresponding to CO_2 masses of about 1 kg/year may be detected.

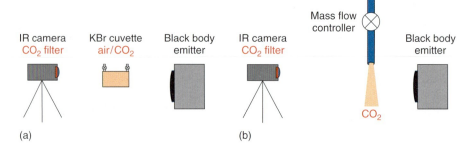

Figure 7.29 Experimental setup for experiments. An IR camera with or without filter detects radiation changes from a background object of known temperature and emissivity.

The principle of two setups are shown in Figure 7.29. MW IR cameras were used to observe a blackbody emitter which served as background of well-defined temperature and emissivity. In some cases, the blackbody emitter was replaced by a large-area hot plate, which was not as homogeneous, but still nicely served as background object. We used both a broadband THV 550 Agema camera (3–5 µm range/PtSi detector/320 × 240 pixels) as well as a broadband FLIR® SC 6000 camera (1.5–5 µm range/InSb detector/640 × 512 pixels). The latter allowed us to insert commercially available room temperature narrowband spectral filters adjusted to the absorption feature of CO_2 gas (here Spectrogon Filter 4235, [14]).

Carbon dioxide was introduced between the camera and the blackbody source either in cells of well-defined length at well-defined pressures (a) or while flowing out of a tube of well-defined diameter with a known flux, defined by mass flow controllers (b), or it could be emitted in unknown concentrations and gas flows due to a variety of processes.

7.6.3.1 Comparison between Broadband and Narrowband Detection

Although broadband MW cameras may detect CO_2 by its absorption [8, 13], it is customary to use narrowband filters. Figure 7.30 depicts a comparison between IR images of a broadband MW camera (3–5 µm) and a narrowband camera (SC6000 with warm filter 4235). Both cameras observe the same CO_2 flow in front of a blackbody source, operated at around 50 °C.

Whereas the broadband camera hardly detects this small CO_2 flow of only 100 ml min^{-1}, the narrowband detection considerably improves the sensitivity of the camera. In addition, the InSb detector of the SC6000 camera is more sensitive than the PtSi detector of the THV 550 camera. Reducing the gas flow allows us to estimate the differences in sensitivity. In the present case for the given conditions ($T_{BB} = 50$ °C, tube inner diameter 6 mm), the difference in the lowest detectable gas flows was around a factor of 20–30, that is, the narrowband camera system allowed us to detect gas flows which were a factor of 20–30

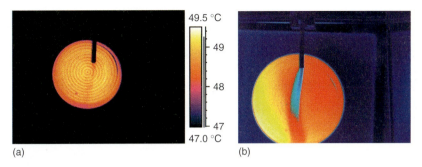

Figure 7.30 Detection of the same flow of CO_2 gas (100 ml min^{-1}, tube inner diameter 6 mm) with a broadband MW camera (THV 550, (a)) and a narrowband camera (SC6000 plus filter, (b)).

smaller than the ones observed with a broadband camera (detection limits, see below). For the practitioner, we note that the smallest detectable gas flows will always be seen as signal changes in life image sequences. This is due to the fact that the image processing within the brain is in general very sensitive to movements in our field of view, that is, image changes as a function of time. In still images, it is much harder to detect such small variations of gas concentrations. All of the following CO_2 experiments were recorded with the narrowband camera system.

7.6.3.2 Detecting Volume Concentration of CO_2 in Exhaled Air

Being the "exhaust gas" of human breathing, CO_2 is directly related to human energy production upon oxygen intake. The average CO_2 concentration in the air, that is, the concentration while inhaling air, is about 380 ppm (in the open air) and around 400–800 ppm in closed rooms (depending on the size of the room, number of people in the room as function of time, ventilation of the room, etc.). The increase of CO_2 concentration in closed rooms is due to breathing. Typical carbon dioxide concentration in exhaled air is around 4–5 vol% (40 000–50 000 ppm), that is, around a factor of 100 larger than typical concentrations in fresh air.

Figure 7.31 depicts qualitative examples for detection of CO_2 concentrations upon exhaling air. Exhaling may occur through the nose (opening diameter around 1 cm, (a)) or through the mouth (opening diameter 4–5 cm, (b)). In both cases, the absorption of background IR radiation due to the colder gas is easily visible.

7.6.3.3 Absorption, Scattering, and Thermal Emission of IR Radiation

Figure 7.32 depicts the influence on scattering of hot objects nearby on IR gas imaging. The background blackbody source was set to 80 °C, the CO_2 gas was at room temperature (about 20 °C), and as additional radiation source we used a soldering iron tip at 400 °C, which could be freely moved in the region close to the gas flow. The orientation of camera, gas tube, and blackbody source was not changed.

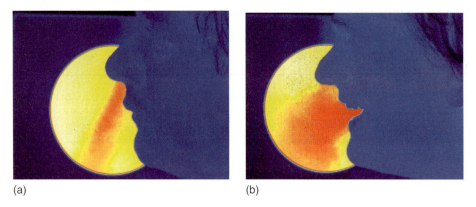

Figure 7.31 The authors (K.-P.M. (a), M.V. (b)) while exhaling through the nose and (while laughing) through the mouth in front of a 50 °C blackbody source.

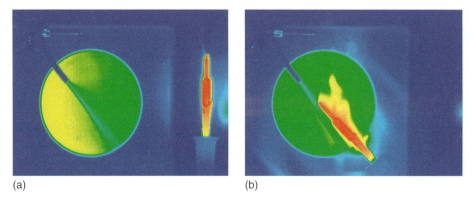

Figure 7.32 Room temperature CO_2 gas flow at 700 ml min^{-1} in front of black body source set to 80 °C. (a) The cold gas leads to attenuation of IR radiation. (b) Scattering of IR radiation from a hot soldering iron at 400 °C, allows simultaneous observation of gas absorption and scattering. The soldering iron was placed several centimeters in front of the gas flow.

In (a), the cold gas (blue) clearly attenuated the IR radiation from the blackbody source (yellow–green). If the additional hot object, the soldering iron (red), is far away (a), scattering contributions from CO_2 are too small to be detected. However, moving the iron tip to a distance of several centimeters from the outflow of cold gas (b), one immediately observes gas features of apparently higher temperature due to the scattering of IR radiation.

In principle, it is also possible to have a small amount of thermal emission from the heated gas close to the soldering iron tip and in practice it may become difficult to separate the scattering from the thermal emission contribution of the gas. In order to estimate the amount of thermal emission, one needs to calculate its emissivity and its optical thickness (Section 7.5). Qualitatively, thermal emission

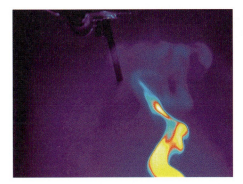

Figure 7.33 CO_2 flow from tube in front of slightly warmer background as well as hot CO_2 combustion product of a lighter flame (the visual flame top was about 10 cm below the lower edge of the IR image).

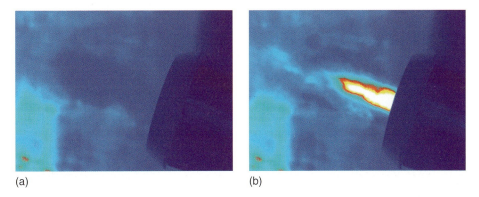

Figure 7.34 Warm (a) and hot (b) CO_2 exhaust gas from combustion in a motorcycle engine. The background was at a temperature of about 80 °C. (Use of motorcycle, thanks to F. Pinno.)

may be observed isolated by studying combustion processes which produce hot CO_2. In the laboratory setup, a cigarette lighter was placed below the blackbody source. It was observed with the IR camera such that the visually observable flame top was about 10 cm below the lower edge of the IR image. Figure 7.33 shows an example: hot CO_2 gas was rising and – while still being hot – emitting IR radiation. Besides, one can still observe the colder gas flowing out of the tube.

An even more impressing visualization of CO_2 combustion products is obtained by studying the exhaust of combustion engines. Figure 7.34b depicts the exhaust pipe of a motorcycle (Suzuki 1100 GSXF, 98 hp) in front of a heating plate which had an inhomogeneous temperature distribution with an average temperature of about 80 °C.

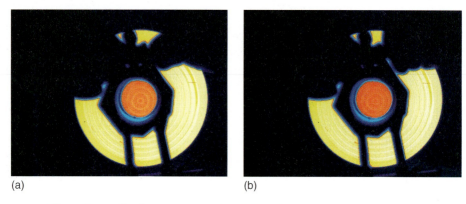

Figure 7.35 Cell with vacuum (a) and filled (b) with room temperature air (CO_2 concentration 400–800 ppm).

Figure 7.34a shows CO_2 absorption in the exhaust after starting the engine and running it at low power. Obviously, the exhaust gases had enough time to cool down within the exhaust pipe. Therefore, only some absorption of IR radiation in front of the warm background was observed. However, while the engine was operated (for a short time) at full power, rather hot exhaust gases were emitted, which showed up in the IR image (b).

7.6.3.4 Quantitative Result: Detecting Minute Amounts of CO_2 in Air

For quantitative measurements of CO_2, we used a cell of 10 cm in length with KBr windows of about 95% transmission. It can be filled with any desired partial pressure. At first, the cell was evacuated and observed while ambient air was streaming in with typical CO_2 volume concentrations in closed rooms of around 400–800 ppm. Figure 7.35 shows the following result: the change in CO_2 content is due to the 10-cm cell while the total measurement distance was about 1 m. The observed change means that we are able to detect minute changes in CO_2 concentrations of 40–80 ppm over measurement distances of 1 m.

The high sensitivity of the measurements even allowed us to detect the additional attenuation due to pressure broadening. Figure 7.36 compares signals of the same partial pressure of CO_2 (51 hPa) for two different total pressures of 51 hPa (pure CO_2) and 1029 hPa. Higher atmospheric pressure results in pressure broadening, which itself leads to larger attenuation of the IR radiation.

7.6.3.5 Quantitative Result: Detection of Well-Defined CO_2 Gas Flows from a Tube

The most important application of any GasFind camera should be its leak detection capability. Figure 7.37 depicts results for varying CO_2 flow in front of a blackbody source operating at around 50 °C. The CO_2 flow (100 vol%) was adjusted using a mass flow controller, the gas was exiting a tube of 6-mm inner diameter several centimeters from the source.

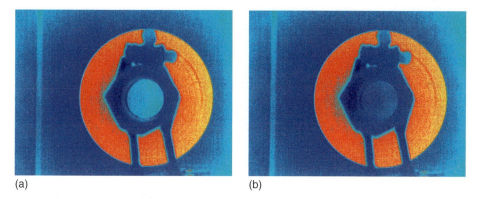

Figure 7.36 Pressure broadening in CO_2 spectra: cell of 10-cm length with KBr windows in front of blackbody source ($T = 35$ °C). The CO_2 partial pressure was 51 hPa in both cases with total pressures of 51 hPa (a) and 1029 hPa (b).

The sequence in Figure 7.37 shows results for mass flows between 5 and 1000 ml min^{-1}. Despite the large density of CO_2 compared to air, small flow rates lead to fumes which are easily driven by currents of the surrounding air. Flow rates between 100 and 500 ml min^{-1} result in nearly laminar flow. For flow rates above 1000 ml min^{-1}, turbulences develop.

The main question concerns the lower limit of detectable gas flow. It depends on a combination of background signal and chosen integration time for the detector (most commercial IR cameras do not allow the user to select integration time). In order to more closely resemble typical outdoor situations, experiments were carried out for blackbody background temperatures only slightly above room temperature ($T_{BB} = 35$ °C). Owing to the corresponding lower radiation signal (with respect to a 50 °C blackbody) longer integration times could be used, yielding a better signal-to-noise ratio. So far, gas flows as low as 1 ml min^{-1} could be detected. In still images, such a small gas flow is sometimes hard to detect; it is easier to observe changes in life images.

The laboratory results suggest that CO_2 with IR cameras is suitable for analyzing gas leaks. The lowest detected gas flows so far of 1 ml min^{-1} would correspond of 0.06 l h^{-1}, 1.44 l/day, or $\approx$0.5 m^3/year. Using the density of CO_2 (about 2 kg m^{-3}), it is therefore possible to detect emissions of as low as about 1 kg/year from individual leaks. Even for much less favorable conditions in industry, lowest detectable gas flows of say 100 ml min^{-1}, that is, $\approx$100 kg/year should be readily observable. Once leaks are localized, the method may be combined with other more sensitive CO_2 spot measurement techniques.

As outlook for future developments, we mention another possibility to increase sensitivity for such gas-sensitive cameras. As is well known from other spectroscopic techniques, it would make sense to use two warm filters to detect the object signal

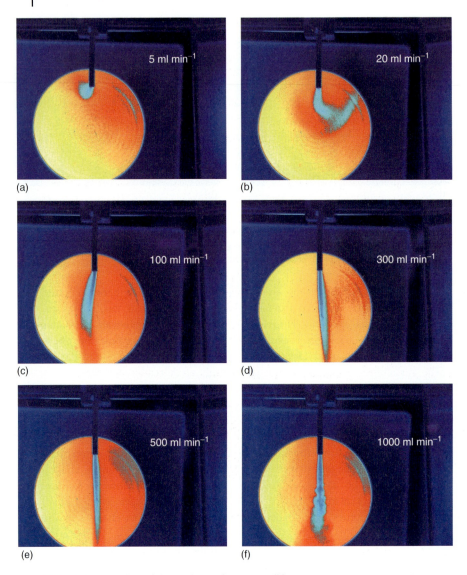

Figure 7.37 Raw data of detected CO_2 flows (in milliliters per minute): 5 (a), 20 (b), 100 (c), 300 (d), 500 (e), and 1000 (f). The SC 6000 camera was operated with an uncooled narrow band filter in front of a black body emitter operated at $T = 50\ °C$.

through the gas, one being off resonance (no absorption) and the other one on resonance (absorption). The ratio of these two signals directly reflects the gas absorption very sensitively. Inserting warm filters in IR cameras should pose no problems. It is already common practice when changing the temperature ranges of IR cameras.

Appendix 7.A: Survey of Transmission Spectra of Various Gases

Here a qualitative survey of spectra of important VOCs and inorganic molecules is given spectra from own calculations own measurements and after data from [2]. They were recorded with a variety of methods (prism and grating spectrometers as well as FTIR spectrometers) by many different groups. Although for many of these spectra, recording conditions would allow quantitative description, we only show qualitative spectra, that is, conditions like gas pressure and optical path length differ from sample to sample such that the transmission at the center of the absorption bands will be similar. More details can be found, for example, on the web [2]. We note that spectra are usually recorded as a function of wavenumber, which is proportional to the frequency or energy of the radiation. Here, we present the data as a function of wavelength since the various camera systems are usually characterized in terms of wavelength ranges. Table 7.A.1, reproduced here for completeness, summarizes all substances whose spectra can be found in this appendix.

Table 7.A.1 Some natural and industrial gases including hydrocarbons and other organic compounds which absorb in the thermal IR region.

Inorganic compounds	Simple hydrocarbons	Simple multiple bond compounds	Benzene compounds
H_2O water vapor	CH_4 methane	C_2H_2 ethylene	C_6H_6 benzene
CO_2 carbon dioxide	C_2H_6 ethane	C_3H_6 propylene	C_7H_8 toluene
CO carbon monoxide	C_3H_8 propane	C_4H_6 butadiene	C_8H_{10} p-xylene
NO nitric oxide	C_4H_{10} butane	**Some alcohols**	C_8H_{10} ethylbenzene
N_2O nitrous oxide	C_4H_{10} isobutane	CH_3OH methanol	$C_{10}H_8$ naphthalene
NH_3 ammonia	C_5H_{12} pentane	C_2H_5OH ethanol	**Hydrocarbons with halogens**
O_3 ozone	C_6H_{14} hexane	C_3H_7OH propanol	CCl_2F_2 Freon 12 (CFC12)
H_2S hydrogen sulfide	C_8H_{18} octane	**Ketones/ethers**	$CHClF_2$ Freon 22 (CFC22)
SO_2 sulfur dioxide	C_8H_{18} isooctane	C_3H_6O acetone	CHF_3 Freon 23 (CFC23)
CS_2 carbon disulfide	$C_{12}H_{26}$ dodecane	C_4H_8O MEK	$CHCl_3$ chloroform
SF_6 sulfur hexafluoride	$C_{16}H_{34}$ hexadecane	$C_6H_{12}O$ MIBK	CCl_4 carbon tetrachloride
–	–	$C_5H_{12}O$ MTBE	$COCl_2$ carbonyl chloride

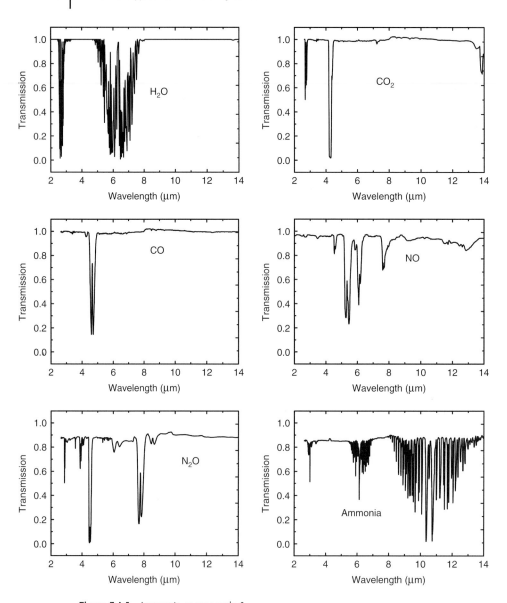

Figure 7.A.1 Inorganic compounds 1.

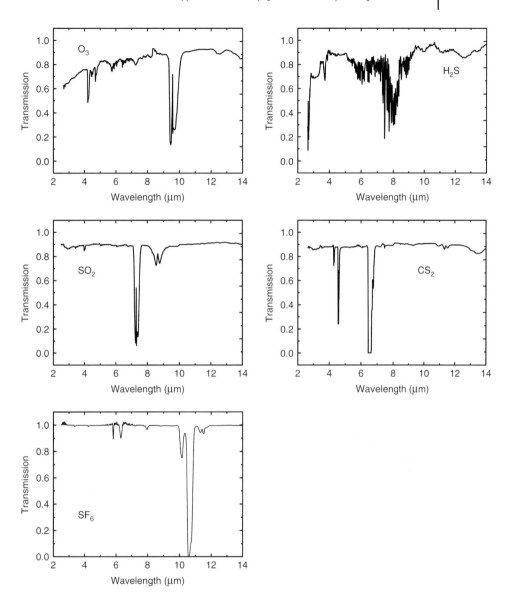

Figure 7.A.2 Inorganic compounds 2.

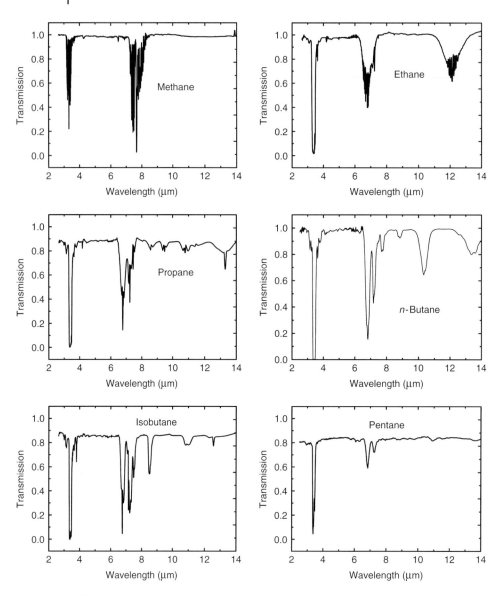

Figure 7.A.3 Simple hydrocarbons 1.

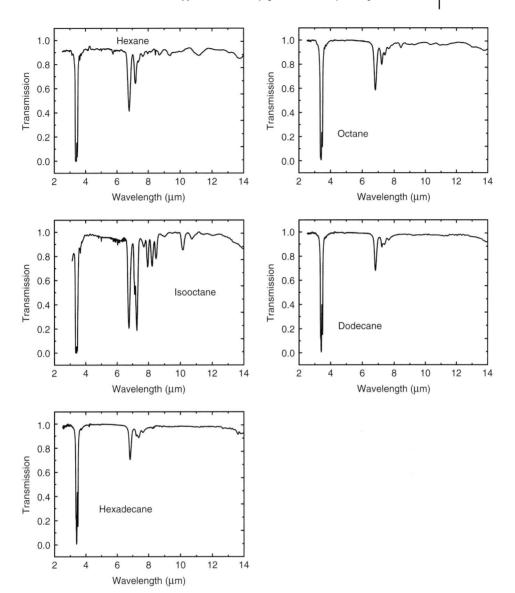

Figure 7.A.4 Simple hydrocarbons 2.

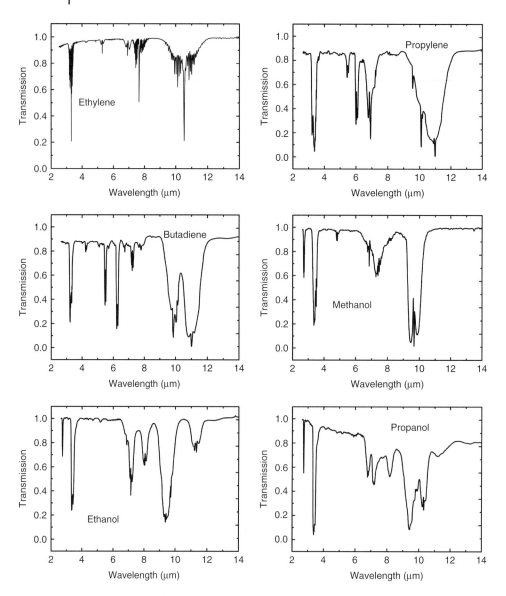

Figure 7.A.5 Simple multiple bond compounds and some alcohols.

Appendix 7.A: Survey of Transmission Spectra of Various Gases | 439

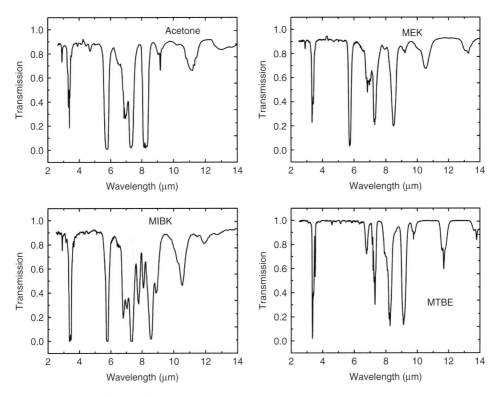

Figure 7.A.6 Some ketones/ethers.

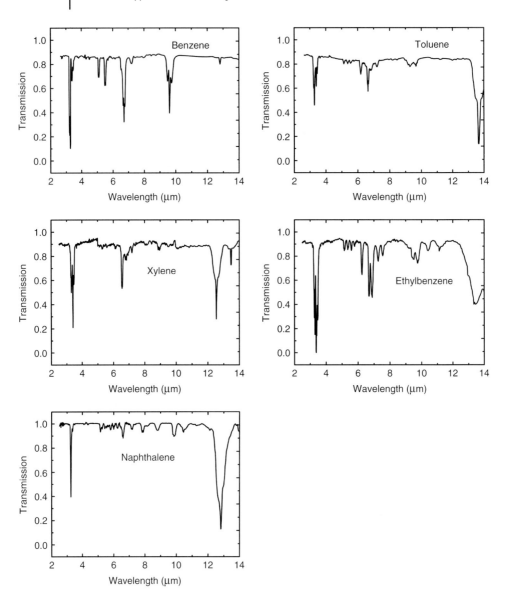

Figure 7.A.7 Some benzene compounds.

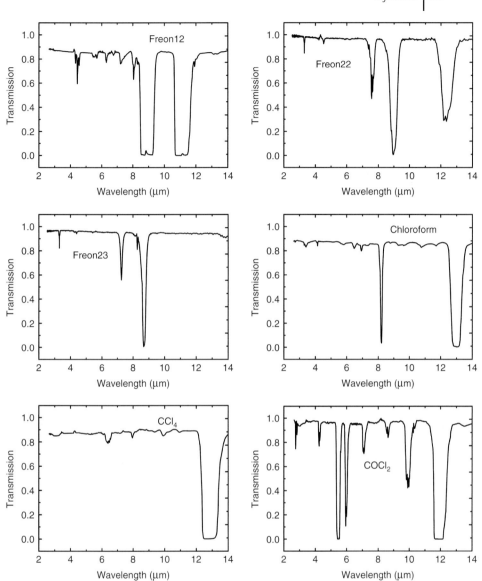

Figure 7.A.8 Some hydrocarbons with halogens.

References

1. Günzler, H. and Gremlich, H.-U. (2002) *IR Spectroscopy – An Introduction*, Wiley-VCH Verlag GmbH.

2. Database of IR Spectra and Properties of Chemical Compounds. http://webbook.nist.gov/chemistry/form-ser.html. (2010)

3. Socrates, G. (2001) *Infrared and Raman Characteristic Group Frequencies, Tables and Charts*, 3rd edn, John Wiley & Sons, Inc.
4. Bell, R.J. (1972) *Introductory Fourier Transform Spectroscopy*, Academic Press, New York.
5. Kauppinen, J. and Partanen, J. (2001) *Fourier Transform in Spectroscopy*, Wiley-VCH Verlag GmbH, Berlin.
6. Richards, A. and Johnson, G. (2005) Radiometric calibration of infrared cameras accounting for atmospheric path effects. Thermosense XXVII, Proceedings of SPIE vol. 5782, pp. 19–28.
7. Strachan, D.C. et al. (1985) Imaging of hydrocarbon vapours and gases by infrared thermography. *J. Phys. E: Sci. Instrum.*, **18**, 492–498.
8. Karstädt, D., Möllmann, K.P., Pinno, F., and Vollmer, M. (2006) Influence of gaseous species on thermal infrared imaging. Inframation 2006, Proceedings vol. 7, pp. 65–78.
9. http://www.flir.com/. (2010).
10. Furry, D., Richards, A., Lucier, R., and Madding, R.P. (2005) Detection of Volatile Organic Compounds (VOC's) with a spectrally filtered cooled mid-wave infrared camera. Inframation 2005, Proceedings, vol. 6, pp. 213–218.
11. Madding, R. and Benson, R. (2007) Sulphur Hexafluoride (SF6) insulating gas leak detection with an IR imaging camera. Inframation 2007, Proceedings vol. 8, pp. 89–94.
12. Vollmer, M. and Möllmann, K.-P. (2009) Perspectives of IR imaging for industrial detection and monitoring of CO_2. Proceedings of the Conference Temperature PTB, Berlin. ISBN: 3-9810021-9-9.
13. Vollmer, M. and Möllmann, K.-P. (2009) IR imaging of gases: potential applications for CO_2 cameras. Inframation 2009, Proceedings vol. 10, pp. 99–112.
14. www.spectrogon.com. (2010).
15. Vollmer, M. and Möllmann, K.-P. (2009) IR imaging of gases: quantitative analysis. Inframation 2009, Proceedings vol. 10, pp. 113–123.
16. Benson, R.G., Panek, J.A., and Drayton, P. (2008) Direct measurements of minimum detectable vapor concentration using passive infrared optical imaging systems. Paper 1025 of Proceedings 101st ACE meeting (held June 2008 in Portland/Oregon) of the Air & Waste and Management Association.
17. Edwards, D.K. (1960) Absorption by infrared bands of carbon dioxide gas at elevated pressures and temperatures. *J. Opt. Soc. Am.*, **50**, 617–626.
18. Burch, D.E. et al. (1962) Total absorptance of CO_2 in the infrared. *Appl. Opt.*, **1**, 759–765.
19. Lapp, M. (1960) Emissivity calculations for CO_2. part 1 of PhD thesis, Cal Technology.
20. Incropera, F.O. and DeWitt, D.P. (1996) *Fundamentals of Heat and Mass Transfer*, 4th edn, John Wiley & Sons, Inc.
21. Schack, K. (1970) Berechnung der Strahlung von Wasserdampf und Kohlendioxid. *Chem.-Ing.-Tech.*, **42** (2), 53–58.
22. DeWitt, D.P. and Nutter, G.D. (1988) *Theory and Practice of Radiation Thermometry*, John Wiley & Sons, Inc., New York.
23. (a) Baehr, H.D. and Stephan, K. (2006) *Wärme – und Stoffübertragung*, 5th German edn, Springer-Verlag; (b) Baehr, H.D. and Stephan, K. (2006) *Heat and Mass Transfer*, 2nd English edn, Springer, Berlin.
24. (a) Vollmer, M. (2005) Hot gases: transition from line spectra to thermal radiation. *Am. J. Phys.*, **73**, 215–223; see also comment by (b) Nauenberg, M. (2007) *Am. J. Phys.*, **75**, 947; reply to comment (2007) *Am. J. Phys.*, **75**, 949.
25. Houghton, J.T. et al. (eds) (2001) *Climate Change 2001, The Scientific Basis, Contribution of Working Group I to 3rd Assessment Report of the Intergovernmental Panel on Climate Change*, IPCC.
26. Solomon, S. et al. (eds) (2007) *Climate Change 2007, The Physical Science Basis, Contribution of Working Group I to 4th Assessment Report of the Intergovernmental Panel on Climate Change*, IPCC.
27. Metz, B. et al. (eds) (2007) *Climate Change 2007, Mitigation of Climate Change, Contribution of Working Group*

III to 4th Assessment Report of the Intergovernmental Panel on Climate Change, IPCC.

28. Yoon, H.W. *et al.* (2006) Flow visualization of heated CO_2 gas using thermal imaging. Thermosense XXVIII, Proceedings of SPIE vol 6205, p. 62050U-1.

29. Tank, V., Pfanz, H., and Kick, H. (2008) New remote sensing techniques for the detection and quantification of earth surface CO_2 degassing. *J. Volcanol. Geotherm. Res.*, **177**, 515–524.

8
Microsystems

8.1
Introduction

Microsystems engineering is likely to be a key technology in the twenty-first century. Microsystems promise to revolutionize nearly every product category. Micro-electro-mechanical systems (MEMSs) are the integration of mechanical, optical, and fluidic elements, sensors, actuators, and electronics through microfabrication technologies. MEMS are made up of components with a characteristic size in the micrometer region. They are also referred to as *micromachines* (in Japan) or *microsystems technology* or MST (in Europe). This is an enabling technology that allows the development of smart products, augmenting the computational ability of microelectronics with the perception and control capabilities of microsensors and microactuators and expanding the range of possible applications. Upon production, every technical product must be characterized such that its properties upon variation of relevant parameters are known.

The concepts of measuring and testing technology known from, for example, microelectronics or precision mechanics can however only be partially applied to microsystems because of the small size and the variety of physical principles used in the microsystem operation. Therefore new testing methods have to be developed to characterize the operation and important performance parameters of microsystems.

For a lot of microsystems, a uniform temperature or a spatial temperature distribution as well as thermal response time are among the most important parameters. Any contact probe for temperature measurement would induce appreciable thermal losses due to thermal conduction which either complicates the analysis or even makes it impossible. Therefore a contactless measurement is needed to analyze these parameters for microsystem components (reactors, sensors, actuators). Obviously, thermography can play an important role for the thermal characterization of different microsystems, to support the research and development work, and to control the system operation without modifying the system performance [1, 2].

In the following, exemplary results are presented for microfluidic systems, sensors, as well as microsystems which utilize an electric-to-thermal energy conversion during operation.

Infrared Thermal Imaging. Michael Vollmer and Klaus-Peter Möllmann
Copyright © 2010 WILEY-VCH Verlag GmbH & Co. KGaA, Weinheim
ISBN: 978-3-527-40717-0

8.2
Special Requirements for Thermal Imaging

The thermography of microsystems poses a number of problems which are usually not encountered when studying macroscopic objects. The specific requirements concern suppression of mechanical instabilities and vibrations, the need for close-up lenses or microscope objectives, and the possibility of high-speed recording.

8.2.1
Mechanical Stability of Setup

Owing to the small size of the microstructures, a very stable mechanical construction of a test bench is necessary to avoid effects of mechanical instabilities or mechanical vibrations on the thermal imaging results. A breadboard or, even better, a vibration insulated optical table with adjustable sample and camera holders is well suited for the microscopic thermal imaging analysis (Figure 8.1).

8.2.2
Microscope Objectives, Close-up Lenses, Extender Rings

Usually a spatial resolution of thermal imaging much better than 1 mm is necessary although in practice the requirement depends on the microstructure size. IR cameras with a standard objective allowing spatial resolutions around 1 mm must be equipped with an additional close-up lens to increase the spatial resolution (Section 2.4.4). For cameras with exchangeable lenses also extender rings can be used (Section 2.4.4.4). The maximum spatial resolution is achieved by using microscope optics. Figure 8.2 demonstrates the improvement of the optical resolution for a miniaturized thermal emitter in a transistor housing using the different additional components. The use of additional optical components for increasing the spatial resolution of the imaging is accompanied by a decreasing working distance (camera objective to object). Therefore the influence of the Narcissus effect for nonblack objects has to be considered [3] (Section 2.4.4.5).

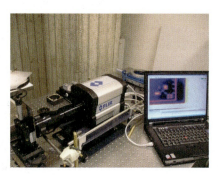

Figure 8.1 Experimental setup for microscopic thermal imaging using an optical table.

8.2 Special Requirements for Thermal Imaging | 447

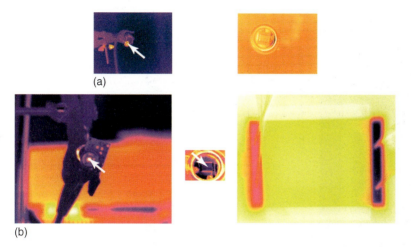

Figure 8.2 Thermograms of a miniaturized infrared emitter (size 2.1 mm × 1.8 mm) in TO-39 housing. Region of interest as indicated by arrows in left and middle thermograms is shown expanded in the right images. (a) MW camera FLIR THV 550 (320 × 240 pixels) with a 24 optics (left) and an additional close-up lens (right). (b) MW camera FLIR SC6000 (640 × 512 pixels) with a 25-mm lens (left), additional extender ring (middle with 160 × 120 pixels), and microscope optics (right).

Before analyzing temperatures at microscopic structures, the spatial resolution of the equipment used needs to be determined to avoid temperature measurement errors. For the determination of the spatial resolution, chromium structures with a well-defined size on a photolithography mask with a glass substrate can, for example, be used. The mask should be heated and the emissivity contrast of the mask can be used for the measurement. Figure 8.3 depicts a measurement across a 34-µm line on the mask. Compared to the visible microscope image (Figure 8.3a, top) the thermogram appears blurred (Figure 8.3a, bottom). Raw signal data across the line were analyzed and a signal plateau was formed by about 6 pixels (Figure 8.3b). If this number of the pixels is compared with the linewidth of 34 µm, one may conclude that for this camera equipment resolution per pixel is around 5–6 µm.

For a more detailed analysis the MTF should be analyzed as described in Section 2.5.4.

8.2.3
High-Speed Recording

Thermal processes such as heat transfer and temperature changes in microsystems are mostly characterized by low time constants in the milliseconds to microseconds range due to the low heat capacitance or thermal mass. For time-resolved thermal imaging, high-speed data acquisition is necessary. The limitation of the time resolution is given by the response time of the IR camera. As described

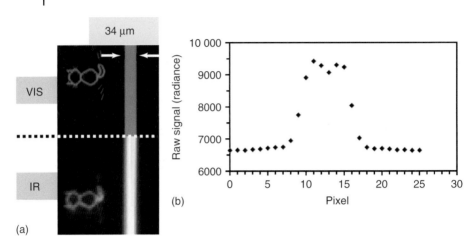

Figure 8.3 Vis microscope image (a) (top) and thermogram (bottom) of a line with a width of 34 μm on a chromium mask. Raw signal profile of the line measured in the thermogram (b) using the SC6000 camera.

in Section 2.5.5, for cameras with photon detectors the sometimes selectable integration time (microseconds to milliseconds) and for cameras with thermal detectors the detector time constant itself (some milliseconds) determine the camera response time. Accurate temperature values and time-dependent temperature changes for transient heat processes can only be measured if the response time of the camera is much smaller than the time constant of the microsystem. Therefore, cameras equipped with bolometer FPA are suitable only to a limited extent for thermal microsystem investigation.

Very accurate temperature measurements by radiation thermometry require an accurate knowledge of the object emissivity. Emissivity determination at microsystems is a complex problem.

Microsystems are made from a large variety of materials which are commonly not well suited for thermal imaging due to the emissivity properties such as, for example, high reflecting materials (metals), semitransparent materials (glass or silicon), or selective emitting materials (plastics). The surface modification by additional colors or emissivity strips as used in standard thermal imaging applications fails because this would modify the physical properties of the microsystems dramatically. If the determination of absolute temperature values is necessary, the emissivity can be determined by tempering the microsystem to known temperature using a climatic exposure test cabinet. The emissivity adjustment at the camera is changed to measure the known object temperature. The object emissivity is given by the correct emissivity adjustment at the camera. The emissivity can also be estimated by infrared spectroscopic measurements using an IR microscope [2].

8.3
Microfluidic Systems

8.3.1
Microreactors

Microreactors are devices in which chemical reactions take place in microchannels with typical lateral dimensions below 1 mm. The microreactors are normally operated continuously and offer many benefits compared to conventional-scale chemical reactors [4] such as large heat exchange coefficients due to large surface-to-volume ratios allowing fast and precise cooling or heating and temperature stabilization, which are important for strong exothermic and dangerous chemical reactions.

Microreactors are fabricated using metals, glass, silicon, or ceramics.

During the design process of a microreactor, first a theoretical model is established considering a lot of parameters such as, for example, the chemistry of the reaction process and the fluidic properties of the reactants.

Owing to the fact that chemical reactions are connected with heat consumption or heat production, the experimental analysis of the reactor temperature, the temperature distribution, and the time-dependent temperature behavior along the reaction channels represents an important tool for the reactor control. It can be used to verify and to improve the design models. For this purpose, exemplary chemical reactions with a limited exothermal behavior are used to characterize the microreactor during operation. The analysis of time and spatially resolved temperatures, for example, by thermal imaging allows to find optimum operation parameters and can help to optimize reactor construction.

Analyzing reaction products without a detailed analysis of reactor temperatures and temperature distribution is usually insufficient to find the optimum internal microreactor structure as well as the optimum operation parameters. Hence the following examples will demonstrate the potential of thermal imaging in microreactor technologies.

8.3.1.1 Stainless Steel Falling Film Microreactor

The design and the operational principle of the falling film reactor are shown in Figure 8.4. It can be used, for example, for gas–liquid reactions [5]. Figure 8.4b illustrates the principle to generate thin films in the micrometer range on a reaction plate with microchannels. The fluid is distributed on the reaction plate and flows downward to the withdrawal zone at the bottom as a fluid film. While the flow direction of the liquid reactant corresponds to the direction of gravity, the gaseous phase can be guided either parallel or antiparallel to the liquid phase.

In order to detect the zone of the chemical reaction and to measure fluid equipartition and temperature profiles at the fluid film on the reaction plate, an FLIR® Thermacam SC 2000 IR camera with a 24° lens was used. To enable thermal process control, the microreactor was equipped with an IR-transparent

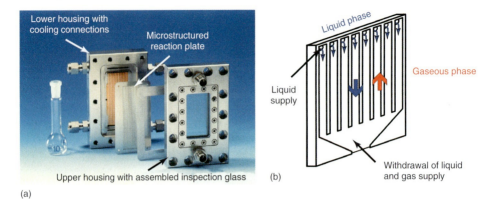

Figure 8.4 (a) Components of a falling film microreactor (total height is 10 cm) and (b) a microstructured reaction plate of falling film reactor (channel width and depth 300 µm × 100 µm or 1200 µm × 400 µm). (Image courtesy for the Vis image: IMM, Mainz, Germany [6].)

inspection window (626-µm-thick Si wafer). In addition, the use of a close-up lens allowed measurements at a high spatial resolution (about 200 µm) sufficient to reveal details smaller than the characteristic dimensions of the microreactor.

One of the most important preconditions for an optimum operation of the reactor is the fluid equipartition in the reaction channels. Mostly, the reaction fluids are transparent in the visible but opaque in the infrared spectral region. Therefore, the homogeneity of the microchannels can be tested by studying the wetting behavior of the reaction plate using isopropanol under nonreacting conditions by thermal imaging in the LW infrared region utilizing the emissivity contrast between the fluid and the reaction plate material. The isopropanol film was heated above room temperature to 30 °C by an integrated micro heat exchanger. The emissivity of the microstructured reaction plate manufactured in high-alloy stainless steel ($\varepsilon \approx 0.5$) differs strongly from that of isopropanol ($\varepsilon \approx 0.9$) in the LW infrared region. Therefore areas wetted with the preheated isopropanol show a bright white color (i.e., larger signal), whereas the dry areas emit less radiation and are indicated in black. This effect was used for easily imaging the fluid equipartition in parallel channels. In Figure 8.5, a sequence of the initial wetting behavior at a volume flow of 250 ml h^{-1} isopropanol is shown for a reaction plate with 15 channels at different times. It is obvious that a very uniform residence time distribution was achieved, since the fluid front has moved the same distance in all microchannels.

Heat management plays an important role for many reactions since most reaction rates strongly depend on the temperature. Thermal imaging allows to record a temperature profile across the whole reaction plate with a single measurement at high spatial resolution sufficient to reveal details smaller than the characteristic dimensions. Figure 8.6a shows an IR image of the temperature distribution of

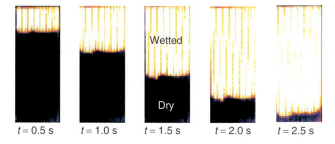

Figure 8.5 Wetting behavior of a reaction plate with channels 1200 × 400 μm² loaded with a volume flow of 250 ml h⁻¹ isopropanol and heated to 31 °C as a function of time.

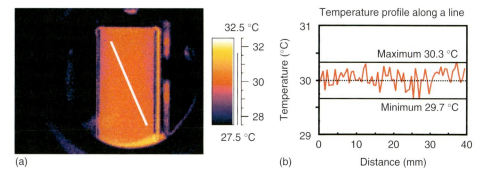

Figure 8.6 (a,b) IR image with a temperature profile of a heated reaction plate, wetted with isopropanol.

the reaction plate, wetted with isopropanol under nonreacting conditions while Figure 8.6b reveals a temperature plot along a line depicted in IR image, extracted from the corresponding data of the IR image, crossing several microchannels. Apart from reflection of environmental radiation at the bottom of the reaction plate, the temperature profile is very homogeneous: The maximum deviation from the set value of 30 °C amounts to ±0.3 °C along the line, and for the whole reaction plate of 27 mm × 65 mm area it amounts to about ±0.5 °C.

The temperature increase and the spatial temperature distribution of the liquid phase during reactor operation can be monitored by thermal imaging analysis of an exothermic test reaction with a gaseous reactant [7, 8]. The investigations were performed in a batchwise manner: the reaction plate was loaded with a defined liquid flow of NaOH under stationary conditions and an over-stoichiometric amount of pure CO_2 was fed into the gas chamber in counterflow to the falling film.

The molar concentration of the NaOH solution was 2.0 M, the liquid load of NaOH was 250 ml h⁻¹, and the reaction plate was heated to 25 °C by the integrated heat exchanger. In Figure 8.7, a sequence of IR images is given which

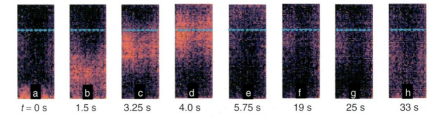

Figure 8.7 Heat release of CO_2 absorption into a 2.0 M solution of NaOH. CO_2 was fed into the chamber from 0 to 6 s. The reaction plate with channels of 1200×400 μm² was loaded with 250 ml h^{-1} and the temperature was set to 25 °C. The IR scale corresponds to $\Delta T = 2$ K from blue to red.

illustrates the heat release of a 2.0 M NaOH solution after opening a valve for CO_2 supply.

The varying position of the reaction zone within the series of images can be explained as follows: As CO_2, which is heavier than air, is fed in from the bottom into the gas chamber, the reaction front rises as the CO_2 level increases in the gas chamber (Figure 8.7a–d). Complete filling of the gas chamber with CO_2 was identified when the reaction front is leaving the observable area of the reaction plate and the reaction is quenched at the bottom (Figure 8.7e). At $t = 5.75$ s the gas chamber is filled with an excess of CO_2. The reaction now only takes place at the entrance port of NaOH. Then at $t = 6$ s the CO_2 supply is stopped. As the CO_2 level in the chamber decreases the reaction front starts moving again downward (Figure 8.7f,g), thereby consuming up the remaining CO_2 in the chamber. Whereas the streaming in of the CO_2 at the beginning of the experiment is a fast process, the consuming up is a slow process of diffusion showing a broad reaction zone with a small temperature peak.

The CO_2 absorption into a 2.0 M solution of NaOH results in a significant temperature increase of about 1.5 °C (Figure 8.8). This temperature increase corresponds to the theoretically calculated data [9].

The results of such analysis allow to optimize and define the flow rates for the chemical reactants in the microreactor in particular, to ensure the chemical reaction in the center of the microstructured reaction plate. Furthermore, the requirements for the reaction plate cooling (for stronger exothermic reactions) can be specified from the temperature increase and the spatial temperature distribution across the reaction plate for stationary operation conditions.

8.3.1.2 Glass Microreactor

Glass is widely used for making chemical reaction vessels. The use of glass is prompted by its unique physical and chemical properties, for example, the excellent resistance to many chemicals, the optical transparency or the mechanical hardness, and high temperature stability. Traditional micromachining technologies known from the silicon-based processes have been adapted to glass processing.

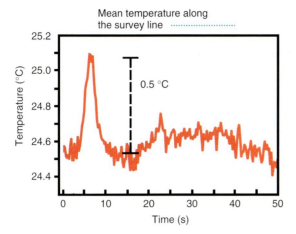

Figure 8.8 Temperature profile of CO_2 absorption into a 2.0 M solution of NaOH from start until complete consumption of all CO_2.

The Fraunhofer ICT developed a glass microreactor for controlled liquid–liquid chemical reactions with strongly exothermic behavior [10]. This reactor, shown in Figure 8.9, is made from photoetchable, chemical resistant, and temperature-stable Foturan® glass.

The reactant liquids are injected through the inputs reactant 1 and 2 (Figure 8.9). The chemical reaction occurs within the mixing and the reaction channels (width approximately 120 μm). A separate cooling system is integrated in the microsystem behind the reaction channels.

Thermal imaging of glass microreactors became possible since Foturan® glass is fortunately semitransparent within the wavelength region 3–5 μm. The raw data signal of an MW IR camera measured for a blackbody with a temperature of 50 °C

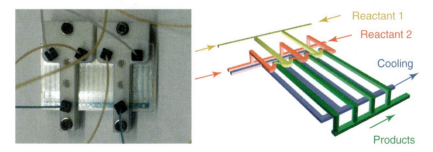

Figure 8.9 Micro glass reactor (total dimension of the depicted area 6 cm × 8 cm; drain outputs blue, reactant 1 inputs gray, reactant 2 inputs yellow, cooling system not connected) for strongly exothermic liquid–liquid reactions. (Image courtesy for the Vis image: Fraunhofer ICT, Pfinztal, Germany [11].)

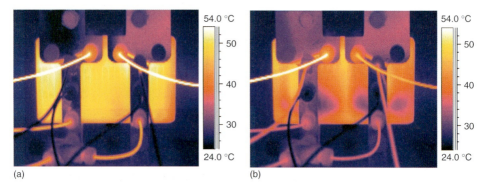

Figure 8.10 Test of the microreactor with H_2O (80 °C) without and with cooling (IR image for $\varepsilon \approx 0.9$) recorded with an AGEMA THV 550 camera.

will drop to 50% if a Foturan® glass plate with a thickness of 1 mm is placed in front of the camera. Therefore thermal processes inside the reactor can be easily observed using a MW IR camera.

During the first tests with hot water ($T \approx 80\,°C$), strong temperature inhomogeneities were found (Figure 8.10). These inhomogeneities were amplified when additional cooling was applied with room temperature water. Such results can be used to optimize the assembly of the reacting and the cooling channels.

For the characterization of the reactor operation within an exothermal chemical test reaction, sulfuric acid (96%) was diluted with water. Figure 8.11 demonstrates that time and spatially resolved thermography is a powerful method for an

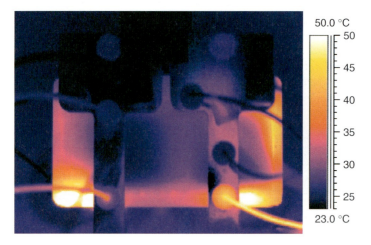

Figure 8.11 The dilution reaction of 96% sulfuric acid with water occurs in the drain due to unfavorable fluid flow at equilibrium.

adjustment and control of the process parameters in chemical microreactors. As shown in Figure 8.11, the flow rates and/or the flow rate ratio of the reactants were unfavorably selected. Therefore the chemical reaction did not occur in the microchannels as desired but it was uncontrolled and without necessary cooling within the drain.

The results of such experiments with varying flow parameters are used to improve the design of the microreactor to ensure a homogeneous and controlled chemical reaction inside the reaction channels.

8.3.1.3 Silicon Microreactor

Silicon is the best-known and best-mastered material in microtechnology. The large mechanical strength, excellent thermal properties connected with a high temperature, and a good chemical stability of silicon combined with the well-established silicon micromachining technologies offer excellent possibilities to fabricate microreactors. Such reactors are suitable for chemical reactions at elevated temperatures and pressures.

Silicon is nonabsorbing in the MW region and is characterized by a flat wavelength dependence of the transmission at about 50% (Figure 1.49). In the LW region absorption features influence the transmission (Section 1.5.4). However, for the typical thickness of a silicon wafer <1 mm the transmission is decreased by only 10% compared to the MW range (Figure 1.49). Therefore temperatures of the chemical reactants inside silicon microreactors can be analyzed by either MW or LW IR cameras.

The silicon microreactor shown in Figure 8.12 was developed for process-controlled strong exothermic liquid–liquid chemical reactions within nine parallel microchannels (channel width approximately 300 μm). The special geometry and the configuration were designed using computer simulations to allow optimum mixing of the reactants.

The silicon microreactor was tested, for example, for the nitration of diethyl urea with N_2O_5. The thermal imaging of this reaction results in an unexpected result. In contrast to expectations, inhomogeneous reactions occur with hot spots along the reaction channels (Figure 8.13). Depending on the pressure, the flow rates, and the temperature of the reactants within the reaction zone, the IR images

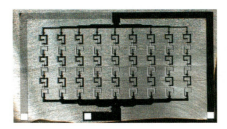
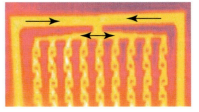

Figure 8.12 Cutaway view of the Si-microreactor and IR image of the reaction channels filled with hotwater (flow directions marked by arrows).

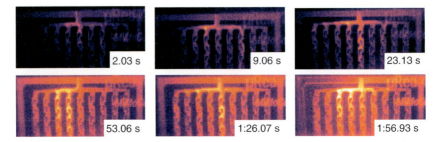

Figure 8.13 Thermal imaging of a nitration reaction of urea in silicon microreactor, recorded with THV 550 MW camera. The numbers indicate the elapsed time after start of the reaction.

show time-dependent localized hot spots like "thunderbolds" in different reaction channels.

So far, the reasons for these unexpected results are unknown. Obviously, Figure 8.13 is, however, a good example to show that time and spatially resolved temperature distributions are extremely valuable tools for reactor optimization. The thermal imaging results led to a complete redesign of the structures and the geometries in order to get a homogeneous and continuous distribution of the chemical reaction across the whole reactor.

8.3.2
Micro Heat Exchangers

Micro heat exchangers are heat exchangers with microchannels for the flowing fluids with typical dimensions below 1 mm. Such microscale thermal devices with very large surface-to-volume ratio are motivated by the fact that the heat transfer coefficient increases if the channel diameter and the channel distance decrease [12, 13]. Micro heat exchangers are fabricated using metals, glass, silicon, or ceramics. Owing to the finite transmission of silicon and glass in the MW region, micro heat exchangers made of these materials can be analyzed by thermal imaging during operation. An example of such a thermal analysis will be demonstrated for a glass micro heat exchanger.

The mgt mikroglas technik AG, Germany has a special know-how in the development of microstructured glass components [14]. One of the latest developments of mikroglas is a micro heat exchanger made from photoetchable Foturan® glass. This type of glass was also used as cover for the glass microreactor.

The assembly of the micro heat exchange system is shown in Figure 8.14. The primary and the secondary circuits are identical meander constructions. The channels of the primary and the secondary circuits are directly on top of each other separated by a Foturan® glass plate with a thickness of 200 µm. Both circuits branch out to five meanders with approximately 700-µm channel width.

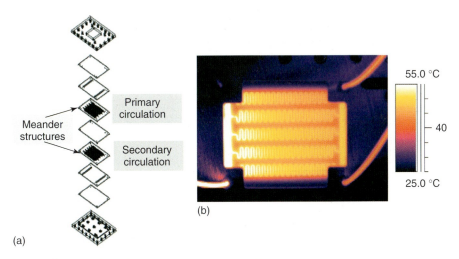

Figure 8.14 Assembly of the micro heat exchange system (a) and (b) an IR image (MW AGEMA THV 550 camera) of the primary circuit with hot water flow inside.

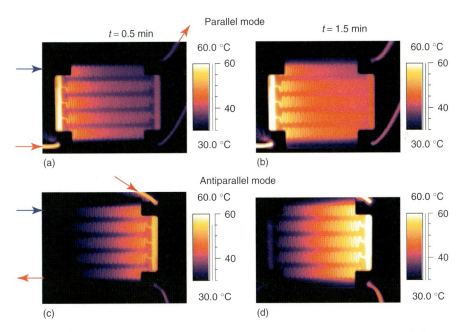

Figure 8.15 IR image of the primary circuit (AGEMA THV 550). Temperature rise of the heat exchanger at parallel flow (a,b) and antiparallel flow (c,d) after 30 s (a,c) and 90 s (b,d). Input (water with $T = 60\,°C$) and output of the primary circuit are marked by red arrows and the cooling water input ($T = 25\,°C$) is marked by a blue arrow. At $t = 0$ the hot water flow was started while the cold water was running continuously.

Applying time and spatially resolved thermography, the thermal processes and the heat transfer in the micro heat exchanger can be studied precisely. The different operation methods (parallel and antiparallel mode, i.e., cocurrent and countercurrent flow) can be characterized very accurately (Figure 8.15).

The IR images show very smoothly varying temperatures across the heat exchanger. Owing to the small temperature gradient, the system operation is remarkably homogeneous. This is shown in Figure 8.16. Obviously all five meanders show about the same temperature gradient especially for antiparallel flow operation.

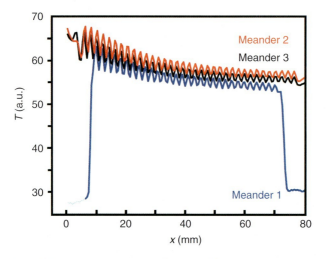

Figure 8.16 Temperature profile (horizontal lines in Figure 8.12 not shown) along selected meanders after about 30 s (antiparallel primary and secondary circuit, noncalibrated temperatures). Meander 1 is partially covered by the housing.

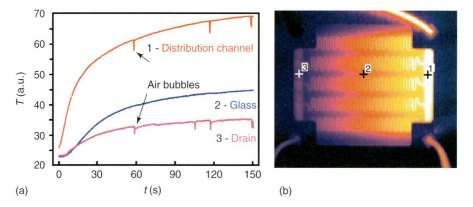

Figure 8.17 (a,b) Time-dependent temperatures at selected test points (antiparallel flow).

Time-resolved thermal imaging offers the possibility for a detailed analysis of thermal processes as shown for antiparallel flow in Figure 8.17. The time-dependent temperatures at selected measuring spots can be used to determine the time response and the heat transfer efficiency. The statistically occurring spikes of the temperatures are caused by small air bubbles within the channels. The corresponding apparent temperature change is caused by the change of emissivity between water and air.

8.4 Microsensors

8.4.1 Thermal IR Sensors

The performance of thermal radiation sensors is determined by the efficiencies and the properties of the two transduction steps from incident radiation power to temperature change and from the temperature change to the electrical output signal (Section 2.5.3). Usually responsivity and time constant measurements for thermal radiation detectors utilize both transduction steps simultaneously. For a more detailed analysis, however, the characterization of each individual transduction step would be desirable. In the first step from energy deposition in the sensor element to temperature increase is of particular interest, because two of the most important thermal parameters, heat capacitance and thermal conductance, can be determined directly from this transduction process. The following examples will demonstrate the potential of thermal imaging for characterization of thermal IR sensors.

8.4.1.1 Infrared Thermopile Sensors

Radiation thermocouples are probably the oldest infrared detectors [15]. They utilize the Seebeck effect [16] for signal generation and consist of alternate junctions of two different materials. *Alternate junctions* are defined as hot and cold junctions. For any temperature difference between the alternate junctions, a voltage is generated that is proportional to the temperature difference. To increase their voltage responsivity, a large number of individual thermocouples are often combined to form thermopiles. Thermopiles exhibit lower responsivities compared to other thermal detectors such as bolometers or pyroelectric detectors. However, thermopiles do not require a bias for operation and exhibit low noise at low frequencies. Thermopiles are frequently used as infrared sensors in pyrometers because of their excellent properties in DC operation not requiring temperature stabilization, in contrast to bolometers (Section 2.5.3).

A typical thermopile used as an IR sensor is shown in Figure 8.18. It consists of 72 thermocouples with one type of junction ("hot junction") in the center of a thin micromachined Si_3N_4 membrane and the other type ("cold junction") connected to the outside silicon substrate [16, 17]. The IR radiation must be absorbed on the thermopile to be detected, so the center is coated with an absorption layer (500-μm

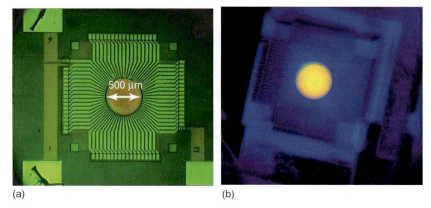

(a) (b)

Figure 8.18 Visible microscope image (a) and IR image (b, homogeneous temperature distribution, emissivity contrast only) of a thermopile (with 72 thermocouples) in stationary thermal equilibrium.

diameter and about 1-μm thickness). The IR image of a homogeneously tempered (sensor temperature slightly higher than ambient temperature) thermopile depicts the emissivity difference between the absorption layer and the surrounding sensor area (Figure 8.18).

The absorption layer should define the radiation-sensitive area. However, incident radiation is also partially absorbed on the uncovered membrane area, which causes an additional contribution to the temperature increase of the membrane. Therefore, the sensor exhibits a larger radiation-sensitive area as defined by the size of the absorption layer. This behavior of an undefined radiation-sensitive sensor area can cause a lot of problems in application of the sensors (Section 2.2).

The Peltier effect is a reverse Seebeck effect, and a thermopile consequently represents a Peltier element if it is operated in the opposite way [18]. If a voltage is applied to the thermopile, one type of junctions is heated up and the other type of junctions is cooled down. The direction of the applied voltage determines what type is heated or cooled, respectively. Figure 8.19 depicts a thermopile detector, operated as such a Peltier element.

The used polarity of the applied voltage causes a heating of the junctions in the center of the membrane. The respective temperature distribution on the radiation-sensitive sensor area is determined by the applied voltage. This behavior allows to study the spatial sensor sensitivity distribution. If a voltage pulse is applied, the sensor time constant can be determined as well.

The spatial sensitivity distribution is usually determined by a laser microprobe using a spatial scan of a focused laser beam. The detector signal is measured for the different laser beam positions at the sensor area. Figure 8.20 depicts a 2D sensitivity profile and a sensitivity false color representation of the spatial sensitivity distribution resulting from such a laser spot measurement. These results can be compared with the thermal imaging results for Peltier element operation of the

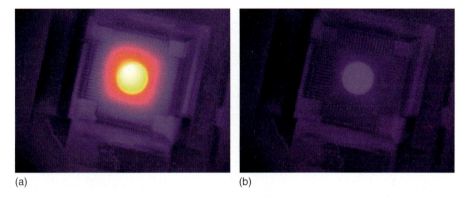

(a) (b)

Figure 8.19 Thermopile operated as a Peltier element: an electrical voltage is applied to the thermopile detector such that the thermocouples in the center are heated (a). For comparison, an image without applied voltage is shown (b). The color scale is different from the one in Figure 8.18.

thermopile (Figure 8.20). Obviously, the thermal imaging gives the same results as the laser spot measurement concerning the spatial sensitivity distribution of the sensor. However, the infrared measurement by thermal imaging is advantageous to the laser spot measurement because the sensor operates in the infrared spectral region where the sensor is used, whereas for the laser spot measurements mostly visible laser radiation is used. Also the IR imaging allows a much faster testing of the sensors.

If the thermopile is operated with a square wave pulsed voltage, the sensor time constant can be determined from the temperature rise and decay during the Peltier element operation (Figure 8.21). For the determination of the time constant, the raw signal data (representing the radiance) have been used. Owing to the fact that there is only a small change of the sensor temperature of some degrees centigrade, one can assume a linear correlation between radiance and temperature. The determined 40-ms time constant corresponds well with the result of frequency-dependent sensitivity measurements using intensity modulation and signal rise/decay measurements with pulses of the incident radiation. The temperature rise and decay are characterized by slightly different time constants. This behavior is caused by the complex thermal conditions within the heated thermopile area [12]. Thermal energy is generated by the electrical heating of the center of the membrane structure. The transient heat transfer in the structure is initiated by convection to the surrounding gas and by the two-dimensional heat conduction into the substrate (outside the membrane structure).

8.4.1.2 Infrared Bolometer Sensors

Microbolometers are very important components of infrared imaging industry (Chapter 2). The excellent achievable detector performance of microbolometers can not only be used for IR thermal imaging but for other applications as well,

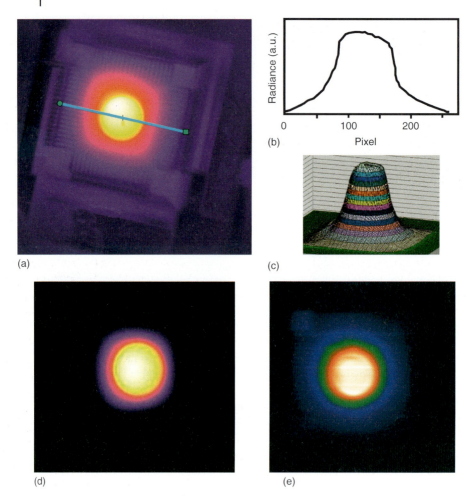

Figure 8.20 IR image of a thermopile operated as a Peltier element with heated thermocouples in the center (a). Comparison of the temperature line profile of the center heated thermopile (b) with the 2D – sensitivity profile from the laser spot measurement (c). False color representation of the temperature distribution (d) due to the Peltier effect with IR imaging and of the spatial sensitivity distribution with laser scanning (e).

for example, for nondispersive gas detection, pyrometry, or IR spectroscopy applications. Therefore, a technology for single micromachined microbolometers and measurement techniques for characterizing the bolometer performance including microthermography have been developed [19].

A single bolometer consists of a thin layer of a material with a high temperature coefficient of resistivity (TCR) with electric contacts (Section 2.5.3). The layer is thermally insulated from the surroundings. Figure 8.22a,b depicts the layout of a single microbolometer.

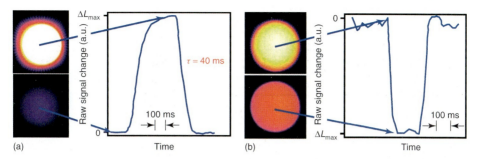

Figure 8.21 Time-resolved measurement of the emitted radiation (raw data signal) for center heating (a) and center cooling of the thermopile (b) due to the Peltier effect during a 250 ms square wave voltage pulse. (Integration time 0.75 ms, 1 kHz frame rate.)

A bulk micromachining technology based on anisotropic silicon etching was used [19]. First a silicon substrate was covered with a thin Si_3N_4 layer forming the membrane for the bolometer structure. A new bolometer material with a high TCR of -2.5 to -3% K^{-1} was developed. For perfect conversion of the incident radiation to heat, an absorbing layer structure (maximum absorbance 0.95, adjustable wavelength for maximum absorbance within 2–14 µm) was evaporated on top of the sensor area. Finally, an inverse pyramid shape cavity of 500 µm × 500 µm was etched into the silicon substrate. KOH is a wet etch which attacks silicon preferentially in the (100) plane, producing a characteristic V-etch, with (111) planes as sidewalls that form a 54.7° angle with the surface [17, 20]. The bolometers are thermally insulated from the silicon substrate by the supporting legs, which simultaneously act as electric contacts. The pixel pitches are 100 and 250 µm. Figure 8.22c,d depicts a microscopic image of the bolometer after complete technological process (c) and the bolometer in a TO-housing (d). The packaging can be completed by mounting of a transistor cap with an infrared transparent window.

The heat capacitance C_{th} and the heat conductance G_{th} of the bolometer structure are important parameters for the performance of the bolometer [21] (Section 2.2.3). Therefore the thermal design of the microbolometer structure has a strong influence on the performance of the detector. The voltage responsivity depends reciprocally on G_{th} and the time constant of the sensor equals the ratio of C_{th} to G_{th} (Figure 2.2).

Within the layout process, a detailed thermal analysis of the structures was done. In order to compare the results of the numerical simulation with the parameters characterizing the processed bolometer structures and for further improvement of the detector performance, the two values C_{th} and G_{th} have to be determined separately. Usually these two parameters are determined experimentally from time constant and sensitivity measurements. In order to increase the accuracy of the determination of these two parameters, an electrical measurement combined with microthermography has been developed utilizing the self-heating process of the

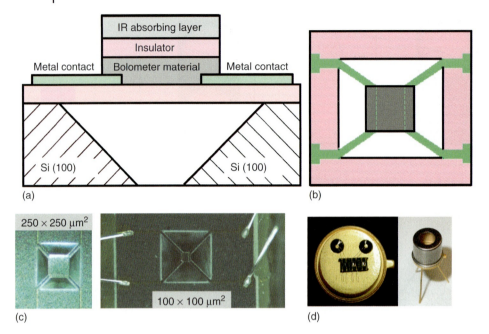

Figure 8.22 Layout (not correctly scaled) of the microbolometer structure (a) and top view on the bolometer (b). Visible microscope image of a micromachined 100 μm × 100 μm bolometer with support legs (c). Mounted microbolometer and complete microbolometer with a housing (d). (Image courtesy: iris GmbH, Berlin, Germany.)

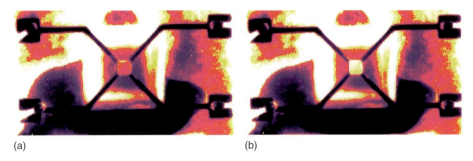

Figure 8.23 IR image of a 100 μm × 100 μm microbolometer electrically heated by an applied voltage pulse (without voltage (a); with applied voltage (b)).

bolometer if a voltage is applied [19]. Figure 8.23 depicts the self-heating effect of the bolometer due to the applied voltage. The thermal conductance can be determined from the ratio of the electric power dissipated in the bolometer structure and the observed temperature change.

The temperature of the bolometer during this self-heating process can be analyzed using microscopic thermal imaging for direct temperature determination or from the changed bolometer resistance with the known temperature-dependent

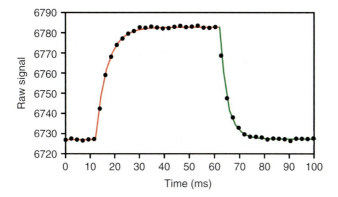

Figure 8.24 Bolometer temperature rise and decay (measured as the raw data camera signal) during an applied 250 ms voltage pulse (FLIR SC6000 camera parameters: integration time 0.8 ms, frame rate 430 Hz). Approximation of the signal rise and decay by an exponential function with a time constant of 4.7 ms (red curve) and 3.4 ms (green curve), respectively.

bolometer resistance. For the 100×100 µm² bolometer depicted in Figure 8.23 a heat conductance of $G_{th} = 4$ µW K^{-1} was determined.

With pulsed heating a time constant of about 4 ms of the microbolometer was determined from transient thermal imaging during the heating and cooling process (Figure 8.24). Similar to the analysis of the miniaturized emitters different time constants for temperature rise and decay were found. The heat capacity was determined from the heat conductance $G_{th} = 4$ µW K^{-1} and the time constant $\tau \approx 4$ ms to be $C_{th} \approx 1.6 \times 10^{-8}$ J K^{-1}.

Furthermore, the thermal images of the electrically self-heated bolometer can be used to analyze the homogeneity of the temperature distribution on the bolometer area and the temperature drop due to thermal conductivity of the supporting legs of the bolometer. The IR image of the self-heated bolometer in Figure 8.25 allows to estimate the spatial sensitivity distribution. Lateral collecting effects increasing the effective detector area can be excluded. These results are confirmed by the results of a laser spot measurement (Figure 8.25).

8.4.2
Semiconductor Gas Sensors

Gas sensors are of prime industrial interest because of their applications in important fields such as security, environment, control, automotive, and domestic applications. A particularly interesting class of gas sensors is based on the resistivity change of metal oxide semiconductor thin films induced by adsorbed gases. The reversible reactions of adsorption and desorption of different gases are controlled by the operational temperature (an integrated heater on the sensor chip allows to vary the temperature between 150 and 900 °C) and the type of metal oxides used. Such sensors allow the analysis of various trace gases such as CO_x, NO_x, NH_3,

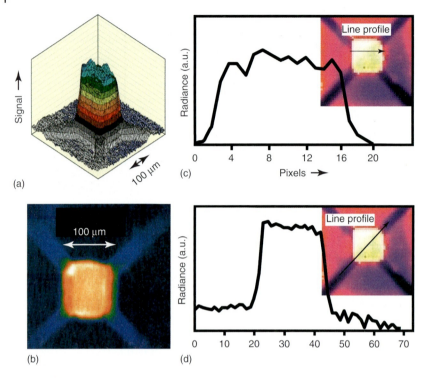

Figure 8.25 Spatial sensitivity distribution determined by a laser spot measurement and thermal imaging of a 100 × 100 μm² bolometer. (a) 2D – sensitivity distribution from a laser spot measurement. (b) False color representation of the spatial sensitivity distribution from the laser spot measurement. (c,d) Thermal image of the heated bolometer with raw signal line profile.

and hydrocarbons down to minimum concentrations in the lower parts-per-million range.

During the last few decades, mostly thick film devices have been used, although their inherent high power consumption is undesirable especially for portable systems. Meanwhile, the technological problems associated with production of thin-film devices have been solved, and the Fraunhofer IPM [22] develops gas sensor arrays for implementation in commercial systems based on the combination of thin-film deposition techniques, CMOS compatible microfabrication, and bulk silicon micromachining. As an example, Figure 8.26 shows such a sensor composed of four quadratic gas-sensitive areas with the heater in the center of the chip.

Similar to the thermopile and bolometer studies, thermal imaging was used to analyze the surface temperature of the sensors, the thermal response time to heat up the sensor to the operational temperature, and the spatial temperature distribution on the chip (Figures 8.26 and 8.27). The temperature distribution across the sensitive sensor area is very homogeneous with $\Delta T \approx 10$ K.

The time response during heating is shown in Figure 8.28. All four sensors show nearly the same behavior during the time-dependent temperature rise with

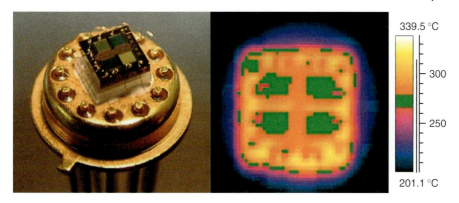

Figure 8.26 Photo of a sensor chip (size approximately 5 mm × 5 mm) and IR image (with a green colored isotherm $\Delta T = 12$ K) of metal oxide gas sensors. (Image courtesy for the Vis image: Fraunhofer IPM, Freiburg, Germany [22].)

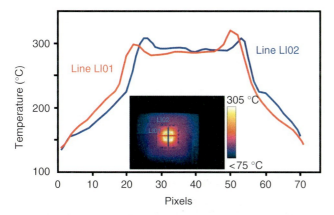

Figure 8.27 Temperature distribution on the chip. The center of the temperature plateau corresponds to the center of the chip.

deviations below 2 K. The heating process shows two time constants (very fast heating up at the beginning and then a slow temperature increase). Because the sensitivity strongly depends on the sensor temperature, these results are now used to optimize the chip structures to get short time constants of the gas sensor.

8.5
Microsystems with Electric to Thermal Energy Conversion

Thermal imaging allows the thermal characterization of miniaturized systems which transform electrical energy into thermal energy, such as, for example,

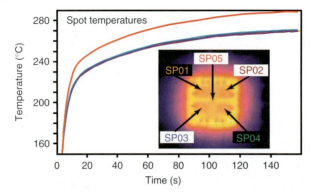

Figure 8.28 Time-dependent temperatures at sensor areas (SP01-SP04) during heating of the chip (SP05-temperature of the heater).

electrically heated radiation emitters, Peltier elements, or cryogenic actuators. For an optimum operation of these systems, the efficiency of energy transfer, the absolute temperature values, and the temperature distribution during operation as well as characteristic thermal time constants are important.

8.5.1
Miniaturized Infrared Emitters

New micromachined thermal infrared emitters in TO-39 housing with protective cap, reflector, or IR-transparent window are available for compact IR spectroscopy applications and nondispersive infrared (NDIR) gas analysis [23]. Typical emitter areas are in the range of several square millimeters. Figure 8.29 depicts two types of miniaturized IR emitters. The miniaturized emitters consist of a resistive heating

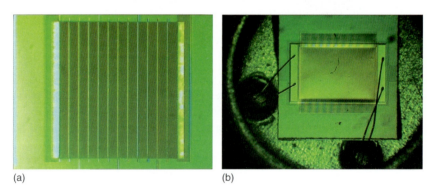

Figure 8.29 Visible microscope images of two types of commercial miniaturized emitters. (Type 1, (a)) Maximum temperature 450 °C, emitter area 2.1 × 1.8 mm^2. (Type 2, (b)) Maximum temperature 750 °C, emitter area 2.8 × 1.8 mm^2.

element on top of a thin insulating membrane suspended by a micromachined silicon substrate. Owing to the low thermal mass of the MEMS structure (heated membrane with a thickness in the micrometer range), these emitters exhibit a time constant in the millisecond range. The maximum emitter temperature is up to 750 °C. The emitters are characterized by wavelength-independent large emissivity values (typically 0.95 in the 2–14 µm range), low electrical power consumption, high electrical to infrared radiation output efficiency, an excellent long-term stability and reproducibility. One of the most important benefits of these emitters is the possibility of fast electrical modulation with high modulation depth (typical 80% at 10 Hz), that is, a chopper wheel for radiation modulation is no longer needed.

Using microscopic and high-speed thermal imaging during emitter operation, the time constant as well as the temperature distribution can be analyzed as shown in Figures 8.30 and 8.31, respectively. The spatial temperature distribution does have an influence on the angular dependence of the emitted radiation.

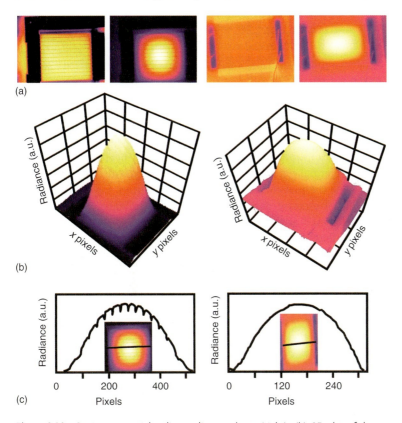

Figure 8.30 Stationary spatial radiance distributions at the emitter surfaces (type 1 left, type 2 right). (a) Thermal images of the emitter surface (stationary temperature distribution) without (left) and with applied voltage (right). (b) 2D-plot of the measured raw data signal distribution on the emitter surface with applied voltage. (c) Line profile of the measured raw data for applied voltage.

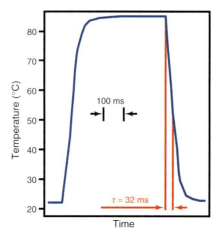

Figure 8.31 Determination of the time constant from the transient temperature signal of the emitter surface temperature for the emitter type 1 after a 250 ms square wave electrical pulse (measurement at 600 Hz frame rate and 0.8 ms integration time, spot temperature measurement at the center of the emitter surface).

The radiance measurements indicate a spatially varying temperature distribution across the emitting area. This behavior is caused by the construction of the miniaturized emitter. The heated area is placed on a membrane area that is connected to the silicon substrate at its border. So the heat generated in the element is transported by thermal conduction to these borders via the membrane material. Owing to the large thermal conductance of the bulk silicon, the temperature at the borders will not increase during heating of the membrane. The maximum temperature is achieved at the center of the membrane. The IR image of the type 1 emitter exhibits additional lines with reduced radiance (Figure 8.30c, left). These lines are also visible in the microscopic image (Figure 8.29) and are the contact lines for current supply to the emitter. Owing to the metal used for establishing these contact lines the emissivity and therefore the emitted radiance are reduced.

A time constant of $\tau = 32$ ms is determined from the measurement of the temperature signal of the emitter surface temperature during voltage pulsed operation of the IR emitter (Figure 8.31). The emitter represents a low-frequency pass and can be characterized by the frequency $f = 5$ Hz for a $\omega\tau = 1$ operation.

8.5.2
Micro Peltier Elements

Micro Peltier coolers for spot cooling or temperature stabilization and microthermogenerators are in high demand for a lot of micro- and optoelectronic applications of the future because of the increasing miniaturization in these technological

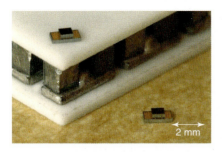

Figure 8.32 Miniaturized thermoelectric elements compared to a "classical" Peltier element. (Image courtesy: Fraunhofer IPM, Freiburg, Germany [22].)

fields. Examples are chip cooling in microelectronics or temperature control for telecommunication lasers.

Peltier coolers and thermogenerators can offer the typical advantages of microsystems such as short response time and small area combined with a high thermoelectric energy conversion. Infineon Technologies AG and Fraunhofer IPM developed the first Peltier devices based on the V–VI-compounds Bi_2Te_3 and $(Bi,Sb)_2Te_3$, which can be manufactured by means of regular thin-film technology, in combination with microsystem technology [24] (Figure 8.32). Nowadays dimensions of as little as 1 mm × 0.5 mm can be achieved.

MicroPelt®, the new generation of thermoelectric components was analyzed by thermography. Some examples are shown in Figures 8.33 and 8.34. Such optimized elements offer many outstanding properties such as cooling densities $>100\,W\,cm^{-2}$, achievable temperature differences $\Delta T > 40$ K, and a time response of the order of 10 ms [25, 26]. Obviously, contactless IR imaging techniques are extremely useful for time and spatially resolved measurements of surface temperatures of these systems.

8.5.3
Cryogenic Actuators

Cryo grippers of the NAISS GmbH Berlin, Germany [27], are a new patented kind of grippers which form a connection between the grippers contact surface and the object under study with freezing vapor (Figure 8.35). This method can be used for any hydrophile material. The emerging connection creates a high holding force without straining the material. Cryo grippers can be used for automatic handling of microprobes, permeable to air, for nonrigid materials and miniaturized components without any tension force.

Time-resolved thermography is a powerful tool for optimizing the handling process (cooling and heating). Gripping is based on water vapor in the vicinity of the material. An integrated nozzle in the holder is used to spray the water vapor on the gripping spots only. The freezing of the vapor is due to a Peltier cooling element. This ensures freezing of small amounts of water within a second. The

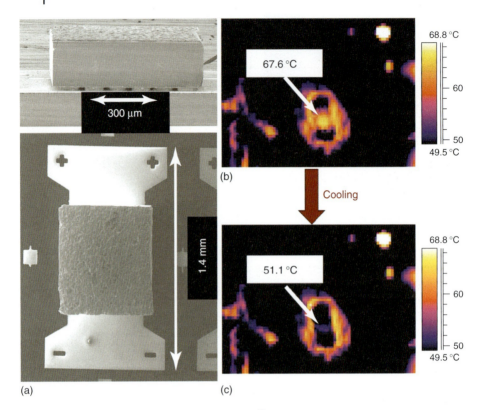

Figure 8.33 Side and top view of a MicroPelt® – Peltier cooler as Vis photos (a) and IR images before (b) and while (c) applying a voltage pulse. (Image courtesy for the Vis image: Fraunhofer IPM, Freiburg, Germany [22].)

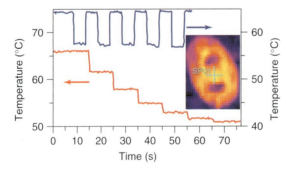

Figure 8.34 Two different time responses of the miniaturized Peltier cooler from a thermography measurement (square pulsed voltage – blue curve, step pulsed voltage – red curve). A typical IR image is shown in the inset.

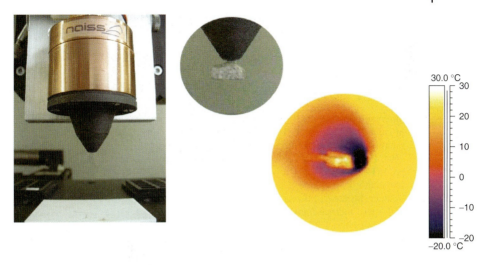

Figure 8.35 Photo of the cryogenic actuator and IR image (recorded at an angle from below the gripper) during cooling. (Image courtesy for Vis image: NAISS GmbH, Berlin, Germany [27].)

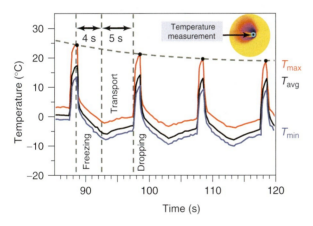

Figure 8.36 Time-dependent temperatures at the cryogenic actuator during cooling and heating.

material will be picked up and dropped off at its destination. For dropping the material, the frozen vapor (i.e., ice) will be liquefied again by heating. Besides melting, the heating also dries by vaporizing the material at the same time. The whole task takes only some seconds.

For the optimization and the functionality of the gripper, a detailed knowledge of the thermal properties at the top of the gripper is necessary. Therefore time-resolved measurements of the temperature distribution were carried out. Figure 8.36 shows the temperatures at the top of the gripper during periodically freezing, transporting, and heating the system. From these results the operating parameters, that is, the current for the Peltier cooler/heater at the top of the gripper with regard to the cycle times, are optimized.

References

1. Möllmann, K.-P., Lutz, N., Vollmer, M., and Wille, Ch. (2004) Thermography of microsystems. Inframation 2004, Proceedings, Las Vegas, Vol. 5, pp. 183–195.
2. Möllmann, K.-P., Pinno, F., and Vollmer, M. (2009) Microscopic and high-speed thermal imaging: a powerful tool in physics R&D. Inframation 2009, Proceedings, Las Vegas, Vol. 10, pp. 303–318.
3. Holst, G.C. (2000) *Common Sense Approach to Thermal Imaging*, SPIE Press, Bellingham.
4. Roberge, D.M., Ducry, L., Bieler, N., Cretton, P., and Zimmermann, B. (2005) *Microreactor Technology: A Revolution for the Fine Chemical and Pharmaceutical Industries?* Chemical Engineering and Technology, Vol. 28, No. 3, Wiley-VCH Verlag GmbH & Co. KgaA, Weinheim.
5. Hessel, V., Ehrfeld, W., Golbig, K., Haverkamp, V., Löwe, H., Storz, M., Wille, Ch., Guber, A., Jähnisch, K., and Baerns, M. (2000) *Conference Proceedings of the 3rd International Conference on Micro-reaction Technology*, Springer-Verlag, Frankfurt a.M., April 18th–21st 1999, pp. 526–540.
6. www.imm-mainz.de. (2010).
7. Hessel, V., Ehrfeld, W., Herweck, Th., Haverkamp, V., Löwe, H., Schiewe, J., Willle, Ch., Kern, Th., and Lutz, N. (2000) *Conference Proceedings of the 4th International Conference on Micro-reaction Technology*, AIChE, Atlanta, March 5th–9th 2000.
8. Wille, Ch., Ehrfeld, W., Haverkamp, V., Herweck, T., Hessel, V., Löwe, H., Lutz, N., Möllmann, K.-P., and Pinno, F. (2000) Dynamic monitoring of fluid equipartion and heat release in a falling film microreactor using real-time thermography. *Proceedings of the MICRO.tec 2000 VDE World Microtechnologies Congress, September 25th–27th 2000, Expo 2000, Hannover, Germany*, VDE Verlag, Berlin und Offenbach, pp. 349–354.
9. Danckwerts, P.V. (1970) *Gas/Liquid Reactions*, McGraw-Hill Book Company, New York, pp. 35, 37, 45, 47, 55.
10. Marioth, E., Loebbecke, S., Scholz, M., Schnürer, F., Türke, T., Antes, J., Krause, H.H., Lutz, N., Möllmann, K.-P., and Pinno, F. (2001) Investigation of microfluidics and heat transferability inside a microreactor array made of glas. Proceedings of the 5th International Conference on Microreaction Technology, 27th–30th May 2001, Strasbourg.
11. www.ict.fhg.de. (2010).
12. Incropera, F.P. and DeWitt, D.P. (1996) *Fundamentals of Heat and Mass Transfer*, 4th edn, John Wiley & Sons, Inc., ISBN: 0-471-30460-3.
13. Brandner, J.J., Benzinger, W., Schygulla, U., Zimmermann, S., and Schubert, K. (2007) Metallic micro heat exchangers: properties, applications and long term stability. ECI Symposium Series, Vol. RP5: Proceedings of 7th International Conference On Heat Exchanger Fouling and Cleaning – Challenges and Opportunities, Engineering Conferences International, Tomar, Portugal, July 1–6, 2007. (eds H. Müller-Steinhagen, M.R. Malayeri, and A.P. Watkinson).
14. www.mikroglas.de. (2010).
15. Seeger, K. (1991) *Semiconductor Physics*, Springer Series in Solid State Science, 5th edn, Springer-Verlag, New York, Berlin and Heidelberg.
16. www.ipht-jena.de/en/. (2010).
17. Madou, M. (1997) *Fundamentals of Microfabrication*, CRC Press, Boca Raton, London, New York, and Washington, DC.
18. Rowe, D.M. (ed.) (2005) *Thermoelectrics Handbook: Macro to Nano*, CRC Press, Boca Raton, London, New York and Washington, DC.

19. Möllmann, K.-P., Trull, T., and Mientus, R. (2009) Single Microbolometer as IR Radiation Sensors Results of a Technology Development Project, Proceedings of the Conference Temperature, PTB, Berlin, ISBN: 3-9810021-9-9.
20. Gerlach, G. and Doetzel, W. (2008) *Introduction to Microsystem Technology*, Wiley-VCH Verlag GmbH.
21. Hudson, R.D. and Hudson, J.W. (eds) (1975) *Infrared Detectors*, John Wiley & Sons, Inc., Dowden, Hutchinson and Ross.
22. *www.ipm.fhg.de*. (2010).
23. *www.leister.com/axetris/*. (2010).
24. *www.micropelt.com*. (2010).
25. Böttner, H. (2002) Thermoelectric micro devices: current state, recent developments and future aspects for technological progress and applications. Proceedings of the 21st International Conference on Thermoelectrics, August 25th–29th, 2002, Long Beach, pp. 511–518.
26. Böttner, H., Nurnus, J., Gavrikov, A., Kuhner, G., Jagle, M., Kunzel, C., Eberhard, D., Plescher, G., Schubert, A., and Schlereth, K.-H. (2004) New thermoelectric components using microsystem technologies. *J. Microelectromech. Syst.*, pp. 414–420.
27. *www.naiss.de*. (2010).

9
Selected Topics in Research and Industry

9.1
Introduction

Applications of IR imaging are manifold in industry as well as in research and development. In this chapter, a number of different topics are discussed, starting with the ever important issue of thermal reflections. It is particularly important in IR imaging of molten or polished metals and ways to overcome the problem of accurately measuring metal temperatures due to the corresponding low emissivities are outlined. In the following sections, specific quality control, safety enhancement, and research applications are presented from different industrial fields such as automobiles, airplanes, and spacecraft. The large field of predictive maintenance (PdM) and condition monitoring (CM) – though present in any kind of industry – is highlighted using examples from power plants, the petrochemical industry, polymer molding, and a variety of electrical utilities.

9.2
Thermal Reflections

Thermal reflections are a common source of problems in interpreting IR thermal images. In particular, not only atomically smooth surfaces such as glass, polished and varnished wood, metals, or wet surfaces but also brick and concrete may easily give rise to reflections of IR radiation from often uncared for sources. Most often people, including the thermographer, are the origin of thermal reflections (Figure 9.1).

If unnoticed, these thermal reflections may give rise to misinterpretations of the object temperature. After analyzing the differences between object IR radiation and thermal reflections theoretically, possibilities to suppress or at least identify such reflections by use of IR polarizers are discussed. Theoretical predictions and experimental results are very promising.

Infrared Thermal Imaging. Michael Vollmer and Klaus-Peter Möllmann
Copyright © 2010 WILEY-VCH Verlag GmbH & Co. KGaA, Weinheim
ISBN: 978-3-527-40717-0

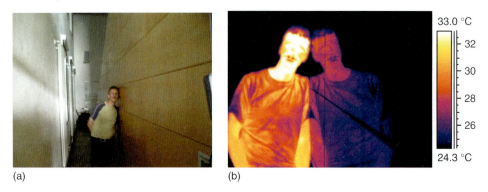

Figure 9.1 Example of thermal reflections of a man standing close to a wall made of varnished wood. (a) Visual image and (b) close-up view with LW IR camera.

9.2.1
Transition from Directed to Diffuse Reflections from Surfaces

In conventional optics, in the visible spectral range, it is common knowledge that flat polished surfaces reflect part of the incident light, whereas the other part is refracted into the material (Figure 9.2). In physics and technology, reflection is used in two different ways: first, and primarily, reflection means specular (i.e.,

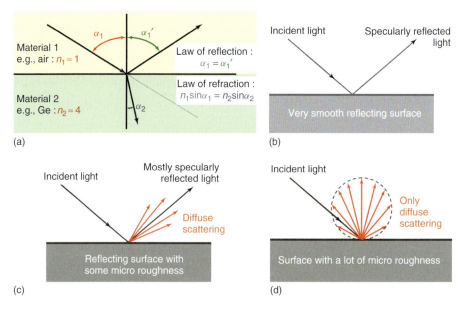

Figure 9.2 Law of specular reflection and refraction (a) and transition from "normal" specular reflection (b) to diffuse scattering (c and d).

directed) reflection, that is, reflection described by the law of reflection:

$$\alpha_1 = \alpha_1' \qquad (9.1)$$

where the angles are defined as in Figure 9.2. The more common case of slightly rough surfaces – which dominate our everyday life – leads to diffuse scattering of incident light as shown in Figure 9.2c,d. For perfect diffusely scattering surfaces, the angular distribution is the one of a Lambertian source (Section 1.3.1.4).

In the following, we refer to specular reflection when we mean conventional mirror reflection. Whenever diffuse scattering is meant, we explicitly say so. Mirrors should have reflectivities R (the portion of reflected light) close to 100% (1.00). Smaller reflectivities ($R < 1$) occur for every boundary between two media. The reflectivity is found in detail by using the law of reflection to find the angle and then Fresnel's equations [1] to compute the reflectivity. In this chapter, all calculations that use the Fresnel equations assume perfectly flat surfaces without any influence of scattering (Figure 9.2a,b). Whenever scattering contributions due to rough surfaces become important, the ideal Fresnel equation results for reflectivity must be modified.

The material input parameter is the index of refraction, which is real for transparent materials and complex for absorbing materials. As an example for transparent materials, Figure 9.3 depicts the reflectivity for light impinging from air onto glass, characterized by an index of refraction $n = 1.5$. Obviously, the reflectivity strongly depends on the polarization of the incident light. The latter is given as the orientation of the electric field of the electromagnetic wave with respect to the plane of incidence, which is defined by the **k**-vector, that is, the propagation direction of the light and the vector normal to the boundary of the surface. Usually, as shown in Figure 9.3, the plane of incidence is the drawing plane.

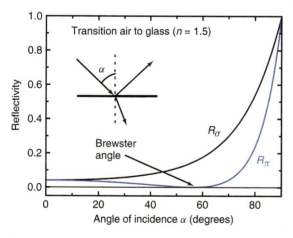

Figure 9.3 Reflectivity computed from Fresnel equations for a transparent material, defined by $n = 1.5$. This can, for example, be the transition from air to glass for visible light.

In this case, normal incidence ($\alpha = 0°$) leads to 4% reflection. Obviously, light polarized parallel (p or π) to the plane of incidence is not reflected at all if the angle of incidence is Brewster's angle, defined by $\tan \alpha = n_2/n_1$. Therefore, the reflected light is polarized perpendicular (s or σ) to the plane of incidence. This fact is used in photography: strong reflections from glass surfaces may easily be suppressed by the use of polarizing filters.

The reflection may not only be calculated for transparent bodies but also for absorbing materials like metals. Theory gives similar results, the main difference being that the materials are characterized by a complex index of refraction. The resulting reflectivity diagrams look similar to those for transparent materials (Figure 9.4); however, the minimum reflectivity does usually not reach zero, that is, the reflected light is only partially polarized. Still, the use of polarizing filters may be useful in partially suppressing reflections.

The situation is very similar when moving from visible light to the thermal IR spectral region. Depending on the materials under consideration and on the wavelength range, one may have nonabsorbing transparent materials such as NaCl ($\lambda < 20$ μm) or absorbing materials such as metals.

Compared to the visible spectral range, the situation regarding reflections may even become worse, as illustrated in Figure 9.5. It depicts an old brass plate, which is covered by oxide.

This plate is an atomically rough surface as can be seen in the visible spectral range: no direct reflection can be seen. Investigating with an LW IR camera in the wavelength range from 8 to 14 μm, one immediately sees reflections from the plate: obviously, the plate is a poor mirror in the visible range, but a good mirror in the IR. This behavior is due to the ratio of surface roughness and wavelength of the radiation. If λ is smaller or of the order of the dimension of the surface roughness, light is scattered diffusely, that is, a good mirror image is not seen. For wavelengths much larger than the roughness dimensions, the radiation is specularly reflected as from a mirror. A classical analogy is that of a soccer ball, which will be reflected most probably according to the law of reflection from a mesh wire if the mesh dimension is much smaller than the ball diameter – now imagine using smaller balls . . . .

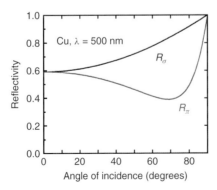

Figure 9.4 Reflectivity from a Cu surface at $\lambda = 500$ nm according to the Fresnel equations ($n \approx 1.12 + i2.6$).

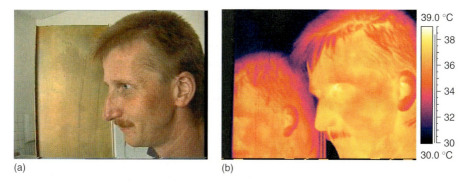

Figure 9.5 An oxidized old brass plate with a lot of surface roughness in the 1-μm scale or below scatters light diffusely for visible light, but specularly, at least, in part for thermal IR radiation of $\lambda \approx 10$ μm.

In Figure 9.5, the IR wavelengths are more than a factor 10 larger than the visible wavelengths and one easily observes the transition from diffuse scattering to specular reflections [2].

Figure 9.6a depicts the specular reflectivity of the brass plate, as measured for an angle of incidence of 20° with respect to an optically polished Au surface for wavelengths from about 1.5 to 25 μm. Measurements were performed with FTIR spectroscopy [3, 4]. Obviously, the directed reflectivity strongly decreases toward the visible spectral range, that is, mostly diffuse scattering takes place in the visible range, explaining why no mirror image is seen. In contrast, at wavelengths of 10 μm, the reflectivity, which already amounts to about 70%, leads to the mirror image seen in Figure 9.5b.

To correlate the low specular reflectivity, that is, the large portion of diffuse scattering, to surface roughness, a small piece of the brass plate was analyzed with conventional microscopy, dark field microscopy, and scanning electron microscopy.

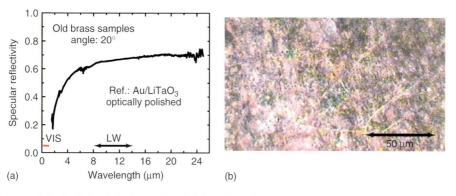

Figure 9.6 Analysis of the brass plate. (a) Specular reflectivity as measured with an FTIR spectrometer and (b) light microscope image of surface of the brass plate.

Figure 9.6b shows one typical example of an area of about 100 µm × 165 µm, magnified with a light microscope. There are some scratches across the surface with widths in the 1–5 µm range and lengths of millimeters or more. In addition, there are many "pointlike" structures in the range of 1–3 µm as well as some larger ones, as is revealed by studying more pictures and also electron micrographs. These structures resemble surface roughness and are responsible for the transition from specular to diffuse scattering.

The consequence of this discussion is as follows: flat and polished surfaces, in particular, all kinds of metal surfaces may easily result in reflections of IR radiation, even so if they are not reflecting in the visible range. Therefore, all analysis of IR thermal images must consider the possibilities of thermal reflections.

If unnoticed, thermal reflections may be misinterpreted as sources of heat on the surfaces of the investigated reflecting bodies. There are many possible thermal sources available as origin for the reflections, for example, the sun for outdoor thermography or moving sources like humans for indoor thermography in the vicinity of objects under consideration.

It is, however, possible to identify thermal reflections and also to suppress them. To achieve this, polarizers for thermal IR radiation in the MW (3.0–5.0 µm) and the LW range (8–14 µm) were used. The principle behind is the same as in the suppression of reflections in visible photography: according to the Fresnel equations, radiation that is either polarized perpendicular or parallel to the plane of incidence will be reflected differently.

9.2.2
Reflectivities for Selected Materials in the Thermal Infrared

Reflectivities have been computed for a number of selected materials the wavelength ranges of IR camera systems. The following examples are based on one set of optical parameters, although sometimes several sets were available [5]. Since all surface properties of materials may appreciably change by oxidation or corrosion, all theoretical examples should only give indications for the reflectivities. If very precise values are needed, each given sample needs to be characterized experimentally.

A quantitative description of how good the suppression by polarizers may be is possible by introducing the parameter z, which is defined for the angle φ_{min} of the minimum of the R_p curve

$$z = (R_s(\Phi_{min}) - R_p(\Phi_{min}))/R_s(\Phi_{min}) \tag{9.2}$$

$z = 1$ would offer the possibility of complete suppression, whereas z close to 0 refers to the case of very small minima, that is, nearly no chance of suppression. In addition, practical requirements favor small Brewster angles that are, however, often not realized. In such cases, it can still be helpful to get at least partial suppression or identification of thermal reflections. Hence, the following discussion focuses on the minimum angles as well as on the parameter z. In the above-mentioned example of Figure 9.3, $z = 1$ at 56.3° and for copper in the visible range (Figure 9.4) $z \approx 0.54$ at around 70°.

9.2.2.1 Metals

The most widely used metals are either iron and iron alloys or aluminum. Figure 9.7 depicts the reflectivity of iron for selected wavelengths in the MW and the LW camera ranges. Similarly, Figure 9.8 gives examples for aluminum. The plots of Figures 9.7 and 9.8 are representative for the MW and the LW regions of thermography since the optical constants are slowly but monotonously increasing functions with wavelength. Obviously, the minima for R_p are at very large angles, that is, nearly grazing incidence, for these (and many other) metals. This is not suitable for practical field work. Hence, a first conclusion is that suppression of thermal reflections will not be applicable to pure metal surfaces. However, even at angles in the range of 40–60°, many metals have z values in the range of several percent. This may be sufficient to at least identify thermal reflections.

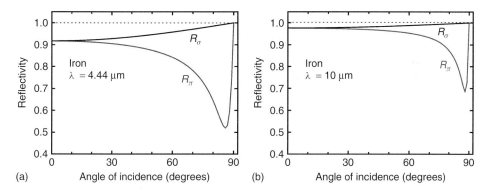

Figure 9.7 Reflectivities versus angle of incidence for iron. (a) $\lambda = 4.44$ μm, $n = 4.59 + i13.8$; $\varphi_{min} = 86°$, $z(86°) \approx 47\%$, $z(60°) = 12.5\%$ and (b) $\lambda = 10.0$ μm, $n = 5.81 + i30.4$; $\varphi_{min} = 88°$, $z(86°) \approx 32\%$, $z(60°) = 3.6\%$.

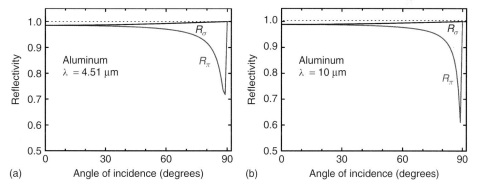

Figure 9.8 Reflectivities versus angle of incidence for aluminum. (a) $\lambda = 4.51$ μm, $n = 7.61 + i44.3$; $\varphi_{min} = 89°$, $z(86°) \approx 28\%$, $z(60°) = 2.2\%$; and (b) $\lambda = 10.0$ μm, $n = 25.3 + i89.8$; $\varphi_{min} = 89°$, $z(86°) \approx 40\%$, $z(60°) = 1.7\%$.

9.2.2.2 Nonmetals

In contrast to metals, other materials of practical use, which are absorbing in the IR, offer much better possibilities. Figure 9.9 illustrates the reflectivities for glass (SiO$_2$). In the MW region, $z = 1$, that is, 100% suppression should be possible at Brewster's angle. Also, an appreciable partial suppression should be possible for angles $>30°$. Because of the absorption maximum of glass in the range from 8 to 10 µm with a peak around $\lambda = 9$ µm, an LW camera would not allow perfect suppression, but again, very satisfying results are already expected for angles $>30°$.

As a final example, Figure 9.10 depicts reflectivity spectra for silicon. This example is motivated by investigations to study the temperatures of silicon wafers *in situ* with IR imaging. However, because of the large real part of the index of refraction of Si of about 3.4, silicon wafers are very good mirrors for thermal radiation; hence, suppression of reflections is essential for correct measurements.

An obvious preliminary conclusion from these theoretical reflectivities is that the use of polarizing filters should be useful in suppressing thermal reflections for many materials, with the exception of metals where the minimum angles are close

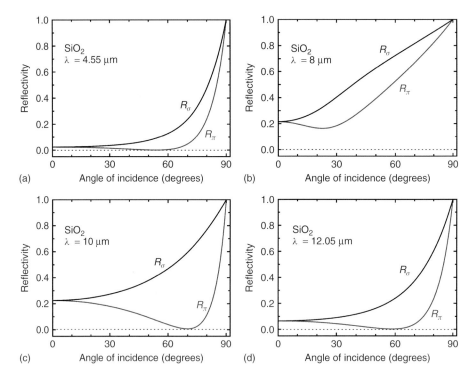

Figure 9.9 Reflectivities versus angle of incidence for glass (SiO$_2$). (a) $\lambda = 4.55$ µm, $n = 1.365 + i0.000256$; $\varphi_{min} = 54°$, $z(54°) \approx 1$; (b) $\lambda = 8$ µm, $n = 0.4113 + i0.323$; $\varphi_{min} = 23°$, $z(23°) \approx 50\%$; (c) $\lambda = 10$ µm, $n = 2.694 + i0.509$; $\varphi_{min} = 70°$, $z(70°) \approx 1$; and (d) $\lambda = 12.05$ µm, $n = 1.615 + i0.267$; $\varphi_{min} = 58°$, $z(58°) \approx 1$.

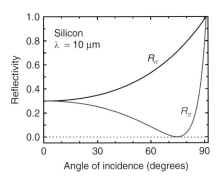

Figure 9.10 Reflectivities for silicon. $\lambda = 10$ μm, $n = 3.4215 + i6.76 \cdot \times 10^{-5}$; $\varphi_{min} \approx 75°, z(75°) \approx 1$. The optical constants are nearly constant across the region of MW and LW thermal IR radiation; hence, results for $\lambda = 4$ μm look exactly the same as for the shown example of 10 μm.

to grazing incidence. A number of practical materials behave in a manner similar to the examples shown, for example, wood used for furniture or other indoor applications is often treated with lacquer, giving a very smooth reflecting surface.

9.2.3
Measuring Reflectivity Spectra: Laboratory Experiments

To verify the theoretical predictions and study the applicability for suppression of thermal reflections, a series of laboratory experiments have been carried out. The objects under study emit unpolarized IR radiation. In addition, there are the disturbing thermal reflections from warm or hot objects in the surroundings. These reflections correspond to partially polarized radiation, which may be eliminated by appropriate polarizers.

9.2.3.1 Polarizers
Polarizers must be of a material that is transparent for IR radiation. In the studies described below, polarizers with an aperture of 50 mm, made of Ge for the LW range and CaF_2 for the MW range, were used (in principle, the Ge polarizer can also be used for the MW range). The polarizing function is due to small metal strips on top of the respective substrates. For the Ge polarizer, they are 0.12 μm wide and made of aluminum with a grating constant of about 0.25 μm. Similar to metal grids for microwave radiation or the Polaroid foils [1], only radiation with the electric field vector pointing perpendicular to the strips will be transmitted. The Ge polarizer is equipped with an AR coating in order to minimize reflection losses.

9.2.3.2 Experimental Setup for Quantitative Experiments
The principal setup for quantitative angularly resolved measurements of reflectivity is shown in Figure 9.11. It was developed to simplify the experiments in keeping the IR source for the thermal reflections, a globar with T from 1000 to 1500 °C, and the detector, that is, an IR camera, at fixed positions while at the same time easily fulfilling the condition of the law of reflection. The whole assembly is mounted on a transparent plate with angular scale. With this assembly, angular measurements in the range between 27 and 85° are possible. For very large samples,

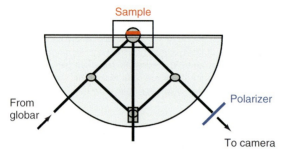

Figure 9.11 Arrangement for letting IR source and IR camera at fixed positions. Thermal radiation from a globar is reflected from the sample and passes a polarizer on its way to the camera. Under operation, the direction of the beam to the camera stays exactly collinear while the sample on the assembly is rotated. The incident radiation from the fixed globar is directed onto the entrance beam of the assembly by mirrors (not shown).

the angular range can be extended to 89°, i.e., grazing incidence that is necessary for metals.

Using this setup, a number of precise measurements have been carried out with, for example, aluminum, iron, SiO_2 (glass), or Si. The results confirm the theoretical expectations according to the Fresnel equations.

9.2.3.3 Example for Measured Reflectivity Curve

For selected materials, wavelength-dependent measurements of the reflectivities for parallel and perpendicular polarization were done. For polished and smooth surfaces, no deviations from the predictions of Fresnel's formulas are expected. If surface roughness comes into play, the influence of diffuse scattering could give rise to discrepancies. Experiments were done with the reflection accessory

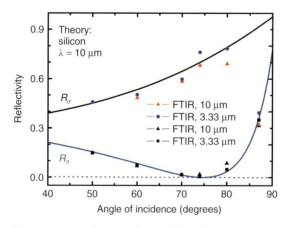

Figure 9.12 Angular dependence of the reflectivities R_p and R_s for silicon at wavelength 10 and 3.33 µm in comparison to theoretical expectations.

of an FTIR spectrometer. For selected angles, reflection spectra were recorded. Therefore, in principle, the spectra for all wavelengths are available. Figure 9.12 depicts some results for a silicon wafer, together with an enlarged view of the theoretical prediction, which is shown in Figure 9.10. The experimental values nicely coincide with the predictions for p-polarization.

9.2.4
Identification and Suppression of Thermal Reflections: Practical Examples

In the following, three applications are described. First, reflections from a polished silicon wafer were studied, second reflections from normal glass (which may be encountered quite often in different thermography applications, e.g., buildings), and third reflections from polished and varnished wood are discussed. More examples are presented in other chapters, for example, metal reflections in Section 9.4.

9.2.4.1 Silicon Wafer

From Figures 9.10 and 9.12, it is expected that thermal radiation polarized parallel to the plane of incidence is completely suppressed at a wavelength of about 10 μm. The absorption coefficient does vary a bit over the wavelength range of LW cameras [5], however, this is not enough to considerably change the plot of Figure 9.10. For MW wavelengths between 3 and 5 μm, it is even lower. Therefore, to first order, silicon should behave as shown in Figure 9.10 for IR radiation all across the MW and the LW ranges.

The setup for this experiment and the one with glass is shown in Figure 9.13. An 8-inch polished silicon wafer was attached to a clean glass plate. It sticks to it immediately via adhesion forces. A human face served as a source of IR radiation, which is reflected from the wafer. A LW IR camera is positioned in a direction corresponding to an angle of 75° with respect to the surface normal of the wafer, that is, close to its Brewster angle. Figure 9.14 depicts the resulting IR images.

Clearly, a perfect mirror image is seen in the visible as well as in the IR if the polarizer transmits the perpendicular component (Figure 9.14a). In particular, the reflections from the wafer and the glass can be studied simultaneously. Rotating the polarizer (Figure 9.14b) by 90°, the thermal mirror image on the wafer is more

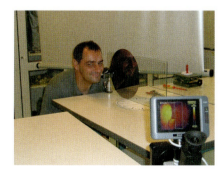

Figure 9.13 Experimental setup with glass plate and silicon wafer (attached to it by adhesion) oriented vertically. The face of a colleague served as an IR source. The angle of incidence (and reflection) was chosen close to 75°.

or less suppressed completely, whereas one may still see a part of the thermal reflection on the glass.

9.2.4.2 Glass Plates

According to Figure 9.9, the angle of 75°, which is the Brewster angle for silicon, is not expected to work similarly well for glass. This is the reason Figure 9.14 still showed the contours of a reflected image on the glass plate. This is an example of how the method works if the angle is not perfect: there is only partial suppression, but this is sufficient to at least identify thermal reflection sources. Still, thermal reflections from glass can also be suppressed more or less completely. Figure 9.15 depicts again the IR image of a person, bending over a horizontal table with a vertically oriented glass plate.

The wafer was removed, that is, only reflections from the glass could occur. Obviously, pronounced thermal reflections from the glass plate can be observed.

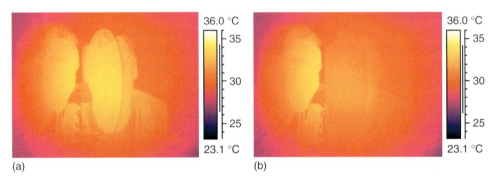

Figure 9.14 Suppression of thermal reflections from the wafer as observed through an IR polarizer oriented (a) perpendicular or (b) parallel to the plane of incidence at an angle of incidence of ≈75°.

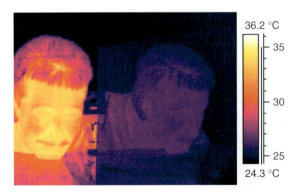

Figure 9.15 Thermal reflections from a glass plate, observed with an LW IR camera without any polarizing filter.

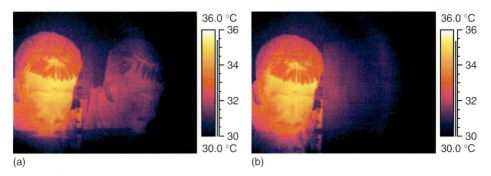

Figure 9.16 Thermal reflections from glass plate of Figure 9.15, observed with an LW IR camera through an IR polarizer oriented perpendicular (a) and parallel (b) to the plane of incidence.

Figure 9.16 shows the same scene through the IR polarizer being oriented either parallel or perpendicular to the plane of incidence. The change in absolute temperature is not relevant here; it is due to the fact that the warm polarizing filter was just placed in front of the camera. First, thermal reflections of the warm detector from the polarizer surface may contribute (the Narcissus effect); second, the IR camera was not calibrated with this filter. Results are, hence, more or less only qualitative.

Figure 9.16 nicely demonstrates the more or less complete suppression of the thermal reflections. According to Figure 9.9, it is not possible to easily attribute a specific Brewster angle for the λ-range from 8 to 14 µm to glass, since the reflectivity curves vary appreciably; therefore, the angle was optimized for these images, and it was probably around 60°. For different angles, the suppression would only be partial, but still sufficient to identify disturbing thermal sources (see example of Si-wafer above).

Glass is used in many applications of thermography, in particular, building thermography, and care should thus be taken to assure that signals, indicated by the camera, are real and not due to reflections.

9.2.4.3 Varnished Wood

Varnished wood has very smooth surfaces, similar to thin films. Hence, specular reflections are to be expected, in particular, for large angles of incidence. Figure 9.1 already demonstrated pronounced thermal reflections from such surfaces. In Figure 9.17, IR images of the same scene were recorded while looking through the Ge polarizer. Similar to the case of the silicon wafer and the glass plate, the reflections can be strongly suppressed if the polarizer is oriented parallel to the plane of incidence (Figure 9.17b).

9.2.4.4 Conclusions

Thermal reflections are a nuisance in IR imaging. They are a common source of disturbing signal contributions to object signals and therefore complicate any

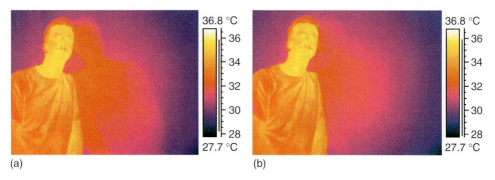

Figure 9.17 Suppression of thermal reflections from varnished wood as observed through an IR polarizer oriented (a) perpendicular and (b) parallel to the plane of incidence.

kind of quantitative analysis. Therefore, it is very important to locate, that is, identify sources of thermal reflections and, if possible, suppress them. So far, no commercial solutions are available. Laboratory experiments have, however, proven the suitability of polarizing filters to at least identify such reflections at nonnormal incidence. The results, demonstrated for several materials, are very encouraging and suggest that the method is applicable to a much wider range of materials.

9.3
Metal Industry

The metal industry deals with all aspects of working with metals to create individual parts, assemblies, or large-scale structures. Metal working generally is divided into three categories, forming, cutting, and, joining. Each of these categories contains various processes. Forming processes modify the work pieces by deforming the object without removing any material. This is done with heat and pressure, with mechanical force, or both. In casting, a specific form of metallic objects is achieved by pouring molten metal into a form and allowing it to cool. Cutting and joining consists of a collection of processes, wherein the material is brought to a specified geometry by either removing excess material using various kinds of tooling or by joining two metal parts by processes like welding. With regard to materials, one often distinguishes the steel industry from the light metal industry (aluminum, magnesium, titanium, etc.) and industries dealing with other nonferrous metals (copper, platinum, noble metals, etc.).

Obviously, the metal industry offers a large variety of different processes, some of them associated with very high temperatures. Therefore, it is natural that attempts were made to measure the associated temperatures with IR thermal imaging. In particular, it seems promising at first glance to measure temperatures of metal molds, for example, in steel or aluminum casting.

9.3.1
Direct Imaging of Hot Metal Molds

Figure 9.18 depicts an example of an aluminum cast. Although contact measurements resulted in large temperatures around 800 °C, IR imaging is not very accurate in determining temperatures. There are several reasons. First, metals have very low emissivities in the thermal IR range (Section 1.4) and small ε-variations can lead to large changes in measured temperature. The variations in emissivity of metal molds are due to ε-dependences on temperature itself, the thickness of the molten slag (which has lower density and lies atop the pure metal), the angle of observation, and so on. In addition, metal molds behave like liquids [6]. In particular, they may exhibit convection cells (like the Bénard convections, Section 5.3.3), which induce temperature fluctuations at the mold surface. All these effects can lead to measurement errors of several hundred Kelvin!

This example indicates that for measurement of temperatures of molten metals, conventional thermography is usually unsuitable. Rather, a method that is independent of emissivity is needed such as ratio thermometry (Section 3.2.2). In pyrometry, the respective instruments are called *two-color pyrometers* [8].

9.3.2
Manufacturing Hot Solid Metal Strips: Thermal Reflections

Once pure and solid metals are available, the machining of the metals still uses high temperatures. As an example, Figure 9.19 depicts a scene from an aluminum factory during the production of aluminum metal strips. In the foreground, a hot pure aluminum metal strip (thickness around 3 cm, width about 2 m) moving with a velocity of about 2 m s^{-1} is seen, which has surface temperatures of around 480 °C. Because of the manufacturing, the aluminum strip surfaces are very flat and have very low emissivities of around $\varepsilon = 0.065$. In the background, colder aluminum

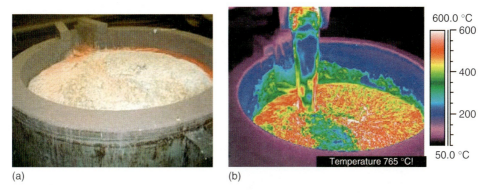

Figure 9.18 Aluminum casting. (a) visible image. (b) Thermal imaging shows huge variations of temperatures of the mold due to various parameters, which have an impact on emissivity. (Images courtesy H. Schweiger [7].)

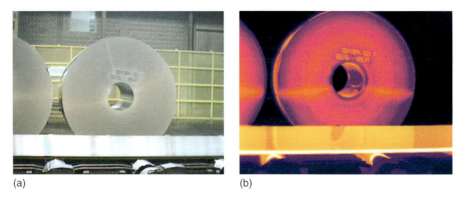

(a) (b)

Figure 9.19 Cold aluminum rolls ($T < 200$ °C) behind a hot ($T = 480$ °C) aluminum strip with very low emissivity ($\varepsilon = 0.065$). Thermal reflections can be seen.

rolls can be seen (thinner strips with thickness usually between 2 and 5 mm at lower temperature of around 200 °C rolled to form a cylinder for transport). Similarly, Figure 9.20 depicts an example of another aluminum strip (width about 0.6 m) produced for research purposes. The strip was supported by several tables to cool down slowly. At the time of the measurement, it still had a temperature of about 200 °C. Both IR images clearly show thermal reflections due to the very flat surfaces.

In Figure 9.19, the reflections of the 200 °C background cylinders show up as apparently higher temperature on the 480 °C aluminum band in the foreground. This seems paradoxical at first glance: why should a colder metal give rise to a larger radiation signal when observed as thermal reflection from a much hotter metal strip? The answer is simple: the emissivity of the front side is increased largely due to the surface roughness, which is induced by rolling the thin metal strips to form the cylinder. The IR image shows another feature: the left-hand side of the cylinder allows a grazing incidence view of the strip surface. At such large angles, emissivities of conductors increase (Figure 1.33), which explains why the perimeter surface of the cylinder emits more radiation than its front. This feature is also apparent in the thermal reflection image from the hot aluminum strip.

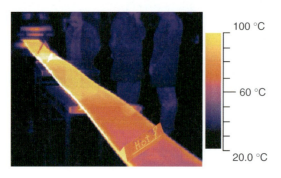

Figure 9.20 Thermal image of a hot and low emissivity aluminum strip (near 200 °C) with strong reflections, demonstrating the influence of inhomogeneous background radiation.

Table 9.1 Signal contributions for an IR camera within the 8–14 μm spectral range for pure aluminum with an emissivity of 0.02 and a background with $T = 25\,°C$.

Al temperature (°C)	Reflected background radiation (%) at 25 °C	Emission (%) from Al surface at $\varepsilon_0 = 0.02$
70	83.4	16.6
200	59.2	40.8
400	34.0	66.0

A similarly surprising feature is shown in Figure 9.20, where thermal reflections of people with skin and clothing temperatures of around 30 °C show up as thermal reflections with larger IR emission from the hot aluminum strip of 200 °C. Again, the people on the image have much higher emissivities compared to the hot metal.

This effect of thermal reflections of cold objects showing up as strong IR radiation sources in thermal reflections from very hot objects can also be understood quantitatively, provided the emissivity ε_0 of the object is known. Besides the emitted object radiation $\varepsilon_0 S_{obj}$, the detected radiation signal S_{det} from highly reflecting surfaces also contains the reflected background radiation $(1 - \varepsilon_0) S_{backgr}$

$$S_{det} = \varepsilon_0 S_{obj} + (1 - \varepsilon_0) S_{backgr} \tag{9.3}$$

S_{obj} is the radiation signal detected from a blackbody object of the same temperature. For aluminum, the emissivities ε_0 are very low, typically $\varepsilon_0 = 0.02$–0.2 depending on the alloy composition, surface roughness, and temperature. Table 9.1 compares the signal contributions of reflected background to object emission in the spectral region of an LW camera (8–14 μm). Surprisingly, the signal from aluminum of 200 °C with $\varepsilon_0 = 0.02$ is still smaller than the radiation of reflected objects of 25 °C. Obviously, background objects of higher temperatures shift the signal ratio to an even more unfavorable value.

9.3.3
Determination of Metal Temperatures if Emissivity is Known

Low emissivities and corresponding reflections from object surfaces are well known theoretically and manufacturers of IR camera systems usually implement a reflectivity correction in order to be able to determine object temperatures. However, such corrections assume a constant ambient temperature of all background objects and precisely known object emissivities. They only work reasonably well if ε_0 is large. For very small ε-values of metals, variations in object emissivity and fluctuation of background temperatures may lead to deviations. If the emissivity is exactly known, the latter problem may be overcome by using a homogeneous background illumination (hemispherical illumination) of known background temperature.

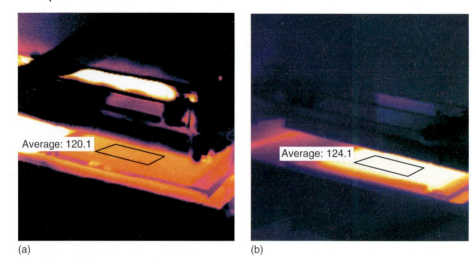

Figure 9.21 Temperature measurement at a 125 °C aluminum strip, background illuminated with black emitters at different temperatures: (a) 30 °C and (b) 140 °C.

Figure 9.21 illustrates this procedure for two different background temperatures in cold aluminum rolling where the emissivity is not changed by the temperature treatment. The aluminum strip (emissivity of the aluminum strip 0.05, pure aluminum alloy with a very smooth and homogeneous surface) was illuminated by two emitters (with emissivity of 0.98, temperatures 30 and 140 °C). For the determination of the strip temperatures, the emitter temperatures have been used as the ambient temperatures. The results fit the aluminum strip temperature of 125 °C, with an uncertainty lower than 5°.

9.3.4
Determination of Metal Temperatures for Unknown Emissivity: Gold Cup Method

The typical situation for industrial applications is more complex (Figure 9.22). In most cases, the emissivity is not known accurately and it is changing within the production process. The object signal is strongly influenced by thermal reflections of background objects (structured, i.e., inhomogeneous background) at different temperatures and changing angles. Because of the strong influence of varying reflections (compare Table 9.1), the object signal may change up to 20%, resulting in large temperature measurement errors.

The "gold cup method" [9] uses a highly reflecting hemisphere with a small hole for the camera (similar to an integrating sphere) placed near to the surface of the object. All radiation sources for thermal reflections are blocked and the multiple reflections inside the cavity increase the apparent emissivity to close to unity. The application of this method requires, however, a very clean environment to avoid a decrease in the reflectivity inside the hemisphere and very small working distances

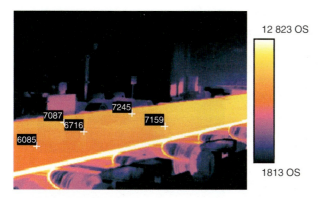

Figure 9.22 Measured object signals at different points of the aluminum strip.

in order to suppress any thermal reflection from outside background. Within most industrial processes, such clean conditions and small distances cannot be obtained.

9.3.5
Determination of Metal Temperatures for Unknown Emissivity: Wedge and Black Emitter Method

There are two possible solutions to solve this problem under industrial conditions with varying emissivities: first, the wedge method [9] can be used for aluminum rolling. In this application, there are highly polished rollers forming the aluminum strip. These rollers and the strip form a wedge (Figure 9.23) acting as a cavity (multiple reflections) and therefore having an increased emissivity [10]. Figure 9.23 demonstrates that, in principle, wedges enhance emissivity and therefore IR emission; however, a quantitative analysis still requires precise knowledge of the emissivity.

To estimate ε-changes induced by a wedge, a test measurement was performed with an aluminum strip of known temperature (provided with a contact measurement). Because of the purity and the smooth and shiny surface of the aluminum strip, the emissivity in the 8–14 μm region was as low as 0.04 (Figure 9.24). The visible image shows the cavity effect with strongly increased absorptance within the formed wedge. The emissivity in the LW region increased from 0.04 to 0.6. The maximum emissivity value is limited by the geometric resolution of the thermal imaging system. Obviously, it is difficult to correctly guess exact emissivity values; therefore, the wedge method is not suitable as a standard method to quantitatively measure metal temperatures.

A second and much better possibility to estimate correct temperatures from low emissivity metals uses two blackbody emitters. The object whose temperature is to be measured is illuminated (approaching hemispherical illumination) with two black emitters that have different temperatures (Figure 9.25). Using Eq. (9.3)

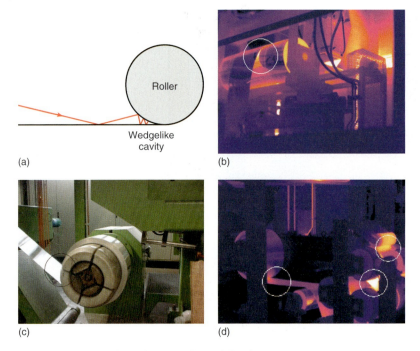

Figure 9.23 Wedges formed by the roller and the aluminum strip. (a,c) Scheme and Vis photo and (b,d) example of IR images.

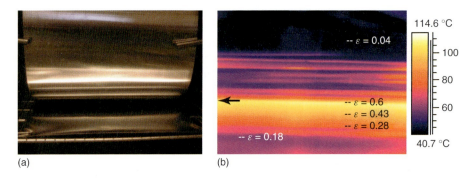

Figure 9.24 Thermal imaging of aluminum using the wedge measurement technique (temperature 110 °C, temperature scale for emissivity setting 0.6), -- ε = calculated emissivity.

with the two different background temperatures, both the emissivity as well as the emitted radiation (object signal) of the measuring object can be determined. Using the calibration curve (temperature versus radiance) of the imager, the object temperature can be determined accurately. For a detailed analysis of radiation thermometry of aluminum, see [11, 12].

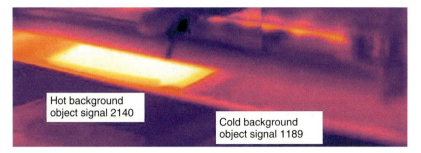

Figure 9.25 Projection of black emitters with different temperatures on the aluminum strip. Analysis of the two signals yield emissivity as well as metal temperature.

Two-color pyrometry (Section 3.2.2) can also accurately measure metal temperatures, provided the emissivities are the same in both chosen wavelength bands. It would make sense to extend the method to thermal imaging systems in the future.

9.3.6
Other Applications of IR Imaging in the Metal Industry

There are many more applications of IR thermal imaging in the metal industry; for an early review of radiation thermometry in the steel and aluminum industry, see [8]. In this chapter, two more examples with IR cameras are mentioned.

First, thermography applications have recently been reported for the equipment in steel works [13] mostly within the PdM program. Examples include machinery, electrical systems including motors and substation conditions as well as furnace and pipeline inspections. In addition, usually, problems in technology or research and development that are due to thermal factors are routinely analyzed by IR.

Another example concerns casting, in particular, lost-foam casting processes. Lost-foam casting is an evaporative-pattern casting process. First, a pattern in the shape of the desired part is made from polystyrene foam such that the final pattern contains mostly air and a few percent of polystyrene. Next, it is coated with a ceramic that forms a barrier so that molten metal does not penetrate. It is then placed inside a flask and backed up with sand (or other molding media), which is compacted. Then, the mold is ready to be poured. The molten metal causes the foam to evaporate, that is, it replaces the foam until the form is completely filled with the desired metal form. The pyrolysis products escape through the porous ceramic coating into the surrounding porous media (sand). After cooling and solidifying, the cast is removed from the flask.

During lost-foam casting, it is important to avoid defect formation such as flaws that may later on lead to failure of the part in service. Cause for such fold formation in aluminum lost-foam casting is the presence of retained foam residues during metal fill and solidification. Therefore, a detailed study using real-time MW IR imaging at 10 frames per second was made to examine the effects of various process parameters on the metal fill and the casting quality [14]. For this purpose, the flask contained a ceramic window for MW IR inspections at mold temperatures of up to 700 °C. It was found that lowering the heat loss rate during the liquid metal cooling should reduce fold formation.

9.4
Automobile Industry

The car industry is a key industry for many countries not only with regard to the large number of employees but also because it is a field where new high-tech innovations are permanently introduced. In this respect, it was only a question of time for IR thermal imaging to bridge the gap between analysis tool for the manufacturing to implementation into the product itself. Applications of thermography in the car industry are, therefore, at least threefold. First, it may be used for condition monitoring (CM) and PdM of all kinds of equipment in the factories (Sections 9.6.1 and 9.7); second, it can be used to check product quality; and third, IR imaging is used as new technology to enhance safety while driving at nighttime. Examples for CM and PdM do not differ from other industries, for example, checking electrical panels, robotic welders, motors, and so on [15, 16], and are not discussed here.

9.4.1
Quality Control of Heating Systems

The car industry offers examples of directly using thermography to check the quality of certain components of cars. Figures 9.26–9.28 give some examples. Figure 9.26 depicts a defective rear-window electrical heater (also called *defogger*), which originally consisted of a row of 12 parallel wires attached to the window. Because of scratching an ice cover from the window, some wires were damaged: the second one from the top and the lowest three wires are clearly not working. Besides detecting new damage to rear-window defoggers of windows, the technique is also used in quality control. Recently, a modern car manufacturer stated a previous failure rate of 1 out of 50 rear-window defoggers. To overcome this, IR thermal inspections were used as a means of quality control and were found to be successful [17].

Similarly, modern cars also have front windshield heaters as shown in Figure 9.27 and it is popular to have seat heaters in cars (Figure 9.28).

Other examples for the use of IR thermal inspections in the car industry include testing of brakes (Section 9.6.6.2; [18]) and, in particular, quality control of tires. Besides regular tires, IR imaging is also regularly used to test the special tires of racing cars.

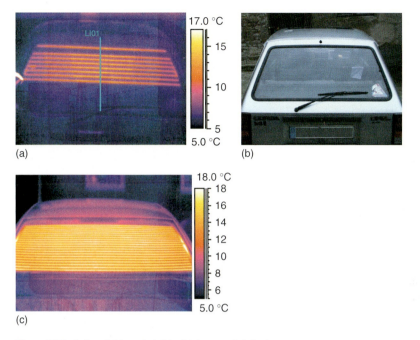

Figure 9.26 Infrared (a) and visible (b) image of defective rear-window heater and (c) example of a properly working heater of a more expensive car (more wires with smaller distances).

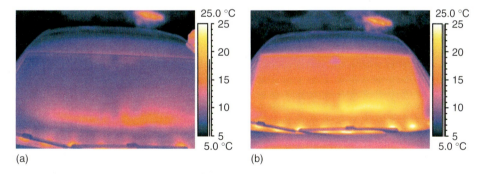

Figure 9.27 Front windshield heaters work using hot air from a fan (eight equidistant openings at bottom) as well as many very small heating wires embedded in the glass, ensuring very fast defogging.

There are a number of different patents associated with car tires and thermography. For example, one European patent application proposed to measure temperatures across the tire profile using high-speed IR imaging [19] and another US patent suggested the forecast of tire wear using high-speed thermography [20].

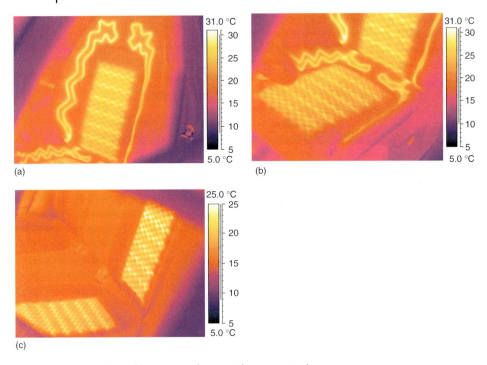

Figure 9.28 (a,b) Front seat heater and (c) rear seat heater in a car.

9.4.2
Active and Passive Infrared Night Vision Systems

A very recent, maybe the most important, innovation of IR imaging in the car industry within the last decades, took place in the field of driving safety. Driving at daytime allows observing the road and potential obstacles ahead for many hundreds of meters, while at nighttime, only the portion of the road that is illuminated by the headlights is visible. Although there are many innovations in the field of the lighting industry, for example, intelligent headlight beams that bend if driving in curves, the visual range while driving at nighttime is drastically reduced. Problems arise in particular if either cold objects like rocks or trees are blocking the road ahead or if warm objects like animals cross the road or people are moving along the road, for example, after a breakdown of their car. If the headlights illuminate such obstacles very late and, hence, if they are seen very late by the driver, allowed reaction time limits in order to avoid accidents can become very small. This limited visual detection ability by drivers can be overcome by using IR thermal imaging. Two different types of such night vision systems are currently available for a certain number of (so far mostly expensive) cars. One system (used, e.g., for Mercedes Benz) uses an active technique while the other (used, e.g., in BMW) utilizes passive IR imaging. The current costs for these night vision systems are around €2000.

Both systems have specific properties: active systems have the advantage of also observing cold objects, whereas passive systems can detect warm obstacles at greater distances.

In the active system, near-IR radiation is emitted by special IR headlights close to the regular visible headlights and an NIR camera behind the windshield detects the scattered near-IR radiation similar to the scattered visible light from objects illuminated by the headlights as detected by the human eye. The IR signals are then displayed as gray-scale images on an 8-inch TFT display in the center of the cockpit (Figure 9.29).

The technique works for distances of up to about 100 m corresponding to the usual range of long distance light, but without blinding drivers of approaching cars. As shown in Figure 9.29, it is easy to detect "hot objects" such as pedestrians. In addition, "cold objects" such as an abandoned car at the side of the road can also be seen in advance.

The passive systems use regular but miniaturized IR cameras (FPA bolometer LW systems 8–14 μm, usually with 324 × 256 pixels, camera size about 6 cm × 6 cm × 7 cm) with a wide-angle lens (36° horizontal and 27° vertical field of view), which can easily detect IR radiation for distances up to 300 m. This system does not

Figure 9.29 Examples of active IR imaging while driving at night with the Mercedes night vision system. (a) View from inside the car and (b) expanded IR image of screen; (c) another night scene. Warm and cold objects can be clearly seen within the range of the IR headlights. The speed is indicated at the lower edge of the screen. (Images courtesy Daimler AG.)

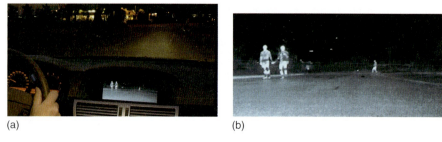

Figure 9.30 (a) Visual image from behind the driver into the cockpit of a BMW 7 and onto the road ahead while driving at night. Two pedestrians on the left side and one on the right side of the road are outside the range of the visible light headlights but they can be clearly seen with IR using the passive night vision system. (b) Enlarged IR image, which is displayed on a screen in the cockpit. (Images courtesy BMW AG, Munich, Germany.)

Figure 9.31 View of a night scene on the road top. (a) Vis image as observed with the eye. (b) Man with dog as observed with night vision. (c) Deer crossing a road as observed with night vision. (Images courtesy FLIR Systems Inc.)

need active illumination but uses the emitted radiation of hot objects Therefore, it may easily identify animals or humans, whereas it may not detect cold objects like rocks on the road. Similar to the active system, images are displayed at a frame rate of about 8 Hz on a monitor in the cockpit. This means that the driver must slightly move the head, but it is assumed that this may become a natural movement similar to checking the rearview mirror. The IR camera optics is equipped with an electrical heating to prevent fog or ice on the lens.

Figures 9.30 and 9.31 depict examples of IR images seen from the passive system.

Last but not the least an indirect application associated with the car industry is that IR thermal imaging can also be used as a method for quality control in the pavement industry [21–25].

9.5
Airplane and Spacecraft Industry

9.5.1
Imaging of Aircraft

National and international air traffic has increased enormously over the past decades. Besides occasional use for airport inspections, the associated industry utilizes IR imaging extensively in various fields. For example, with regard to PdM (Section 9.6), thermography has been used as a nondestructive testing method in general [26–28], for engine analysis, electrical and hydraulics inspections, for detecting bleed air leaks as well as studying composite materials, for example, of propellers [29]. Other applications are inspections of aircraft fuselage using lock-in thermography [30] (Section 3.5.4).

In a research-oriented context, thermography has also been used to study new propulsion systems [31].

Here, we want to mostly show qualitative images of aircraft from the outside. The features within thermal images of airplanes depend on the type of propulsion system. Propeller-driven planes as well as helicopters will have a characteristic thermal feature due to the exhaust plume from the engines. Similarly, the different exhaust plumes of turbines are characteristic for jet planes. If recorded during takeoff or landing, one may easily spot thermal reflections from the runway. Whatever kind of airplane is investigated, they all will usually cover an extremely large range of temperatures or object radiances starting from relatively cold surroundings (e.g., sky) to the ambient temperature hull of the plane to the very hot engines or exhaust plumes. This large range of object temperatures usually leads to radiance differences exceeding more than an order of magnitude. Therefore, an image cannot be recorded by a camera with a given set of parameters (integration time, filters) without suffering from either low-signal nonlinearities or high-signal saturation effects of the detectors. Therefore, airplanes are ideal examples for the use of superframing (Section 3.3) in order to have a very large dynamic range of several orders of magnitude in radiance within a single image. In addition, planes are usually fast-moving objects, which means that high-speed thermal imaging is needed with small integration times and high frame rates in order to get sharp still images. Figures 9.32–9.34 depict several examples.

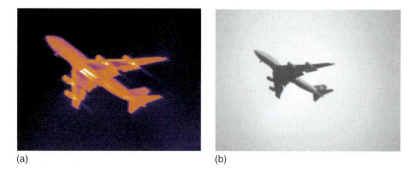

(a) (b)

Figure 9.32 Plane preparing for landing near Frankfurt/Main airport in Germany. The hot spots at the planes underbelly are due to the opening for the carriage. They seem hottest in the image since the hot engine exhaust plumes which are observed from the front are optically thin. (Courtesy SIS Schönbach Infrarotservices.)

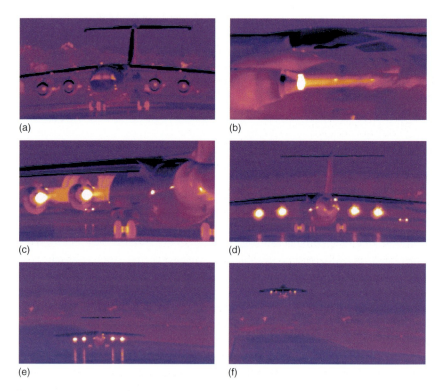

(a) (b)
(c) (d)
(e) (f)

Figure 9.33 Qualitative IR image sequence of a plane during takeoff recorded with a high-speed Agema 900 camera. The exhaust gas is still optically thin when observed from the side. The hot engines give rise to thermal reflections from the runway. (Courtesy of FLIR Systems AB.)

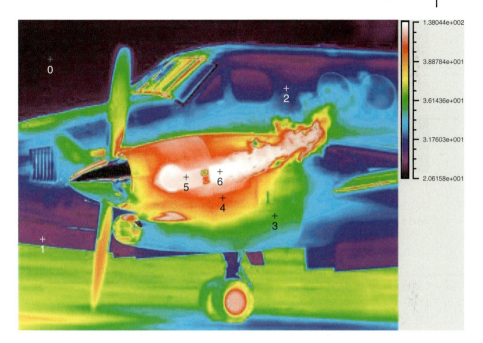

Figure 9.34 Small propeller-driven plane before takeoff. Because of the large temperature differences, the image was recorded using superframing (Section 3.3). Temperatures increased from spot 1 to spot 6 being in the range between 26 and 140 °C. Larger temperature cannot be seen because the exhaust gas is not yet optically opaque. (Image courtesy A. Richards, FLIR systems Inc.)

9.5.2
Imaging of Spacecraft

Nondestructive testing of aerospace structures is a challenging task owing to the exceptional demands on materials and processes used [26]. Any spacecraft that is flying into space and returning to earth needs to be equipped with a heat shield. This is essential for a safe reentry of the vehicle into the atmosphere since the frictional forces between a fast-moving object and air of the atmosphere can easily lead to temperatures of several thousand degrees. A corresponding natural phenomenon is the occurrence of meteor showers and shooting stars – the sometimes spectacular night sky phenomenon when small meteorites enter the earth's atmosphere, heat up very fast, and finally evaporate upon emission of light.

Spacecraft like Apollo or Sojus used techniques with heat shields for single use; on the other hand, multiple-use vehicles such as the space shuttles needed heat shields, which could be repaired, if necessary. After the accident of the space shuttle Columbia in 2003, efforts were made to perform on-orbit inspections and potential damage assessments of the outer shell of the shuttle using IR cameras [32–34]. For this purpose, a commercial IR camera had to be turned into a space-hardened camera ready for extra vehicular activities in space. The first successful inspections

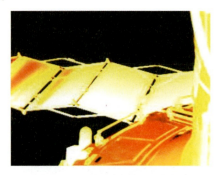

Figure 9.35 One of the first qualitative IR images recorded in space during space shuttle mission STS 121 with a commercial modified IR camera. Besides investigating shuttle heat shield samples, the astronauts Sellers and Fossum also directed the camera toward an array of radiator fins of the international space station ISS for later analysis on earth. (Image courtesy NASA.)

in space – also of the International Space Station (ISS) (Figure 9.35) – took place during the mission STS 121 of the space shuttle Discovery [35].

During this mission, the IR camera was also used to inspect an array of predamaged reinforced carbon-carbon samples – resembling parts of the heat shield – which were studied in order to test new adhesive repair techniques.

Very recently, researchers at NASA reported the use of IR thermal imaging to study the heat patterns on the surface of the space shuttle during reentry into the Earth's atmosphere [36]. So far, three missions of the shuttles Atlantis and Discovery (Figure 9.36) have been investigated for a project called *Hypersonic Thermodynamic Infrared Measurements* (*HYTHIRM*). The project was developed for inspection and protection of shuttle missions in response to the Columbia accident in 2003 when damage to the shuttle's wing compromised its heat-resistant shield, causing it to lose structural integrity and break apart during reentry.

Some goals of the project are to create 3D surface-temperature maps during reentry, compare them to measurements from thermal sensors on the shuttle's underbelly surface, and model these temperature maps using computational fluid dynamics. Once it is understood when and where peak temperatures occur, one may design the type of material for and the size of a protection system for the successor of the space shuttle whose missions are planned to end after STS 134.

Because of the very high temperatures, imaging of the shuttle upon reentry into the earth's atmosphere within the HYTHIRM project was done by using a camera with filters for near-IR radiation. The camera was on board an aircraft flying as close as 37 km to the space shuttle, thereby acquiring data for about 8 min. Calibration of sensors allows, in principle, to calculate surface temperatures, once corrections for the varying atmospheric path from shuttle to camera is taken into account. The research focused on the underbelly of the shuttle, which is covered by about 10 000 thermal protective tiles. The highest temperature areas were found to be near the nose and along the leading edge of each wing. As the shuttle enters the atmosphere, it pushes air molecules out of the way. Thereby, a boundary layer – acting as protective layer – forms around the shuttle with temperatures between about 1100 and 1650 °C. Outside this protective boundary layer, temperatures can rise to around 5500 °C.

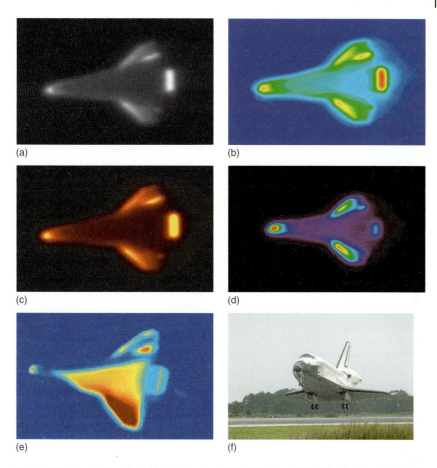

Figure 9.36 Thermal and visible images of space shuttles Atlantis and Discovery. (All images courtesy NASA.) (a,b) The Discovery at mission STS 128. Original data result in black-and-white images (a). In the false color image (b), blue refers to lowest and red to largest temperatures. (c,d) The Discovery, during boundary-layer experiment at mission STS 128 (c) and the Atlantis at mission STS 125 with hot spots at nose and wings while traveling at 8 km s^{-1} (d). (e,f) The Discovery, at mission STS119 while traveling at about 4.8 km s^{-1} (e) and Vis image of space shuttle discovery upon landing of missions STS121 (f).

Obviously, any damage to the tiles or bumps on the surface of the shuttle must be avoided since they can cause breaks in the boundary layer. In this case, the extreme heat from the outside may flow to the surface, leading to critical damages. To study such effects, artificial tiny bumps were added to the wing of Discovery. Similarly, heat shield materials for future spacecraft were tested. Figure 9.36 depicts some qualitative results of the space shuttle thermal imaging for the shuttles Atlantis and Discovery recorded during the three missions STS119, STS125, and STS128. Quantitative results have not yet been published.

9.6
Miscellaneous Industrial Applications

9.6.1
Predictive Maintenance and Quality Control

In the preceding chapters of this book and in the foregoing sections of this chapter, exemplary applications of thermography in various fields have been presented. However, thermography is nowadays so widely used in industry that it is no longer possible to give examples from all application areas. It is, however, possible to identify the fundamental application fields of IR thermal imaging in industry. Thermography is, on the one hand, mostly used as a tool in CM within PdM programs and, on the other hand, it offers some unique possibilities in the field of quality control.

PdM programs are major tools in industry that are used to reduce costs on the long term. Their most important part is CM. In general, PdM techniques should help determine the condition of in-service equipment in order to predict when maintenance should be performed in order to avoid system failure. The goal is that repair or maintenance tasks are performed only when needed. Ideally, maintenance may also be scheduled in advance when it is also most cost effective, for example, during regular shutdown periods of the machinery. This method is much cheaper than time-based maintenance programs, where parts are maintained or replaced not because they may need it, but because a predefined time interval has passed.

To monitor when equipment or parts of it loses optimum performance, PdM program use CM. In CM, inspections of equipment are performed while they are in service, that is, CM should not affect the regular system operation while at the same time it should give reliable information on significant changes of equipment parameters that are indicative of a developing failure.

Besides qualitative visual inspections that are cheap and quite reliable if performed by experts, the most usual CM techniques – depending on the equipment to be tested – are the following nondestructive testing methods:

- vibrational analysis (in particular, of equipment with rotating parts, using seismic, piezoelectric, or eddy current transducers, combined with fast Fourier transform (FFT) analysis);
- oil inspections (spectrographic analysis of chemical composition of oil);
- ultrasound (mechanical applications, high-pressure fluid flows); and
- thermography.

In recent years, IR analysis with thermography has become an important tool, since high surface temperatures of parts are often indicative of a developing failure of components, for example, of degrading electrical contacts and terminations. It may also detect a multitude of other thermal anomalies like missing or rooting thermal insulation around tubes, or leakages of tubes and valves. In all tests, a deviation from a reference value (e.g., a specific vibration behavior, oil quality, temperature value) must occur to identify impeding damages.

Quality control is another important factor in any industry. It should ensure that the industrial products have properties as designed. Quality control, therefore, deals with failure testing in design and production of products. Obviously, all kinds of destructive as well as nondestructive tests methods may be applied. And again, thermography is one of the nondestructive test tools, which has the advantage of being contactless while measuring the thermal properties of products either during or after manufacturing.

Applications of thermography in the fields of PdM, CM, and quality control have been reported [37], for example, for fossil [38], gas [39], and nuclear power plants [40, 41], electrical power converter system with capacity of 3 GW [42], chemical and petroleum industry [43], coal mines [44], the sand mining industry [45], the sugar industry [46], the paper industry [47–49], a crankshaft production line [50], the inspection of marine vessels [51, 52], or pipelines [53, 54], refineries [55, 56], big mail distribution centers [57, 58], solar panel inspections using lock-in techniques [59], wind energy rotor blade inspections [60], or cooled food in super markets/grocery stores [61].

Each application would deserve a section on its own. Here, we present only three examples: first, PdM of tubes and valves in a power plant; second, detection of the levels of liquids in large storage tanks in the petrochemical industry; and third, an example of quality control in the production line of bicycle helmets using polymer welding techniques. The wide field of electrical applications is discussed in the subsequent section.

9.6.2
Pipes and Valves in a Power Plant

Power plants are very complex systems, which have a large variety of equipment ranging from electrical to fluid dynamic components. For most parts of the equipment, thermal signatures do provide important information for CM systems. Therefore, thermography is already in common use as an established technique for power plant PdM programs. Many examples for early detection of problems have been reported [40, 41]. They include detection of hot spots in a cable connection for a large chiller, a hot bus bar connection in a critical panel, a complicated compressor issue, hot spots on a terminal board, remote monitoring for pump seal leakage, leak detection of critical valves, evaluation of loading on compressors, and the study of hot spots in generator equipment, switchyard equipment, and motor control centers. Because of the complexity of power plants, it has proven important to prepare careful documentation of detected irregularities in the form of case histories, which enter, for example, a component health database. This can also help to judge the severity of an identified anomaly and give hints to the most important question "how long will a component showing irregularities last?"

Figure 9.37 is an example of remote monitoring of a pump seal leakage from a high-energy feed pump. Recording IR images of leakage from remote drain piping eliminates the need to enter the room, providing a safer means of monitoring the equipment condition. If temperature remains constant in the piping profile, the

510 | *9 Selected Topics in Research and Industry*

 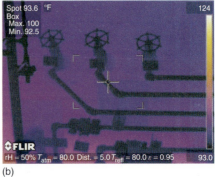

(a) (b)

Figure 9.37 IR images of a seal leak (a) and of a companion pump without leak (b). (Images courtesy Michael J. Ralph, Exelon Nuclear.)

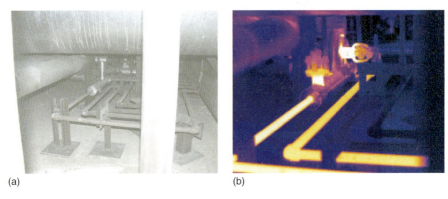

(a) (b)

Figure 9.38 Photo of the drain valve arrangement (a) and IR image of hot pipes from leaking drain valves (b). (Images courtesy Michael J. Ralph, Exelon Nuclear.)

leakage should also remain relatively constant. The repair can be deferred until the end of a two-year cycle.

Figure 9.38 depicts an IR image and the corresponding photo of drain lines with leaking valves from the main steam lines. This leakage had minimal effect on plant efficiency, but it was causing concern to the waste-processing equipment, which had to address the inputs of high-temperature water. It was imperative to accurately identify which of potentially dozens of valves were leaking in order to make the repair during a scheduled refuel outage.

9.6.3
Levels of Liquids in Tanks in the Petrochemical Industry

The petrochemical industry makes intensive use of IR thermal imaging not only for the detection of volatile organic compounds with GasFind cameras (Chapter 7) but also for other equipment [55, 56]. As an example, the process conditions within oil field production vessels could be successfully studied with IR thermal imaging [62, 63].

In particular, thermography was used to monitor the crude oil conditions in various storage, treatment, and transfer tanks used in oil production fields and it was shown that the levels of liquids and solids in tanks and vessels could be easily detected with thermography. Thermography is useful for locating levels in tanks and silos since other methods are often unreliable. The need for precise information is often critical, for example, to verify if there is still enough space in a tank to be filled.

The primary focus in the example presented below was to ensure that foaming, solids, and other crude oil process conditions are known and properly handled to optimize the cost of handling, processing, and transferring. The technique has been demonstrated for the production of gunbarrel tanks, water storage tanks, oil storage tanks, oil heaters, and production vessels.

A *gunbarrel* is a large tank that is used to separate oil from water as it flows from the production wells. Such tanks typically range from 500 barrels up to 10 000 barrels (1 barrel corresponds to a volume of ≈ 159 l). They are made from either steel or fiberglass and are usually painted black. Gunbarrels are located within a tank battery that also includes a number of oil and water storage tanks and a unit to monitor the oil being sold to the pipeline companies. If the contaminants in the oil (water, sediment, and other matter) exceed a preset limit, usually 1%, the oil is diverted back to a storage tank for treatment. This oil will go through another treating process to be within acceptable limits before it can be introduced back into the sales tank. The gunbarrel is a vital part of the crude oil sales and water treating process. The crude from the producing wells goes directly into the gunbarrel where the gas, oil, water, and small amounts of solids go through a separation process. The oil separates and floats upward, and the gas also goes to the top of the gunbarrel. The heavier water and solids settle in the bottom.

IR thermal imaging can then be used to identify all these layers from outside a gunbarrel tank [63]. The layer between the oil and water is called the *interface pad*. It is a normal part of the separation process, but sometimes the paraffin, asphaltenes, and iron compounds can become suspended within the interface pad. When this happens, the pad becomes thicker and can become hard, inhibiting the separation process and causing a number of operational problems. By using IR to locate these pads and their thickness, one can take measures to break it up before it gets too thick and hard. For example, chemicals can be added to break up the suspensions within the gunbarrel. Another method is to shut down operations to the gunbarrel and physically remove the contents, which usually requires personnel to go in and scrape out the solids, load them into trucks, and dispose of the waste materials.

Oil and water storage tanks are another area where IR has become an important CM tool. Water storage tanks are usually associated with water injection plants and receive the water from the crude oil batteries and gunbarrels. The water is then reinjected into the producing formations to enhance oil production. Some problems associated with water tanks are oil that has not been separated from

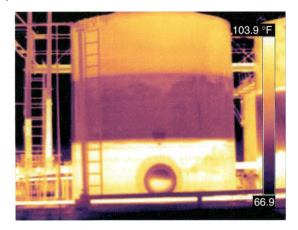

Figure 9.39 Locating levels of liquids in tanks. An about 1.5 m, a cool oil layer pad (5 feet) is floating on top of warmer water. The tank has a diameter of about 4.7 m and a height of 4.9 m with a capacity of 500 barrels (79.5 m³). The oil layer corresponds to about 23.9 m³ of oil. (Image courtesy Danny Sims, Chevron USA.)

the water, and, as with the gunbarrels, solids and other unwanted compounds. IR thermography can be used to locate these levels of "bottoms" – settlements and oil layers floating on the water, which can be recovered and sold. Storage tanks with a large accumulation of bottoms are also more prone to tank bottom corrosion problems.

Figure 9.39 is a thermal image of a water storage tank with substantial oil floating on the water. It provides the data needed to understand what is going on inside these tanks and vessels, instead of just making an educated guess. This knowledge allows prioritization of maintenance efforts on these tanks and to go right to the most critical problems first, minimizing the potential environmental risks of tank overflows and leaks. It also helps not only in identifying those tanks and vessels that are high priority risks, but it also allows deferral of maintenance on tanks that do not need attention.

Obviously, one needs thermal contrast in order to record such an image [64]. It is due to the fact that materials in a tank or silo, whether solids, liquids, or gases, behave in different ways when subjected to a thermal transition. Gases typically change temperature much more easily than liquids due to their much lower heat capacity. Consider, for example, air and water. If the same amount ΔQ of thermal energy is added to equal volumes of air and water, the resulting temperature differences differ appreciably. From the general relation $\Delta Q = cm\Delta T$, where c is the specific heat capacity and $m = \rho V$ is the mass (ρ: density, V: volume) of the gas, liquid, or solid; it follows that

$$\frac{\Delta T_{\text{air}}}{\Delta T_{\text{water}}} = \frac{c_{\text{water}} \cdot \rho_{\text{water}}}{c_{\text{air}} \cdot \rho_{\text{air}}} \tag{9.4}$$

From known specific heats and densities (air: $c_p \approx 1$ kJ (kg · K)$^{-1}$, $\rho = 1.293$ kg m^{-3}; water: $c = 4.182$ kJ (kg · K)$^{-1}$, $\rho = 1000$ kg m^{-3}), it follows that air temperature differs by a factor of more than 3000 from that of water. Now imagine a tank with a liquid and a gas on top. If the tank is heated from the outside, for example, due to solar load, the gas can easily increase its temperature, that is, it will adapt its temperature to that of the tank wall. In contrast, the liquid below will show a much smaller temperature rise due to its large thermal heat capacity. This obviously results in a thermal signature on the outside wall at the interface between liquid and gas. If different liquids with different thermal heat capacity are present, the same mechanism leads to thermal signatures at the liquid interface.

In tanks, one may also have solids like sludge. Since liquids and solids have similar densities, the temperature rises due to the same thermal input energy are not as pronounced as compared to the gases. In this case, the thermal contrast is in addition affected by the different methods of heat transfer (Section 4.2). Within solids, heat is transported by conduction, whereas in liquids, convections also play a major role. If heated form the wall of the tank, the various heat transfer mechanisms inside the solid and liquid give rise to the observed thermal signature.

How was the thermal contrast achieved in the example discussed here? The weather conditions in Texas provide a lot of sunshine, which means solar load on the tank walls. In Figure 9.39, the gases on top of the oil layer (upper third of image) have low heat capacity, which means that the wall temperature is mainly due to solar load. Below the cold oil cools the outside wall and at the bottom, the warm water heats the wall. Because of the large heat capacity of the fluids, quasi-stationary conditions are achieved.

9.6.4
Polymer Molding

Shape and block molding of expanded polymers such as polystyrene, polypropylene, and various blends therefrom are nowadays widely used in manufacturing many everyday life products, ranging from simple packaging materials or picnic supplies to technical industrial products, for example, automotive bumpers, bicycle helmets, or aircraft structural parts. Besides creating products directly, such foams are also used, for example, in metal–foam casting processes (Section 9.3.6).

Manufacturing of foam-molded parts requires precise molding control such as precise amounts of steam energy delivered to the mold and later extracted in the subsequent cooling cycles. Accomplishing this "energy balance" is critical in molding good quality parts. Besides quality control, one also wants to reduce excessive energy use.

Characterization of freshly molded parts is possible with IR thermal imaging. The parts will have varying moisture content and thermal gradients from the manufacturing process. Quantitative analysis of the gradients can lead to process

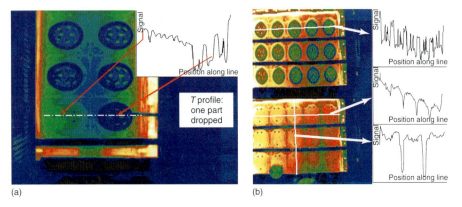

Figure 9.40 Thermal images of two thermal presses at the point of bicycle helmet ejection. (Images courtesy G.V. Walford.)

improvement and the identification of defect areas in the manufactured parts. In particular, thermal gradients are often directly related to mechanical property variations at the respective locations.

Figure 9.40 depicts examples of the manufacturing of protective bicycle helmets [65]. These are molded in presses. The press halves separate after completion of the process and the frame slides out to allow retrieval of the freshly molded helmets. Figure 9.40 shows two images of the frames with helmets at the time of helmet ejection, showing significant temperature gradients for the molds and the helmets (line plots at the right).

The IR image of the helmet shown in Figure 9.41 was recorded 60 s after its ejection from the mold. To find out whether the observed temperature gradient has an impact on helmet properties, an X-ray mass profile was made of the part. The helmet was inspected in slices by scanning the cross-section of the helmet in several sections. These are also plotted in Figure 9.41.

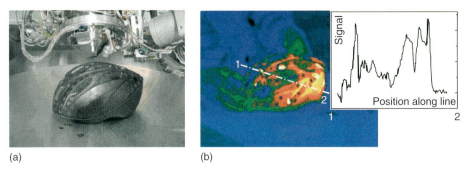

Figure 9.41 Visible (a) and thermal image (b) of a helmet, removed from the press. The helmet was subsequently analyzed using X-ray profiling; see inset at thermal image. (Images courtesy G.V. Walford.)

The higher mass profile on one side of the helmet coincides with the thermal behavior in the press. This may imply a bead filling issue on one side that when coupled with a differing steam distribution across the helmet provides differing fusion and density profiles [65]. Such features can lead to a nonuniformity of the mechanical properties of the helmet. In this case, the asymmetry of the helmet was readily observable with thermal imaging. Such nonuniformities can have significant impact upon the end application be it helmets or other mold parts, since mechanical properties are correspondingly modified from the original design specification.

9.6.5
Plastic Foils: Selective Emitters

Accurate temperature measurements are crucial to any process in the plastics industry where temperature is a relevant factor. There is large variety of applications for thermal imaging in extrusion, coating, thermoforming, laminating, and embossing. However, plastics offer problems concerning IR imaging due to the fact that they are selective emitters (Section 1.4). Thin plastic foils are transparent to IR radiation (Figure 9.42). The transmittance depends on the material and

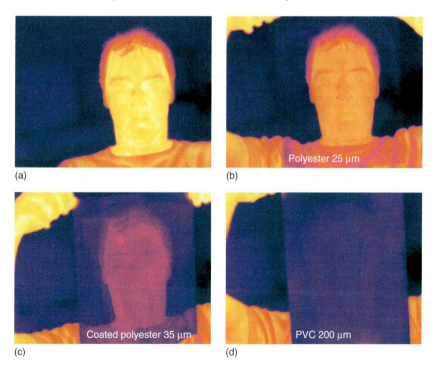

Figure 9.42 Demonstration the transmittance of various plastic foils in the LW IR spectral region using thermal imaging.

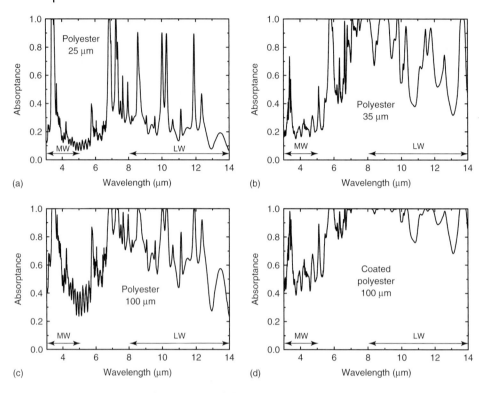

Figure 9.43 Spectral dependence of absorptance (emissivity) of polyester foils with different thickness (measured with Bruker IFS 66 Fourier transform spectrometer).

the thickness [6]. The behavior of the plastic foils can be understood from their transmission spectra $T(\lambda)$.

Thin plastic foils are typically selective emitters. Their spectral emissivity equals the spectrally dependent absorptance $\alpha(\lambda)$. Reflectances are very small (order of %) and neglected here. Figure 9.43 gives examples of absorptance spectra of several foils, computed from the transmission spectra according to $\alpha(\lambda) \approx 1 - T(\lambda)$.

As is obvious from Figure 9.43, emissivity of thin plastic foils (such as PVC, polyester, polyethylene, polypropylene, nylon, etc.) is strongly changing within the spectral ranges of the MW– and LW–thermal imaging systems. Such foils are selective emitters. In addition, they are not opaque, that is, an IR camera partially looks through such a foil. As a consequence, any broadband thermal imaging gives inaccurate temperature values, particularly if other hot objects are behind the foil.

Even if background radiation is negligible, a quantitative analysis is tricky. One may, for example, think of correcting the changing emissivities by averaging the absorptance in the spectral range of the imaging system. It is not possible to calculate simply an average emissivity because the spectral distribution of

9.6 Miscellaneous Industrial Applications | 517

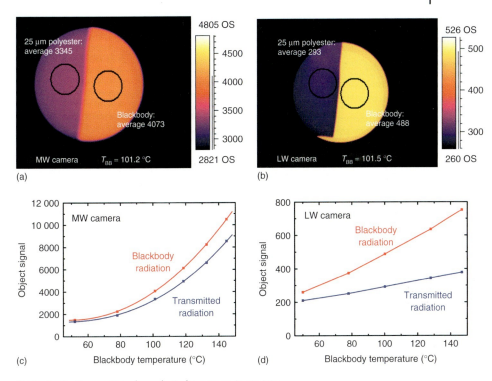

Figure 9.44 Temperature-dependent absorptance/emissivity of 25-μm polyester foil within MW and LW spectral range (object signals measured at different emitter temperatures).

the emitted light follows Planck's law (Section 1.3.2). Therefore, an "effective emissivity" within the spectral bands used for thermal imaging depends on the object temperature [66], as has been shown for selectively transmitting IR windows.

An experiment was performed to illustrate this effect. One half of a heated blackbody source was covered with a 25-μm polyester foil (only during the measurement to prevent a temperature increase of the foil). The camera radiometric signals with and without foil were recorded and compared for a MW (3–5 μm) and a LW (7–14 μm) camera (Figure 9.44). For the MW camera, a nearly constant ratio between the two object signals with and without foil is observed, indicating that an "effective emissivity" value may be used to characterize the foil absorption. This is due to the fact that in the MW region, the Planck curves are still rising and the main effect of the foil is just to make the detected spectral region a little bit narrower, due to the absorption peak around 3.43 μm.

This constant ratio no longer holds for the LW region, since the shift of the Planck curve with temperature is now affected by the absorption bands around 8.5, 10, and 12 μm. We conclude that MW cameras are better suited to measure temperatures of such thin plastic foils.

From the spectra of Figure 9.43, it is, however, also clear that LW cameras are well suited to analyze temperatures of coated plastics (and also of foils with thickness > 250 μm) since emissivity is high with only a minor wavelength dependence.

The best choice to measure temperature of thin plastic foils is to use a filter, adapted to a narrow absorption band in the spectrum. The wavelengths of the absorption bands depend on the type of plastics and can be determined by an IR transmittance measurement.

9.6.6
Line Scanning Thermometry of Moving Objects

Quite often, technological, industrial, or everyday life processes of moving objects need to be considered. Depending on the speed of the objects, different techniques are used. The best solution would be very high-speed IR cameras. If either those are not available or – more importantly – if less expensive solutions are needed for mass production equipment, IR line scanners are the first choice.

9.6.6.1 Line Scans of Fast-Moving Objects with IR Camera

Consider the following task: measure the transient temperature distribution of the disk of a brake system of a car wheel on a brake test bench while assuming the outer edge of the disk to have an initial velocity of 180 km h^{-1}. For a disk diameter of about 32 cm (circumferences 1 m), the outer edge velocity of 180 km h^{-1}, that is, 50 m s^{-1}, corresponds to 50 revolutions per second. If the temperature distribution of the wheel is to be measured, the integration time needs to be quite small. Within 10 μs, a point on the circumference would move only about 0.5 mm, whereas 100 μs integration time would already lead to 5 mm, that is, a smearing out of the image. The needed integration time also depends on the signal-to-noise ratio. Fortunately, the fact that brakes can reach very high temperatures (very high levels of emitted thermal radiation) counterbalances the low emissivity of the brakes, which are made from metal.

The need of small integration times explains why bolometer cameras cannot be used to analyze such a problem (Section 2.4.2). Figure 9.45 depicts the result as measured with a bolometer camera. Obviously, only radial symmetry is observed, whereas any nonuniformity along the circumference is smeared out.

To measure the temperature distribution with higher time resolution, high-speed cameras are needed. All such cameras must operate with very small integration times in order to achieve high speed. As discussed in Section 2.4.2 (Figure 2.32), high speed must go along with a reduction in image size and highest speeds are usually achieved with single lines or linelike rectangles with one side being much longer than the other side. For the case of the brake wheels, an Agema 900 camera was used with its fastest mode of image generation, the line scan mode, which gives 3500 lines s^{-1}. Figure 9.46 depicts the scheme of the measurement: the camera records line scans while the wheel rotates. For the rotating brake wheel, this corresponds to 70 lines per revolution.

9.6 Miscellaneous Industrial Applications | 519

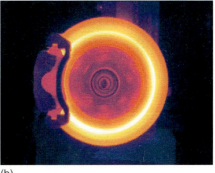

(a)   (b)

Figure 9.45 Vis and IR bolometer camera image of a rotating wheel while braking. (Courtesy SIS Schönbach Infrarotservices.)

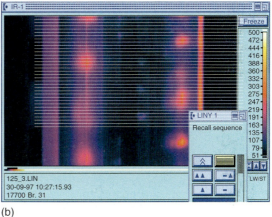

(a)   (b)

Figure 9.46 The camera measures line scans while the wheel is rotating (a). The line scans (b) just give raw data that need to be transformed back taking into account the wheel size and speed. For this purpose, the camera was triggered by the wheel, that is, there is a correlation between each scan and position of the wheel. (Courtesy SIS Schönbach Infrarotservices.)

It should be noted that because of the metallic material, a very careful compensation of reflected background radiation as well as measurement of the emissivity of the brake wheel was necessary.

Figure 9.47 depicts an example of a reconstructed IR image (right) of the wheel, which is due to a large number of individual line scans (left). In comparison to the bolometer image (Figure 9.45), it is obvious that the wheel is heated nonuniformly. Such investigations help characterize the heat dissipation of brakes and, hence, modify the brakes, if needed.

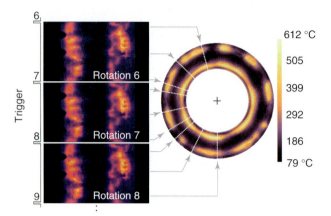

Figure 9.47 Reconstructed IR image of the rotating brake wheel from a large number of individual line scans. It clearly reveals structures. (Courtesy SIS Schönbach Infrarotservices.)

9.6.6.2 Line Scanning Thermometry of Slow-Moving Objects

Figure 9.48 depicts a typical industrial example. Some products, here windshields, are transported with a certain velocity during production. The task is to analyze product quality during the movement. In the example of Figure 9.48, windshields enter the bending station and it should be tested whether the bending process introduces any defects in them.

There are two possible solutions to the problem. First, one may use a high-speed IR camera as in the above example. Since the lateral dimension of the windshield must be recorded with high spatial resolution, one would again end up with detection areas like either lines or elongated rectangles. Although possible, this is a very costly method in comparison to the second possibility, which is to use line scanners (Section 2.4.1, Figure 2.19) from the very beginning. As a matter of fact, line scanners are ideal candidates for monitoring of industrial processes with linear velocity: the lateral dimension is recorded along lines with short integration

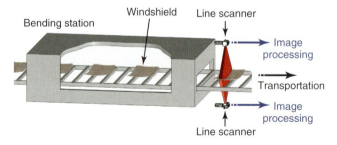

Figure 9.48 Measurement of the temperature distribution of windshields at the exit of the bending station for automotive glass production. (Image courtesy: Raytek GmbH Berlin.)

times. The longitudinal dimension (parallel to velocity) is automatically recorded by the time sequence of the line scans.

Consider, for example, commercial line scanners that are available typically with up to 1024 pixels. The largest available pixel numbers are clearly an advantage compared to the maximum number of 640 from typical 640 × 480 pixel cameras (due to higher cost, one megapixel IR cameras are still rather the exception than the rule). The operating wavelength for the glass measurements is about 5 μm. This is achieved by using broadband detectors with a narrow filter (Section 3.2.1). Glass is opaque with a high emissivity value at this wavelength, which enables studies of the surface-temperature distribution of the windshields. Line scanners have the advantage of a very large FOV of about 90°, which means that short distances between scanner and windshield are possible. In contrast, most standard lenses for IR cameras offer at maximum a 45° FOV.

Let us consider an example with a scanner made of 256 pixels, which allows, for example, a 150-Hz frame rate, that is, a time of 6.6 ms between two scans (1024 pixels would allow, e.g., 40 Hz). Assuming a velocity of 1 m s^{-1} of the windshields on the band, two subsequent line scans correspond to a distance of 6.6 mm (25 mm for 1024 pixels). Figure 9.49 depicts how the windshield is scanned. The sequential readout given by the scan speed leads to scan lines that are slightly rotated with respect to the transverse object direction. The distance between two consecutive scans results in a line grid, which is superimposed on the object area and which forms a two-dimensional slightly rotated IR image of the object. Its transverse spatial resolution is given by the number of pixels and the longitudinal spatial resolution is defined by the distance between two scan lines (6.6 mm in the above example). The tilt angle depends on the ratio of scan speed and speed of object.

Figure 9.50 depicts an IR image of a windshield with a defect recorded with this method. The IR image was first treated with the advanced image-processing techniques, which allow to get rid of the tilt angle and noise. Obviously, such IR images can detect any extended inhomogeneities of the temperature distribution, for example, due to fissures, scratches, and so on, unless they are exactly parallel to the scan lines and are missed due to the line distance (in the example, 6.6 mm). Point defects of smaller sizes would be difficult to detect. The windshield is still warm due to the manufacturing process. Any fissures of other defects can therefore show up in the IR image since such defects can lead first to inhomogeneities in emissivity and second change the heat transfer mechanisms like conduction in the vicinity of the defect. Images such as Figure 9.50 can then be analyzed online

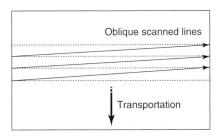

Figure 9.49 Line scans of a moving object lead to grid lines that are tilted with respect to the object.

Figure 9.50 Thermal image of a windshield showing a line-type defect. (Image courtesy: Raytek GmbH Berlin.)

using software tools of image processing (Section 3.2.2) adapted to the problem. Thereby, automated alarms could be installed if a certain defect level is detected.

9.7
Electrical Applications

There are numerous examples of IR thermal imaging in the field of electrical applications [67]. They range from low-voltage indoor surveys, study of components on electrical assemblies or electrical circuit boards and electronic components [68–71] electric motors (e.g., [72]) via oil-filled circuit breakers [73], transformers, and substations [74, 75] to the study of high-voltage lines, sometimes only observed from helicopters [76]. For any quantitative outdoor inspections, potential wind speed effects on temperature readings of the components under study must be taken into account [77].

Since thermography inspections of electrical equipment makes sense only if components are under full load, covers of electrical panels, and so on must usually be removed [78] unless special IR windows are used [79, 80], which allow to study the objects from a safe distance. In any case, whenever high-voltage and/or high-power equipment under load is investigated, special personal protection precautions must be taken to avoid arc flash hazards [81–84].

In the following, just a few special topics that are relevant for industry are discussed. Some other examples on electrical heating within wires, the microwave oven, thermoelectric effects, or eddy currents have already been given (Chapter 5).

9.7.1
Microelectronic Boards

Because of the ongoing miniaturization of all electrical components, in particular in the microelectronic boards of computers, the problem of malfunctioning of whole boards becomes important. Although the voltage and also the currents are very low, they still produce heat and many components malfunction above a critical temperature. The old 486 processors still operated uncooled and sometimes at

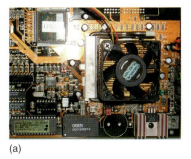 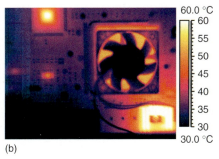

Figure 9.51 Microelectronic board with Pentium 100 processor below a fan. The heat sources are well separated on this board. For demonstration purposes, the fan was not operated.

working temperatures above 80 °C. Starting with the Pentium series, processors were air cooled by fans. Still, it is necessary to separate all heat-producing elements on a board from each other in order to avoid overheating of individual parts. To test or design new boards, it is essential to operate the boards under working conditions and analyze, for example, IR images. Figure 9.51 depicts an example of a board containing a fan-cooled Pentium 100 processor. Upon switching the computer on, two IC chips and another processor on the board got heated, however, typically only to 15 °C above room temperature.

As can be seen in the visible image, the board is made of different materials; therefore, the IR images usually reflect temperature as well as emissivity contrast. If emissivities are quite high ($\varepsilon > 0.9$), small variations in their values are not critical. However, whenever small emissivities are present, in particular those of hot objects, the quantitative analysis must be performed very carefully. To overcome such problems, it was recently suggested that electronic boards be covered with a thin coating of known emissivity, which does not change the electric properties of the components [68]. In this case, the boards will show homogeneous emissivity and hence allow sensitive relative temperature measurements across them.

Besides the design of microelectronic boards for computers, IR imaging may also be used to study the behavior of simpler electronic boards. As an example, Figure 9.52 depicts the electronic board of a small loudspeaker of a type, which is usually connected to computers. The visible image shows the board that is mounted on the back cover of the speaker, which was opened for the experiment. Playing music obviously led to a heating up of four diodes on the board, whereas the power transistors were sufficiently cooled by the metallic cooling fins.

9.7.2
Old Macroscopic Electric Boards

Electric switchboards are related to typical industrial applications of engineers, since the regular testing and surveying of electrical switches is essential for

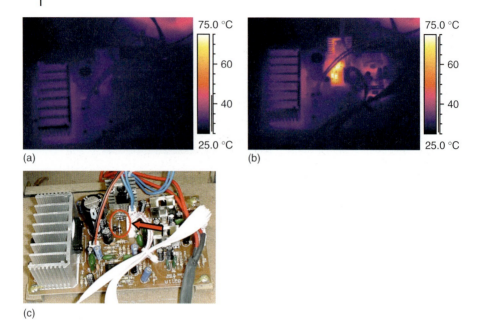

Figure 9.52 Visible (c) and IR images of the electronic board of a speaker, which is connected to a computer. First, the speaker was turned on, but operated without emitting any sound (a), and then the speaker was used to play music which was clearly audible (b). Obviously, the four diodes on the board (red circle in Vis image) were heating up to about 75 °C during operation of the speaker.

reducing service shutdown times and, hence, for guaranteeing a high productivity of machines in factories. Therefore, a student project [85] dealt with an electrical switch assembly to simulate malfunctioning and subsequently identify the relevant parts. To visualize typical defects occurring in low-voltage applications ($U < 1$ kV), we prepared an old electric fuse board previously used in industry. Several typical failures were included artificially for student teaching. Figure 9.53 depicts the visual image as well as an expanded view of the upper section. Several different defects can be clearly seen, such as a broken connection, a deformed (squeezed) wire connection, a loose connection (loose screw) with transition resistance as well as a very old oxidized 16-A fuse. Whenever such anomalies are observed in IR thermal imaging, one must have a closer look, which usually identifies and explains the problem immediately.

9.7.3
Substation Transformers

A very important application concerns high-voltage power lines and their connection to transformer units. Figure 9.54 depicts an example of a 115–23 kV substation transformer unit [86]. The distance between the camera and the transformer, which was run at about 40% of the maximum allowed load, was determined with a laser

 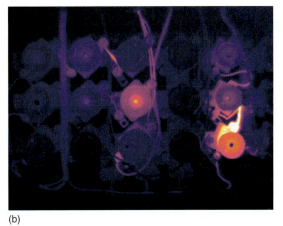

(a) (b)

Figure 9.53 Visible (a) and IR image (b) of (very) old electrical equipment. The board was modified to include three typical thermal features of malfunctioning. Top row of fuses: broken wire within insulation gives resistance, heating up the wire. Middle row: an old oxidized fuse heats up. Bottom row to the right: loose connection giving a transition resistance.

range finder to be around 11 m. At least two of the phases show problems. Figure 9.54 also illustrates a typical problem of outdoor electrical applications, encountered when studying high-voltage equipment. Since the distances between observer (camera) and the targets are often above 10 m – with no opportunity for the thermographer to get closer – and targets (like the 3/4-inch cables in Figure 9.54) are sometimes just a few centimeters in size, the spatial resolution of the camera can impose problems for a quantitative analysis. In the example presented, three different measurements of the same targets were made at about the same time and from the same distance under the same environmental and load conditions. Three different cameras (high, medium, and low number of pixels) were used, which differed only by their detector arrays (high: 640 × 480 pixels, medium 320 × 240 pixels, and low 160 × 120 pixels). Since the field of view determined by the standard optics was similar, the instantaneous field of view of the cameras differed (0.65, 1.3, and ≈2.6 mrad, respectively). As a consequence, the number of pixels is directly related to the spatial resolution (Section 2.5.3).

A cable of 2 cm at a distance of 11 m corresponds to an angular size of about 1.8 mrad or 0.1°. As mentioned in Chapter 2, the minimum object size should be about 2–3 times the IFOV in order to get reliable temperature measurements. This condition is only fulfilled for the high-resolution camera with IFOV = 0.65 mrad. Therefore, it is to be expected that the medium- and low-resolution cameras will show appreciably lower temperatures, since the background temperature of the clear sky behind the cables was well below 0 °C and the detectors received signals not only from the hot cable but also form the much colder background.

The two problem areas with thermal anomalies of the jumpers from the bushings were detected with all cameras. The jumpers from the bushings to the feed are stranded cables where some of the strands have broken over time or are not well-connected electrically. This causes overloading of the remaining good conductors resulting in significant heating. Preliminary conclusions indicate that one may detect thermal anomalies with all cameras.

However, the question arises, whether these anomalies impose a problem that needs to be dealt with. Obviously, one must quantitatively estimate the temperature increases of the problem areas. The maximum temperature of the left-most phase in Figure 9.54 was analyzed for all three images. Not surprisingly, the temperature rise with respect to ambient temperature differs appreciably for the three cameras showing 50 °C (low), 95 °C (medium), and 117 °C (high) for 40% load of the transformer.

This 40% load situation was observed in early June when the weather was cool and electrical demand in the area was low. Obviously, later in summer, the use of

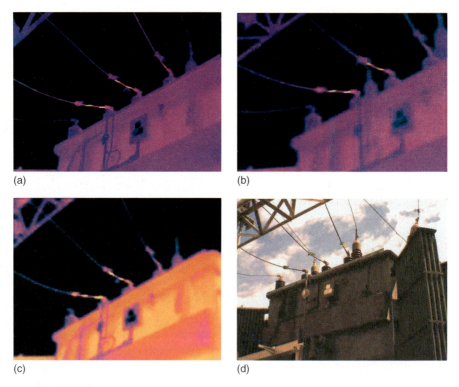

Figure 9.54 Thermal images of a 115–23 kV substation transformer low side. (a) High resolution (640 × 480 pixels), (b) medium resolution (320 × 240 pixels), (c) low resolution (160 × 120 pixels), and (d) visible image. Significant thermal anomalies can be clearly observed with all cameras, quantitative analysis however, results in large differences (for details, see text). (Images courtesy Infrared training center.)

air conditioners would lead to much higher demand. Therefore, the temperature rises were also estimated for 100% load (for load correction [77]). It was found that temperature differences would significantly increase to 211 °C (low), 402 °C (medium), and even 494 °C (high). Whether a temperature rise is critical or not depends on the criteria [87], for example, those of the American Electric Power Research Institute (EPRI). In this case, the problem severity according the EPRI criteria was already critical at 40% load for the medium- and high-resolution measurements! For the estimate of 100% load, results indicated a very high probability of major failure.

The example of Figure 9.54 nicely demonstrates the importance of exactly knowing the spatial resolution of the camera used, in particular, if the distance to small targets is quite large. For the conditions presented here, a cheap and simple low-resolution camera of 160 × 120 pixels is useless.

It should also be mentioned that whenever surface temperatures are measured from high-voltage electrical components, small measured values at the outside surface may correspond to large internal temperature rises. This can, for example, be the case for oil circuit breakers [88]. In this case, power loss calculations including convective and radiative heat transfer can be based on measured surface temperatures in order to estimate the total power loss and therefrom the internal electrical resistance can be calculated.

9.7.4
Overheated High-Voltage Line

Figure 9.55 depicts a number of 130-kV lines under load. Three cables are clearly overheated as can be seen in direct comparison with the other (blue color) cables which were much colder. It turned out that the load upon these three cables was occasionally enlarged without taken into account, whether they are run with load

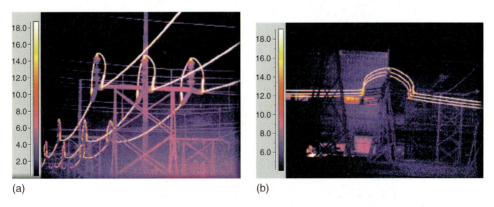

Figure 9.55 About 130-kV high-voltage lines that are clearly overheated. (Images courtesy J. Giesecke.)

above the specified limit. The diameter of the cables needs to be increased too to reduce their resistance, that is, the electrical power consumption by the cables themselves.

9.7.5
Electric Fan Defects

Transformer units at substations are critical components. Their housings are usually cooled from their sides by a series of powerful fans. Figure 9.56 depicts an array of nine fans, which cool an area of about $4\,m^2$ of a transformer housing. The center fan in the image suffered from a severe malfunctioning. The fan motor was overheated and only allowed for a strongly reduced cooling operation as can be seen by the elevated transformer wall temperatures. When this unit was exposed to extremely high solar load, the reduced cooling sometimes resulted in problems, that is, the fan motor had to be exchanged or repaired immediately.

9.7.6
Oil Levels in High-Voltage Bushings

A very important application of IR imaging is monitoring oil levels in bushings of HV power equipment [89, 90]. This is due to the fact that it indirectly deals with the most costly piece of equipment in the power grid, the power transformer. Bushings are devices that allow electrical current to pass through a barrier and provide an electrical connection on each side while providing electrical insulation between the center conductor and the ground. The inside conductor is surrounded by electrical insulation. In higher voltage bushings, it consists of concentric layers of insulation and layers of conductor foils. These layers of insulation and conductors form a concentric capacitor between the high-voltage center core and the bushing flange at ground potential. In most modern bushings above 26 kV, paper provides the skeleton for the insulation system. The paper is impregnated with mineral oil to provide more insulation.

One of the most important failure modes of HV bushings are oil leakages. Whatever be the reason for the leakage, moist air may enter the bushing and

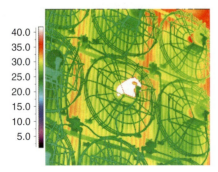

Figure 9.56 Housing of a transformer unit, cooled by a series of electric fans. The center fan is malfunctioning leading to reduced cooling power. (Image courtesy J. Giesecke.)

replace the oil. Since the paper will become void of oil, the insulation ability of the bushing decreases. Eventually, electrical discharges (corona discharges) can develop, which may eat like worms through the paper, thus causing the foil layers to short out. In the worst case, the dry insulation may flash and lead to losses in the million dollar range (the cost of high-voltage transformer units). Therefore, it is of the utmost importance to regularly check the oil levels in such high-voltage transformer bushings.

IR thermal imaging is a very useful method in determining the oil levels in such bushings in PdM programs. The method for detecting oil levels is simple. In any transformer, there are energy losses that lead to a temperature increase in the transformer coils. These heat up the oil in the tank of the transformer and, via thermal conduction, the oil filling in the attached bushing. Because of the thermal conductivity through the wall of the bushing and heat capacitance of the oil with respect to the one of air above it, the bushing oil level can be seen. So the bushings will appear warmer where the oil is and cooler where there is no oil. The IR image (Figure 9.57) shows two bushings: the one on the right being full of oil and the other one being only partially filled. The oil-filled bushing is warmer (around 31 °C) due to the heat conduction from the warm transformer tank full of oil up the bushing all the way to the top. In contrast, the bushing with low oil level is much colder, here only around 15 °C, since there is no oil in the bushing to conduct the heat from the transformer to the bushing. Only the very lowest part of the bushing on the left shows signs of heat being conducted thus marking the location of the oil level. After verifying the low oil level, the transformer was immediately removed from service for repair in order to avoid possible catastrophic failure.

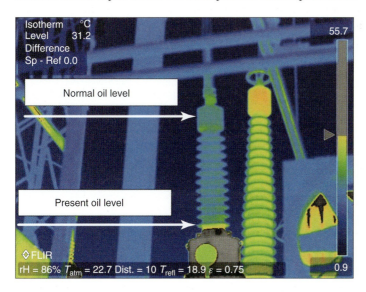

Figure 9.57 Checking oil levels in bushings of HV power equipment. For details, see text. (Image courtesy M.B. Goff.)

References

1. Hecht, E. (1998) *Optics*, 3rd edn, Addison-Wesley.
2. Vollmer, M., Henke, S., Karstädt, D., Möllmann, K.P., and Pinno, F. (2004) Identification and suppression of thermal reflections in infrared thermal imaging. Inframation 2004, Proceedings vol. 5, pp. 287–298.
3. Bell, R.J. (1972) *Introductory Fourier Transform Spectroscopy*, Academic Press, New York.
4. Kauppinen, J. and Partanen, J. (2001) *Fourier Transform in Spectroscopy*, Wiley-VCH Verlag GmbH, Berlin.
5. Palik, E.P. (ed.) (1985) *Handbook of Optical Constants of Solids*, vol. 1, Academic Press; vol. 2, (1991).
6. Möllmann, K.P., Karstädt, D., Pinno, F., and Vollmer, M. (2005) Selected critical applications of thermography: convections in fluids, selective emitters and highly reflecting materials. Inframation 2005, Proceedings vol. 6, pp. 161–173.
7. http://www.thermografie-schweiger.de. (2010).
8. De Witt, D.P. and Nutter, G.D. (1988) *Theory and Practice of Radiation Thermometry*, Jhon Wiley & Sons, Inc., New York.
9. Holst, G.C. (2000) *Common Sense Approach to Thermal Imaging*, SPIE Optical Engineering Press, Washington, DC.
10. Vollmer, M., Henke, S., Karstädt, D., Möllmann, K.P., and Pinno, F. (2004) Challenges in infrared imaging: low emissivities of hot gases, metals, and metallic cavities. Inframation 2004, Proceedings vol. 5, pp. 355–363.
11. Schmidt, V., Möllmann, K.-P., Muilwijk, F., Kalz, S., and Stoppiglia, H. (2002) Radiation thermometry on aluminium. VIR[*]-Conference 2002, Ijmuiden, vol. 78, October 2002, pp. 897–901.
12. Gärtner, R., Klatt, P., Loose, H., Lutz, N., Möllmann, K.-P., Pinno, F., Muilwijk, F., Kalz, S., and Stoppiglia, H. (2004) New aluminium radiation thermometry. VIR[*]-Conference 2004, Brussels, vol. 80, June 2004, pp. 642–647.
13. Papp, L. (2004) Using IR surveys in DUNAFERR steelworks complex diagnostic systems. Inframation 2004, Proceedings vol. 5, pp. 371–386.
14. Nolan, D., Zhao, Q., Abdelrahman, M., Vondra, F., and Dinwiddie, R.B. (2006) The effect of process characteristics on metal fill and defect formation in aluminum lost foam casting. Inframation 2006, Proceedings vol. 7, pp. 51–64.
15. Calmes, F. (2000) Infrared inspections of robotic welders in automotive assembly. Inframation 2000, Proceedings vol. 1, pp. 25–23.
16. Sinclair, D. (2004) Infrared cameras enhance productivity and safety at GM. Inframation 2004, Proceedings vol. 5, pp. 281–286.
17. Predmesky, R. and Ruane, T. (2004) Using infrared cameras for process control. Inframation 2004, Proceedings vol. 2, pp. 27–29.
18. Royo, R. (2004) Characterization of automation brake thermal conditions by the use of infrared thermography. Inframation 2004, Proceedings vol. 5, pp. 259–266.
19. Infrared detection of standing waves. European patent application 1255102, see http://www.freepatentsonline.com/EP1255102.pdf. (2010).
20. Tire wear forecasting method and apparatus. US patent 6883962. (2005).
21. Putman, B.J. and Amirkhanian, S.N. (2006) Thermal segregation in asphalt pavements. Inframation 2006, Proceedings vol. 7, pp. 233–244.
22. Steyn, W.Jvd.M. (2005) Applications of thermography in pavement engineering. Inframation 2005, Proceedings vol. 6, pp. 81–89.
23. Monem, T.A., Olaufa, A.A., and Mahgoub, H. (2005) Asphalt crack detection using thermography. Inframation 2005, Proceedings vol. 6, pp. 139–150.
24. Amirkhanian, S.N. and Hartman, E. (2004) Applications of infrared cameras in the paving industry, Inframation 2004, Proceedings vol. 5, pp. 91–99.

25. Philipps, L., Willoughby, K., and Mahoney, J. (2003) Infrared thermography revolutionizes hot-mix asphalt paving. Inframation 2003, Proceedings vol. 4, pp. 213–221.
26. Crane, R.L., Astarita, T., Berger, H., Cardone, G., Carlomagno, G.M., Jones, T.S., Lansing, M.D., Russell, S.S., Walker, J.L., and Workman, G.L. (2001) Aerospace applications of infrared and thermal testing, in *Nondestructive Testing Handbook,* Infrared and Thermal Testing, vol. 3, 3rd edn, Chapter 15 (ed. P.O. Moore), American Society for Nondestructive Testing, Inc., Columbus. pp. 489–526.
27. Spring, R.W. (2001) An overview of infrared thermography applications for nondestructive testing in the aerospace industry. Inframation 2001, Proceedings vol. 2, pp. 191–196.
28. Genest, M. and Fahr, A. (2009) Pulsed thermography for nondestructive evaluation (NDE) of aerospace materials. Inframation 2009, Proceedings vol. 10, pp. 59–65.
29. Figg, J. and Daquila, T. (2001) Applying the benefits of infrared thermography in naval aviation. Inframation 2001, Proceedings vol. 2, pp. 207–214.
30. Tarin, M. and Kasper, A. (2008) Fuselage inspection of Boeing-737 using lock-in thermography. *Proceedings of the Thermosense XXX, Proceedings of SPIE vol. 6939*, SPIE Press, Bellingham, electronic file: 693919-1 to 693919-10.
31. Holleman, E., Sharp, D., Sheller, R., and Styron, J. (2007) IR characterization of Bi-propellant reaction control engines during auxiliary propulsion systems tests at NASA's White Sands test facility in Las Cruces, New Mexico. Inframation 2007, Proceedings vol. 8, pp. 95–107.
32. Gazarek, M., Johnson, D., Kist, E., Novak, F., Antill, Ch., Haakenson, D., Howell, P., Jenkins, R., Yates, R., Stephan, R., Hawk, D., and Amoroso, M. (2005) Infrared on-orbit RCC inspection with the EVA IR camera: development of flight hardware from a COTS system. Inframation 2005, Proceedings vol. 6, pp. 273–284.
33. Gazarik, M., Johnson, D., Kist, E., Novak, F., Antill, C., Haakenson, D., Howell, P., Pandolf, J., Jenkins, R., Yates, R., Stephan, R., Hawk, D., and Amoroso, M. (2006) Development of an extra-vehicular (EVA) infrared (IR) camera inspection system. Proceedings of the Thermosense XXVIII, Proceedings of SPIE vol. 6205, p. 62051C.
34. Elliott Cramera, K., Winfreea, W.P., Hodgesb, K., Koshtib, A., Ryanc, D., and Reinhardt, W.W. (2006) Status of thermal NDT of space shuttle materials at NASA. Proceedings of the Thermosense XXVIII, Proceedings of SPIE vol. 6205, p. 62051B.
35. ITC Newsletter, vol. 7, issue 9, p. 2, see *http://itcnewsletter.com* (September 2006).
36. Sauser, B. *http://www.technologyreview.com/computing/23533/*, published by MIT, Monday, Image Credit: NASA/HYTHIRM team (28 September 2009).
37. Schultz, C. (2004) Developing and implementing an infrared predictive maintenance program. Maintenance Technology, May 2004, pp. 28–32.
38. May, K.B. (2003) Predictive maintenance inspections – boilers in fossil power plants. Inframation 2003, Proceedings vol. 4, pp. 161–166.
39. Updegraff, G. (2005) Infrared thermography program at a gas-fired turbine generation plant. Inframation 2005, Proceedings vol. 6, pp. 293–300.
40. Ralph, M.J. (2004) Power Plant Thermography- wide Range of Applications. Inframation 2004, Proceedings vol. 5, pp. 241–248.
41. Ralph, M.J. (2009) The continuing story of power plant thermography. Inframation 2009, Proceedings vol. 10, pp. 253–258.
42. Didychuk, E. (2006) Predictive maintenance of HVDC converter stations within Manitoba Hydro. Inframation 2006, Proceedings vol. 7, pp. 409–422.
43. Bales, M.J., Berardi, P.G., Bishop, C.C., Cuccurullo, G., Grover, P.E., Mack, R.T., McRae, T.G., Pelton, M.W., Seffrin, R.J., and Weil, G.J. (2001) Chemical and petroleum applications of infrared and thermal testing, in *Nondestructive Testing Handbook*, Infrared and Thermal Testing, vol. 3, 3rd edn, Chapter 17

44. Massey, L.G. (2007) IR keeps West Virginia coal miners safe. Inframation 2007, Proceedings vol. 8, pp. 327–334.
45. Blanch, M. (2008) Thermal imaging extracting faults in the sand mining industry. Inframation 2008, Proceedings vol. 9, pp. 45–48.
46. Blanch, M. (2008) How sweet it is using infrared thermography. Inframation 2008, Proceedings vol. 9, pp. 49–51.
47. Thon, R.J. (2002) Troubleshooting paper machine problems through thermal imaging. Inframation 2002, Proceedings vol. 3, pp. 191–194.
48. Baird, L.W. and Bushee, R.L. (2003) Paper mill predictive maintenance utilizing infrared. Inframation 2003, Proceedings vol. 4, pp. 17–24.
49. Thon, R.J. (2007) Using thermography to reduce energy costs in paper mills. Inframation 2007, Proceedings vol. 7, pp. 553–555.
50. Bremond, P. (2004) IR Imaging Assesses Damage in Mechanical Parts. Photonics Spectra (February 2004), pp. 62–64.
51. Allison, J.N. (2005) Applying infrared imaging techniques to marine surveying ... identifying moisture intrusion in a wood cored FRP yacht. Inframation 2005, Proceedings vol. 6, pp. 1–13.
52. Allison, J.N. (2008) Infrared thermography as part of quality control in vessel construction. Inframation 2008, Proceedings vol. 9, pp. 11–16.
53. Ershov, O.V., Klimova, A.G., and Vavilov, V.P. (2006) Airborne detection of natural gas leaks from transmission pipelines by using a laser system operating in visual, near-IR and mid-IR wavelength bands. Thermosense XXVIII, Proceedings of SPIE vol. 6205, p. 62051G.
54. Miles, J.J., Dahlquist, A.L., and Dash, L.C. (2006) Thermographic identification of wetted insulation on pipelines in the arctic oilfields. Thermosense XXVIII, Proceedings of SPIE vol. 6205, p. 62051H.
55. Bonin, R.G. (2002) Infrared applications in the petrochemical refinery. Inframation 2002, Proceedings vol. 3, pp. 13–17.
56. Whitcher, A. (2004) Thermographic monitoring of refractory lined petroleum refinery equipment. Inframation 2004, Proceedings vol. 5, pp. 299–308.
57. Susralski, B.D. and Griswold, Th. (2000) Predicting mechanical systems failures using IR thermography. Inframation 2000, Proceedings vol. 1, pp. 81–87.
58. Pearson, J. and Pandya, D.A. (2006) IR thermography assessment enhanced compressed air system operations & indirectly slashed $26 K utility bills at New Jersey international and bulk mail center. Inframation 2006, Proceedings vol. 7, pp. 117–121.
59. Tarin, M. (2009) Solar panel inspection using lock-in thermography. Inframation 2009, Proceedings vol. 10, pp. 225–237.
60. Aderhold, J., Meinlschmidt, P., Brocke, H., and Jüngert, A. (2008) Rotor blade defect detection using thermal and ultrasonic waves. Proceedings of the 9th International German Wind Energy Conference DEWEK, held 26th – 27th November 2008, in Bremen, http://08.dewek.de/; see also Fraunhofer Gesellschaft/Germany: http://www.vision.fraunhofer.de/en/20/projekte/459.html.
61. Moore, S. (2007) Thermography improves operations at grocery stores. Inframation 2007, Proceedings vol. 8, pp. 181–186.
62. Sims, D. (2001) Using infrared imaging on production tanks and vessels. Inframation 2001, Proceedings vol. 2, pp. 119–125.
63. Sims, D. (2004) Monitoring the process conditions in oil field production vessels with infrared technology. Inframation 2004, Proceedings vol. 5, pp. 273–280.
64. Snell, J. and Schwoegler, M. (2004) Locating levels in tanks and silos using infrared thermography. Proceedings of the Thermosense XXVI, Proceedings of SPIE vol. 5405, pp. 245–248.
65. Walford, G.V. (2008) Use of infrared imaging techniques to understand

and quantify expanded polymer molding operations and part quality. Inframation 2008, Proceedings vol. 9, pp. 493–503.
66. Madding, R.P. (2004) IR window transmittance temperature dependence. InfraMation 2004, Proceedings vol. 5, pp. 161–169.
67. Bosworth, B.R., Eto, M., Ishii, T., Mader, D.L., Okamoto, Y., Persson, L., Rayl, R.R., Seffrin, R.J., Shepard, S.M., Snell, J.R., Teich, A.C., Westberg, S.-B., and Zayicek, P.A. (2001) Electric power applications of infrared and thermal testing, in *Nondestructive Testing Handbook*, Infrared and Thermal Testing, vol. 3, 3rd edn, Chapter 16 (ed. P.O. Moore), American Society for Nondestructive Testing, Inc., Columbus. pp. 527–570
68. Bennett, R. (2009) Normalizing the E-values of components on a printed circuit board. Inframation 2009, Proceedings vol. 10, pp. 157–162.
69. Fergueson, D.E. (2006) Applications of thermography in product safety. Inframation 2006, Proceedings vol. 7, pp. 203–206.
70. Fishbune, R.J. (2000) IR thermography for electronic assembly design verification. Inframation 2000, Proceedings vol. 1, pp. 211–217.
71. Wallin, B. and Wiecek, B. (2001) Infrared Thermography of electronic components, in *Nondestructive Testing Handbook*, Infrared and Thermal Testing, vol. 3, 3rd edn, Chapter 19 (ed. P.O. Moore), American Society for Nondestructive Testing, Inc., Columbus. pp. 659–678
72. Radford, P. (2008) Ship's stearing gear motor problem diagnosed using IR technology. Inframation 2008, Proceedings vol. 9, pp. 373–376.
73. Madding, R.P. (2001) Finding internal electrical resistance from external IR thermography measurements on oil-filled circuit breakers during operation. Inframation 2001, Proceedings vol. 2, pp. 37–44.
74. Leonard, K. (2006) What if? Inframation 2006, Proceedings vol. 7, pp. 123–127.
75. Maple, C. (2003) Secure Plant infrastructure through remote substation monitoring. Inframation 2003, Proceedings vol. 4, pp. 151–159.
76. Brydges, D. (2005) Aerial thermography surveys find insulator and other problems. Inframation 2005, Proceedings vol. 6, pp. 193–202.
77. Madding, R.P. (2002) Important measurements that support IR surveys in substations. Inframation 2002, Proceedings vol. 3, pp. 19–25.
78. Gierlach, J. and DeMonte, J. (2005) Infrared inspections on electric panels without removing covers: can the inspection be completed correctly? Inframation 2005, Proceedings vol. 6, pp. 203–212.
79. www.iriss.com, free download of The Ten Things You Need To Know About Infrared Windows. (2010)
80. Robinson, M. (2007) Infrared Windows: Where do I start? Inframation 2007, Proceedings vol. 8, pp. 291–307.
81. Newton, V. (2008) Arc flash and electrical safety for thermographers. Inframation 2008, Proceedings vol. 9, pp. 321–332.
82. Theyerl, M.N. (2008) Arc Flash concerns & detecting early signs. Inframation 2008, Proceedings vol. 9, pp. 447–457.
83. Androli, D. (2006) A thermographers common sense approach to arc flash safety. Inframation 2006, Proceedings vol. 7, pp. 207–214.
84. Woods, B. (2005) Reducing arc flash hazards through electrical system modifications. Inframation 2005, Proceedings vol. 6, pp. 15–21.
85. Möllmann, K.-P. and Vollmer, M. (2007) Infrared thermal imaging as a tool in university physics education. *Eur. J. Phys.*, **28**, S37–S50.
86. Madding, R.P., Orlove, G.L., and Lyon, B.R. (2006) The importance of spatial resolution in IR thermography temperature measurement – three brief case studies. Inframation 2006, Proceedings vol. 7, pp. 245–251.
87. Giesecke, J.L. (2006) Adjusting severity criteria. Inframation 2006, Proceedings vol. 7, pp. 33–35.
88. Madding, R.P., Ayers, D., and Giesecke, J.L. (2002) Oil circuit breaker

thermography. Inframation 2002, Proceedings vol. 3, pp. 41–47.
89. Goff, M.B. (2001) Substation equipment (bushings). Inframation 2001, Proceedings vol. 2, pp. 113–117.
90. Goff, M.B. (2008) The secrets of bushing oil level. Inframation 2008, Proceedings vol. 9, pp. 167–174.

10
Selected Applications in Other Fields

10.1
Medical Applications

10.1.1
Introduction

Thermal imaging is increasingly becoming accepted by physicians as a supplementary diagnostic as well as a monitoring tool for various diseases. This is based on the fact that diseases as well as injuries lead to changes in body surface temperature, which are recorded by sensitive IR cameras.

Surface temperature of an extremity reflects the result of a complex combination of central and local regulatory systems [1]. As a living organism, the human body tries to maintain homeostasis, that is, an equilibrium of all systems within the body, for all physiologic processes, which (amongst others) leads to dynamic changes in heat emission. They depend on internal (e.g., blood flow, hormones, food intake in particular alcohol, smoking, exercise, emotion) and external (e.g., room temperature, humidity, clothing, cosmetics, jewelry) conditions. This makes it inevitable to develop standard operating procedures in order to be able to interpret thermal imaging results that might influence the physician's decision on further steps through the diagnostic process and treatment.

The most important factor seems to be arterial blood flow as surface temperature increases with intensified blood flow, whereas augmented venous blood flow correlates with a decrease (e.g., physical exercise leads to a dilation of veins in order to cool down core body temperature) [2]. This physiologic process is maintained by many factors: the activation of the sympathetic nervous system and hormones/catecholamines seem to play an important role as activation of the sympathetic nervous system – partly controlled by hormones and catecholamines – leads to vasoconstriction. In a healthy person, skin temperature is symmetrical. Nevertheless, acral temperature varies more than the temperature in the area of the trunk, where it is usually higher than in more distal parts [3].

The major advantage of medical IR imaging next to being an easy-to-handle tool in diagnostics and monitoring is that the progress of a disease and/or the effect of therapeutic measures can be visualized for patients. Especially in the beginning of a

Infrared Thermal Imaging. Michael Vollmer and Klaus-Peter Möllmann
Copyright © 2010 WILEY-VCH Verlag GmbH & Co. KGaA, Weinheim
ISBN: 978-3-527-40717-0

treatment, patients may not notice sufficient relief. In this case, IR images can help to demonstrate initiated changes and increase compliance for intensified and/or time-demanding efforts of the medical staff. This is best illustrated for pressure ulcers in patients with diabetes: because of the underlying disease, perception of healing is impaired by polyneuropathies. Here, IR thermograms can visualize positive effects of treatment before obvious macroscopic improvement. Advantages of thermography with modern camera systems are the high spatial resolution as well as fast data recording.

Standard conditions in thermographic imaging have been discussed copiously [4]. However, first of all because of lack of time, second because of patients' incompliance, and third because of lack of knowledge, they are not followed regularly. For example, room temperature as well as humidity and medication, which may have an impact on surface blood flow, should be included in the report. For standardization, the following five major conditions should be fulfilled in order to avoid common errors [5]:

1) proper room conditions
    a. should have a minimum size of 9 m^2
    b. should have a constant room temperature (23.5 °C)
    c. should have a humidity of 45–60%
    d. should avoid turbulent airflow, that is, no heat emitters, ventilators, or air conditioning close to the patient
    e. if necessary, shades should be available in order to reduce the effect of sunlight
    f. should have a proper selection of material on walls and ceiling to avoid thermal reflections
2) proper preparation of the patient
    a. acclimatization for at least 15 min in a proper room (resting in a natural position, light clothing without feeling too hot or too cold), taking off clothes from areas to be examined
    b. no intake of food, alcohol, nicotine, coffee, or tea 2 h before examination
    c. no physical exercise 24 h before examination
    d. no deodorants, crèmes, perfume, no solutions, or shaving at the body surface areas to be monitored
    e. no acupuncture, transcutaneous electrical nerve stimulation, or invasive procedures before examination
    f. taking off jewelry 4 h before examination
    g. no excessive tanning or sun burns 7–10 days before examination
3) standardized equipment
    a. calibrated IR camera, low NETD
    b. camera switched on at least 1 h before examination in the same room (see switch on behavior and thermo shock, Section 2.4.6)
4) proper recording and storage of acquired images
    a. avoid grazing incidence of radiation on camera, select viewing angle close to 0° since $\varepsilon = \varepsilon(\text{angle})$ (Section 1.4)

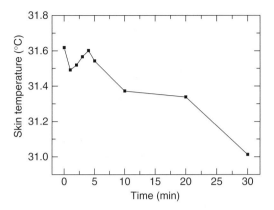

Figure 10.1 Decrease in skin temperature (hands) over a period of 30 min after an acclimatization period of 15 min. (Graph courtesy K. Agarwal.)

  b. files should include gender, age, diagnosis, current medication, size, weight, room conditions, and date
5) representation of images
  a. standardized temperature scale (color palette, level, span), often a span 23–36 or 24–38 °C is chosen.

In multiple books on dermatology or physiology, "normal skin temperature" is defined to range from 32 to 34 °C. In addition, despite proper acclimatization, skin temperature continues to drop at least 30 min into the examination (Figure 10.1). Even if all of the above-mentioned items are followed strictly, the human body is a highly dynamic organism, that is, even under the same conditions, surface temperature may vary intra- (0.4–11 °K) and interindividually (up to 12.3 °K) without any obvious cause, the largest spread occurring at the extremities (Figure 10.2). This renders interpretation of thermograms even more sophisticated [6].

For each application, different aspects are important; therefore, here only a few case studies of medical applications are discussed in more detail, that is, pain management, acupuncture, and breast cancer, whereas others are only mentioned briefly.

10.1.2
Diagnosis and Monitoring of Pain

Objectively determining pain intensity is a key intention in pain management. Unfortunately, so far no tool has been developed that sufficiently fulfills this task; hence, it is extremely complex to conclude if treatment offered to a patient is helpful or not. In particular forms of pain, that is, neuropathic and sympathetically maintained pain, blocks of the sympathetic nervous system are performed in order to diminish pain intensity. They often result in significant relief, though there is no way of proving if pain improvement results from blocking other structures

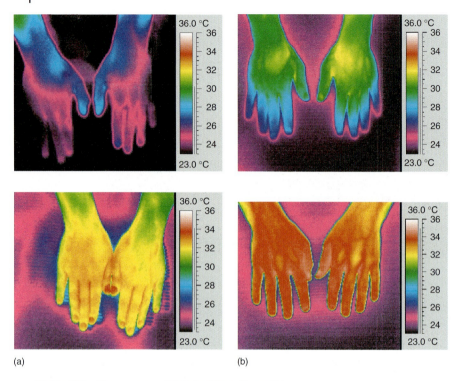

Figure 10.2 Thermograms of four healthy subjects ((a) female, (b) male, all age 25) after acclimatization over 15 min at standardized conditions. (Images courtesy K. Agarwal.)

than the sympathetic nervous system, for example, spinal nerves or if it is just a psychological effect. As sympathetic fibers influence blood flow by varying size of vessels, a rise in the skin temperature is expected when performing sympathetic blocks, because variation of skin perfusion correlates with surface temperature. The configuration of changes in skin temperature provides us with precious information on the success of the intervention besides the patient's subjective impression. Therefore, it was proposed [7–9] that IR thermography can obviously be utilized as a diagnostic tool with extraordinary sensitivity and repeatability, although it has been rarely applied in that manner so far. Two examples of pain treatment are discussed in more detail [10].

First, the complex regional pain syndrome (CRPS) is a painful syndrome usually affecting distal extremities of the body, manifesting itself with a wide variety of symptoms. The most outstanding feature is unbearable pain, including spontaneous pain, allodynia, hyperpathia, and hyperalgesia. Usually, the affected extremity displays changes in color and/or temperature (vasomotor disturbances), edema, alterations in transpiration, hair and nail growth (sudomotor disturbances), and muscular atrophy and/or dysfunction (motortrophic disturbances). In some cases, a specific initiating event can be identified, for example, trauma or surgery,

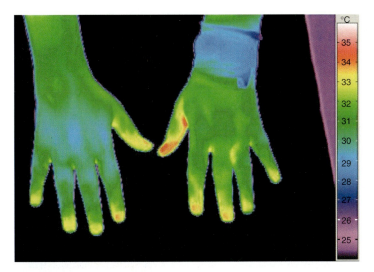

Figure 10.3 Thermogram prior to sympathetic block in a female patient with CRPS; note surface temperature on the right hand is about 2 K lower than in the unaffected left hand. (Images courtesy K. Agarwal.)

but sporadically it is observed after stroke, myocardial infarction, infection, or even without an obvious inciting event. The course of the disease varies interindividually. If proper treatment is not available, or in some cases despite treatment, it may lead to persisting diminished function of the affected limb, which deeply impairs the quality of life of the patient.

Temperature variations are commonly considered as a major diagnostic criterion in the diagnosis of CRPS; hence, IR thermography can be utilized as a diagnostic tool in CRPS with extraordinary sensitivity and repeatability, though it is not specific (Figure 10.3). Skin temperature is a superior predictor of sympathetic activity as a good correlation is found between skin temperature and skin sympathetic nerve activity.

This malady might be treated with blocks of the sympathetic nervous system, which in return have an impact on skin perfusion (Figure 10.4). Pain usually can be reduced within a few hours after injecting the local anesthetic, but the complete treatment is very time consuming; patients often lose patience. To increase compliance to continue therapy, IR images are very useful.

Second, osteomyelitis is a painful infection of the bone and marrow. In adults it primarily affects digits of the feet and is most often associated with diabetes mellitus. Very often it is accompanied by neuropathic pain, which summarizes the collection of an assortment of pain states with a common feature: symptoms are suggestive of dysfunction and/or lesion of nerves.

Treatment may be sophisticated and time consuming, often requiring prolonged antibiotic therapy for weeks or months and sometimes it may also require surgical intervention [10]. The lack of efficacy of antibiotic treatments may correlate

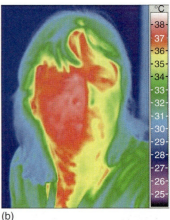

(a)   (b)

Figure 10.4 Thermograms 2 h after sympathetic block of the hands in the same patient; the block was performed on the right side and, hence, perfusion of that hand and that side of the face led to an increase in skin temperature by $\Delta T \approx 10$ K as opposed to $\Delta T \approx 3$ K contralaterally. All thermograms were recorded using the above-mentioned standardized procedure. (Images courtesy K. Agarwal.)

with diminished blood flow in the affected area. This can be easily detected using thermography. In particular, it may depict physiologic changes that cannot be demonstrated by ultrasound, computer tomography (CT), or magnetic resonance imaging (MRI). A healthy human usually presents a symmetric surface temperature that correlates with tissue blood flow, which is influenced by sympathetic activity. In the treatment of sympathetically maintained pain, thermography is used to prove the efficacy of sympathetic chain blocks in order to determine whether neuroablation is indicated and may be effective. Figures 10.5–10.7 depict IR images of a young male who suffered from osteomyelitis in his jaw for more than five years after an accident. He had undergone initial surgery including bony reconstruction of the jaw and had received multiple doses of antibiotics; still healing would not progress as desired and he suffered from neuropathic pain in the affected area that started spreading over his whole face. As sufficient blood supply is inevitable for delivery of satisfactory amounts of antibiotics to the affected site, the primary aim in treating this patient was to increase perfusion in the area involved.

Figure 10.5 shows the patients face and hands prior to treatment. On the basis of the hypothesis that the whole disease might be sustained by the sympathetic nervous system, he received blocks of the stellate ganglion and the thoracic sympathetic chain. Figure 10.6 shows thermograms during continuous infusion of a local anesthetic. The increased blood flow led to higher skin temperature on the right side of the face as intended. In addition, the same applies to the hand, which proves that the effect is most probably sustained by the sympathetic nervous system. Figure 10.7 depicts IR images after a 10-day infusion treatment with local anesthetics: facial blood flow is augmented all over, whereas the perfusion of the

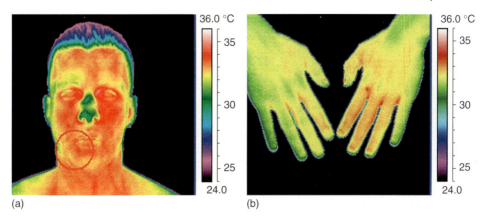

Figure 10.5 IR images of patient before treatment. (a) Facial image, the circle indicates the location of the infection. (b) Dorsal IR image of the hands. (Images courtesy K. Agarwal.)

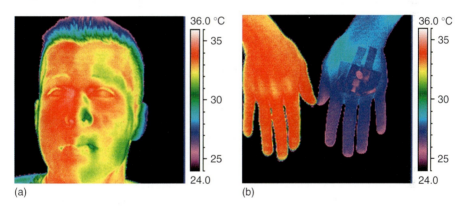

Figure 10.6 IR images of patient during treatment. (a) Facial image, indicating the increased blood flow on the right side of the face. (b) Palmar IR image of the hands. The pink dots and dark blue shades on the dorsal surface of the hand originate from an intravenous access that has been taped to the skin. (Images courtesy K. Agarwal.)

hands has returned to the preinterventional level. The patient was still feeling well 15 months after the treatment.

10.1.3
Acupuncture

Acupuncture is an ancient Chinese approach to various kinds of diseases; in particular, it has been adapted in western society to the treatment of pain. It is increasingly applied in therapeutic concepts for treating multiple illnesses especially

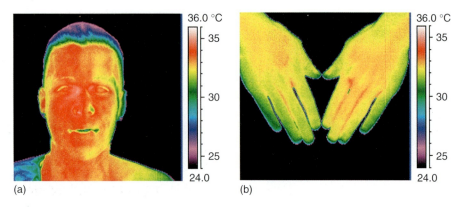

Figure 10.7 IR images of patient after treatment. (a) Facial image, the blood flow is augmented all over. (b) Perfusion of the hands had returned to the preinterventional level. (Images courtesy K. Agarwal.)

due to growing demand by patients themselves as they expect this treatment to be less harmful and a softer method than western medicine, avoiding chemicals as in regular medication and pills. Compared to classical western medicine, acupuncture is considered to be a nontraditional approach in Europe and America. The demand for acupuncture has also grown in cases where western medicine may not offer a solution if no morphologic correlates can be found in laboratory specimen and/or other diagnostic procedures like imaging techniques.

The most important effective structural elements of acupuncture are acupoints. According to traditional Chinese medicine (TCM), their distribution defines the system of channels, also called *meridians*, through which "qi" flows. No generally accepted proof for meridians has so far being given. There is no exact translation for "qi" into western medicine, though it can be attributed to "vital energy." In acupuncture, needles are inserted into acupoints. In this case, patients often state a strange sensation that may comprise a feeling of warmth, cold, strong emotions, or relaxation referred to as the *de-qi-sensation*. So far, scientists have not been able to attribute this feeling to a specific bodily reaction and overall no scientific proof for the efficacy of acupuncture has been found yet. There are several theories of why and how acupuncture might work. Because of a lack of scientific evidence, traditional medicine often does not accept nor believe in symptom alleviation by acupuncture. Recent research revealed that at least at the point Hegu a specific effect occurs: needling of this acupoint results in a massive increase in skin temperature, whereas needling of the skin or muscle (the so-called sham-needling or false needling) or no needling leads to a decrease (Figure 10.8). The study was performed with 50 healthy volunteers as a randomized single-blinded placebo-controlled cross-over clinical trial [11].

Visualization of acupoint distribution as well as the course of the meridians by measuring the skin surface temperatures is being discussed controversially. As meridian-like structures could be found by some researchers, others accused

10.1 Medical Applications | 543

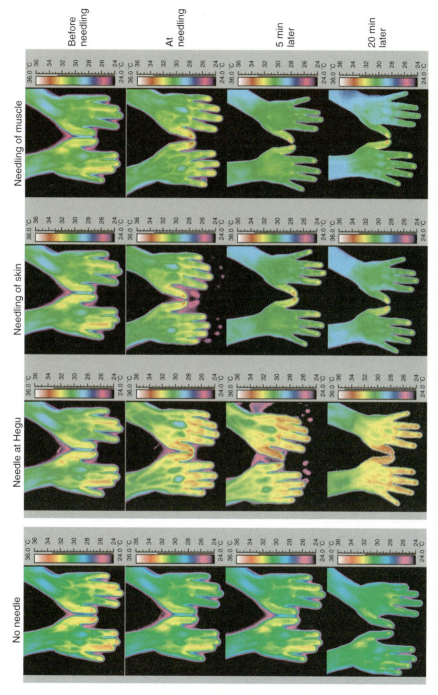

Figure 10.8 Thermograms of the hands before, during, and after needling at Hegu, the skin and muscle. (Images courtesy K. Agarwal.)

that inferior techniques would produce these phenomena. Litscher [12] could not visualize structures according to meridians as described in TCM. With the aid of various IR cameras with diverging properties like resolution and with different methods of stimulation (i.e., Moxibustion, acupuncture, LASER), they were clearly able to objectively quantify technical reflection artifacts from the moxa-cigar on the skin of healthy probands that appeared to be meridian-like at first sight. No thermal reflection phenomena were present and also no meridians could be detected as soon as stimulation without heat was utilized. Other authors describe channel-like pathways not known to TCM. Considering all the facts discovered so far, it is clear that an underlying anomalous dispersion has to be insinuated.

10.1.4
Breast Thermography and Detection of Breast Cancer

Because of its noninvasive, nonradiating, and passive nature, thermographic imaging is again becoming more accepted in early detection of anomalies of the breast. Usually, clinical breast exams, mammograms, and ultrasound are utilized in diagnosing pathologies of the breast. Although conventional breast-imaging techniques routinely include mammography and ultrasound, growing interest in other approaches has drawn increasing attention to exploiting the anatomic and physiologic basis for understanding breast cancer. With the aid of thermography, tumors in the breast are to be identified due to elevated surface temperature caused by the increased blood flow as a response to augmented cellular activity.

In detail, breast thermography is based on the premise that before the growth of abnormal cells is possible, a constant blood supply must be circulated to the growth area. Thermography measures the breast surface temperatures and, hence, the heat generated by the microcirculation of blood in the breast during this process.

The chemical and blood vessel activity in both precancerous tissue and the area surrounding a developing breast cancer is almost always higher than in the normal breast [13, 14]. Since cancerous masses are highly metabolic tissues, they need an abundant supply of nutrients to maintain their growth, which is achieved by increasing blood supply to their cells. The resulting increase in regional surface temperatures of the breast can be detected with IR breast imaging.

In 1956, thermography was introduced into screening but was abandoned 20 years later as it was not on par with previously mentioned methods even though no single tool provides excellent sensitivity and specificity. Breast thermography was later approved by the FDA in 1982 as an adjunctive diagnostic screening procedure for the detection of breast cancer. Protocols have also been standardized concerning patient preparation, image collection, and reporting [13].

Being a passive procedure, thermography application especially in younger men and women is favored as cumulating hazardous effects of radiation might be minimized. An assemblage that incorporates thermography may enhance accuracy. It is desired that through new approaches including regression analysis, highly precise diagnosis using thermography techniques, and high-resolution digital camera systems can be achieved [15]. In the past few years, many studies on

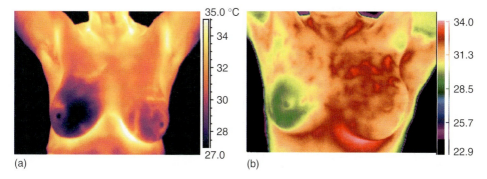

Figure 10.9 Two IR images of screening the breast under standard procedures. (a) Temperature differences of around 2 K suggest additional diagnostic tools and closer observation. In this case, the mammogram was negative. (b) Example of thermogram of patient with regular negative annual mammography examination. Re-examination after this high-risk thermography finding confirmed breast cancer which led to mastectomy of left breast. (Images courtesy A. Mostovoy.)

breast thermography have been reported not only in the medical but also in the IR community [13, 14, 16–18].

Figure 10.9 depicts two examples of IR images of two high-risk patients. Such images are evaluated following standard procedures which include analyzing the symmetry of thermal patterns between breasts, the consistency of thermal patterns with normal anatomy, and the quantitative temperature differences between breasts [13]. In Figure 10.9, both patients show temperature anomalies in the range of 2 K or higher. All images are analyzed using specialized software and the images of the breasts are rated in five risk categories from TH-1 to TH-5, where TH-1 indicates the lowest possible risk of developing breast disease, while TH-5 indicates the highest risk. Any abnormal thermographic scan result of the breast clearly demonstrates abnormal areas of heat. This should be considered as an alert that something might be wrong with physiology of the breast. It could be an infection, inflammatory disease, trauma, or cancer.

As a consequence, thermal imaging is definitely not a stand-alone screening examination. It is simply a method to detect physiological changes associated with the presence or increased risk for the development of breast cancer. Breast thermography has been studied for over 50 years, and over 800 peer-reviewed breast thermography studies have been published. In the index-medicus database, well over 250 000 women have been included as study participants. As a result, breast thermography has an average sensitivity and specificity of 90% [13].

10.1.5
Other Medical Applications

10.1.5.1 Raynaud's Phenomenon
In Raynaud's phenomenon episodic, long-lasting recurrent vasospasms occur primarily in the extremities in response to cold or emotional stress, though other

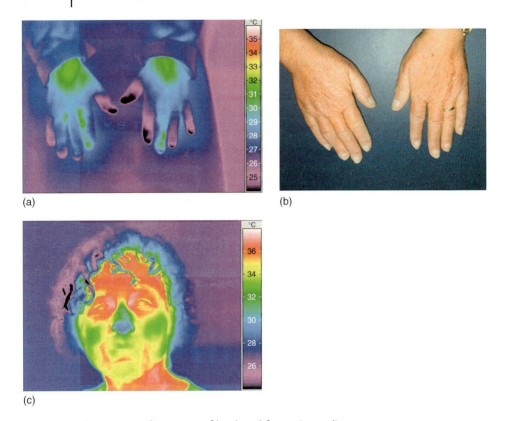

Figure 10.10 Thermogram of hands and face in Raynaud's Phenomenon; note that clinically on inspection no pathology can be found in the picture of the hands. In the thermogram of the face, a markedly reduced temperature ($\Delta T = 7.6$ K) of the nose can be identified. (Images courtesy K. Agarwal.)

acral parts such as the ears, chin, nose, nipples, and tongue may be affected, too. Patients do have reversible discomfort and color changes (pallor, cyanosis, and erythema). The disease is common in cold climates. Young women are more often affected than other groups. Pathophysiology may result from exaggerated α_2-adrenergic response triggering vasospasms. Sometimes an association with rheumatic conditions can be recognized. Diagnosis is clinical only. IR thermography can reveal clinical unapparent Raynaud's syndrome as skin sometimes looks normal upon inspection, though patients report of pain and/or paresthesias (Figure 10.10).

10.1.5.2 Pressure Ulcers

Pressure ulcers (bedsores, decubiti) occur as tissue is compressed between bony prominences and solid underground, leading to skin defects ranging from simple

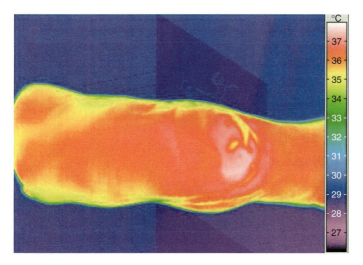

Figure 10.11 Pressure ulcer on the right heel. (Images courtesy K. Agarwal.)

erythema to skin loss with extensive soft-tissue necrosis. Older patients or people suffering from impaired circulation, immobilization, malnutrition, and incontinence are prone to this complication, and bacterial superinfection is a serious risk. The clinical appearance of the lesion usually suffices to diagnose an ulcer, though depth and extent may be intricate to verify. Constant and repetitive assessment is compulsory for successful management of the disease. Serial photographs, that is, conventional pictures as well as IR thermograms, aid in documenting the course of healing (Figure 10.11).

Quite a few other successful medical thermography applications have been reported. As selected examples, we mention first evaluating the coronary flow during the operation of revascularization of the myocardium [19], second using IR imaging as a tool in medical emergencies to quickly gain triage information, that is, help to decide which victims, for example, of car accidents need help first [20], and third development of new expanded medical imaging by registering and merging 2D IR imagery with 3D magnetic resonance imaging techniques [21].

Finally, new medical applications of IR cameras are sometimes also due to topical issues. One example is the outbreak of severe acute respiratory syndrome (SARS) a couple of years ago. In 2003, the outbreak of the SARS resulted in 813 deaths worldwide in half a year, with officially 8437 people reported as being infected [22]. SARS is transmitted by respiratory droplets or by contact with respiratory secretions of patients. It can spread out very rapidly through public transport such as trains and airplanes. Therefore, soon after the outbreak, authorities arranged for screening procedures of travelers, looking for fever symptoms [23]. The goal was to prevent spreading of the disease by identifying the few infected people with fever out of thousands without fever by screening, for example, the large number of passengers arriving at airports from abroad. Because of the large numbers, traditional oral thermometers or spot IR thermometers did not work. Rather IR imaging systems were used [22, 23] to find potentially febrile persons. Because of the political pressure of stopping a developing pandemic, large numbers of IR camera systems and IR spot pyrometers were needed, which momentarily gave a push to the industry.

10.2
Animals and Veterinary Applications

Animals like humans vary in response to painful changes, be it illness or injury, in their bodies. Most often, increased or decreased blood flow to the respective body regions lead to temperature changes from normal. Therefore, many diseases and/or structural changes due to injuries can be seen indirectly via thermal anomalies or asymmetries on the animal surface. The important issue is that the body of an animal – even if not showing acute signs of pain – does usually develop a thermal signature or asymmetry and often such temperature changes are the first indication of a problem. Thermal imaging is the proper tool to identify changes in heat patterns and use this information in diagnosis [24].

Consider, for example, horses that suffer pain in the leg: some become obviously lame, some may experience minor changes but both require significant methods of examination to try and detect a cause for the change. In contrast other horses will mask their pain and show no visible obvious change in their gait [25]. In the latter case, IR cameras are valuable tools when used to evaluate and detect such lameness issues, which otherwise may take up to two weeks before showing up with obvious signs. If working together with a veterinarian, thermal imaging allows to make immediate changes, possibly preventing possible disastrous effects for the animal.

However, whenever thermography is used for veterinary purposes, it must be kept in mind that the method should usually be combined with other examination methods like X-ray scans, if needed. For example, thermal anomalies should lead to additional methods such as X-ray, CT scan, and/or MRI and the information from all methods should be combined for a diagnosis.

Consider, for example, X-rays: they are helpful in identifying bone changes and some changes in soft tissue, but they do not have the capability to evaluate soft tissue changes to the extent of thermal imaging. Thermal imaging shows physiological changes such as increased or decreased blood circulation, lack of or diminished nerve function as well as swelling. It can, therefore, give additional, partially complementary information to other examination techniques. Since it is a noninvasive technique, it has the additional advantage that it reduces stress of the owner and the animal.

Nowadays, animal thermography is usually divided into equine thermography and studies of other animals. This is most probably due to the fact that horses are quite expensive animals, in particular, if used in sports.

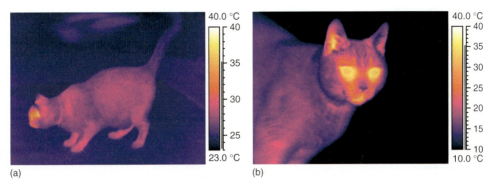

Figure 10.12 (a) IR images of a cat inside a house (21 °C) or (b) after staying for a while outside at air temperatures below 10 °C.

10.2.1
Images of Pets

Most owners of IR cameras have probably recorded some images of animals – mostly healthy pets (Figure 10.12). Therefore, such IR images have no relation to veterinary applications. They are usually fun to watch, but they may also point out differences between various animals, for example, in body temperatures or of thermal insulation due to fur or bare skin, and so on.

Regular observations of one's pet can also help to detect illnesses or injuries, since comparison with "healthy" IR images will reveal thermal anomalies or asymmetries. In this respect, pets of a thermographer may probably get the best possible benefit of thermal veterinary imaging. The examples given below discuss some cases where thermography has proven to be a versatile and successful diagnostic tool.

10.2.2
Zoo Animals

Many major cities of the world have zoos with a large variety of animals in captivity. Obviously, veterinary care for zoo animals is of primary concern and thermography has proven to be a valuable tool in diagnosis, documentation, and monitoring in particular for very large animals such as elephants or camels [26, 27].

Figure 10.13 shows an example of IR images of two giraffes [25]. The front giraffe (Figure 10.13a) had a known injury to the left rear leg. The IR image clearly shows a thermal anomaly in the form of lower temperature compared to the other legs. It is due to the lack of blood circulation in the leg.

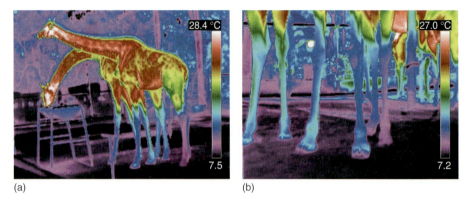

Figure 10.13 (a) IR images of giraffes, one of them having a known injury in the left rear leg. (b) The dark blue-purple color in the leg and the purple-lavender in the foot indicate lower temperatures due to a lack of circulation or nerve conduction. (Images courtesy of Peter Hopkins, United Infrared, Inc. [28].)

10.2.3
Equine Thermography

Equine thermography was one of the first applications of thermography to animals and is nowadays a fast growing field of veterinary applications of IR imaging [25]. It is extremely useful to study thermal asymmetries. As an example, Figure 10.14 shows a routine scan done on a race horse scheduled to leave for the track seven

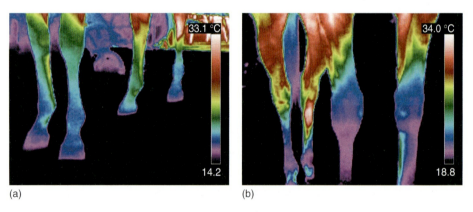

Figure 10.14 Routine scans of race horses. (b) Abnormal hot spot in leg (second leg from left) later confirmed as fracture of cannon bone. (a) Lower temperature of left rear leg hoof (first leg from right) indicated shifting weights due to the cannon bone injury. (Image courtesy Peter Hopkins, United Infrared, Inc. [28].)

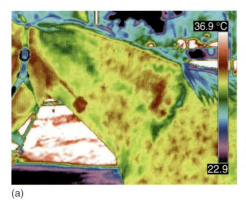 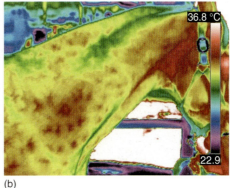

(a) (b)

Figure 10.15 Scans of horse necks from (a) and (b) indicated a thermal asymmetry, which was later confirmed as lower cervical neck injury (C6/7 fracture). (Image courtesy Peter Hopkins, United Infrared, Inc. [28].)

days after the recording. The IR scan identified an abnormal hot spot that was confirmed later as a fracture in the cannon bone. In addition, the rear leg hoof temperature was cooler. This was a consequence of the horse shifting its weight due to the previously unknown injury.

Another example is shown in Figure 10.15. The IR images of the horse were recorded after an emergency evacuation took place during a barn fire. The behavior of the horse changed after the incident and it was suspected that an injury may have taken place during the evacuation. The thermal scan shows an asymmetry in the neck, possibly indicating a lower cervical neck injury. Later it was confirmed via a radiograph that a C6/7 fracture had in fact been present.

Thermography is also used to detect laminitis, one of the most serious horse diseases [29].

10.2.4
Others

A variety of other animals have been studied using thermography. The examples range from nice shots of animals still living in their natural environment like, for example, whales, elephants, leopards, or seals [30, 31] to purposeful studies like, for example, rescue, recovery, and rehabilitation of birds following oil spills [32]. The detection of birds or bat in flight [33] can raise some important question about the time response of the detectors (Chapter 2). Slow cameras with microbolometers detectors usually show streaks behind flying birds. These must of course not be interpreted as air, warmed during the flyby of the bird. Rather, they represent the effect of detector integration time discussed in Section 2.5.5. An example of a flying bird is shown in Figure 10.16, which depicts a hummingbird. Hummingbirds can practically stay in the air while rapidly flapping their wings in order to get access

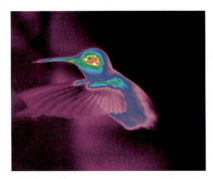

Figure 10.16 A Hummingbird in flight, recorded with 430 frames per second with a MW camera. (Image courtesy A. Richards, FLIR systems Inc.)

to nectar within flowers. Therefore, images must be recorded with high-speed cameras.

Another typical veterinary application is the dairy and poultry industry [27]. For example, it has been demonstrated that IR imaging can be used as an early identifier of bovine respiratory disease in bison or for detection of mastitis in dairy cows. In the poultry industry, thermography has also been used to study the feather cover, which in itself is an important factor determining animal welfare.

Another important application of animal thermography in certain parts of the world is the field of termite and pest detection [34, 35]. For example, termites do more damage in the United States each year than fires and storms combined [35]. The most destructive termite species in the world is the Subterranean Formosan termite because they can secrete an acid that dissolves, for example, concrete, steel, lead, copper, glass, and of course also wood. One colony can consume more than 1 kg of wood per day, which means that wooden homes can be destroyed very fast if infected. If moisture sources are available within the structure, nests are built above ground somewhere in walls. This is where thermography comes in. Termites create irregular heat patterns due to two facts. On the one hand, the mud tubes they construct have high moisture content and IR imaging techniques to detect moisture can be applied. On the other hand, heat is released from the digestive system of the termites in the form of CO_2. This leads to thermal changes on the surfaces of walls, ceilings, and floors, creating irregular temperature patterns that can be detected with IR imaging.

Finally, another fascinating example of insect temperature study using thermography was reported recently [36]. The Japanese giant hornet is a predator of bees and wasps and often several hornets combine for a mass attack of a bee nest. Unlike European honey bees, the Japanese honeybees have developed an ingenious strategy to respond to a hornet attack. Since individual counterattacks are useless, they use heat produced by a large group of bees to kill hornets that try to enter the nest. If warned of an attack, typically many worker bees wait inside the entrance and once a hornet tries to enter, it is quickly engulfed within a ball of about 500 workers. Whereas the temperature of a typical worker without defending is about 35 °C, the temperature within the ball quickly rises to 47 °C (Figure 10.17). This temperature is lethal to the hornet (upper lethal temperature of Japanese hornets

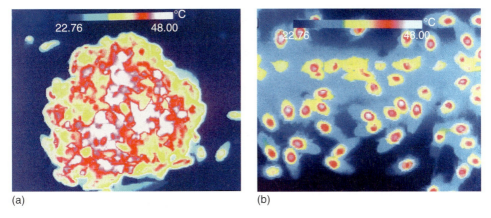

(a) (b)

Figure 10.17 Hot defense ball of many bees, raising the inside temperature to about 47 °C (a) and individual defender bees crawling around the nest entrance. Their thoraxix temperature is already elevated (b). (Images courtesy M. Ono, Tokyo.)

is 44–46 °C) but not to the bees (upper lethal temperature of Japanese honeybees is 48–50 °C).

10.3
Sports

In sports, there are many processes that involve heat transfer. Quite often, these processes are also fast, and require high-speed data acquisition. Examples include all kinds of game activities that involve balls that are either hit by some racket, by hands, by feet, or by something else like in tennis, squash, soccer, volleyball, billiard, and so on. For these activities, IR thermal imaging may record the temperature changes following inelastic collision processes. The corresponding information may have varied uses, depending on the sport. For example, one may directly determine whether a tennis serve is in the field or outside. In the following, first a number of applications involving inelastic collisions as well as frictional forces are presented. Then, other potential benefits from IR imaging in sports are discussed.

10.3.1
High-Speed Recording of Tennis Serve

An introduction to the collision of a tennis ball with the floor was already briefly discussed in Section 5.2.4. Here, we discuss the problem in more detail. Modern tennis balls are pressurized rubber balls covered with felt. Their masses are around 57–58 g, with diameters of about 6.7 cm. They are quite elastic and new balls are required to bounce back to a height of at least 1.35 m if dropped from a height of 2.5 m. Energy conservation requires that all of the initial potential energy of the ball is transferred into other kinds of energy. Just before hitting the floor, the ball gains maximum kinetic energy, which is about as large as the initial potential energy (friction due to the surrounding air is negligible for these small speeds). The bouncing back, however, starts with less kinetic energy, which is the reason, why later on, the ball comes to a stop at a smaller height (at least 54% of the original height). The difference in energy before and after the collision is due to the deformation of the ball upon hitting the floor. Thereby, part of the kinetic energy is transferred into deformation energy. Because the deformation is not purely elastic, a part of this energy is transferred into thermal energy, that is, ball and floor are heated up a little bit. Energy conservation, thus, means that potential energy is transferred into the sum of kinetic energy and thermal energy.

The resulting temperature increase is quite small when a ball is just dropped from a height of 2.5 m. However, the service of top 10 tennis players can reach about 250 km h^{-1}. If a tennis ball of this speed hits the court surface (be it grass or artificial), the inelastic collision leads to an enormous deformation of the tennis ball. Figure 10.18 depicts a visual image of a tennis ball first before and second while hitting the floor with a speed of only about 86 km h^{-1}. During the inelastic collision, upon hitting the floor the ball is compressed appreciably to at maximum of about half of its original volume during the contact time of ≈ 4 ms with the floor. After hitting the floor, the velocity of the bouncing ball drops by about a factor of 2.

This difference in the kinetic energies of the ball before and after the collision is transferred into thermal energy, that is, one may expect the ball as well as the floor to heat up. Figure 10.19 depicts two thermal images recorded with an LW IR microbolometer camera operated at 50-HZ frame rate. The first image shows the situation directly after the ball has hit the floor. Since the ball was too fast, it just left a comet-like streak behind, similar to the examples of the falling balls discussed in Section 2.5.5 (if the ball with initially 86 km h^{-1} would have transferred 40% of its kinetic energy into thermal energy it would have left the image field of view

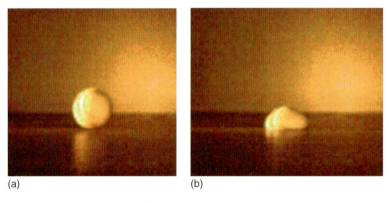

(a) (b)

Figure 10.18 Visual high-speed camera images of a tennis ball hitting the floor with a speed of about 86 km h^{-1}.

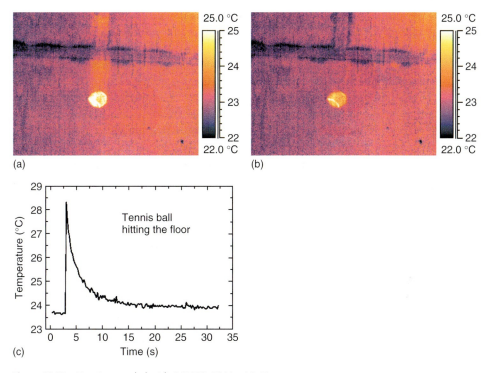

Figure 10.19 Tennis recorded with SC2000, 50 Hz. (a) First image after ball hits floor, (b) 1.7 s after ball has hit floor, and (c) measured surface temperature of contact spot on floor as function of time.

within 25 ms after floor contact, which roughly equals the 25 ms corresponding to the frame rate).

With these images, it is not possible to quantitatively analyze the whole collision process, since first only the heating up of the floor is recorded while the ball itself has left the scene with undefined temperature. Second, the heating is a transient phenomenon that depends on heat conduction within ball and floor as well as on the respective heat capacities. However, it is possible to at last analyze the floor heating as a function of time. Figure 10.19c shows the maximum temperature of the hot spot on the floor as a function of time. The initial temperature rise in this case was about 5 K and the time constant was several seconds. Top players can easily reach ΔT values of 10–15 K, with an easy to observe signal for more than a minute. Therefore, the IR detection of the collision of ball with floor is easily detectable, which means that it can in principle be used to determine the location of the serve in the field of the opponent. This idea was tested already more than 10 years ago at the ATP finals 1996 in Hannover/Germany. The side lines of the court in our experiment were just roughly marked with white chalk on a carpet, which gives only a little bit of emissivity difference. In a real tennis court, the lines can be made visible using the large emissivity contrast of court and metals by

556 | *10 Selected Applications in Other Fields*

having thin metal threads woven into the line material. Unfortunately, the method, although being available as instant replay in near real time, was not chosen as aid for referees in tennis – probably because of the costs of using several IR cameras.

Figure 10.19b shows the IR image about 1.7 s after the floor contact. It nicely demonstrates the different thermal properties of the hairy felt part and the rubber lines on the tennis ball surface. On the one hand, the smooth surface of the rubber lines allowed for a much better thermal heat transfer; on the other hand, the thermal diffusion is also larger. As a result, the rubber lines have gained a higher temperature than the rest of the ball.

To also observe the tennis ball itself, a high-speed IR camera (SC 6000) was used to observe the collision. Figure 10.20 depicts three images of a slightly slower ball recorded with a frame rate of 400 Hz and an integration time of 0.75 ms. Because of small integration time, more or less sharp and focused images result. The first image represents the maximum deformation of the ball. On its upper surface, one can still see the structure of the string grid with which it was hit by the racket. The second and third images show the situation after the collision with the floor. Now the floor has also heated up. One may even observe a slight rotation of the tennis ball in the thermal image. On the one hand, such experiments directly visualize the

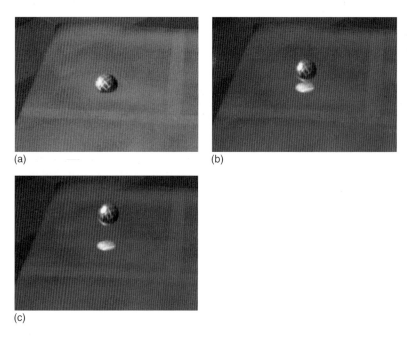

Figure 10.20 IR images of tennis ball while hitting the floor (a) and while bouncing up again (b,c) recorded with a frame rate of 400 Hz and integration time of 0.75 ms. The position on the ball that was hit by the racket, in particular the heated lines from contact with the string network made of nylon or polyester synthetic materials, is seen as well as the warming of the contact spot on the floor.

kinematics of inelastic collision processes; on the other hand, they allow to study the energetics.

10.3.2
Squash and Volleyball

Similar to tennis, squash consists of a repetition of inelastic collisions of a ball with walls and rackets. Squash balls consist of small rubber-like balls that are a few centimeters in diameter and filled with air. When dropped from a height of 1 m, they do not bounce very well, that is, they are quite inelastic. The near permanent collisions during a game will obviously lead to heating of the ball. Assuming steady-state conditions, that is, hitting the ball at a given frequency with similar strength each time will lead to an equilibrium of heat transfer to the ball due to the inelastic collision processes and heat transfer from the ball due to convection and radiation and to a small extent also conduction to the walls during the short contact times. Figure 10.21 depicts a squash ball before starting to play and after a few minutes of playing. The ball temperature has risen by about 20 °C and the ball is nearly homogeneously heated all across its surface.

Volleyballs have a diameter of about 21 cm and a mass of typically 270 g. Several types of inelastic collisions are observable in volleyball games, for example, while hitting the ball for service or for attack with a single hand, while defending with both arms, or while the ball is hitting the floor. During an attack, the hand hits the ball and deforms it, which already leads to heating of the ball. Second, when such a fast attack ball (velocities up to 30 m s^{-1}) hits the floor, it also leads to heating of the floor similar to the tennis ball.

The available kinetic energy of a volleyball is similar to that of a tennis ball: the mass is about a factor of 4 higher, whereas the maximum velocity is about a factor of 2 lower. Hence, we may assume a similar amount of energy, which may be transformed into heating the ball and the floor. However, because of its much larger diameter, the contact area of a volleyball with the floor is much larger and hence the thermal energy is transferred to a larger area. Overall, a smaller temperature rise is to be expected. Figure 10.22 shows results of a ball hitting the floor with high speed, again recorded with the high-speed IR camera SC 6000. Similar to the tennis ball, the floor heats up appreciably and one can readily see the geometrical structure of the ball surface on the floor due to different heat transfer rates.

10.3.3
Other Applications in Sports

There are many more IR imaging applications related to sports. For example, sport shoes have been tested using thermography, or sport injuries, which result in thermal asymmetries or anomalies, may be studied. Here, one more example is briefly discussed, dealing with the study of heat-loss patterns in athletes [37]. During exercises and any kind of sports activity, working muscles produce heat,

558 10 Selected Applications in Other Fields

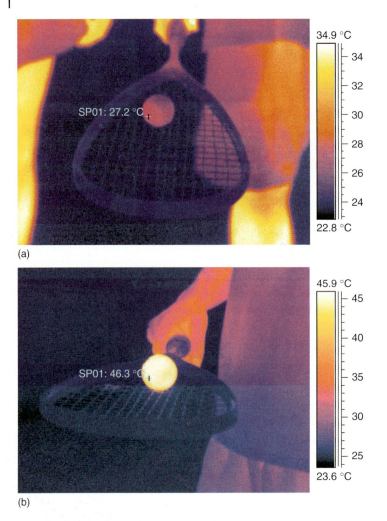

Figure 10.21 The temperature of a squash ball increases by about 20 °C after playing for some minutes.

that is, the body temperature increases. This heating up is counteracted by various thermoregulatory mechanisms, all of them dissipating this excess heat to the external environment. Two mechanisms are particularly important: first sweating with cooling via the heat of vaporization and second direct radiative cooling from the body surface to the environment. These cooling mechanisms are essential for the athlete to avoid heat illness, which may lead to decreased performance or in the worst case to severe injuries or even death. In the period between 1995 and 2008, 33 high school, college, and professional football player fatalities have been attributed to heat stroke!

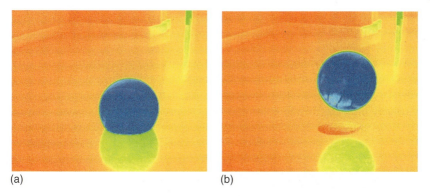

Figure 10.22 Two snapshots of a volleyball hitting the floor. Upon impact it is deformed appreciably which led to a temperature increase of floor and ball. In addition, thermal reflections of the previously frosted ball can be seen.

A particularly critical sport with regard to heat dissipation by the athlete is professional American football, because players must wear extensive padding on chest, shoulder, and legs, gloves, and other equipment, decreasing the effectiveness of many of the body's methods for eliminating excess heat. Therefore, besides part of the arms and legs, the face itself can serve as a heat dissipating surface, after the helmet is removed during breaks of the game.

Results of a recently published investigation [37] used IR imaging to analyze heat patterns of a venous plexus in the face for 53 National League Football players during 12 regular league games. These facial regions have shown to be effective for heat dissipation due to increased blood flow when body temperature rises during exercise. Players were filmed while resting on the sideline during the brief recovery periods after exertion, yielding 1858 images. IR tracking of the skin temperatures proved to be a useful technique for the understanding of physiological heat exchange patterns in the cheek region after intense exertion. Characteristic trends were discernable during the postexertion period. For example, many players typically display a period of significant cutaneous temperature increase in the cheek region across the initial 0–3 min interval following exertion.

10.4
Arts: Music, Contemporary Dancing, Paintings, and Sculpture

Surprisingly many applications of IR thermal imaging have already been reported in the field of arts. Two examples, one in music and another in contemporary dancing, are discussed in more detail and others concerning paintings and other artwork are briefly discussed.

10.4.1
Musical Instruments

Musical wind instruments produce sound when warm air from the musician's breath is blown into the instrument. Therefore, the instrument's temperature will change once the ambient temperature differs from body temperature, which is usually the case. In the following, brass instruments are discussed. Playing brass instruments indoors, for example, in concert halls, leads to, after short times of several minutes, an effective temperature of the air in the inner cavity, which determines the intonation of the instrument. Intonation is a measure for deviations of the fundamental frequencies of tones from those of a reference scale. Usually, one refers to the equally tempered scale, where one octave is divided into 12 half-tone steps of equal interval size (defined by the frequency ratios). For quantitative analysis, one octave is divided into 1200 intervals of 1 cent each, that is, each half-tone interval is divided into 100 cents.

The average temperature indoors is relatively stable and each musician can vary the intonation of his instrument to some extent by blowing techniques; therefore, there are seldom any problems of intonation changes in an orchestra, which relate to the ambient temperature. The situation, however, changes outdoors at low temperatures, say below 10–15 °C. Again, the instrument will be characterized by an effective temperature that determines the intonation. However, musical pauses, which occur quite frequently for brass instruments during a concert (e.g., around christmas time), lead to a cooling of the inner cavity air. Therefore, the intonation changes during these breaks [38]. If the temperature differences are too large, the necessary corrections done by the musicians using the strength of the lips may become tedious. These problems become more severe for large brass instruments, which, on the one hand, have a large heat conductivity of the metallic surfaces and, on the other hand, they are too big to be kept warm by pressing them against the body and/or by holding them below clothes.

Intonation changes were measured while, simultaneously, the temperature across the first upper slide of a trombone was measured using IR thermal imaging. As a result, it was found that the warming up of the instrument due to warm airflow through the mouth piece leads to a change in intonation of about 3 cents K^{-1}. The change occurs for all natural tones similarly, that is, the instrument does not change its characteristics during the temperature changes. Even at low temperatures of 5 °C, the process of warming up takes less than 10 min, that is, the instrument can be characterized as stable with respect to intonation after this time. To correct the tuning, a trombonist either corrects the length of his instrument by its tuning slide or in the worst case by using different lip strength.

Figure 10.23 depicts two IR images obtained while a trombone was being played. Obviously, a temperature gradient evolves and there is a pronounced effect at the crook of the main slide. After some minutes of continuous playing, steady-state conditions are achieved, characterized by nearly body temperature at the mouthpiece and a much smaller gradient at a higher temperature level.

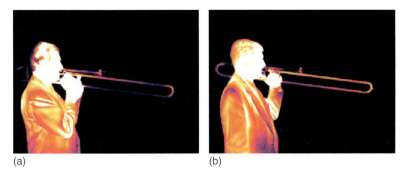

Figure 10.23 Two IR images while playing the trombone. At the beginning, quite a large temperature gradient can be observed, whereas after some minutes of playing, steady-state conditions are reached.

10.4.2
Contemporary Dancing

Pointed spot lights often suffer from the problem that although people or objects are illuminated with high light flux as desired, they usually also produce disturbing shadows. In addition, stage technicians must often manually direct the spots on the chosen points in real time. To overcome both drawbacks, a new illumination technique was invented by Milos Vujković [39, 40] based on IR thermal imaging. Using IR–visible feedback loops that are coupled to spotlights for a target-based illumination, it could be demonstrated that any kind of shadow upon illumination can be avoided. Furthermore, this innovative technique of personalized spot lights does not require any manual adjustments of the spots. The technique was demonstrated for contemporary dancing, enabling real time movements of dancers on stage while only their bodies remain fully illuminated [40].

In general, feedback describes a situation when an output signal from a phenomenon is used to somehow have an influence on the same phenomenon in the future. If the respective signals are used in a causal way, that is, if the output signal (or part of the output signal) of a phenomenon is used as an input signal to create a changed new output signal, one speaks of feedback loops. Such feedback loops are well known from many different areas in science, industry, and technology. Typical examples range from electronic circuits, where feedback loops are used for the designs of amplifiers, oscillators, and logic circuit elements to mechanical, audio, and video feedback phenomena.

In arts, optical feedback is sometimes used to achieve special effects. The feedback occurs when an optical signal, for example, the image on a television screen, a computer monitor, or a video beamer is observed by a camera, whose output is fed into the monitor/beamer (Figure 10.24). The feedback loop creates an infinite series of images of the same object, all within the same image. The actual appearance of the resulting image depends on camera and monitor settings, for example, light amplification, contrast, distance, angle, and so on.

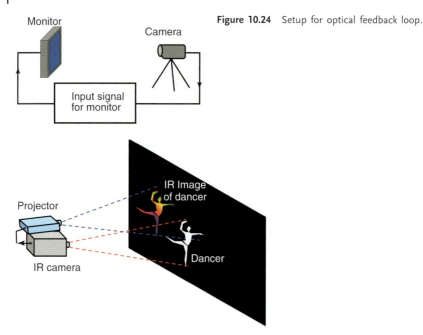

Figure 10.24 Setup for optical feedback loop.

Figure 10.25 Scheme for an IR-optical feedback loop. An IR camera records an image of a dancer, which serves as an input signal of a projector.

Using IR cameras rather than cameras in the visible spectral range, a very similar IR–visible feedback loop can be generated. The idea is to record a scene with an IR camera, for example, a performing dancer (Figure 10.25). The life signal of the camera can be fed into a video beamer, which projects the IR image onto a screen, which is behind the dancer. If the screen would be spatially separated from the dancer, one would just project typical IR images (Figure 10.26) onto the screen.

If the projector is aligned as close as possible to the IR camera (it is best to use a telephoto lens such that the distances of camera plus projector are large compared

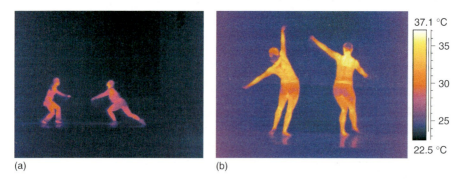

Figure 10.26 Two examples of IR images of contemporary dancing.

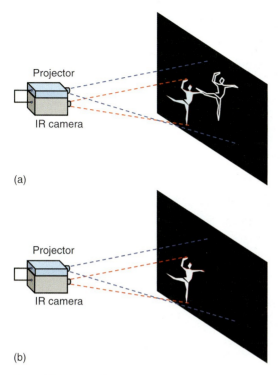

Figure 10.27 (a,b) Two useful alignments for the IR-visual feedback loop (for details, see text).

to the size of the dancers), that is, if both are nearly in collinear alignment, the output signal of the video projector can be superimposed onto the dancers. By adjusting the telephoto lens of the video projector, it is possible to adjust the size of the projected image to that of the dancers (Figure 10.27). For demonstration purposes, the size of the projected image can be chosen slightly larger than that of the dancer (Figure 10.27a). In principle, the size can, however, be reduced such that only the dancer is illuminated (Figure 10.27b).

The idea behind this optical-IR feedback loop seems so simple that one may try to realize this illumination by a visual camera. This would, however, raise the following problem. A visual light camera does need a certain light level for operation, which would automatically mean that the background would also be visible and amplified by the feedback loop. This would lead to bad contrast between the dancer and the background. The use of an IR camera, however, easily overcomes this problem, since it detects the surface temperature of the dancer, which is much higher than that of the background. The contrast can be made very large by simply changing the temperature span of the IR image (Figure 10.28). If this feedback signal is used, the background remains very dark and only the warm skin and clothes of the dancers remain visible. The result is a whitish projection, which fits nearly perfectly to the geometrical form of the dancing bodies.

Figure 10.28 IR image of Figure 10.26b with different temperature span in order to reach saturation in the image of the dancers. This image can be used for the optical feedback signal to the projector.

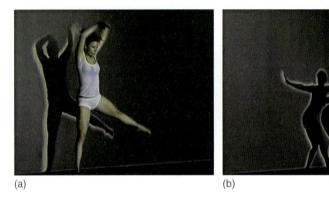

(a) (b)

Figure 10.29 (a,b) Some photos of dancers on stage which were illuminated with a slightly larger projected IR image of themselves. For more details on art aspect, see [39].

Figure 10.29 depicts several still images from a video sequence, which refer to the situation in Figure 10.29a. The projected image was chosen to be slightly larger than the dancer. This allows to simultaneously observe the edge of the projected image as well as the moving shadow of the dancer. Movie sequences, analyzed in slow motion, reveal that the projection can be made very fast such that the projection follows even very rapid movements of the dancers in real time.

In the experiments [40], an SC 6000 research IR camera (wavelength range 1.5–5 µm) with 640×512 pixels was used. It provides a 30-Hz analog output signal, which could be directly used as an input for the video projector. The IR-optical feedback loop can be easily used even for fast body movements during contemporary dancing and any kind of IR camera with 30 Hz output may be used for these experiments, although it is desirable to have at least 640×512 pixels. A number of improvements for the future were suggested, for example, using several cameras simultaneously from the front and the sides, using cameras with higher frequency analog output, or including a computer between IR camera and projector for manipulations of the camera signal.

10.4.3
Other Applications in the Arts

In recent times, quite a lot of sophisticated equipment is used for restoration of paintings in museums. Some of the techniques also allow one to look beneath the various layers of paint. One particularly interesting technique, IR reflectography, uses SW IR cameras like the NIR cameras operating in the range from 1 to 2.5 µm wavelength [41]. IR reflectography utilizes the fact that in certain lighting situations, oil paints can appear transparent to the IR camera because they allow specific IR wavelengths to pass through them and be reflected off of the canvas. Graphite sketches on the canvas do not reflect the IR light and thus can be seen as dark areas on the canvas. The opaqueness of paint layers increases with paint thickness. If the paint is not too thick, the optimum wavelength to see through the paint while maintaining the ability to detect the graphite lines against a white background has been shown to be around 2 µm. These images are referred to as *infrared reflectographs*, since the image is made using reflected light, rather than thermally self-emitted radiation (thermographs). In this way, art historians study the thought process of the artist, that is, see where artists may have changed their minds between the stage of sketching a scene and painting it. In the 1960s, cameras with PbS detectors (peak responsivity at 2 µm) were used to study the underdrawings of a sixteenth century painting by using an active mode with the painting illuminated with a tungsten lamp.

In a recent investigation, a Phoenix NIR camera with a 320 × 256 InGaAs (sensitive from 0.9 to 1.7 µm) focal plane array was used. Lighting of the painting in the museum was by incandescent tungsten bulbs. An underdrawing test panel was made in order to optimize the camera settings and filter selection. This test panel consisted of an ordinary primed artist canvas with vertical black stripes drawn by a charcoal pencil. These stripes were then covered with different colors of oil paint. The InGaAs IR camera was successful at imaging an underdrawing present below the specific chosen original painting attributed to Loyd Branson. The underdrawing was oriented in the landscape direction, whereas the top painting was oriented in the portrait direction. Besides paintings, IR thermal imaging was also used as a nondestructive test method to study historical wall paintings, frescos, or other indoor and outdoor stone materials constituting cultural heritage [42–46].

Very similar techniques are used in forensic sciences [47]. Criminologists often use NIR imaging techniques to examine documents for possible alterations. Figure 10.30 shows examples of several documents that have been altered by covering some letters with either correcting fluid or ink, for example, from a ballpoint pen. The ink and correcting fluid neither absorb nor reflect SW IR radiation; therefore, they are more or less transparent to the radiation of 1.6 µm wavelength. The underlying paper does, however, reflect IR as well as Vis radiation. The printer toner finally absorbs SW IR light, which leads to the image contrast.

Similarly, it is possible to look through thin layers of paint with SW IR radiation as is demonstrated for metal containers, whose serial numbers were covered by paint, Figure 10.31. In the Vis range, it is not possible to detect the numbers,

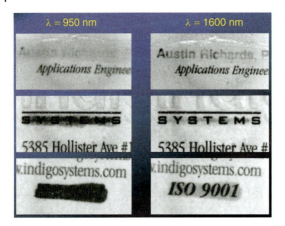

Figure 10.30 NIR imaging – questioned documents. (Courtesy A. Richards, FLIR Systems Inc.)

Figure 10.31 NIR imaging-looking through paint. (Courtesy A. Richards, FLIR Systems Inc.)

whereas SW radiation between 1.3 and 2.5 µm readily detects the hidden numbers below the paint.

10.5
Surveillance and Security: Range of IR Cameras

10.5.1
Applications in Surveillance

Not surprisingly, IR thermal imaging had and has a lot of applications in the field of security and surveillance. Here modern camera systems have many advantages compared to classical NIR image intensifier systems [48]. IR imaging is used extensively for the protection of critical infrastructure, that is, industrial assets such as airports, power plants, petrochemical installations, or warehouses. In particular, it is used for perimeter monitoring. Nowadays, it is even possible to construct sophisticated "thermal fences" around any facility by using a series of IR cameras with overlapping fields of view and appropriate software tools. Whenever some object of elevated temperature crosses such a line, an automated alarm is triggered.

Figure 10.32 Monitoring a secured parking lot with a surveillance camera. Attempts of car theft can be detected.

All kinds of IR cameras can be used as regular surveillance tools, the only difference to other imaging applications being the fact that these cameras must be operated in any weather situation, be it extremely warm or cold, be it sunshine, rain, or fog. Therefore, surveillance cameras have mostly a special design.

Figures 10.32 and 10.33 depict two typical examples. Surveillance cameras may be used to monitor large and dark parking lots (Figure 10.32) or the fenced perimeter of an industrial facility during nighttime (Figure 10.33).

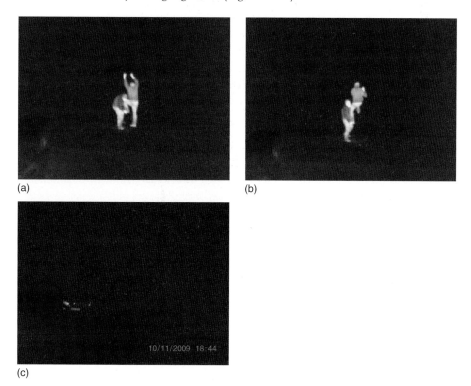

Figure 10.33 Monitoring a fenced property with a surveillance camera. In contrast to the IR images (a,b), the visual camera image (c) shows no features since the fence is not illuminated.

Besides security applications, IR imaging is also used when searching for missing persons, for example, disoriented people, fugitives, or even shipwrecked people at sea [49].

Surveillance cameras are available with either cooled MW detectors or uncooled microbolometer LW detectors and either 320×240 or 640×480 pixels and a variety of lenses and fields of view. To avoid ice on lenses in wintertime, cameras are usually equipped with a lens heater (which is also standard for the PathFindIR cameras in cars (Section 9.4.2). This also helps to avoid water films on the lens in situations of high humidity.

10.5.2
Range of IR Cameras

There is one major question that is usually raised in surveillance issues, which is not as important in regular thermal imaging applications (with the exception of passive IR in cars, Section 9.4.2): is it possible to see through haze, fog, and rain and what is the resulting range of an IR camera? This question is related to quite a few aerospace applications. For example, safety issues during landing at airports under unfavorable weather conditions like fog have raised the question of whether IR cameras can improve the visibility and thus help approaching aircraft during landing [50].

A similar question has been addressed from a different perspective: what is the influence of long atmospheric paths of tens to hundreds of meters through a hazy atmosphere with respect to radiometric calibrations of MW IR cameras? This problem was investigated by comparing experimental data to atmospheric MODTRAN modeling results [51]. For the MW range, the two major attenuating species are CO_2 and H_2O vapor. The aerosol contribution is usually modeled by the type of aerosol (see below) and the visual range. It was found that deviations occur between experiment and theory, which means that quantitative radiometric measurements can usually not be performed over distances of hundreds of meters to 1 km.

The question of what is the actual range of an IR camera does, however, not require any quantitative radiometric data; hence, much larger distances may be expected. However, a number of factors will have an influence on actual ranges of IR cameras:

1) spectral range, type of detector (cooled or uncooled), and sensitivity;
2) lens and size of object;
3) temperature of the target and the background;
4) atmospheric transmission for ideal clear sky conditions;
5) humidity of the air;
6) aerosol contribution in the atmosphere: haze;
7) additional atmospheric properties: fog, rain.

The first three factors are common to all applications. The range is related to the reduction in apparent contrast [52, 53]. In the visible part of the spectrum, the visual range is defined by a reduction of apparent contrast radiance between the target and background to 2%. The range for IR cameras is related to the MRTD and the NETD (Section 2.5). In a model [50], the IR range was determined by analyzing the temperature difference of a bar pattern versus background. The *range* was then defined as the distance when this radiation temperature difference equals the NETD.

The attenuation itself is defined by the attenuation due to the molecular species in the atmosphere (Chapter 7), in particular, the strongly varying humidity. In addition, various kinds of aerosols may scatter and/or absorb IR radiation. For easier modeling, only six climate types were defined: tropical climate, midlatitude summer, midlatitude winter, subarctic summer, subarctic winter, and US standard. Each climate could in addition be combined with different types of aerosols. The following aerosol types were analyzed: rural, urban, maritime, advection fog, radiative fog, and desert.

Each aerosol type is characterized by size and material contributions with well-defined absorption and scattering properties of radiation. The concentration of these aerosols is usually characterized by the respective visual range or meteorological optical range [54].

Rather than speaking of aerosols in the air, one often refers to a hazy atmosphere. Haze refers to the fact that dust, smoke, and other dry particles (from traffic, industry, wildfires, etc.) can obscure the clarity of the sky thereby reducing visibility. Besides haze, the world meteorological organization also classifies horizontal obscuration in the atmosphere by the categories fog, mist, haze, smoke, volcanic ash, dust, sand, and snow.

Usually, dry haze refers to small particles made of dust or salt particles with typical sizes of the order of 0.1 µm. These particles interact with radiation mostly according to Rayleigh scattering, which favors shortwavelength scattering. If the relative humidity of the air increases above 75–80%, these tiny aerosol particles of the dry haze act as condensation nuclei, thereby producing wet haze. It is composed of small liquid droplets with sizes in the micrometer range. They are able to scatter light much more efficiently than dry haze, thereby reducing the range. If the nuclei grow much larger, fog may result with droplet sizes in the range of 10 µm, which is similar to droplet sizes in clouds. Again the scattering efficiency of these droplets increases with size, which further reduces the range.

IR radiation can have a much larger range than visible light under certain circumstances. The situation is easiest for dry haze with scattering particles of, say, diameters below 500 nm. For this size range, visible light is appreciably scattered. The dominant Rayleigh scattering mechanism decreases strongly with increasing wavelength (scattering is proportional to $1/\lambda^4$). Therefore, IR radiation in the SW range of 0.9–1.7 µm is much less affected. This means that NIR can give much sharper images through such haze than visible light. Examples where NIR radiation was used to detect objects in horizontal distances at sea level of around 50 km are shown in Figure 10.34. Obviously, the respective visible light produced only very fuzzy images – if they could be observed at all.

Figure 10.34 Example of SW IR image ((a), recorded with InGaAs camera and telescope optics) versus Vis images (b) (regular color film) of oil platform in a distance of 47 km. (Images courtesy A. Richards, FLIR Systems Inc.)

In principle, the attenuation in the atmosphere due to Rayleigh scattering decreases for the even longer wavelengths of MW and LW cameras, which is why one may suspect even larger ranges for these systems. This is, however, not the case. The major reason is that NIR long ranges are usually observed during daytime, when sunlit objects are observed at great distances. Therefore, the detected radiation consists of scattered radiation from the sun in this wavelength range, which is still very large compared to contributions in the MW or the LW IR spectral region (compare Figure 6.44). The LW cameras, for example, will detect a comparably negligible amount of scattered sun light. To see long range objects, they need a thermal signature, for example, that the objects are much warmer than the surroundings. For the observed objects in Figure 10.34, this is not the case. And their small temperature differences with respect to the surrounding only give small object signals, which are attenuated due to aerosol scattering and absorption by water vapor and by carbon dioxide.

As a consequence, NIR cameras are the best option for long ranges in dry haze environments during daylight conditions. During nighttime, they do, however, also depend on thermal object radiation. Because of their spectral range, only very hot objects will be detectable at large ranges.

For dry and wet haze, that is, fog situations, model calculations with MODTRAN, an atmospheric radiative transfer code, have been performed to characterize the range of IR cameras with respect to the visual range [50]. Parameters for the model are the climate and aerosol type, humidity, visibility, the geometry, and lengths of the atmospheric path as well as temperature and emissivity of the object and background. Transmission spectra are computed from the UV to the millimeter wavelength region with spectral resolution of $1\,\text{cm}^{-1}$ for the above-mentioned climate models and differing locally prevailing aerosols. Typically 25 atmospheric

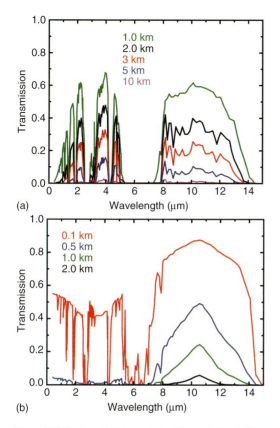

Figure 10.35 Spectral transmission for predefined climate and aerosol models (a) Midlatitude summer climate and rural aerosol, visual range 1220 m (b) Midlatitude winter, radiative fog, visual range 610 m. Image after [50], Courtesy Deutsches Zentrum für Luft- und Raumfahrt (DLR).

species are included as well as pressure, temperature, and mixing ratios for 50 atmospheric layers from sea level to 120 km altitude.

In general, maritime aerosols have the lowest ranges since the usually have larger particles such as, for example, rural or urban aerosols. For fog situations, the range also depends very sensitively on the type of fog, that is, its size distribution, and concentration of the droplets. Figure 10.35a depicts an example of reasonable visual range in the visible, due to rural aerosol particles, only, that is, pure dry aerosols that are typically smaller than 1 μm. The transmission in the visible band is significantly lower than in both IR bands due to the strong influence of Rayleigh scattering. Path ranges of more than 5 km still show reasonable transmission, that is, they are still detectable in both IR bands. As a consequence, the IR range is expected to be larger by at least a factor of four compared to the visual range.

Figure 10.36 Example of looking through slightly foggy atmosphere with (a) Vis and (b) IR camera. (Images courtesy FLIR systems Inc.)

Figure 10.35b shows the results for radiative fog in midlatitude winter [50]. The visual range was assumed to be around 0.61 km. Results for the IR show similar low transmission for the MW range, whereas LW IR radiation shows a much larger transmission. In this case, the LW camera is expected to have a range of about a factor of four larger than in the Vis or the MW IR. Finally, very dense fogs with visibilities below 300 m or 90 m will have similar transmission spectra for VIS, MW, or LW IR. Radiation does just not penetrate through this dense fog in all spectral bands (Vis, NIR, MW, and LW), that is, the atmosphere itself is the limiting factor for the range, which is the same in the VISD and the IR.

As an example, Figure 10.36 depicts Vis and IR images while looking at distant objects through a slightly foggy atmosphere. The IR image is obviously quite clear and even allows to identify the pedestrian, which is not easily possible for the still image in the visible spectral range.

These transmission spectra have been used to actually determine the range according to the NETD criterion mentioned above. The temperature differences between object and background were varied between 1 and 100 K. Figure 10.37 depicts results assuming an NETD of 0.15 K of the camera used (lower values would just lead to a lowering of the respective horizontal line).

The detection range increases with temperature difference between the object and the background and has a value of about 2.4 km for $\Delta T = 10$ K, which is a factor of four larger than for the visual range. Lowering the NETD would also lead to longer ranges.

These ideal, sensor-independent results must still be corrected for the sensor properties, in particular, the limited spatial and radiometric resolution of the camera due to the optics, the detector, and the system noise.

Finally, it is mostly the apparent angular size of the object (having an influence on the MTF, see Chapter 2) that will determine the achievable range. Figure 10.37b shows results for a $\Delta T = 10$ K difference of the object versus background as a function of distance. The limited spatial resolution is dominant as can be seen by

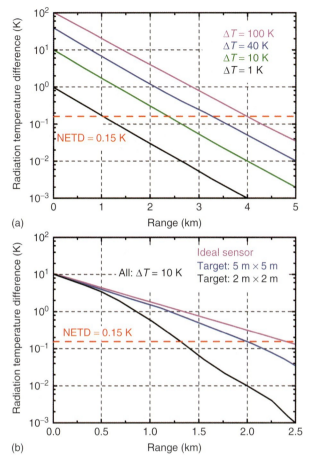

Figure 10.37 Ranges for LW IR cameras (details see text) for the case of Figure 10.36b with midlatitude winter, radiative fog, visual range 610 m. (a) Ideal situation, only due to atmospheric conditions for $\Delta T = 1$, 10, 40, and 100 K. (b) Comparison of ideal (MTF $= 1$) to realistic conditions (MTF < 1) for $\Delta T = 10$ K. Images after [50], Courtesy Deutsches Zentrum für Luft- und Raumfahrt (DLR).

comparing the results for different target sizes. The smaller targets will become invisible at much shorter distances.

Overall, these model results suggest that fog penetration is higher in the LWIR compared to that in the MWIR band. LW systems have the additional advantages of being less affected by CO_2 absorption and they are cheaper than MW cameras.

Finally, rainfall consists of even larger droplets with sizes in the millimeter range. Their increased scattering contribution and absorption properties significantly reduce the range for both LWIR and MWIR. Depending on the density of rain droplets, the signal decreases strongly and the range is seldom above 500 m.

Another rain-related question is whether raindrops (or dirt) on the lenses will be important. First of all, if small particles are on a lens, they are out of focus and hence will not affect image quality (similar to normal photography in the Vis range). Second, in most cameras, the lenses also have heaters to prevent condensation.

10.6
Nature

10.6.1
Sky and Clouds

Nearly everybody owning an IR camera has probably directed it once toward the sky. Optically thick clouds allow to measure cloud temperatures. Clear skies, however, pose a problem. The measured apparent temperature is useless, since the atmosphere is not optically thick in the IR spectral range. Figure 10.38 depicts a spectrum of the vertical atmosphere in the near IR range. Obviously, both in the LW and MW ranges, transmission has finite values due to scattering and absorption losses. As a consequence, the vertical atmosphere is not optically thick, that is, an IR camera detects radiation from the atmosphere, but from the much colder background, that is, space, as well. The small finite absorption contribution correlates with not only small but also finite emissivity values. Therefore, the measured radiance contribution from a certain direction of the clear sky is the sum of the background radiance from outer space (3 K) (see Figure 1.22) and the

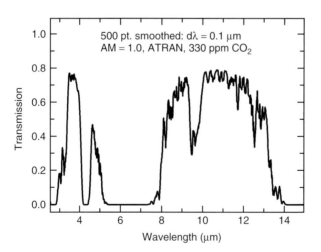

Figure 10.38 Transmission spectrum of the vertical atmosphere in the thermal IR spectral range. Spectrum courtesy A.Krabbe/D. Angerhausen DSI Universität Stuttgart, Germany, computed with ATRAN (S. Lord 1992).

emission of atmospheric gases and aerosols. The latter depends on the temperature of the gases/aerosols, which themselves depend on the height in the atmosphere. These emission contributions, therefore, lead to measured "temperatures," which are well above the expected 3 K background radiation of the universe. Besides, the typical measurement ranges are well above the minimum sky temperatures, which may be measured with IR cameras. Obviously, sensitive detectors outside of the atmosphere may indeed get rid of the atmospheric emission and hence directly observe IR radiation from the space. This is the fascinating field of astronomical IR spectroscopy.

To make the situation more complicated, the apparent sky temperature is also a function of zenith angle. This is due to the so-called air mass: radiation that penetrates the atmosphere not from the zenith sky but under well-defined zenith angles will pass a longer way within the atmosphere, thereby interacting with more matter. The total amount of matter is included in the dimensionless parameter air mass, which can reach the value of about 38 near the horizon (Figure 10.39) [55].

This means that radiation that penetrates the atmosphere along a tangential path from the horizon to an observer must pass 38 times the amount of matter compared to the vertical, that is, the zenith direction. Consequently, it has a much higher chance of being scattered, absorbed, or emitted. Therefore, IR radiation from a direction closer to the horizon will be increased due to a higher atmospheric contribution. Overall, the apparent sky temperature will decrease from horizon toward the zenith (Figure 10.40). Since the horizon paths will include more contributions from lower parts of the atmosphere, the measured temperatures do not directly correlate with the air mass dependence. A practical problem of the user consists in the proper choice of the distance and emissivity. Distances used in the camera software usually refer to horizontal paths ending at opaque objects. The atmosphere is, however, first not opaque, second the paths are not horizontal, and third the maximum distance is usually limited to, say, 10 km. Furthermore, what

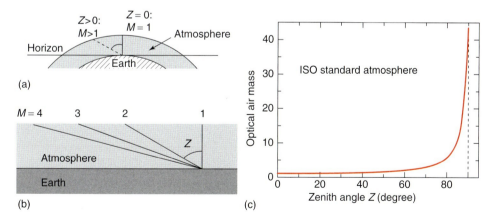

Figure 10.39 (a,b) Planar and spherical geometry for explanation of air mass and (c) resulting air mass as a function of zenith angle.

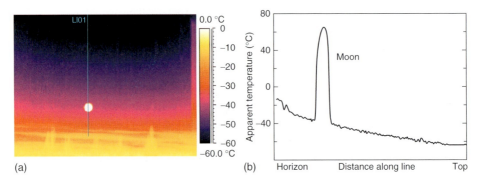

Figure 10.40 View of horizon and sky with setting moon at a cold winter morning ($T_{air} = -10$ °C). The apparent temperature of the sky drops with increasing angle above horizon. The exact values depend on the chosen distance and emissivity (here arbitrarily $\varepsilon = 0.95$ and 2000 m).

is the kind of emissivity that should be used for the atmosphere? Another problem is that humidity and temperatures vary with height quite appreciably, for example, T amounts to about -60°C at 0 km height. Rather than discussing these problems in detail, results should be interpreted qualitatively. The lowest detected apparent temperatures are easily below the lowest temperature range of the conventional IR cameras, for example, the lower limit of about -60°C was reached in Figure 10.40 for the top of the line plot.

Figure 10.40 also depicts an image of the moon. The moon or any other hot object observed through the atmosphere also gives rise to an apparent temperature, which is due to different radiance contributions. First, the IR camera detects contributions from the atmospheric emission and second, contributions from the moon radiance, which are attenuated while passing through the atmosphere. The latter also again depends on the air mass, that is, the zenith angle [56].

Low temperatures of clear skies during nights also have applications with relevance to IR imaging. On the one hand, they can also be used for cooling purposes [57] and, on the other hand, clear night sky radiative cooling will have an impact on outdoor building thermography (Chapter 6). For the respective radiative transfer rates, it is often useful to define effective sky temperatures, which resemble average values, defined by the overall energy transfer [58]. The effective sky temperature depends primarily on dew point temperature and cloud cover.

10.6.2
Wildfires

Unplanned wildfires occur all over the world, mostly in hot and dry seasons. They can have devastating impacts on the environment by consuming thousands of hectares of productive timber, destroying wildlife habitat, and filling the air with smoke and tons of carbon dioxide.

Wildfires are due to either natural (as a result of lightning activity) or human causes and require immediate and effective suppression actions. In some regions like southern Europe (e.g., Italy, Spain, and Greece), the United States (e.g., California), Canada (e.g., Alberta), or Australia, wildfires are well-known regular events. For example, the province of Alberta alone suffers from an average of about 1380 of these devastating events annually [59]. This resulted in the burning of more than 107 000 ha in the decade between 1998 and 2007, alone. Lightning accounted for 49% of these wildfires; the remaining 51% were preventable human caused wildfires.

Therefore, detection and fast reporting of wildfires as well as protective actions for the land, the people, and wildlife are very important. In the following, an exemplary program from the Alberta Sustainable Resource Development, Forestry Division is presented, which uses a multiple IR approach to support wildfire suppression activities. This program is based on 32 years of thermal IR operations in the field. It uses first handheld IR cameras (primary use of IR scanning, for example, from helicopters), second low- and high-altitude surveys (gyro-stabilized gimbal mounted on a helicopter or fixed at wing aircraft for high altitudes), and third satellite-based systems (using the moderate resolution imaging spectroradiometer, MODIS).

Overall IR imaging helps to first detect holdover fires that are a result of winter burning such as site clearings, power and pipelines construction, brush piles, and so on, second to determine the most effective location for air tanker drops in heavy smoke conditions, and third to obtain hotspot locations and perimeters of ongoing wildfires. Overall, the allocation of limited fire fighting resources is based upon the following priorities: human life, communities, sensitive watersheds and soils, natural resources, and important infrastructure.

A typical use of handheld IR imaging is to use a camera within a helicopter by removing the helicopter door. The operator is required to wear a safety harness. He scans a wildfire and marks the identified spots with marking material and the use of GPS coordinates. Such handheld scanning works best for wildfires less than 1000 ha, or when hotspots need to be located along portions of the fire line.

Low-altitude IR scanning involves the use of a light helicopter (Bell 206) with a gyro-stabilized gimbal mounted IR camera under the aircraft. These (more expensive) systems allow a more flexible scanning of larger areas of typically 1000–0000 ha. The spatial resolution of the hotspot ranges from one inch to one foot for a maximum altitude of 1000 ft above ground level. Approximately 100 km of perimeter can be scanned each day, depending upon the number of hotspots and speed of the aircraft.

In general, IR imaging does not always work the same way due to several reasons:

- Rainfall causes the ground to cool down through evaporation. This cooling reduces the heat signature and can drive hotspots into the ground where it can smolder within organic material. Hotspots with sufficient energy can resurface,

creating a potential risk. By scanning immediately following a rainfall, one may miss such hotspots. Therefore, depending on the amount of moisture, IR surveys following rain fall should wait for some dry days before trying to find resurfacing hot spots.
- The time of a scan may have an impact on the result due to first solar gain and second solar reflectance. Both effects can lead to a false-positive identification of hotspots. Therefore, one should observe the same spot while changing the observation angle of the target.
- Imaging is not possible through solid objects (such as tree trunks, but also windows in helicopters), which are opaque in the IR.
- IR radiation can penetrate smoke to a much larger extent than visible light. Therefore, IR imaging is able to identify perimeters and/or the head of wildfires through smoke from a safe distance. However, smoke may change the detected IR signal due to first absorption of the hotspot signal behind the smoke, second, emission of an own IR signal depending on smoke temperature and emissivity (which depends on its optical thickness), and third scattering of IR radiation. The importance of these effects depends on the ratio of size of smoke particle (typically larger than Vis and smaller than IR wavelengths) and wavelength of relevant radiation. For smokes of wildfires, LW camera signals are less susceptible to absorption and scattering.

The temperature range of hotspots found on wildfires ranges from above 800 °C for flaming combustion to below 100 °C for hot areas buried below a lot of white ash and maybe observed through smoke. The latter spots are very difficult to detect with other means.

Figure 10.41 depicts an example (Vis and IR image) of a low-altitude inspection, detecting hot spots of a wildfire.

(a)

(b)

Figure 10.41 Hotspots of a wildfire observed with low-altitude IR imaging (b) and corresponding visible image (a). The colored line on the Vis image resembles the retardant used to slow down advancement of fire. Courtesy of Alberta Sustainable Resource Development.

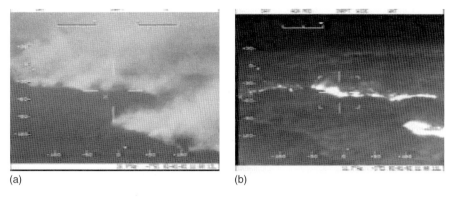

Figure 10.42 Head of a wildfire in thermal infrared (b) and visible image (a). IR radiation penetrates through smoke and allows to identify the head oft the fire front. Courtesy of Alberta Sustainable Resource Development.

High-altitude IR scanning is done by mounting a gyro-stabilized gimbal and IR camera on the nose or wing of an aircraft. The spatial resolution of the method, which is useful for wildfires greater than 10 000 ha, ranges between 0.5 and 30 m on the ground. The images can precisely locate the head and flank of a wildfire thus effectively supporting the drops of retardant and water by air tankers, which are often hindered by poor visibility due to heavy smoke (Figure 10.42). Because of the velocity of the moving aircraft, images seem sometimes to be slightly fuzzy, which does, however, not affect the information about hot spots or fire heads, etc.)

The largest scale scans of hotspots are performed using the MODIS and Aqua MODIS imaging spectroradiometer instruments aboard Earth orbiting satellites. They view the entire Earth's surface every 1–2 days and cover the province of Alberta twice a day. Fire detection data are derived from the data and a variety of products related to hotspot detection are available daily [60]. The data have a spatial resolution of approximately 1 km; hence, satellite data can give overviews of what is happening on a large scale.

Of course, IR imaging techniques, in particular if used from helicopters and planes, are expensive. For example, thermal IR operations cost the Alberta Government approximately $1.6 million for each fire season [59]. This must however be seen in perspective to the huge benefits.

10.6.3
Geothermal Phenomena

10.6.3.1 Geysirs
Geothermal phenomena belong not only to the most spectacular but also to the most dangerous phenomena in nature. A safe observation of geothermal physics is possible in National Parks such as Yellowstone or in Iceland, where there are spectacular geyser eruptions taking place at well-defined time intervals. Geysir

Figure 10.43 Eruption of a geyser in Iceland. Image courtesy Infrared Training Center.

eruptions are observed, if hot water is in a cavern under high pressure due to the water column which extends from the cavern up to the surface, ending, for example, in a small pond. Because of geothermal heat, the water temperature in the cavern rises above the boiling temperature, which is higher than 100 °C due to the high pressure. Whenever the temperature exceeds a critical limit, boiling starts and gas bubbles form in the cavern. On their way up the channel to the surface their volume expands rapidly with decreasing pressure, thereby ejecting the water column on top of it with high velocities. Figure 10.43 depicts such a Geyser eruption with the hot water ejection as observed in Iceland. The fountain is quite high as can be seen by the spectators in the foreground, which were of course at a safe distance from the hot water.

10.6.3.2 Infrared Thermal Imaging in Volcanology

The value of ground-based thermal radiometry in volcanology has been recognized for decades [61]. Mostly because of the development of powerful radiometric imagers in the form of handheld IR cameras, however, it is only recently that this method has become routine including data capturing from helicopters and from the sea [62–67].

In recent years, thermal IR measurements of active volcanic features have been used to

1) recognize magma movements within the uppermost volcanic conduits;
2) detect the upward movement of shallow feeder dikes;
3) track eruptive activity, for example, within the craters even through the thick curtain of gases;
4) distinguish and measure the thermal and rheological properties of active basaltic lava flows, lava tubes, and silicic lava flows as well as sea entry thermal anomalies;

5) track the development of the structures on the lava flow field such as channels, tubes, tumuli, and hornitos;
6) analyze the evolution of fumarole fields;
7) study active eruption plumes, strombolian activity, and persistent degassing;
8) obtain effusion rate for active lava flows;
9) detect potential failure planes on recently formed cinder cones and fractures developing just before flank collapse at active volcanoes; and
10) analyze active lava lakes.

Thus thermal imaging has contributed to a detailed eruption chronology and understanding of the lava flow and eruption processes operating during eruption. Thermal mapping has proven essential during effusive eruptions, since it distinguishes lava flows of different age and concealed lava tubes' path, improving hazard evaluation. In particular, the 2002/2003 Etna and Stromboli eruptions were analyzed in detail with thermal imaging with more than 100 000 thermal images recorded before, during and after eruptions. The analysis of these images, combined with visible images, as well as data on seismicity, ground deformation, and gas geochemistry collected routinely during the eruption, allowed improved quantification of eruptive parameters such as effusion rate and maximum temperature at the bottom of the summit craters, qualitative tracking of lava flow features, fissures, and vents in situations where traditional, ground-based measurements were difficult and dangerous, and in general a greater understanding of eruptive phenomena.

On Etna, thermal images recorded monthly on the summit of the volcano revealed the opening of fissure systems several months in advance. After the onset of the 2002 flank eruption, daily thermal mapping allowed to monitor a complex lava flow field spreading within a forest, below a thick plume of ash and gas. At Stromboli, helicopter-borne thermal surveys allowed to recognize the opening of fractures along Sciara del Fuoco, one hour before the large failure that caused severe destruction on the island on 30 December 2002. This was the first time ever that a volcanic flank collapse has been monitored with a thermal camera. In addition, the exceptional explosive event of 5 April 2003 at Stromboli was observed from helicopter with a thermal camera recording images immediately before, during, and after the huge explosion. As a consequence of these investigations, it is argued that a more extended use of thermal cameras in volcano monitoring, both on the ground, from fixed positions and from helicopters, will significantly improve the understanding of volcanic phenomena and hazard evaluations during volcanic crisis. It is expected that if thermal imaging is applied to daily tracking of volcanic features, it can contribute significantly to volcano surveillance and hazard evaluation.

Figures 10.44 and 10.45 depict some typical examples of thermal images from lava flows. Fresh lava flows at Mt. Etna can have inside temperatures of up to 1085 °C [67] and outer surface temperatures of up to 900 °C (maximum temperatures at other volcanos can be higher, as is, e.g., the case in Hawaii). Figures 10.46 and 10.47 show explosive Strombolian activities with ash plume. The ash plumes are optically

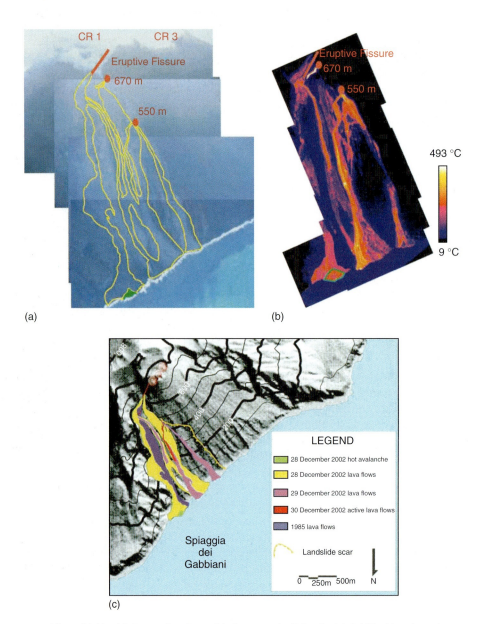

Figure 10.44 (a) Composite photo, (b) thermal image, and (c) map, view from north, of the eastern half of the Sciara del Fuoco, Stromboli volcano in southern Italy, showing the 28, 29, and 30 December 2002 lava flows, the 1985 lava flow, the hot avalanche on the Spiaggia dei Gabbiani beach, and part of the 30 December 2002 landslide scar. Images courtesy Sonia Calvari, Istituto Nazionale di Geofisica e Vulcanologia, Catania.

10.6 Nature | 583

(a)

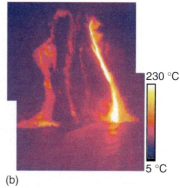
(b)

Figure 10.45 (a) View from north of the Sciara del Fuoco (Stromboli volcano), showing the lava flow field erupted from the 500-m elevation vent. Thermal image (b) of active (yellow) and inactive (red-purple) single flow units entering the sea on 3 January 2003, one may easily detect thermal reflections from the sea water surface. Images courtesy Sonia Calvari, Istituto Nazionale di Geofisica e Vulcanologia, Catania.

Figure 10.46 Explosive strombolian activity and ash column detected by thermal and visible images. An example is shown of the 2750 m vent, with digital photos and overlaps of the corresponding thermal images of the 19 November 2002 transitional activity with the ash plume bent-over close to the vent by the wind. Image courtesy Letizia Spampinato, Istituto Nazionale di Geofisica e Vulcanologia, Catania.

thick, but sometimes they allow to estimate the maximum temperatures. For example, temperatures of about 900 °C are typically measured at 250 m distances. However, for any quantitative analysis of volcanic features, it is required to take into account the special properties of the surroundings, for example, strong absorption due to SO_2 and aerosol absorption within the line of sight [65]; see also spectrum of SO_2 in the Appendix 7.A.

Figure 10.47 Explosive strombolian activity from the SW crater at Stromboli volcano. Image courtesy Sonia Calvari, Istituto Nazionale di Geofisica e Vulcanologia, Catania.

References

1. Jessen, C. (2003) in *Lehrbuch der Physiologie*, 4th edn (ed. S. Klinke), Thieme, Stuttgart.
2. Bennaroch, E. (2007) Thermoregulation: recent concepts and remaining questions. *Neurology*, **69**, 1293–1297.
3. Uematsu, S. (1988) Quantification of thermal asymmetry. Part 1: normal values and reproducibility. *J. Neurosurg.,break* **69** (4), 553–555.
4. Leroy, P. and Filasky, R. (1992) in *Evaluation and Treatment of Chronic Pain*, 2nd edn (ed. G. Aronoff), Williams and Wilkins, Baltimore, London, Hong Kong, and Munich, pp. 202–212.
5. Cockburn, W. (2006) Common errors in medical thermal imaging. Inframation 2006, Proceedings Vol. 7, pp. 165–177
6. Agarwal, K., Lange, A.-Ch., and Beck, H. (2007) Thermal imaging in healthy humans, what is normal skin temperature? Inframation 2007, Proceedings Vol. 8, pp. 399–403.
7. Agarwal, K., Spyra, P.-S., and Beck, H. (2008) Monitoring treatment effects in Osteomyelitis with the aid of infrared thermography. Inframation 2008, Proceedings Vol. 9, pp. 1–6.
8. Agarwal, K., Spyra, P.-S., and Beck, H. (2007) Thermographic imaging for interventional pain management. Inframation 2007, Proceedings Vol. 8, pp. 391–394.
9. Agarwal, K., Spyra, P.-S., and Beck, H. (2007) Thermal imaging provides a closer look at stellate blocks in pain management. Inframation 2007, Proceedings Vol. 8, pp. 391–394.
10. Agarwal, K., Spyra, P.-S., and Beck, H. (2008) Monitoring effects of sympathicolysis with infrared-thermography in patients with complex regional pain syndromes (CRPS). Inframation 2008, Proceedings Vol. 9, pp. 7–10.
11. Agarwal-Kozlowski, K., Lange, A.C., and Beck, H. (2009) Contact free infrared thermography for assessing effects of acupuncture: a randomized single-blinded placebo-controlled cross-over clinical trial. *Anesthesiology*, **111** (3), 632–639.
12. Litscher, G. (2005) Infrared Thermography Fails to Visualize Stimulation-induced Meridian-like Structures, BioMedical Engineering,

online *http://www.biomedical-engineering-online.com/content/4/1/38*, accessed 2010.
13. Mostovoy, A. (2008) Breast thermography and clinical applications. Inframation 2008, Proceedings Vol. 9, pp. 303–307.
14. Mostovoy, A. (2009) Clinical applications of medical thermography. Inframation 2009, Proceedings Vol. 10, pp. 23–27.
15. Kennedy, D.A., Lee, T., and Seely, D. (2009) A comparative review of thermography as a breast cancer screening technique. *Integr. Cancer Ther.*, **8** (1), 9–16.
16. Cockburn, W. (2007) Functional thermography in diverse medical practice. Inframation 2007, Proceedings Vol. 8, pp. 215–224.
17. Bretz, P. and Lynch, R. (2007) Me lding Three emerging technologies: pharmacogenomics, digital infrared and argon gas, to eliminate surgery, chemotherapy and radiation in diagnosing and treating breast cancer. Inframation 2007, Proceedings Vol. 8, pp. 225–234.
18. Bretz, P., Lynch, R., and Dreisbach, Ph. (2009) Breast cancer in tough economic times – is there a new diagnostic and treatment paradigm? Inframation 2009, Proceedings Vol. 10, pp. 435–448.
19. Brioschi, M.L., Vargas, J.V., and Malafaia, O. (2004) Review of recent developments in thermographic applications in health care. Inframation 2004, Proceedings Vol. 5, pp. 9–17.
20. Brioschi, M.L., Silva, F.M.R.M., Matias, J.E.F., Dias, F.G., and Vargas, J.V.C. (2008) Infrared imaging for emergency medical services (EMS): using an IR camera to identify life-threatening emergencies. Inframation 2008, Proceedings Vol. 9, pp. 549–560.
21. Brioschi, M.L., Sanches, I., and Traple, F. (2007) 3D MRI/IR imaging fusion: a new medically useful computer tool. Inframation 2007, Proceedings Vol. 8, pp. 235–243.
22. Wu, M. (2004) *Proceedings of the Thermosense XXVI (2004), Proceedings of the SPIE Vol. 5405*, SPIE Press, Bellingham, pp. 98–105.
23. Tana, Y.H., Teoa, C.W., Onga, E., Tanb, L.B., and Sooa, M.J. (2004) *Proceedings of the Thermosense XXVI (2004), Proceedings of SPIE Vol. 5405*, SPIE Press, Bellingham, pp. 68–78.
24. Harper, D.L. (2000) The value of infrared thermography in the diagnosis and prognosis of injuries in animals. Inframation 2000, Proceedings Vol. 1, pp. 115–122.
25. Hopkins, P. and Bader, D.R. (2009) Mammals communicate, infrared listens utilizing infrared imaging for injury identification. Inframation 2009, Proceedings Vol. 10, pp. 179–188
26. Reese, J. (2006) Applications of infrared imagery in the monitoring and diagnosis of zoo animals in anchorage, Alaska. Inframation 2006, Proceedings Vol. 7, pp. 145–150.
27. Church, J.S., Cook, N.J., and Schaefer, A.L. (2009) Recent applications of infrared thermography for animal welfare and veterinary research: everything from chicks to elephants. Inframation 2009, Proceedings Vol. 10, pp. 215–224.
28. *www.unitedinfrared.com*, *www.equineir.com*. (2010).
29. West, M. (2007) The odd couple … the professor and the thermographer. Inframation 2007, Proceedings Vol. 8, pp. 281–289.
30. Allison, J.N. (2007) Infrared goes on safari. Inframation 2007, Proceedings Vol. 8, pp. 353–359.
31. Whitcher, A. (2006) Hunting seals with infrared. Inframation 2006, Proceedings Vol. 7, pp. 155–164.
32. Reese-Deyoe, J. (2008) Practical applications of infrared thermography in the rescue, recovery, rehabilitation and release of wildlife following external exposure to oil. Inframation 2008, Proceedings Vol. 9. pp. 389–402.
33. Schwahn, B. (2009) Infrared thermal imaging to prevent bat mortality at wind farms. Inframation 2009, Proceedings Vol. 10, pp. 385–392.
34. Rentoul, M. (2007) The practice of detecting termites with infrared thermal imaging compared to conventional techniques. Inframation 2007, Proceedings Vol. 8. pp. 15–19.

35. Bruni, B. (2004) Three ways the pest professional can use infrared thermography. Inframation 2004, Proceedings Vol. 5, pp. 109–119.
36. Ono, M., Igarashi, T., and Ohno, E. (1995) Unusual thermal defense by a honeybee against mass attack by hornets. *Nature*, **377**, 334–336.
37. Garza, D., Rolston, B., Johnston, T., Sungar, G., Ferguson, J., and Matheson, G. (2009) Heat-loss patterns in national football league players as measured by infrared thermography. Inframation 2009, Proceedings Vol. 9, pp. 541–547.
38. Vollmer, M. and Wogram, K. (2005) Zur Intonation von Blechblasinstrumenten bei sehr niedrigen Umgebungstemperaturen. *Instrumentenbau-Zeitschrift - Musik International*, **59** (3-4), 69–74.
39. Coram Populo! – A Mystery, International Dance Theatre Production, (2008), project proposal by Miloš Vujković, Arthouse Tacheles Berlin; contact: milosh@tacheles.de.
40. Vollmer, M., Vujković, M., Trellu, Y., and Möllmann, K.-P. (2009) IR feedback loops to spotlights: thermography and contemporary dancing. Inframation 2009, Proceedings Vol. 10, pp. 89–97.
41. Dinwiddie, R.W. and Dean, S.W. (2006) Case study of IR reflectivity to detect and document the underdrawing of a 19th century oil painting. Proceedings of the Thermosense XXVIII, Proceedings of SPIE Vol. 6205, p. 620510 /1 - 12.
42. Grinzato, E. and Rosina, E. (2001) Infrared and thermal testing for conservation of historic buildings, in *Nondestructive Testing Handbook*, Infrared and Thermal Testing, Vol. 3, 3rd edn, Chapter 18.5 (ed. P.O. Moore), American Society for Nondestructive Testing, Inc., Columbus, p. 624–646.
43. Humphries, H.E. (2001) Infrared and thermal testing for conservation of fine art, in *Nondestructive Testing Handbook*, Infrared and Thermal Testing, Vol. 3, 3rd edn, Chapter 18.6 (ed. P.O. Moore), American Society for Nondestructive Testing, Inc., Columbus, pp. 647–658.
44. Poksinska, M., Wiecek, B., and Wyrwa, A. (2008) Thermovision investigation of frescos in Cistercian monastery in Lad (Poland). Proceedings of the 9th International Conference on Quantitative InfraRed Thermography in Krakow, QIRT 2008, pp. 653–658.
45. Poksinska, M., Cupa, A., and Socha-Bystron, S. (2008) Thermography in the investigation of gilding on historical wall paintings. Proceedings of the 9th International Conference on Quantitative InfraRed Thermography in Krakow, Poland, QIRT 2008, pp. 647–652.
46. Magyar, M. (2006) Thermography applied to cultural heritage. Inframation 2006, Proceedings Vol. 7. pp. 325–332.
47. Richards, A. (2001) *Alien Vision*, SPIE Press, Bellingham.
48. Rogalski, A. and Chrzanowski, K. (2006) Infrared devices and techniques, in *Handbook of Optoelectronics*, vol. **1** (eds J.R. Dakin and R.G.W. Brown), Taylor and Francis, New York, p. 653–691.
49. Zieli, M. and Milewski, S. (2006) Thermal images and spectral characteristics of objects that can be used to aid in rescue of shipwrecked people at sea. Inframation 2006, Proceedings Vol. 7, pp. 279–286.
50. Beier, K. and Gemperlein, H. (2004) Simulation of infrared detection range at fog conditions for enhanced vision systems. *Civ. Aviat. Aerosp. Sci. Technol.*, **8**, 63–71.
51. Richards, A. and Johnson, G. (2005) *Proceedings of the Thermosense XXVII (2005)*, Proceedings of SPIE Vol. 5782, SPIE Press, Bellingham, pp. 19–28.
52. Duntley, S.Q. (1948) The reduction of apparent contrast by the atmosphere. *J. Opt. Soc. Am.*, **38**, 179–191.
53. Jha, A.R. (2000) *Infrared Technology*, John Wiley & Sons, Inc.
54. Mason, N. and Hughes, P. (2001) *Introduction to Environmental Physics*, Taylor and Francis.
55. Vollmer, M. and Gedzelman, S. (2006) Colors of the sun and moon: the role of the optical air mass. *Eur. J. Phys.*, **27**, 299–309.
56. Vollmer, M., Möllmann, K.-P., and Pinno, F. (2010) Measurements of sun and moon with IR cameras: effect of air mass. Inframation 2010, Proceedings Vol. 11, to be published.

57. Möllmann, K.-P., Pinno, F., and Vollmer, M. (2008) Night sky radiant cooling – influence on outdoor thermal imaging analysis. Inframation 2008, Proceedings Vol. 9, pp. 279–295.
58. Martin, M. and Berdahl, P. (1984) Characteristics of infrared sky radiation in the United States. *Solar Energy*, **33**, 321–336.
59. Simser, S. (2008) Utilization of thermal infrared technology on wildfires in Alberta. Inframation 2008, Proceedings Vol. 9, pp. 417–428.
60. *http://activefiremaps.fs.fed.us* (2010).
61. Birnie, R.W. (1973) Infrared radiation thermometry of Guatemalan volcanoes. *Bull. Volcanol.*, **37**, 1–36.
62. Calvari, S., Lodato, L., and Spampinato, L. (2004) *Proceedings of the Thermosense XXVI (2004), Proceedings of SPIE Vol. 5405*, SPIE Press, Bellingham, pp. 199–203.
63. Calvari, S., Spampinato, L., Lodato, L., Harris, A.J.L., Patrick, M.R., Dehn, J., Burton, M.R., and Andronico, D. (2005) Chronology and complex volcanic processes during the 2002–2003 flank eruption at Stromboli volcano (Italy) reconstructed from direct observations and surveys with a handheld thermal camera. *J. Geophys. Res.*, **110**, B02201, doi: 10.1029/2004JB003129.
64. Calvari, S., Spampinato, L., and Lodato, L. (2006) The 5 April 2003 vulcanian paroxysmal explosion at Stromboli volcano (Italy) from field observations and thermal data. *J. Volcanol. Geotherm. Res.*, **149**, 160–175.
65. Sawyer, G.M. and Burton, M.R. (2006) Effects of a volcanic plume on thermal imaging data. *Geophys. Res. Lett.*, **33**, L14311.
66. Spampinato, L., Calvari, S., Oppenheimer, C., and Lodato, L. (2008) Shallow magma transport for the 2002–2003 Mt. Etna eruption inferred from thermal infrared surveys. *J. Volcanol. Geotherm. Res.*, **177**, 301–312.
67. Calvari, S., Coltelli, M., Neri, M., Pompilio, M., and Scribano, V. (1994) The 1991-1993 Etna eruption: chronology and lava flow-field evolution. *Acta Vulcanolo.*, **4**, 1–14.

Index

a
aberrations   118
active thermal imaging   217ff
– lock in thermography   226
– nondestructive testing of composites   229
– pulsed phase thermography   234
– pulsed termography   221
– thermal waves   219
acupuncture, see medical applications
adiabatic processes   300
advanced methods   157ff
aircraft fuselage   229
air mass   575
airplanes   503ff
alcohol, see ethanol
ambient temperature   98
ammonia, see NH_3
animals, see also veterinary applications   548ff
– bees   552
– horses   550
– hummingbird   552
– pets   549
– zoo animals   549
antireflection coatings, see optical material properties
arts   559ff
– contemporary dancing   561
– musical instruments   560
– paintings   565
atmospheric temperature   98
atmospheric transmission   53ff, 98, 571, 574
attenuation/transmission of IR radiation   52
– by absorbing slabs   57
– by nonabsorbing slabs   55
automotive industry   498ff
– heating systems   498
– night vision systems   500
autoshutter   112

b
background noise limited detection   86ff
bad pixel correction, see image formation
band emission, see blackbody radiation
bandwidth Δf, see detectors, performance parameters
Bénard Marangoni convection   296
bicycle   283ff
– helmets   514
Biot number   250
blackbody calibration standards, see emissivity
blackbody cavities   47ff, 320
– absorption vs. emission   323
blackbody radiation   21ff
– band emission   26
– Planck's law representations   22–25
– Stefan–Boltzmann law   25
blackbody radiator, see blackbody radiation
black emitter method   495
blower door test   386
bolometer, see also detectors, thermal   78, 461
bottles/cans   305
Bouguer's law   408
breast thermography, see medical applications   544
breathing   427
Brewster angle   14, 480
buildings   329ff
– blower door test   386
– carports   377
– energy standards   336
– floor heating system   347

Infrared Thermal Imaging. Michael Vollmer and Klaus-Peter Möllmann
Copyright © 2010 WILEY-VCH Verlag GmbH & Co. KGaA, Weinheim
ISBN: 978-3-527-40717-0

buildings (contd.)
– general rules   335
– half timbered structures   338ff
– inside thermal insulation   344
– interpretation problems   333
– moisture   358
– night sky radiant cooling   375
– quantitative imaging   391
– shadows   366
– solar load   361
– solar reflections   369
– structural defects   352
– thermal bridges   348ff
– view factor effects   373ff
– wind   356
– windows   381ff
bushing   528

c

calibration of IR camera   129ff
camera systems   101ff
– comparison   114
camera system performance   137ff
– spatial resolution (IFOV, SRF)   142, 525
– temperature accuracy   137
– temperature resolution (NETD)   139
– time resolution   150
carbon capture and storage (CCS)   424ff
car industry, see automotive industry
carport   377
cavity blackbody radiators, see also blackbody cavities   49
cheese   302
close-up lens   446
clouds   574
CO_2 401, 407, 424ff
color palette   333
condition monitoring (CM)   508ff
conduction, see heat transfer
convection of water, see also heat transfer   295

d

D*, see detectors, performance parameters
detectors   73ff
– cooling   90, 116ff
– performance parameters   74ff, 75
– photon detectors   83ff, 91–93
– temperature stabilization   116ff
– thermal detectors   77ff
detector noise
– Johnson–Nyquist noise   79
– shot noise   85
– temperature fluctuation noise   79

dew point   260
diffuse reflection   12, 318, 478ff
directed reflection, see specular reflection
dispersion   11
dynamic range   132, 186

e

Eddy currents   309
electrical boards   523
electrical circuits   308
electric fans   528
electromagnetic waves   6ff
– polarization   8
– spectrum   8
emissivity   32ff
– and blackbody calibration standards   47ff
– and Kirchhoff's law   34
– angular dependence   38, 322
– definitions   33
– measurements   45
– object classification   33
– of cavities   49, 320ff
– of gases   416
– parameters   35ff
ethanol   421
evaporative cooling   298
exhaust gases   429
extender rings   122, 446

f

feature extraction, see image processing
feedback   561
filters, see also imaging with filters   69, 126, 158ff
focal plane array (FPA)   82, 103ff
field of view (FOV)   76, 120ff
floor heating, see buildings
Fourier number   264
frame rate   103, 150ff
Fraunhofer   398
Fresnel's equations   13
– Al   483
– Cu   480
– Fe   483
– glass   479, 484
– Si   485
FTIR spectrometry, see also multi spectral imaging   178

g

gases, compilation of spectra   433ff
gases, theory   397ff
– absorption   403ff
– calibration curves   413

- CO_2 400, 407, 424ff
- emission   403ff, 416ff
- H_2O 400ff
- hydrocarbons   401
- measurement conditions   414
- molecular spectra   397ff
- quantitative analysis   408ff
- scattering   403ff
gases, measurements   326, 417ff
- broadband vs. narrowband   426
- GasFind cameras   417ff
- inorganic compounds   420ff
- organic compounds   417ff
gasoline   422
gas detection   417
gas sensors   465
geometrical optics   10ff
- laws of reflection and refraction   11
geometrical transformations, see image processing
geysirs   579
gold cup method   494
gray objects, see emissivity, classification

h

heat diffusion equation   262
heating systems   292
heat of vaporization/condensation, see latent heats
heat transfer   239ff
- comparison electrical/thermal circuits   254
- conduction   240
- convection   243, 356
- one dimensional wall   252
- radiation   244
- windows   257
HgCdTe, see MCT
hidden structures   340ff, 565
- below paint   565
- half timbered house   338ff, 565
high speed imaging/recording   189ff, 447
high voltage lines   527
Hough space, see image processing, pattern recognition
hyperspectral imaging, see multi spectral imaging

i

IFOV, see camera system performance
image formation   102ff
- bad pixel correction   113
- fill factor   105

- NUC   108ff
- pixel numbers   105ff
image processing   190ff
- digital detail enhancement (DDE)   203
- feature extraction   212
- geometrical transformations   209
- image building   193
- image fusion   192
- image subtraction   196, 388
- noise reduction   207
- pattern recognition   215
- spatial derivatives   202
- segmentation   212
- time derivatives   198
image quality   145
imaging with filters   158ff
index of refraction n 10ff
inelastic collisions   286, 288
InGaAs   93
InSb   92
integration time   150ff, 189
IR emitters   468
IR camera
- parameters   4
- signal contributions   3, 98ff
IR gas detection   417ff
IR spectral regions   9,10

k

Kirchhoff's law, see emissivity

l

Lambert – Beer law, see Bouguer's law
Lambertian radiator, see radiometry   479
latent heats   245
Leslie cube   39, 322
level   132, 331
light bulbs   276
line scanning   103, 518ff
- principle   104
- rotating wheel   519
- windshields   520
Lock In thermography, see active thermal imaging

m

MCT   92
measurement process   97ff
medical applications   535ff
- acupuncture   541
- breast thermography   544
- pain   537
- pressure ulcers   546

medical applications (*contd.*)
– Raynaud's phenomenon   545
– standard conditions   536
MEMS   445
metal industry   490ff
– hot molds   491
– solid Al strips   491
microelectronic board   522
microscope objective   446
microsystems   445ff
– cryogenic actuators   471
– IR emitters   468
– measurement requirements   446
– micro heat exchangers   456
– micro-electrical-mechanical-systems (MEMS)   445
– microreactors   449ff
– microsensors   459
– Peltier elements   470
microwave oven   312ff
moon   576
motorbike   283ff
multi spectral imaging   176ff
MTF, MRTD, MDTD, see image quality

n

narcissus effect   124, 446
NETD, see detectors, performance parameters; camera system performance
NEP, see detectors, performance parameters; camera system performance
Newtons law of cooling   271ff
NH_3 420
Night vision systems   500ff
noise reduction, see image processing
nondestructive testing, see active thermal imaging
nonuniformity correction (NUC), see image formation

o

oil level, see bushing
optical components IR camera   118ff
optical filters, see filters
optical material properties   51ff
– antireflection coatings   67ff
– thin film coatings   64ff
– transmission experiments   316
– transmission spectra of materials   57ff
optical thickness   408

p

pattern recognition, see image processing
Peltier elements   116ff, 470
Peltier effect   311
photoelecric effect,   83
photon detectors, see detectors
photoconductors, see detectors, photon detectors
photodiodes, see detectors, photon detectors
pixel numbers, see image formation
Planck's law   22ff
plastic foils   515ff
polarizer   485
polarization   485ff
polymer molding   513
power plant pipes/valves   509
predictive maintenance (PdM)   508ff
PtSi   94
pulse thermography, see active thermal imaging
pulsed phase thermography, see active thermal imaging

q

QWIP, see also detectors, photon detectors   94ff

r

radiation, see also heat transfer
radiometry   14ff
– excitance (emissive power)   15
– irradiance   15
– Lambertian radiator   18–19, 38
– radiance   17
– radiant intensity   17
– radiant power (energy flux)   15
– radiation transfer   19–21
– spectral densities of radiometric quantities   16
– view factor   19–21
range of IR cameras   568ff
ratio thermography, see two color thermography
Rayleigh Jeans law   166
reflectivity   485ff
– glass   484, 487
– Si   485, 487
– wood   489
responsivity, see detectors, performance parameters
ROIC   82, 104
rotational-vibrational molecular spectra   403
R-value   255ff, 344, 391ff

s

Schottky barrier, see detectors, photon detectors

Seebeck effect   311
segmentation, see image processing
selective emitters   34, 62, 326, 515ff
SF_6 420
shadows   366
signal noise limited detection   85
sky   574
sliding friction   282ff
software tools   136ff
solar cell inspections   230
solar constant   29
solar load   361
solar reflections   369
solid angle   17
space shuttle   505ff
span   132, 331
specific heat   262
spectral camera response   118
specular reflection   12, 318, 478
sports   553ff
– American football   559
– squash   557
– tennis serve   287, 553ff
– volleyball   557
SRF, see camera system performance
Stefan–Boltzmann law   25
Stirling cooler   117
structural defects   352ff
substation transformers   524
superframing   185ff
surveillance & security   566ff
switch on behavior   133

t
tank levels   510ff
teaching physics   281ff
– electromagnetism   308
– mechanics   281
– optics/radiation physics   316
– thermal physics   291
temperature accuracy, see camera system performance
temperature scales   5, 6
tennis serve, see also sports   287, 553
thermal bridges   258, 348ff
thermal conductivity   241, 292ff,
– of solids   293
– of water   294
thermal detectors, see detectors
thermal diffusivity, see also active thermal imaging/thermal waves   219, 262
thermal effusivity, see also active thermal imaging/thermal waves   219
thermal equilibrium   248
thermal mismatch, see also active thermal imaging/thermal waves   220
thermal penetration depth, see also active thermal imaging/thermal waves   221
thermal reflections   14, 320, 477ff, 491
– Fresnel equations   479ff
– identification and suppression   487
– measurements   485ff
thermal shock   134
thermal time constants   267ff, 278, 364, 461, 465
thermal waves, see active thermal imaging
thermoelectricity   310
thermopile sensors   459
thin films, see optical material properties
transformers, see substation transformers
transmission spectra, see optical material properties and see atmospheric transmission
transient effects   261ff
two color thermography   163ff
– applications   174
– comparison to single band method   173

u
U-value   255ff, 344, 391ff

v
vehicle tires   290
veterinary applications   548ff
– equine thermography   550
– zoo animals   549
viewing angle   38
view factor   19, 373ff
visualization   2, 281
VOC's   417
volcanology   580

w
walking   289
walls, see heat transfer
wavelength, λ 6ff
wedge method   495
Wien approximation   167ff
Wien displacement laws   22, 23
wildfires   576ff
windows, see also heat transfer   381ff